SPICE FOR
MICROELECTRONIC
CIRCUITS

THIRD EDITION

By SEDRA / SMITH

GORDON W. ROBERTS
McGill University

AND

ADEL S. SEDRA
University of Toronto

SAUNDERS COLLEGE PUBLISHING
Harcourt Brace Jovanovich College Publishers
Fort Worth Philadelphia San Diego New York Orlando Austin San Antonio
Toronto Montreal London Sydney Tokyo

Copyright© 1992 by Saunders College Publishing

All rights reserved. No part of this publication may be reproduced or transmitted in any form or by any means, electronic or mechanical including photocopy, recording or any information storage and retrieval system, without permission in writing from the publisher.

Requests for permission to make copies of any part of the work should be mailed to: Permissions Department, Harcourt Brace Jovanovich Publishers, 8th Floor, Orlando, Florida 32887.

Printed in the United States of America.

The following device models are produced by MicroSim:

D1N4148   Q2N2222A   Q2N696   Q2N3904

© 1985–1990 by MicroSim Corporation. Used by permission.

Roberts/Sedra: Spice for Microelectronic Circuits, 3/e

ISBN 0-03-052617-5

345 021 987654321

Preface

Today, most, if not all, microelectronic circuit design is carried out with the aid of a computer-aided circuit analysis program such as Spice. Spice, an acronym for *Simulation Program with Integrated Circuit Emphasis,* is considered by many to be the de-facto industrial standard for computer-aided circuit analysis for microelectronic circuits, mainly because it is used by the majority of IC designers in North America today. It is reasonable to say that to master electronic circuit design, one must also develop a fair amount of expertise in a circuit analysis program such as Spice. It is therefore our aim in this book to describe **how** Spice is used to analyze microelectronic circuits and, more importantly, outline **how** Spice is used in the process of design itself.

It is our view that electronic circuit design begins by assembling various known subcircuits together in a systematic manner assuming rather simple mathematical models of transistor behavior. Keeping the mathematical model of the transistor simple enables the designer to quickly configure an electronic network and determine using hand analysis whether the resulting circuit has potential for meeting required specifications. Once satisfied, the designer can use a more complex model for the transistors in conjunction with Spice to better judge the behavior of the overall circuit as it will appear in integrated form. If the circuit fails to meet specification, the designer can resort back to a simpler computer model, preferably the same one that was used during the initial design, and identify the reason for the discrepancy. In this way, the designer is in a position to decide where the shortfall lies; whether it be in the designer's own understanding of circuit operation or whether additional circuitry is required to circumvent inherent problems caused by the nonidealities of the devices. Examples throughout the text will emphasize the importance of this approach.

There is a tendency for new designers of electronic circuits to be overwhelmed by the analysis capability of a circuit analysis program such as Spice, and ignore the thought-process provided by a hand analysis using simple models for the transistors. Instead, they usually begin their designs directly with complex transistor models, falsely thinking that the results generated by the computer will provide the necessary insight into circuit operation when the circuit fails to perform as required. Experience has shown that this generally leads to poor designs because most of the design effort is spent blindly searching for ways to improve

Preface

the design using a brute-force hit-and-miss approach. It is our intention in this book to avoid this pitfall and teach the reader what not to do with Spice. This is accomplished by keying each example of this text to those presented in **Microelectronic Circuits**, Third Edition, by Sedra and Smith, where a complete hand analysis is provided. In this way, the insight provided by a hand analysis is readily available to our readers. To allow the reader to quickly locate the hand analysis in Sedra and Smith, each example of this text that has a corresponding hand analysis will be denoted by the appropriate example number in a square bold box located in one of the corners of the schematic that illustrates the example.

Spice, developed in the early 1970's on main-frame computers, is now being used by undergraduates in engineering schools all across North America. Although other programs for computer-aided circuit analysis exist, and are being used by various groups, none has found as much wide-spread use as Spice. This largely stems from the generous distribution policies of the Electronics Research Laboratory of the University of California, Berkeley, during the early stages of the program's development. Until recently, Spice was largely limited to main-frame computers on a time-sharing basis. However, today, one can find various versions of Spice for personal computers (PCs).

There are various Spice-like simulators built for the PC with very similar characteristics; however, *PSpice*, developed and distributed by the MicroSim Corporation, provides a version free of charge to students and their instructors that run on IBM PCs or compatible with at least 512 kilobytes of resident memory. Although limited to circuits containing no more than 10 transistors or 20 electrical nodes (whichever takes precedence), this simplified version is usually more than adequate for the types of circuit problems facing students at the undergraduate level. Hence, PSpice enables the integration of computer-aided circuit analysis into the undergraduate curriculum at reasonable costs, ie. the cost of a PC. In this text all circuit examples will be simulated using the student version of PSpice unless they exceed the circuit size limit. In such cases, which there are few, we will resort to the professional version of PSpice purchased from the MicroSim Corporation or Spice version 2G6 distributed by the University of California.

The student version of PSpice can be obtained from the MicroSim Corporation by writing directly to them at the following mailing address:

> MicroSim Corporation
> 23175 La Cadena Drive
> Laguna Hills, CA 92653
> USA

Although we make direct reference to the text by Sedra and Smith, we present material in such a way that the book can be used as a stand alone text. In this way, this book can also serve as a tutorial on computer-aided circuit analysis using Spice.

The organization of this book is as follows:

Chapter 1 provides an introduction to electronic circuit simulation using Spice. A brief description of the capabilities of Spice and the computer concept of electrical and electronic elements are outlined. Moreover, this chapter illustrates the role that computer-aided circuit simulation plays in the process of circuit design.

Chapter 2 demonstrates how Spice can be used to simulate the ideal and nonideal behavior of op amp circuits. Various models of op amps are introduced to assist the user in investigating the effect of op amp behavior on closed-loop circuit operation. Additional Spice commands are also introduced.

Chapters 3 to 5 present simulation details for circuits containing semiconductor diodes, zener diodes, bipolar junction transistors (BJTs), metal-oxide-semiconductor field-effect transistors (MOSFETs), junction field-effect transistors (JFETs) and metal-semicondu-ctor field-effect transistors (MESFETs). The main objective of these chapters is to understand how to simulate circuits containing active devices and to calculate the quiescent point of each circuit from which the small-signal model of the circuit can be determined. Most Spice results are compared to those computed by hand analysis.

Chapter 6 investigates both the large and small-signal operation of differential and multistage amplifiers using Spice. Various attributes of current-mirror and current-source circuits are also investigated using Spice.

Chapter 7 investigates the frequency response behavior of various amplifier circuits using Spice. Spice is ideally suited for frequency-response type calculations. The accuracy of the method of short- and open-circuit time constants for estimating the 3 dB bandwidth of wideband amplifiers is investigated with Spice.

Chapter 8 deals with the topic of feedback. Stability issues are also investigated with Spice.

Chapter 9 investigate the DC and transient behavior of various types of output stages.

Chapter 10 presents several circuit simulation studies of analog integrated circuits. This includes a detailed investigation of the 741 bipolar op amp, and a CMOS and a BiCMOS op amps. The ideal behavior of analog-to-digital and digital-to-analog converters is also presented.

Chapter 11 investigates the frequency response behavior of various types of active-RC filter circuits and LC tuned amplifiers. In addition, the reader is exposed to the use of computer-aided circuit design to fine-tune the behavior of a circuit.

Preface

Chapter 12 investigates the nonideal behavior of various types of signal-generator and waveform-shaping circuits. Many of the analyses involve the use an op amp macromodel of a commercial op amp circuit.

Chapters 13 and 14 deal with bipolar and MOS digital circuits.

All of the chapters have in-depth problem sets that are intended to be solved using Spice. In most cases, the student version of PSpice is sufficient to carry out these exercises.

We would like to acknowledge the support and help of many people during the preparation of the text. In particular we would like to thank our students Philip Crawley, Michael Toner, Andrew Bishop, Jean-Charles Maillet, Xavier Haurie, Stuart Banks and Antoine Chemali for debugging many of the exercises at the end of each chapter and highlighting unclear areas of the text. We would also like to acknowledge the useful feedback provided by many of the students of the 1991 Analog Microelectronics course at McGill University as they struggled through initial drafts of the text. The help of Jacek Slaboszewicz and Pierre Parent with the questions about TEXwere much appreciated. Finally, we would like to thank Mark Moraes for supplying the drawing program *xpic* that was used to draw most of the figures in this text.

Furthermore, the authors would like to acknowledge the assistance of a number of individuals who made the text possible. We are grateful to our developmental editor Alexa Barnes, and our editor, Barbara Gingery. We would also like to acknowledge the help of Mei Sum Kwan. Finally, we like to thank our families for much encouragement and support.

Gordon W. Roberts
Adel S. Sedra

Table of Contents

Preface .... *iii*

Chapter 1 Introduction To Spice 1
   1.1 Computer Simulation Of Electronic Circuits 1
   1.2 An Outline Of Spice 4
       1.2.1 Types Of Analysis Performed By Spice 5
       1.2.2 Input To Spice 6
             Circuit Description 9
             Analysis Requests 16
             Output Requests 18
             A Simple Example 20
       1.2.3 Output From Spice 21
   1.3 Output Post-Processing Using Probe 25
   1.4 Examples 27
       1.4.1 Example 1: DC Node Voltages Of A Linear Network 29
       1.4.2 Example 2: Transient Response Of A 3-Stage Linear Amplifier . 32
       1.4.3 Example 3: Setting Circuit Initial Conditions During A Transient Analysis 35
       1.4.4 Example 4: Frequency Response Of A Linear Amplifier 39
   1.5 Spice Tips 39
   1.6 Bibliography 41
   1.7 Problems 41

Chapter 2 Operational Amplifiers 47
   2.1 Modeling An Ideal Op Amp With Spice 47
   2.2 Analyzing The Behavior Of Ideal Op Amp Circuits 48
       2.2.1 Inverting Amplifier 48
       2.2.2 The Miller Integrator 50
       2.2.3 A Damped Miller Integrator 53
       2.2.4 The Unity-Gain Buffer 57
       2.2.5 Instrumentation Amplifier 59
   2.3 Nonideal Op Amp Performance 64
       2.3.1 Small-Signal Frequency Response Of Op Amp Circuits 65
       2.3.2 Modeling The Large-Signal Behavior Of Op Amps 67
             Limited Current-Output Of The Front-End Transconductance Stage 69
             Output Saturation 71

Table of Contents

    Frequency Response 71
    A PSpice Large-Signal Op Amp Subcircuit 72
 2.4 The Effects Of Op Amp Large-Signal Nonidealities On Closed-Loop Behavior 73
  2.4.1 DC Transfer Characteristic Of An Inverting Amplifier 73
  2.4.2 Slew-Rate Limiting 75
 2.5 Other Op Amp Nonidealities 77
  2.5.1 Common-Mode Gain 77
  2.5.2 Input And Output Resistances 80
  2.5.3 DC Problems 81
 2.6 Spice Tips 83
 2.7 Bibliography 85
 2.8 Problems 85

Chapter 3 Diodes 94

 3.1 Describing Diodes To Spice 94
  3.1.1 Diode Element Description 94
  3.1.2 Diode Model Description 95
 3.2 Spice As A Curve Tracer 98
  3.2.1 Extracting The Small-Signal Diode Parameters 100
  3.2.2 Temperature Effects 101
 3.3 Approximating Ideal Diode Behavior 102
 3.4 Voltage Regulation Using A String Of Diodes 104
 3.5 Zener Diode Modeling 108
    Voltage Regulation Using A Zener Diode 109
 3.6 Rectifier Circuits 115
  3.6.1 Half-Wave Rectifier 115
  3.6.2 Full-Wave Peak Rectifier 119
  3.6.3 A Voltage Regulated Power Supply 127
 3.7 Limiting And Clamping Circuits 132
    A Diode Limiter Circuit 132
    A DC Restorer Circuit 133
    Voltage Doubler Circuit 135
 3.8 Spice Tips 136
 3.9 Problems 137

Chapter 4 Bipolar Junction Transistors (BJTs) 143

 4.1 Describing BJTs To Spice 143
  4.1.1 BJT Element Description 144
  4.1.2 BJT Model Description 145
  4.1.3 Verifying NPN Transistor Circuit Operation 146
 4.2 Using Spice As A Curve Tracer 149

4.3		Spice Analysis Of Transistor Circuits At DC	151
	4.3.1	Transistor Modes Of Operation	151
		Active Region	151
		Saturation Region	153
		Cutoff Region	154
	4.3.2	Computing DC Bias Of A PNP Transistor Circuit	154
	4.3.3	DC Operating Point Of A Multiple Transistor Circuit	157
4.4		BJT Transistor Amplifiers	160
	4.4.1	BJT Small-Signal Model	160
	4.4.2	Single-Stage Voltage-Amplifier Circuits	162
		Another Example:	165
4.5		DC Bias Sensitivity Analysis	168
		Sensitivity To Component Variations	168
		Sensitivity To Temperature Variations	172
4.6		Basic Single-Stage BJT Amplifier Configurations	174
		The Common-Emitter Amplifier	175
		The Common-Base Amplifier	182
		The Common-Collector Amplifier	185
4.7		The Transistor As a Switch	185
4.8		Spice Tips	187
4.9		Bibliography	189
4.10		Problems	189

Chapter 5 Field-Effect Transistors (FETs)   194

5.1		Describing MOSFETs To Spice	194
	5.1.1	MOSFET Element Description	194
	5.1.2	MOSFET Model Description	196
	5.1.3	An Enhancement-Mode N-Channel MOSFET Circuit	198
	5.1.4	Observing The MOSFET Current – Voltage Characteristics	200
5.2		Spice Analysis Of MOSFET Circuits At DC	203
	5.2.1	An Enhancement-Mode P-Channel MOSFET Circuit	204
	5.2.2	A Depletion-Mode P-Channel MOSFET Circuit	207
	5.2.3	A Depletion-Mode N-Channel MOSFET Circuit	209
5.3		Describing JFETs To Spice	212
	5.3.1	JFET Element Description	212
	5.3.2	JFET Model Description	213
	5.3.3	An N-Channel JFET Example	214
	5.3.4	A P-Channel JFET Example	217
5.4		FET Amplifier Circuits	219
	5.4.1	Effect Of Bias Point On Amplifier Conditions	219
	5.4.2	Small-Signal Model Of The FET	221
	5.4.3	A Basic FET Amplifier Circuit	225

5.5	Investigating Bias Stability With Spice	230
5.6	Integrated-Circuit MOS Amplifiers	235
	5.6.1 Enhancement-Load Amplifier Including The Body Effect	235
	5.6.2 CMOS Amplifier	239
5.7	MOSFET Switches	242
5.8	Describing MESFETs To PSpice	247
	5.8.1 MESFET Element Description	247
	5.8.2 MESFET Model Description	250
	5.8.3 Small-Signal MESFET Model	251
	5.8.4 A MESFET Biasing Example	253
5.9	Spice Tips	255
5.10	Bibliography	256
5.11	Problems	257

Chapter 6 Differential And Multistage Amplifiers ... 261

6.1	Input Excitation For The Differential Pair	261
6.2	Small-Signal Analysis Of The Differential Amplifier: Symmetric Conditions	267
6.3	Small-Signal Analysis Of The Differential Amplifier: Asymmetric Conditions	275
	Input Offset Voltage	275
	Input Bias And Offset Currents	277
6.4	Current-Mirror Circuits	279
	Current-Gain Accuracy:	279
	Output Resistance:	282
	Minimum Output Voltage:	285
	Input Current Range:	285
6.5	A High-Performance Current Mirror	287
6.6	Current-Source Biasing In Integrated Circuits	287
6.7	A CMOS Differential Amplifier With Active Load	290
6.8	GaAs Differential Amplifiers	298
6.9	A BJT Multistage Amplifier Circuit	304
6.10	Spice Tips	312
6.11	Bibliography	314
6.12	Problems	314

Chapter 7 Frequency Response ... 323

7.1	Investigating Transfer Function Behavior Using PSpice	323
7.2	Modeling Dynamic Effects In Semiconductor Devices	326
7.3	The Low-Frequency Response Of The Common-Source Amplifier	330
7.4	The High Frequency Analysis Of A Common-Source Amplifier	332
7.5	High-Frequency Response Comparison Of The Common-Emitter and Cascode Amplifiers	335

7.6	High-Frequency Response Of The CC-CE Amplifier	341
7.7	Frequency Response Of The Differential Amplifier	344
7.8	The Effect Of Emitter Degeneration On Differential Amplifier Characteristics	352
7.9	Spice Tips	353
7.10	Problems	354

Chapter 8 Feedback ... 358

8.1	The General Feedback Structure	358
8.2	The Four Basic Feedback Amplifier Topologies	359
	8.2.1 Voltage-Sampling Series-Mixing Topology	359
	Checking Basic Feedback Theory Assumptions:	363
	8.2.2 Current-Sampling Series-Mixing Topology	367
	8.2.3 Voltage-Sampling Shunt-Mixing Topology	374
	8.2.4 Current-Sampling Shunt-Mixing Topology	380
8.3	Determining Loop Gain With Spice	388
	An Alternative Method:	392
8.4	Stability Analysis Using Spice	394
8.5	Investigating The Range Of Amplifier Stability	401
8.6	The Effect Of Phase Margin On Transient Response	408
8.7	Frequency Compensation	410
8.8	Spice Tips	413
8.9	Bibliography	416
8.10	Problems	416

Chapter 9 Output Stages And Power Amplifiers ... 423

9.1	Emitter-Follower Output Stage	423
9.2	Class B Output Stage	428
	Power Conversion Efficiency	429
	Transfer Characteristics And A Measure Of Linearity	432
9.3	Class AB Output Stage	436
9.4	Short-Circuit Protection	439
9.5	Spice Tips	440
9.6	Problems	442

Chapter 10 Analog Integrated Circuits ... 446

10.1	A Detailed Analysis Of The 741 Op Amp Circuit	446
	DC Analysis Of The 741	448
	Gain And Frequency Response Of The 741	454
	Slew-Rate Limiting Of The 741	457
	Noise Analysis Of The 741 Op Amp	459
	A Summary Of The 741 Op Amp Characteristics	461
10.2	A CMOS Op Amp	462

Table of Contents

10.3	Two Different Technology Versions Of A Folded-Cascode Op Amp	468
10.4	Data Converters	476
	R-2R Ladder D/A Converter	476
	3-Bit A/D Flash Converter	479
10.5	Spice Tips	483
10.6	Bibliography	484
10.7	Problems	484

Chapter 11 Filters And Tuned Amplifiers — 487

11.1	The Butterworth And Chebyshev Transfer Functions	487
11.2	Second-Order Active Filters Based On Inductor Replacement	491
11.3	Second-Order Active Filters Based On The Two-Integrator-Loop Topology	494
11.4	Single-Amplifier Biquadratic Active Filters	500
11.5	Tuned Amplifiers	502
11.6	Spice Tips	507
11.7	Bibliography	507
11.8	Problems	507

Chapter 12 Signal Generators And Waveform – Shaping Circuits — 516

12.1	Op Amp-RC Sinusoidal Oscillators	517
12.1.1	The Wien-Bridge Oscillator	518
12.1.2	An Active-Filter-Tuned Oscillator	522
12.2	Crystal Oscillators	526
12.3	Multivibrator Circuits	529
	A Bistable Circuit:	529
	Generation Of A Square-Wave Using An Astable Multivibrator:	533
	The Monostable Multivibrator:	534
12.4	A Nonlinear Waveform-Shaping Circuit	536
12.5	Precision Rectifier Circuits	538
	A Half-Wave Rectifier Circuit:	539
	A Buffered Peak Detector:	541
	A Clamping Circuit:	542
12.6	Spice Tips	543
12.7	Bibliography	544
12.8	Problems	545

Chapter 13 MOS Digital Circuits — 553

13.1	NMOS Inverter With Enhancement Load	553
	Dynamic Operation	556
13.2	NMOS Inverter With Depletion Load	562
13.3	The CMOS Inverter	564

	Dynamic Operation	565
13.4	A Two-Input CMOS NOR Gate	570
13.5	CMOS SR Flip-Flop	572
13.6	A Gallium-Arsenide Inverter Circuit	577
13.7	Spice Tips	580
13.8	Problems	580

Chapter 14 Bipolar Digital Circuits ... 583

14.1	Transistor-Transistor Logic (TTL)	583
14.2	Emitter-Coupled Logic (ECL)	592
14.3	BiCMOS Digital Circuits	602
14.4	Bibliography	606
14.5	Problems	606

Appendix A Device Model Parameters ... 611

A.1	Diode Model	611
A.2	BJT Model	612
A.3	JET Model	613
A.4	MOSFET Model	613
A.5	MESFET Model	617
A.6	Bibliography	618

Appendix B Spice Options ... 620

Index ... 623

Chapter 1 Introduction To Spice

This chapter provides an introduction to electronic circuit simulation using Spice.[†] This is accomplished by outlining the basic philosophy of circuit simulation, and why it has become so important for todays electronic circuit design. This will then be followed by a brief description of the capabilities of Spice and the computer conception of electrical and electronic elements. Several examples will be used to illustrate these concepts.

Although the title of this chapter is called **Introduction To Spice** it could just as easily be called an **Introduction To PSpice**. In many of the following discussions we make specific reference to the program Spice by name, but it should be made clear here that these discussions apply equally well to PSpice. However, statements made about PSpice do not, in general, apply to the program Spice.

1.1 Computer Simulation Of Electronic Circuits

Traditionally, electronic circuit design was verified by building discrete prototypes, subjecting the circuit to various input simulii (input signals, temperature changes, power supply variations, etc.) and measuring its response using appropriate laboratory equipment. Although, somewhat time-consuming, practical experience is directly gained from the prototype. This, in turn, provides the necessary insight into circuit behavior from which to judge the manufacturability of the design.

[†] Spice is an acronym for Simulation Program with Integrated Circuit Emphasis.

1 Introduction To Spice

The design of an integrated circuit (IC) requires a different approach to verifying its intended operation. Due to the minute dimensions associated with the IC, a breadboarded version of the intended circuit will bear little resemblance to its final implementation. The parasitic components that are present in an IC are entirely different from the parasitic components present in the breadboard. As a result, signal measurements obtained from the breadboard would not usually provide an accurate representation of the signals appearing on the IC. To further complicate matters, one might be inclined to suggest that one measure the appropriate signals directly on the IC itself; although this is sometimes done, it requires extreme mechanical and electrical measurement precision. Not to mention that it is limited to specific types of measurements (ie. it is very difficult to measure currents). Furthermore, an IC implementation does not lend itself easily to circuit modifications. As a result, circuit modifications must be made at the IC mask level prior to circuit fabrication. Due to the processing delay time associated with IC development, this approach results in unreasonable times (on the order of weeks) between executing the modification and observing its effect.

Computer programs that simulate the electrical performance of an electronic circuit provide a simple means of verifying the intended operation prior to circuit construction. Not only does this provide a cost effective approach to design verification, it also provides a means for quickly verifying new ideas that could ultimately lead to improved circuit performance. It was precisely the introduction of such computer programs that revolutionized the electronics industry, leading to the development of today's high-density monolithic circuit schemes (ie. VLSI).

Spice, the de-facto industrial standard for computer-aided circuit analysis, was developed in the early 1970's at the University of California, Berkeley. Although other programs for computer-aided circuit analysis exist, and are being used by many different electronic design groups, none has found as much wide-spread use as Spice. Until recently, Spice was largely limited to main-frame computers on a time-sharing basis. Today, however, one can find various versions of Spice for personal computers (PCs). In general, these programs use slightly different algorithms than Spice for performing the circuit simulations; but, many of them adhere to the same input format description as Spice, elevating the Spice-input syntax into a computer-like language.

Commercially supported versions of Spice can be considered to be divided into two types: main-frame versions and PC-based versions. Generally, main-frame versions of Spice are intended to be used by sophisticated integrated-circuit designers who require large amounts of computer power to simulate complex circuits. Commercial versions of Spice include *HSpice* from Meta-Software and *IG-Spice* from A. B. Associates. There are also many other companies now supporting main-frame version of Spice but are too numerous to mention here. PC-based versions of Spice simply allow circuit simulation to be performed on a low-cost computer system. One such version that is of interest to us in this text is the program *PSpice*, from MicroSim Corporation. We shall have more to say about this program in a moment.

Although Spice was originally intended for analyzing integrated circuits, its underlying concepts are general and can apply to any type of network that can be described in terms of a basic set of electrical elements (ie. resistors, capacitors, inductors, and dependent and independent sources). Today, Spice is often used for such applications as the analysis of high-voltage electrical networks, feedback control systems, and the effect of thermal gradients on electronic networks.

At this juncture we should make perfectly clear the role that computer-aided circuit simulators play in the cycle of design. A design usually begins with a set of specifications (eg. frequency response, step-response, etc.) from which a circuit is to be found that meets all of the proposed specifications. It is then the objective of a designer to configure an electronic circuit that satisfies the given specifications. It is precisely this task which intrigues most circuit designers because, at this time, computers have not yet acquired the intelligence to perform this task. It is therefore left up to the designer to accomplish this, relying on his or her knowledge of electronic circuit design. By developing a simple understanding of circuit design – one that utilizes an approximate method of analysis – different designs can be configured and quickly analyzed by hand to determine if they have the potential for meeting the proposed specifications.

Subsequently, once a design is found that is thought to meet the required specifications, the designer proceeds to analyze the circuit using more complex models of device behavior with a circuit simulator such as Spice. On completion of Spice, the behavior of the design is checked against the required specifications to determine whether the circuit meets the desired requirements. If the circuit fails to meet specification, the designer can resort back to a simpler computer model, preferably the same one that was used during the initial design, and identify the reason for the discrepancy. In this way, the designer is in a position to decide where the shortfall lies; whether it be in the designer's own understanding of circuit operation or whether additional circuitry is required to circumvent inherent problems caused by the nonidealities of the devices. Such a thought process is depicted in the flow-chart shown in Fig. 1.1.

The importance of basic circuit understanding is essential in the design process because if after a computer simulation the performance of a circuit is not adequate the designer has

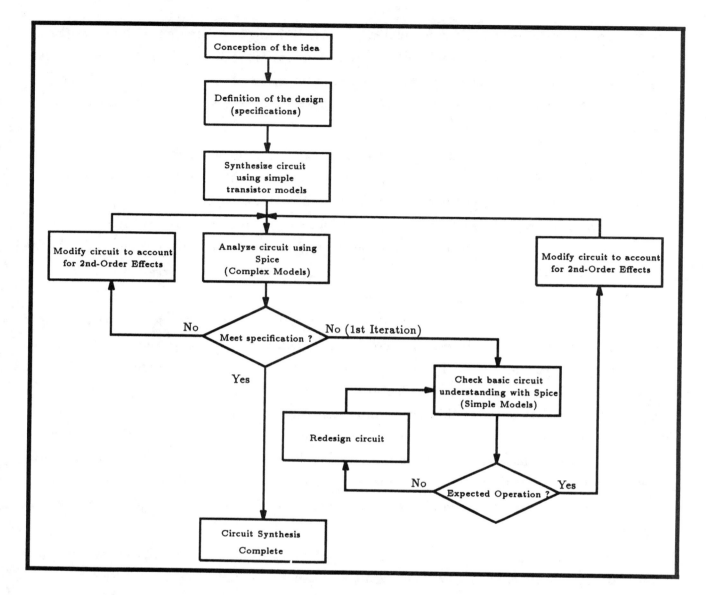

Figure 1.1 Illustrating the role of circuit simulation in the process of circuit design.

some idea about the components of the design that may be altered to improve its circuit performance. Otherwise, the designer must rely on altering components and circuit structure using a brute-force hit-and-miss approach. Such an approach that usually results in a lot of wasted effort and probably no improvement to the circuit.

1.2 An Outline Of Spice

Spice simulates the behavior of electronic circuits on a digital computer and tries to emulate both the signal generators and measurement equipment found on one's laboratory bench. This would include such measurement instruments as multimeters, oscilloscopes, curve-tracers and frequency spectrum analyzers. In the following we shall outline the analysis available in Spice and the way in which to describe a circuit to Spice. By no means is the following description complete. It is only meant to introduce the reader to Spice in a gradual manner. More details on using Spice to simulate electronic circuits are included throughout this text.

1.2.1 Types Of Analysis Performed By Spice

Spice is a general purpose circuit simulator capable of performing three main types of analysis. These are nonlinear DC, nonlinear transient, and linear small-signal AC circuit analysis.

Nonlinear DC analysis or simply **DC analysis** calculates the behavior of the circuit when only a DC voltage or current is applied to the circuit. In most cases, this analysis is performed first before any other analysis is undertaken. We shall commonly refer to the results of this analysis as the DC bias or operating-point characteristics.

The second analysis is probably the most important analysis type, that being **Transient Analysis**. This analysis computes the various voltages and currents in the circuit with-respect-to time. This analysis is most meaningful when time-varying input signals are applied to the circuit; otherwise this analysis generates results identical to the DC analysis.

The third and final major analysis type that Spice can perform is a small-signal **AC analysis**. This analysis begins by linearizing the circuit around the DC operating point and subsequently calculates the linearized network variables as function of frequency. This, of course, is equivalent to calculating the sinusoidal steady-state behavior of the circuit assuming that the signals applied to the network have infinitesimally-small amplitudes.

Spice is also capable of performing other types of analysis, however these are generally viewed as special cases of the above three analysis types. We shall briefly outline these below.

DC Sweep allows a series of DC operating points to be calculated while sweeping or incrementally changing the value of an independent current or voltage source. This analysis is largely used to determine the DC large-signal transfer characteristic of a given circuit. A related analysis is the **Transfer Function** analysis. This analysis computes the small-signal DC gain from a specified input to a specified output (ie. one of the following: voltage gain, transconductance, transresistance, or current gain), and the corresponding input and output resistance.

In a similar manner as DC Sweep, **Temperature Analysis** allows a series of analyses to be performed while varying the temperature of the circuit. Because the characteristics of many passive and active devices depend on temperature, this facility provides a useful

```
            Title Statement
                Circuit Description
                    Power Supplies / Signal Sources
                    Element Descriptions
                    Model Statements
                Analysis Requests
                Output Requests
            .END
```

Figure 1.2 Suggested format for a Spice input file.

tool for investigating the effect of temperature variation on circuit operation. Any of the above main analysis types can be performed in conjunction with Temperature Analysis, thus providing important insight into the circuit's temperature dependencies.

Often circuit designers are interested in determining the components of a given circuit which affect circuit performance more critically than others. This is helpful in determining the appropriate manufacturing tolerances required by the circuit components. This type of analysis is referred to as **Sensitivity Analysis**, and depending on the version of Spice one has access to, there are are usually two sensitivity analyses available. The first, which we shall refer to as **DC Sensitivity**, is used to compute changes in the DC operation of the circuit subject to infinitesimally-small changes in the values of various circuit components. This analysis is available in most, if not all, versions of Spice. The second sensitivity analysis is called **Monte Carlo Analysis**. This analysis simply performs multiple runs of selected analysis types (DC, AC and transient) using a pre-determined statistical distribution for the values of various components. This analysis is rather complicated and is beyond the scope of this text.

Finally, the dynamic range of a circuit can be calculated through the **Noise** and **Fourier Analysis** procedures. More specifically, the noise analysis begins by calculating the noise contribution of each element, injecting its effect back into the circuit and calculating its total effect on the output node in a mean-square sense. The Fourier analysis computes the Fourier series coefficients of the circuits voltages or currents with respect to the period of the input excitation(s).

1.2.2 Input To Spice

In order for Spice to perform its circuit simulation function, the circuit to be analyzed is described to Spice by a sequence of lines entered into a computer file via a computer

Power-of-Ten Suffix Letter	Metric Prefix	Multiplying Factor
T	tera	10^{+12}
G	giga	10^{+9}
Meg	mega	10^{+6}
K	kilo	10^{+3}
M	milli	10^{-3}
U	micro	10^{-6}
N	nano	10^{-9}
P	pico	10^{-12}
F	femto	10^{-15}

Table 1.1 Scale-factor abbreviations recognized by Spice.

terminal. This computer file is commonly referred to as the **Spice input deck or file**. Each line is either a statement describing a single circuit element or a control line setting model parameters, measurement nodes, and analysis types. A simplified outline of the various Spice statements will be given later in this section. The first line in the Spice input deck *must* be a "**title**" and the last line *must* be an "**.END**" statement. The former command line is used to identify the output generated by Spice and the latter indicates an end to the Spice input file. The order of the remaining lines is arbitrary. Based on the authors experience, the format shown in Fig. 1.2 is recommended for layout of the Spice input files; however, we need to stress that this arrangement is arbitrary and we will sometimes deviate from it in subsequent examples. To improve the readability of the input file, one usually sprinkles **comments** throughout the file. This is useful for others, and usually for one-self, to understand the rationale of the present simulation, and identify the various components of the given design. A comment is added into the Spice input file by inserting an "*" as the first character of the comment line.

The format of each statement is of the free-format type. That is, the various words used in each statement can be separated by either arbitrary-sized spaces (of course, limited by the line length) or commas, or both. For lines longer than 80 characters (ie. the screen width), they can be continued on the next line by entering a + (plus sign) in the first column of the new line. In the original version of Spice, all letters had to be given in upper case; however, more recent versions of Spice make no distinction between upper and lower case. In our future examples we will mix the case type at our convenience. A number can be represented as either an integer or floating point, using either decimal, scientific notation or engineering

Spice Suffix	Units
V	volts
A	amps
Hz	hertz
Ohm	ohm (Ω)
H	henry
F	farad
Degree	degree

Table 1.2 Element dimensions.

scale factors. The recognized scale factors are listed in Table 1.1. Not included in this table, but recognized by Spice, is the suffix MIL which is equivalent to 1/1000th of an inch. In addition, the dimensions or units of a given value can also be appended to any element value to clarify its context. The allowed suffix types are listed in Table 1.2.

One word of caution about attaching the dimensions of Farads to a capacitor value; Spice, unfortunately, uses the same letter (F) to denote a scale factor of 10^{-15} (femto) – see Table 1.1. One must therefore be careful not to confuse these two suffixes in a Spice input file. Placing a single suffix F on the value of a capacitor indicates that the value of the capacitor is to be expressed in femto-farads not farads. Thus 1 F is 10^{-15} farads while 1 is one farad.

Before creating the Spice input deck, a clearly labeled circuit diagram should be drawn with all nodes distinctly numbered with nonnegative integers between 0 and 9999. The ground node must be labeled node "0" (zero). In addition, all components must be uniquely labeled. To illustrate this, consider the network shown in Fig. 1.3(a). Here we have a linear network consisting of various resistors, capacitors and sources (both dependent and independent) having specific numerical values. To prepare this circuit for a Spice simulation, each element is assigned a unique name, say for sake of illustration, that the 1 Ω resistor is assigned the name R_1, the load resistor of 10 Ω is given the name R_{load}, the 2.65 mF capacitor is assigned the name C_1, the voltage-controlled voltage source is assigned the name E_1, and the input sinusoidal voltage source is assigned the name v_i. The beginning letter, (eg. R, C, E, V) of each of these elements has a special meaning that will be explained in a moment. Subsequently, each node of the circuit is assigned a positive integer value with the ground node labeled with the number 0. Here we assigned the numbers 1, 2, and 3 to the non-grounded nodes.

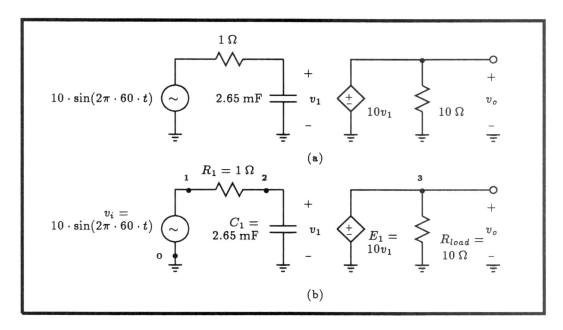

Figure 1.3 Preparing a network for Spice simulation. (a) Schematic drawing of a linear network. (b) Each element is uniquely labeled and each node is assigned a positive number with the ground reference point assigned the number 0.

Circuit simulation using Spice is made up of three major components: a detailed circuit description, a set of analysis types and various output requests. A brief glimpse of this was presented in Fig. 1.2. In the following we shall outline the basic syntax of the various commands that make up these three major components of Spice.

Circuit Description

Each circuit element is described to Spice by **an element statement** that contains the element name, the circuit nodes to which it is connected, and its value. Spice knows about four general classes of network elements. These are: passive elements, independent and dependent sources, and active devices (ie. diodes and transistors).

As a standard convention of Spice, the first letter of an element description signifies the element type, eg. R for resistor, D for diode, etc. As a quick reference, in Table 1.3 we list the key letter of each element type available in Spice. This letter is then followed by some alphanumeric name limited to 7 characters to uniquely identify that element from all others. The subsequent information that follows the element type and name depends on the nature of the element. This we describe in the following.

Passive Elements: In Fig. 1.4 we depict the Spice single-line descriptor statement for an arbitrary resistor, capacitor and inductor. As mentioned above, the first field (or a set of characters separated by blank spaces) of each statement describes its type and provides a

1st Letter Representation	Element
B	GaAs Field-Effect Transistor (MESFET)
C	Capacitor
D	Diode
E	Voltage-Controlled Voltage Source (VCVS)
F	Current-Controlled Current Source (CCCS)
G	Voltage-Controlled Current Source (VCCS)
H	Current-Controlled Voltage Source (CCVS)
I	Independent Current Source
J	Junction Field-Effect Transistor (JFET)
K	Coupled Inductors
L	Inductor
M	MOS Field-Effect Transistor (MOSFET)
Q	Bipolar Transistor (BJT)
R	Resistor
V	Independent Voltage Source

Table 1.3 Basic element types in Spice.

unique name for each element. This is followed by two fields that describe how each element is connected to the rest of the network via node numbers. Although these elements are bilateral, each element is assigned a positive and negative terminal. This is a convention used to assign direction to the current flowing through each device as denoted in Fig. 1.4, but more importantly, it is used to specify the polarity of the initial condition associated with the energy storage devices. The fourth field is used to specify the value of the passive element. Resistance is specified in Ohms, capacitance in Farads and inductance in Henries. These values are usually positive but can also be assigned a negative value (in which case the elements are not passive). For either the capacitor or inductor, an initial (time-zero) voltage or current condition can be specified in its fifth field, as depicted in Fig. 1.4.

For the circuit example shown in Fig. 1.3, the element statements for passive elements R_1, C_1 and R_{load} would appear in the Spice deck as follows:

```
R1      1 2 1Ohm
C1      2 0 2.65mF
Rload   3 0 10Ohm
```

For easy reading, we have attached on the end of each parameter value the dimensions of each element.

1.2 An Outline Of Spice

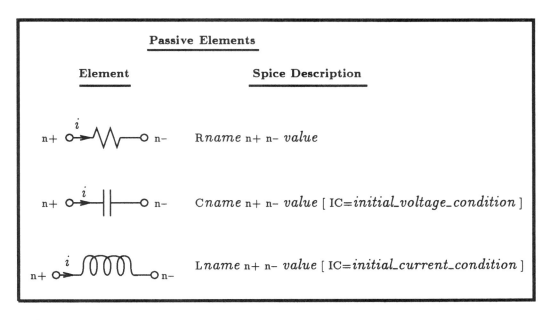

Figure 1.4 Spice descriptors for passive elements. Fields surrounded by [ ] are optional.

Independent Sources: There are three types of independent sources that can be described to Spice. This includes a DC source, a frequency-swept AC generator and various types of time-varying signal generators. In addition, the independent signal associated with any one source can be either voltage or current. To simplify the discussion, we present a shortened summary of the description used to represent these various sources to Spice in Fig. 1.5. Also listed beside each source description is the type of analysis that would be most appropriate for the source type.

With regards to the Spice description of each of these signal sources, the first field begins with the letter V or I, depending on whether it is a voltage or a current source, respectively, followed by a unique 7 character name. The next two fields describe the nodes to which the source is connected to the rest of the network. It is important to respect the order of the nodes because of the signal polarity associated with the source. For example, in the case of a voltage source, the first node is connected to the positive side and the second node to the negative side. As far as the polarity (sign) of the current through a voltage source is concerned, the convention in Spice is as follows: Current flowing into the positive terminal of the source is taken as positive. In the case of a current source, a positive current is pulled from the positive node (n+) and returned to the negative node (n−). The next field specifies the nature of the signal source, ie. DC, AC or time-varying. The remaining fields are then used to specify the characteristics of the signal waveform generated by the particular source. The subsequent signal level parameters associated with the DC and AC sources should be

11

1 Introduction To Spice

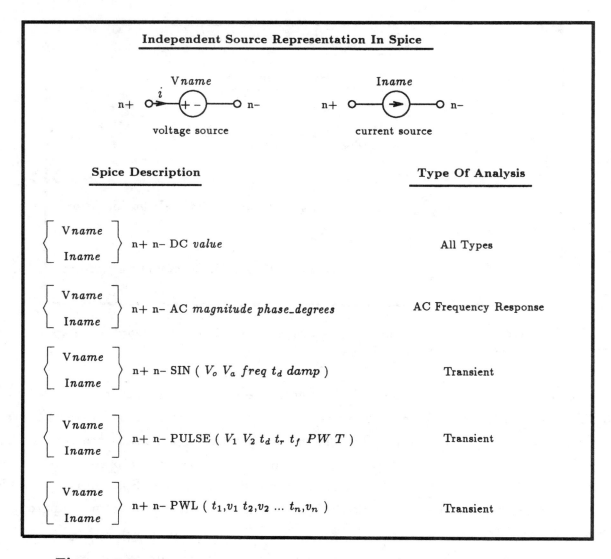

Figure 1.5 Independent sources and their Spice descriptions. Also shown is the analysis type for which it is normally used. One exception is for DC sources which are commonly used to set bias conditions in all types of circuits.

obvious from the Spice descriptions listed in Fig. 1.5. The signal level of the DC source is specified by the field labeled *value*. The peak amplitude and the phase (in degrees) of the AC source are simply specified in the fields labeled *magnitude* and *phase_degrees*, respectively. If the field labeled by *phase_degrees* is left blank, Spice will assume that the phase is zero.

In addition to DC and AC sources, one can also describe several different types of time-varying signal sources as listed in Fig. 1.5. Included in this list are element statements describing a sinusoidal signal (denoted by the SIN flag), a periodic pulse signal (PULSE) and an arbitrary waveform consisting of piece-wise linear segments (PWL). Obviously, due to the time-varying nature of these signals, these sources can only be used in transient

1.2 An Outline Of Spice

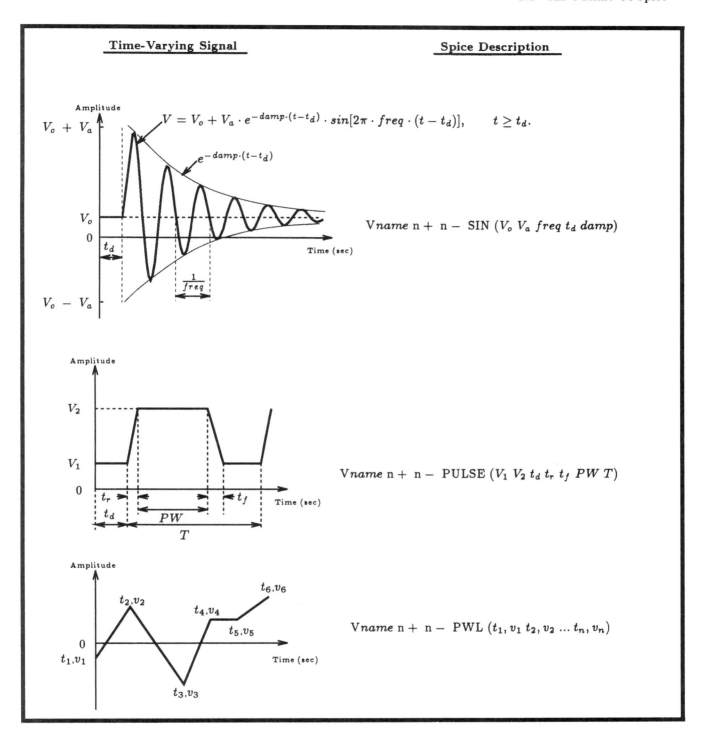

Figure 1.6 Various time-varying signals available in Spice and the corresponding element statements. Top curve: damped sinusoid; Middle curve: periodic pulse waveform; Bottom curve: piecewise linear waveform.

analysis studies. The syntax of these time-varying sources is more complex than that for the DC or AC sources described above, so to simplify matters, we list in Fig. 1.6, the corresponding waveform with the appropriate signal-determining parameters superimposed on each waveform. Although we express these waveforms in terms of voltage, it should be readily apparent that similar waveforms can be described for current sources.

For the circuit example of Fig. 1.3, the input sinusoidal voltage source v_I described by $10 \cdot \sin(2\pi \cdot 60 \cdot t)$ would have the following Spice description:

Vi 1 0 SIN (0V 10V 60Hz 0 0).

In many cases, the delay time t_d and the damping factor *damp* are both zero, so we will commonly shorten the above Spice statement to the following:

Vi 1 0 SIN (0V 10V 60Hz).

This is acceptable to Spice.

Linear Dependent Sources: There are four dependent sources that Spice knows about: voltage-controlled voltage source (VCVS), voltage-controlled current source (VCCS), current controlled voltage source (CCVS) and current-controlled current source (CCCS). These can be either linear or nonlinear, however, here we are only concerned with the linear ones. In Fig. 1.7 we depict these four dependent sources. The relationship between the input and output variables is clearly evident for each dependent source. Beside each controlled source in Fig. 1.7 we list the corresponding statement used to describe the element to Spice. The names of each dependent source begins with a unique letter (ie. E, G, H and F) followed by a unique 7 character name. This is exactly the same as for the passive elements described above.

Each dependent source is a two-port network with either the voltage or current at one port (terminals denoted as n+ and n−) being controlled by the voltage or current at the other port (terminals denoted as nc+ and nc−). In the case of a voltage-controlled dependent source, the controlling voltage is derived directly from the network node voltage variables. This is unlike that of a current-controlled source which must sense a current through a short circuit that is described to Spice using a zero-valued voltage source, ie. V*name* nc+ nc− 0. In comparison, a voltage-controlled source is completely described to Spice with one statement whereas a current-controlled source requires two statements. Remember to keep this in mind when working with current-controlled dependent sources.

The gain factor associated with the input and output variables is specified in the field labeled *value*. The dimensions of this gain factor will depend on the type of the dependent source.

As an example, the voltage-controlled voltage source in the circuit of Fig. 1.3 can be

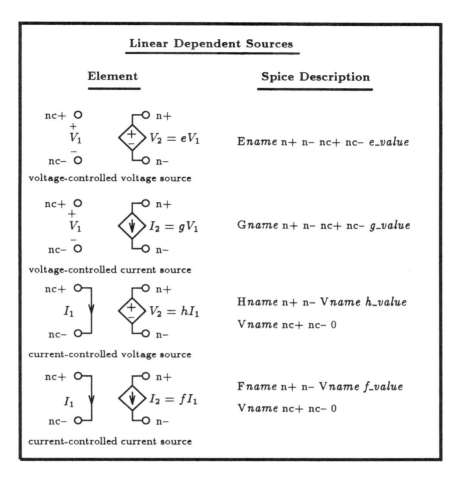

Figure 1.7 Linear dependent sources. Notice that the CCVS and the CCCS are both specified using two Spice statements, unlike the other two dependent sources.

described to Spice as follows:

E1 3 0 2 0 10.

Active Devices: Thus far our discussion on the elements describable to Spice has been limited to very basic electrical elements. The real computational strength of Spice lies in its capability to simulate the behavior of various types of active or electronic devices. This would include such devices as diodes, bipolar transistors, field-effect transistors, and more recent versions have been extended to include Gallium Arsenide transistors. Active devices are described to Spice in much the same manner as that used for the above electrical elements; a Spice description statement indicating the device type and name followed by the manner in which it is connected to the rest of the network. The subsequent fields, however, make reference to a specific model statement found on another line of the Spice input deck rather than list a set of device parameters. The model would then contain the parameters of the device and the nature of the device model, eg. *npn* bipolar transistor. Most active device

1 Introduction To Spice

Analysis Requests	Spice Command
Operating Point	.OP
DC Sweep	.DC *source_name start_value stop_value step_value*
AC Frequency Response	.AC DEC *points_per_decade freq_start freq_stop* .AC OCT *points_per_octave freq_start freq_stop* .AC LIN *total_points freq_start freq_stop*
Transient Response	.TRAN *time_step time_stop* [*no_print_time max_step_size*] [UIC] .IC V($node_1$)=*value* V($node_2$)=*value* ....

Table 1.4   Main analysis commands.

models are quite sophisticated and consist of many parameters. So this approach has the advantage that more than one device can reference the same model, simplifying data entry into the Spice input file.

At this time we are not ready to discuss the modeling of active devices so we shall defer this discussion until Chapter 3 where we will begin our study with diode circuits.

Analysis Requests

Once a circuit has been described and entered into the Spice input file, one must then specify the analysis that Spice should perform. As already mentioned, Spice has the capability of performing three main analyses: DC operating-point, AC frequency response, and transient response. The syntax of the Spice commands required to execute these analyses are listed in Table 1.4. Also included in this Table is the DC sweep command. Notice that each of these commands begins with a " · ". This dot signifies to Spice that the subsequent line is a command line requesting that Spice take action instead of trying to recognize that line as part of the circuit description. We shall see other dot commands in the next section and in later chapters of this text. Below we shall describe the meaning of the syntax used in these commands so that we can get on with simulating electronic circuits using Spice.

The command specifying a DC operating point calculation is simply .OP. This causes Spice to solve for the DC operating point of the given circuit. This includes finding all the DC node voltages and the currents and power dissipation of all voltage sources (both dependent and independent types). This command will also cause Spice to print the results of the calculation into the output file (discussed below). In many circuit simulations, one wants to vary the level of some DC source to determine the DC transfer characteristic. Rather than perform this analysis by repeated application of the .OP command while varying the level of some DC source, Spice provides a DC Sweep command (.DC) that performs this calculation automatically. The syntax of this command requires that one specify the name of the DC source (*source_name*) that will be varied beginning with the value marked by *start_value* and increased or decreased in steps of *step_value* until the value *stop_value* is reached. One can also vary the temperature of the circuit this way by simply replacing the name of the source in the field labeled *source_name* by TEMP.

The AC frequency response command (.AC) causes Spice to perform a linear small-signal frequency response analysis. Spice automatically calculates the DC operating-point of the circuit, thereby calculating the small-signal equivalent circuit of all nonlinear elements. The linear small-signal equivalent circuit is then analyzed at various frequency points beginning at *freq_start* and ending at *freq_stop*. Points in between are either spaced logarithmically by decade (DEC) or octave (OCT) with the number of points in a given frequency interval specified by the fields labeled *points_per_decade* or *points_per_octave*, respectively. Alternatively, one can specify a linear frequency sweep (LIN) between the frequency limits using a total number of points specified by the field marked *total_points*. One usually uses a linear frequency sweep when the bandwidth of interest is narrow and a logarithmic sweep when the bandwidth is large.

Finally, the transient response command (.TRAN) causes Spice to compute the various network variables as a function of time over a specified time interval. The time interval begins at time $t = 0$ and proceeds in linear steps of *time_step* seconds until *time_stop* seconds is reached. Although all transient analysis must begin at $t = 0$, one has the option of delaying the printing or plotting of the output results by specifying the *no_print_time* in the third field enclosed by the square brackets. This is a convenient way of skipping over the transient response of a network and viewing only its steady-state response. In order to have Spice avoid skipping over important waveform details within the time interval specified by *time_step*, the field designated by *max_step_size* should be chosen to be less than or equal to the *time_step* value. The origins of *max_step_size* are rather involved and interested readers should consult the *PSpice Users' Manual*. For most, if not all examples of this text, we chose the *max_step_size* equal to the *time_step*.

Prior to the start of any transient analysis, Spice must determine the initial values of the circuit variables. This is usually determined from a DC analysis of the circuit; however,

1 Introduction To Spice

if the optional UIC (use initial conditions) parameter is specified on the .TRAN statement, then Spice will skip the DC bias calculation and instead use only the "IC=" information supplied on each capacitor or inductor statement (see Fig. 1.4). All elements that do not have an "IC=" specification are assumed to have an initial condition of zero.

Initial conditions can also be set using an .IC command. This command simply clamps specific nodes of the circuit at the user-specified voltage levels during the DC bias calculation. The DC solution is then used as the initial conditions for the transient analysis. The syntax of the .IC statement is listed under the .TRAN command in Table 1.4. Note that this command is not used with the UIC flag of the transient analysis command.

Spice can perform many variations of the above mentioned analysis, as mentioned previously. However, it is felt that at this time introducing these additional commands would only burden our readers with details that will not be used until later chapters of this text. Therefore we shall defer discussion of these additional commands until a more convenient time.

Output Requests

The number of network variables generally associated with any circuit simulation is usually quite large. It would then be impractical to pass all the data associated with the network variables over to the user on the completion of Spice. Instead, Spice provides a display feature that enables the user to specify which network variables they want to see and the format that they would like to see it in. This is much like taking a measurement probe and placing it at some node in a given circuit. The types of formats include a printing method which lists in tabular form a set of network variable as a function of the independent variable associated with the analysis, or a graphical display (of a rather low quality) of the selected network variables. To invoke either the print or plot process, one specifies in the Spice input file one or more of the commands listed in Table 1.5 which we explain below.

To print a set of network variables, a print statement (.PRINT) must be inserted into the Spice input file. In addition, one must also specify the type of analysis (ie. DC, AC, or TRAN) for which the specified outputs are desired. Subsequently, the user specifies a list of voltage or current variables (denoted as *output_variables*). Generally, a voltage variable is specified as the voltage difference between two nodes, say $node_1$ and $node_2$, as V($node_1, node_2$). When one of the nodes is omitted, the omitted node is assumed to the ground node (0). In Spice, less freedom is available to specify branch currents. In fact, only those currents flowing through independent voltage sources can be observed. This would be specified by writing I(*Vname*) in the list of output variables where *Vname* is the name of

Output Requests	Spice Command
Print data points	.PRINT DC *output_variables*
	.PRINT AC *output_variables*
	.PRINT TRAN *output_variables*
Plot data points	.PLOT DC *output_variables* [(*lower_plot_limit*, *upper_plot_limit*)]
	.PLOT AC *output_variables* [(*lower_plot_limit*, *upper_plot_limit*)]
	.PLOT TRAN *output_variables* [(*lower_plot_limit*, *upper_plot_limit*)]

Notes:
1. Spice *output_variables* can be a voltage at any node V(*node*), the voltage difference between two nodes V($node_1$, $node_2$), or the current through a voltage source I(*Vname*).

2. AC output_variables can also be:

   Vr, Ir : real part
   Vi, Ii : imaginary part
   Vm, Im : magnitude
   Vp, Ip : phase
   Vdb, Idb : decibels

3. PSpice provides a greater flexibility for specifying output_variables.

Table 1.5 Spice output requests.

the independent voltage source through which the current is flowing. If a particular branch current is to be observed and no voltage source exists in series with the branch of interest, then one simply adds a zero-valued voltage source in series with this branch and request that the current flowing through this source be printed or plotted.

The variables printed for a DC analysis are the network node voltages or branch currents computed as a function of the level of a particular DC source in the network.

For an AC analysis, the output variables are sinusoidal or phasor quantities as a function of frequency and are represented by complex numbers. Spice provides several different ways in which to access these results; in the form of real and imaginary numbers or in magnitude and phasePhase: angle form. In the case of the magnitude and phase form, the magnitude can also be expressed in terms of dB's when it is convenient. To access a specific variable type, a suffix is appended to the letter V or I in the output variable list according to that seen in Table 1.5.

The results of a TRAN analysis are the network node voltages or branch currents computed as a function of time.

1 Introduction To Spice

Instead of printing a list of points in the output file, Spice provides a graphical feature which will generate a simple line plot from the list of output variables as a function the independent variable. The syntax for the plot command is identical to that specified for the print command except that the word .PRINT is replaced by .PLOT. Also, the range of the y-axis given by (*lower_plot_limit, upper_plot_limit*) can be specified as an optional field on the .PLOT command line — See Table 1.5.

There are no restrictions on the number of .PRINT or .PLOT commands that can be specified in the Spice input file. This is a convenient way of controlling the number of data columns appearing in the output file.

A Simple Example

For the simple circuit of Fig. 1.3 let us consider creating a Spice input file that would be used to compute the transient response of this circuit for 3 periods of the input 10 V, 60 Hz sinusoidal signal. Based on the above outline, the Spice input file for this circuit would appear as follows:

```
Transient Response Of A Linear Network

** Circuit Description **
* input signal source
Vi 1 0 SIN ( 0V 10V 60Hz )
* linear network
R1     1 2 1Ohm
C1     2 0 2.65mF
Rload 3 0 10Ohm
E1 3 0 2 0 10

** Analysis Request **
* compute transient response of circuit over three full
* periods (50 ms) of the 60 Hz sine-wave input with a 1 ms
* sampling interval
.TRAN 1ms 50ms 0ms 1ms

** Output Request **
* print the output and input time-varying waveforms
.PRINT TRAN V(3) V(1)
* plot the output and input time-varying waveforms
* set the range of the y-axis between -100 and +100 V
.PLOT   TRAN V(3) V(1) (-100,+100)

* indicate end of Spice deck
.end
```

Here the first line begins with the title: "Transient Response Of A Linear Network," followed by a circuit description, analysis request and several output request statements. The final statement is an .END statement. Many comments are sprinkled throughout this file in order to improve its readability. The transient analysis statement,

.TRAN 1ms 50ms 0ms 1ms,

is a request to compute the transient behavior of this circuit over a 50 ms interval using a 1 ms time-step. Furthermore, the results of the analysis will be stored in resident memory beginning at time $t = 0$ and will later be available for printing or plotting. The last field of this statement specifies that the maximum step size is to be limited to 1 ms; the same value as that used for the time-step. In almost all cases in this text, we shall set the maximum step size equal to the time-step. Finally, to observe the output response behavior, we request that the voltage at the output (node 3), together with the voltage appearing across the input terminal (node 1), be both printed and plotted. In the case of the plot command, we are further requesting that the two node voltages be plotted on the same graph with the $y-axis$ having a range varying between −100 and +100 V.

1.2.3 Output From Spice

Once the Spice input file is complete, the Spice computer program is executed with reference to this file, and the results will ultimately be found in what's known as the **Spice output file**. One is then at liberty to examine the contents of this file to observe the results of the different analyses requested in the Spice input file. If other analysis is needed, one must edit the original Spice input file by either altering existing element statements or by adding additional analysis commands and re-executing the Spice program. Obviously, this process can be repeated until one obtains all the information that they require.

For the example presented above, the results found in Spice output file appear as follows:

```
******* 11/19/91 ******* Student PSpice (Dec. 1987) ******* 10:24:00 *******

Transient Response Of A Linear Network

****        CIRCUIT DESCRIPTION

******************************************************************************

** Circuit Description **
* input signal source
Vi  1 0 SIN ( 0V 10V 60Hz )
* linear network
R1     1 2 1Ohm
C1     2 0 2.65mF
Rload  3 0 10Ohm
E1     3 0 2 0 10

** Analysis Request **
* compute transient response of circuit over three full
* periods (50 ms) of the 60 Hz sine-wave input with a 1 ms
```

1 Introduction To Spice

```
* sampling interval
.TRAN 1ms 50ms 0ms 1ms

** Output Request **
* print the output and input time-varying waveforms
.PRINT TRAN V(3) V(1)
* plot the output and input time-varying waveforms
* set the range of the y-axis between -100 and +100 V
.PLOT  TRAN V(3) V(1) (-100,+100)

* indicate end of Spice deck
.end
```

******* 11/19/91 ******* Student PSpice (Dec. 1987) ******* 10:24:00 *******

Transient Response Of A Linear Network

****   INITIAL TRANSIENT SOLUTION   TEMPERATURE =   27.000 DEG C

****

NODE	VOLTAGE	NODE	VOLTAGE	NODE	VOLTAGE	NODE	VOLTAGE
(1)	0.0000	(2)	0.0000	(3)	0.0000		

VOLTAGE SOURCE CURRENTS
NAME   CURRENT

Vi   0.000E+00

TOTAL POWER DISSIPATION   0.00E+00 WATTS

******* 11/19/91 ******* Student PSpice (Dec. 1987) ******* 10:24:00 *******

Transient Response Of A Linear Network

****   TRANSIENT ANALYSIS   TEMPERATURE =   27.000 DEG C

****

TIME	V(3)	V(1)
0.000E+00	0.000E+00	0.000E+00
1.000E-03	6.504E+00	3.652E+00
2.000E-03	2.120E+01	6.745E+00
3.000E-03	3.923E+01	8.920E+00
4.000E-03	5.645E+01	9.842E+00
5.000E-03	6.896E+01	9.382E+00
6.000E-03	7.398E+01	7.604E+00
7.000E-03	7.010E+01	4.758E+00
8.000E-03	5.739E+01	1.244E+00
9.000E-03	3.731E+01	-2.445E+00
1.000E-02	1.247E+01	-5.790E+00
1.100E-02	-1.381E+01	-8.322E+00
1.200E-02	-3.792E+01	-9.686E+00
1.300E-02	-5.657E+01	-9.689E+00
1.400E-02	-6.716E+01	-8.331E+00
1.500E-02	-6.825E+01	-5.803E+00

1.2 An Outline Of Spice

```
 1.600E-02  -5.971E+01  -2.460E+00
 1.700E-02  -4.276E+01   1.228E+00
 1.800E-02  -1.977E+01   4.744E+00
 1.900E-02   6.007E+00   7.594E+00
 2.000E-02   3.095E+01   9.377E+00
 2.100E-02   5.155E+01   9.843E+00
 2.200E-02   6.492E+01   8.927E+00
 2.300E-02   6.917E+01   6.757E+00
 2.400E-02   6.371E+01   3.638E+00
 2.500E-02   4.931E+01   8.124E-03
 2.600E-02   2.798E+01  -3.623E+00
 2.700E-02   2.716E+00  -6.745E+00
 2.800E-02  -2.292E+01  -8.920E+00
 2.900E-02  -4.535E+01  -9.842E+00
 3.000E-02  -6.140E+01  -9.382E+00
 3.100E-02  -6.883E+01  -7.604E+00
 3.200E-02  -6.659E+01  -4.758E+00
 3.300E-02  -5.500E+01  -1.244E+00
 3.400E-02  -3.568E+01   2.445E+00
 3.500E-02  -1.136E+01   5.790E+00
 3.600E-02   1.456E+01   8.322E+00
 3.700E-02   3.844E+01   9.686E+00
 3.800E-02   5.692E+01   9.689E+00
 3.900E-02   6.740E+01   8.331E+00
 4.000E-02   6.842E+01   5.803E+00
 4.100E-02   5.983E+01   2.460E+00
 4.200E-02   4.283E+01  -1.228E+00
 4.300E-02   1.982E+01  -4.744E+00
 4.400E-02  -5.972E+00  -7.594E+00
 4.500E-02  -3.093E+01  -9.377E+00
 4.600E-02  -5.154E+01  -9.843E+00
 4.700E-02  -6.491E+01  -8.927E+00
 4.800E-02  -6.917E+01  -6.757E+00
 4.900E-02  -6.371E+01  -3.638E+00
 5.000E-02  -4.965E+01  -2.800E-06


******* 11/19/91 ******* Student PSpice (Dec. 1987) ******* 10:24:00 *******

Transient Response Of A Linear Network

****        TRANSIENT ANALYSIS              TEMPERATURE =    27.000 DEG C

*******************************************************************************


  LEGEND:

*: V(3)
+: V(1)

  TIME        V(3)
(*+)---------  -1.0000E+02  -5.0000E+01   0.0000E+00   5.0000E+01   1.0000E+02
             - - - - - - - - - - - - - - - - - - - - - - - - - -
  0.000E+00  0.000E+00 .             .            X            .            .
  1.000E-03  6.504E+00 .             .            .+*          .            .
  2.000E-03  2.120E+01 .             .            . +    *     .            .
  3.000E-03  3.923E+01 .             .            . +         *.            .
  4.000E-03  5.645E+01 .             .            .  +         . *          .
  5.000E-03  6.896E+01 .             .            .  +         .       *    .
  6.000E-03  7.398E+01 .             .            .  +         .         *  .
  7.000E-03  7.010E+01 .             .            .+           .        *   .
  8.000E-03  5.739E+01 .             .            +            . *          .
  9.000E-03  3.731E+01 .             .            +.           *            .
  1.000E-02  1.247E+01 .             .            + .  *       .            .
```

```
1.100E-02 -1.381E+01 .         .         * +.        .         .
1.200E-02 -3.792E+01 .         .    *    + .         .         .
1.300E-02 -5.657E+01 .         *         + .         .         .
1.400E-02 -6.716E+01 .       * .         + .         .         .
1.500E-02 -6.825E+01 .       * .         + .         .         .
1.600E-02 -5.971E+01 .        *.         +.         .         .
1.700E-02 -4.276E+01 .         . *       +           .         .
1.800E-02 -1.977E+01 .         .       * .+          .         .
1.900E-02  6.007E+00 .         .         . X         .         .
2.000E-02  3.095E+01 .         .         . +    *    .         .
2.100E-02  5.155E+01 .         .         .  +        . *       .
2.200E-02  6.492E+01 .         .         .  +        .    *    .
2.300E-02  6.917E+01 .         .         .  +        .      *  .
2.400E-02  6.371E+01 .         .         . +         .    *    .
2.500E-02  4.931E+01 .         .         . +         .*         .
2.600E-02  2.798E+01 .         .         +. *        .         .
2.700E-02  2.716E+00 .         .        +  .*        .         .
2.800E-02 -2.292E+01 .         .   *    + .          .         .
2.900E-02 -4.535E+01 .        .*         + .         .         .
3.000E-02 -6.140E+01 .         *         + .         .         .
3.100E-02 -6.883E+01 .       * .         + .         .         .
3.200E-02 -6.659E+01 .       * .         +.         .         .
3.300E-02 -5.500E+01 .        *.         +          .         .
3.400E-02 -3.568E+01 .         .    *    .+         .         .
3.500E-02 -1.136E+01 .         .       * . +        .         .
3.600E-02  1.456E+01 .         .         . +  *     .         .
3.700E-02  3.844E+01 .         .         .  +     * .         .
3.800E-02  5.692E+01 .         .         .  +       .  *      .
3.900E-02  6.740E+01 .         .         .  +       .     *   .
4.000E-02  6.842E+01 .         .         .  +       .     *   .
4.100E-02  5.983E+01 .         .         . +        .   *     .
4.200E-02  4.283E+01 .         .         . +        * .        .
4.300E-02  1.982E+01 .         .         +.    *    .         .
4.400E-02 -5.972E+00 .         .         X .        .         .
4.500E-02 -3.093E+01 .         .    *    +.         .         .
4.600E-02 -5.154E+01 .         *         + .         .         .
4.700E-02 -6.491E+01 .       * .         + .         .         .
4.800E-02 -6.917E+01 .      *  .         + .         .         .
4.900E-02 -6.371E+01 .       * .         +.         .         .
5.000E-02 -4.965E+01 .         *         +          .         .
- - - - - - - - - - - - - - - - - - - - - - - - - - - - - - - - - - - - - - -

JOB CONCLUDED

TOTAL JOB TIME        5.82
```

The output file from Spice contains four separate parts: (1) a replica of the Spice input file or circuit description, (2) the initial conditions for the transient analysis, (3) the results of the transient analysis in tabular form generated by the .PRINT command, and (4) a graphical plot of the transient results produced by the .PLOT command. As is evident from the graphical results, the output waveform (V(3)), denoted by star-signs (*), almost completes three period of oscillation, and lags behind the input voltage waveform (V(1)), denoted by plus-signs (+), by about 2 ms or 45 degrees. The transient portion of the output waveform is rather short, less than one complete cycle of the input signal of 60 Hz. The amplitude of the output voltage can be seen to be approximately 70 V. Exact values that correspond to points on the waveform can be read from the two columns of numbers positioned on the left hand side of the graphical plot; the first column denotes the time axis, and the second column denotes the corresponding values of the output voltage. At time $t = 4$ ms, the value of the output voltage is seen to be 56.45 V. To determine the voltage level of the input signal at this time, one must refer to the table of numbers computed by the .PRINT command. On doing so, one would find at $t = 4$ ms the input voltage is 9.842 V. The value of both the input and output voltage levels at other times can be found in exactly the same way.

1.3 Output Post-Processing Using Probe

To improve the accessibility of the information generated by Spice, commercial vendors of Spice are making available post-processing facilities that allow the users of Spice to graphically display their results on a computer monitor. This not only provides easier access to the results computed by Spice, it also generates a higher-quality graph than the line-plot produced by Spice. Thus, more detail in the plotted waveform becomes visible. Furthermore, cursor facilities are made available that enable the user to access any point of the graph and determine its numerical value. This relieves the user from having to search a long table of numerical values in the Spice output file to find specific values of a particular waveform.

In this text we make specific use of the *Probe* facility available with the PSpice program supplied by the MicroSim Corporation. Probe is meant to function like a software version of an oscilloscope. Probe enables the user to look at various results of a simulation using an interactive graphics processor. In addition, Probe has many built-in computational capabilities that allow for an interactive investigation of circuit behavior after a completed PSpice simulation. For example, one can compute and graphically display the instantaneous power dissipated by a transistor by multiplying its collector current as a function of time by the corresponding collector-emitter voltage. The mathematical commands available in Probe are powerful, including such mathematical functions as integration and differentiation, and a quick summary of these is given in Table 1.6. The variable x used in the argument of each function represents any Spice recognizable output variable, but also includes additional network variables created by PSpice. For easy reference, we list in Table 1.7 the types of variables created by PSpice and recognized by Probe.

In order to use Probe, one must include in the Spice input file a **.Probe** statement. This statement causes PSpice to create the necessary data file for later use by Probe. The contents of this data file will include **all** the network variables associated with the present simulation, eg. the results of DC sweep, AC frequency response and transient response. Another useful feature of Probe is that hardcopies of any graphical result can easily be created for future

Probe Command	Available Functions
abs(x)	$\|x\|$
sgn(x)	+1 (if $x > 0$), 0 (if $x = 0$), -1 (if $x < 0$)
sqrt(x)	$x^{1/2}$
exp(x)	e^x
log(x)	$ln(x)$ (log base e)
log10(x)	$log(x)$ (log base 10)
db(x)	$20 log(\|x\|)$ (log base 10)
pwr(x,y)	$\|x\|^y$
sin(x)	$sin(x)$ (x in radians)
cos(x)	$cos(x)$ (x in radians)
tan(x)	$tan(x)$ (x in radians)
atan(x)	$tan^{-1}(x)$ (result in radians)
arctan(x)	$tan^{-1}(x)$ (result in radians)
d(x)	derivative of x with respect to the X-axis variable.
s(x)	integral of x over the range of the X-axis variable.
avg(x)	running average of x over the range of the X-axis variable.
rms(x)	running RMS average of x over the range of the X-axis variable.
min(x)	minimum of x.
max(x)	maximum of x.

Table 1.6 Probe mathematical functions.

reference. It is, in fact, these plots that we shall use to illustrate the results of most of the simulations performed in this text.

To illustrate the simplicity of using Probe, consider adding a .Probe statement to the above listed Spice deck and then re-run the Spice deck through PSpice. On completion, the results of the PSpice simulation are stored in a special file for later use by Probe. Invoking the Probe program enables one to view the simulation results directly on the screen of the computer monitor. For example, in Fig. 1.8 we display an actual view of the computer screen as seen by the user of Probe. Here we have plotted the voltage waveform associated with node 3 of the circuit shown in Fig. 1.3. Seen in the lower righthand corner of the screen are the x and y axis values that correspond to the point at which the two cursors lie on the voltage waveform. For instance, in the case of the first cursor, C1, it is positioned on the second negative peak of the waveform corresponding to a time of 31.734 ms and a voltage value of –69.079 V. Likewise, the second cursor, C2, is positioned on the second positive peak of the waveform corresponding to a time of 22.734 ms and a voltage value of 69.961 V.

Voltage Variables		
Node Voltage	Voltage across a two-terminal element	Voltage at transistor terminal x †
V(*node*) V(*node₁*, *node₂*)	V(*element_name*)	Vx(*trans_name*)

Current Variables	
Current through a two-terminal element	Current into transistor terminal x †
I(*element_name*)	Ix(*trans_name*)

† x can be any one of the following transistor terminals:
    BJT (Q): C (collector) B (base) E (emitter) S (substrate)
    FET (B, J, M): D (drain) G (gate) S (source) B (bulk, substrate)

‡ AC suffixes can also be appended – see Table 1.5.

Table 1.7 Variables generated by PSpice and recognized by Probe.

Below the co-ordinates of these two cursors is the relative distance the two cursors are apart. From this difference, we see that the peak-to-peak amplitude of this waveform is 139.043 V.

Two or more traces can be added to the same graph, as is shown in Fig. 1.9, and the two cursors can be used to access information on either graph. Probe allows other variants of these graphical features; see the *PSpice Users' Manual* for more details.

1.4 Examples

To further illustrate the capabilities of Spice, let us explore four different circuit examples. The first example will involve calculating the DC node voltages of a linear network, the second example will explore the transient behavior of a three-stage linear amplifier subject to a sine-wave input, the third example will illustrate how circuit initial conditions are established

1 Introduction To Spice

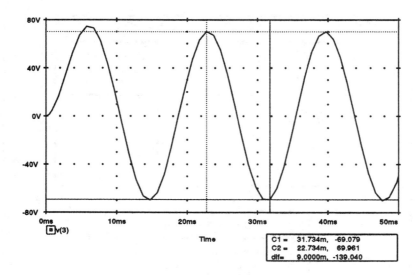

Figure 1.8 Screen display seen by user of the Probe facility of PSpice. Two cursors are superimposed on the waveform in order to read values directly off the waveform.

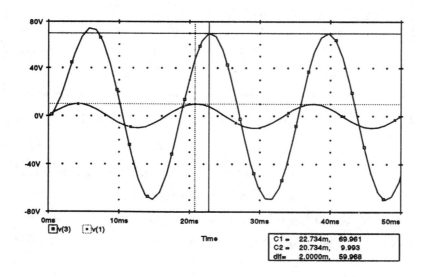

Figure 1.9 Screen display seen by user of the Probe facility of PSpice. Two waveforms are present on the same graph with separate cursors placed on each.

during a transient analysis, and finally, the last example of this chapter will compute the frequency behavior of a linear amplifier.

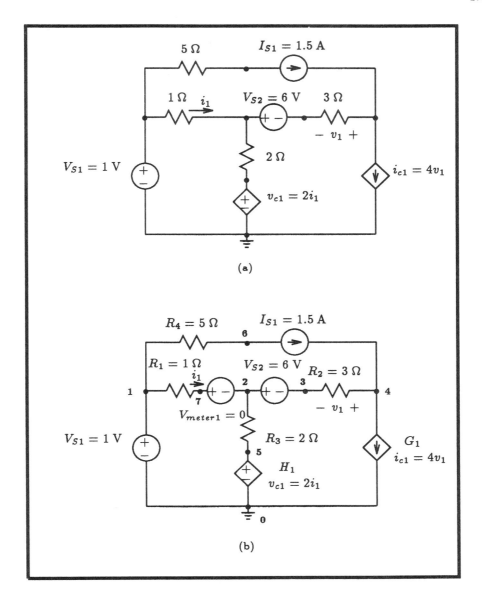

Figure 1.10 (a) Resistive network with dependent sources. (b) Each node is assigned a non-negative integer number and each element is assigned a unique name. Also, a zero-valued voltage source is placed in series with R_1 to monitor the current denoted by i_1.

1.4.1 Example 1: DC Node Voltages Of A Linear Network

Our first circuit example is shown in Fig. 1.10(a) which displays a rather complicated network of resistors and sources. Sources V_{S1}, V_{S2}, and I_{S1}, are independent DC sources whose values are given on the circuit schematic. The remaining two sources, v_{c1} and i_{c1}, are dependent sources where v_{c1} is a current-controlled voltage source (CCVS) and i_{c1} is a voltage-controlled current source (VCCS). In the case of v_{c1}, the voltage generated by this source is proportional to the current that flows through the 1 Ω resistor, designated by i_1.

```
Resistive Network With Dependent Sources

** Circuit Description **
* signal sources
Vs1 1 0 dc 1V
Vs2 2 3 dc 6V
Is1 6 4 dc 1.5A
* resistors
R1 1 7 1ohm
R2 3 4 3ohm
R3 2 5 2ohm
R4 1 6 5ohm
* CCVS with ammeter
H1 5 0 Vmeter1 2
Vmeter1 7 2 0
*VCCS
G1 4 0 4 3 4

** Analysis Requests **
* compute DC solution
.OP

** Output Requests **
* by default the ".OP" command prints all node voltages
.end
```

Figure 1.11 Spice input deck for the circuit shown in Fig. 1.10(b).

On the other hand, the current generated by i_{c1} is proportional to the voltage appearing across the 3 Ω resistor.

The first step in preparing the circuit schematic of Fig. 1.10(a) for a Spice simulation is to identify each element of the circuit by assigning it a unique name, and then, label each node of the circuit with some non-negative integer. It is also necessary to label the ground node of the circuit as node 0. The results of this labeling process are shown in the circuit diagram displayed in Fig. 1.10(b). Further, due to the presence of the CCVS, a zero-valued voltage source must be placed in series with the 1 Ω resistor (R_1) in order to sense the current through it. Thus, we have added a zero-valued voltage source V_{meter1} in series with R_1. Recall form Table 1.7 that this is a necessary requirement for describing current-controlled sources to Spice.

Our next step is to describe each element of the network to Spice using the element statements described in the previous sections. The results of this are listed in the Spice input deck shown in Fig. 1.11. The first line of the Spice input deck is the title of this particular example. In this case, "Resistive Network With Dependent Sources." This is then followed

```
****        SMALL SIGNAL BIAS SOLUTION        TEMPERATURE =   27.000 DEG C
*****************************************************************************

    NODE    VOLTAGE     NODE    VOLTAGE     NODE    VOLTAGE     NODE    VOLTAGE

(    1)     1.0000  (    2)     .8462   (    3)    -5.1538  (    4)    -4.8077

(    5)     .3077   (    6)    -6.5000  (    7)     .8462

    VOLTAGE SOURCE CURRENTS
    NAME         CURRENT

    Vs1         -1.654E+00
    Vs2         -1.154E-01
    Vmeter1      1.538E-01

    TOTAL POWER DISSIPATION   2.35E+00   WATTS

****        OPERATING POINT INFORMATION      TEMPERATURE =   27.000 DEG C
*****************************************************************************

**** VOLTAGE-CONTROLLED CURRENT SOURCES

NAME         G1
I-SOURCE     1.385E+00

**** CURRENT-CONTROLLED VOLTAGE SOURCES

NAME         H1
V-SOURCE     3.077E-01
I-SOURCE     2.692E-01
```

Figure 1.12 DC node voltages of the circuit shown in Fig. 1.10(b). Also shown are the voltages and currents associated with the independent and dependent sources.

by a series of lines describing the circuit to Spice beginning with the comment statement: ** Circuit Description **. Notice that each circuit element description in the Spice input deck corresponds directly with an element of the circuit shown in Fig. 1.10(b). Following this circuit description, we list the analysis command: .OP. This tells Spice to compute the DC node voltages of the circuit. Normally, we would then follow this command, and any other analysis request command, by a series of output requests; however, by default, the .OP command prints all the node voltages into the Spice output file. The final statement is an ".end" statement, signifying an end to the Spice input file.

Submitting this input file to Spice for execution, would on completion, result, in part, in the Spice output file shown in Fig. 1.12. The only part not shown is a description of the input circuit which was already given in Fig. 1.11. The results that are shown consist of two parts: a small-signal bias solution and operating point information. The small-signal bias solution refers specifically to the voltages on each node of the circuit relative to node 0 and the current flowing through each independent voltage source. Notice that the current supplied by V_{S1}

1 Introduction To Spice

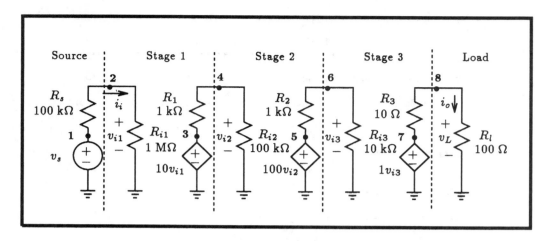

Figure 1.13 A three stage amplifier with input signal and load.

and V_{S2} is negative. Thus according to the Spice convention (positive current flows from the positive terminal of the voltage source to its negative terminal) both currents are actually flowing away from the positive terminal of each source. Also calculated, and included in the output file, is the total power dissipated by the circuit. The second part of output file refers to the DC operating point information of the dependent sources. Specifically, the controlled signal (not the controlling signal) of each dependent source is listed in the output file. In the case of dependent voltage sources, both the controlled voltage and the current it supplies to the circuit are given. The current that controls this source is the current that flows through Vmeter1 (1.538E-01) seen listed in the "SMALL SIGNAL BIAS SOLUTION".

1.4.2 Example 2: Transient Response Of A 3-Stage Linear Amplifier

Our next example is shown in Fig. 1.13 which consists of a three-stage linear amplifier fed by a signal source having a source resistance of 100 kΩ. This example was first encountered in section 1.5 of the text by Sedra and Smith where they computed by hand the overall voltage gain, $A_v = v_L/v_s$, to be 743.6 V/V, the current gain, $A_i = i_o/i_i$, to be 8.18×10^6 A/A, and the power gain, $A_p = A_v \cdot A_i$, to be 98.3 dB. Here we shall compute the same gains by using the transient analysis capability of Spice and the graphical post-processing features of Probe, and compare the results with those found by hand.

We begin our analysis by creating the Spice circuit description shown in Fig. 1.14 for the circuit of Fig. 1.13. All nodes have been pre-labeled except for the ground node which is assumed to be node 0. The input generator is a 1 volt time-varying sinusoidal voltage source of 1 Hz frequency with zero voltage offset. Using the .TRAN statement we are instructing PSpice to calculate the time response of the circuit from $t = 0$ to $t = 5$ s in 10 ms time steps. Instead of specifying a specific output request, we will simply utilize Probe, the graphical

```
Transient Response Of A 3-Stage Linear Amplifier

** Circuit Description **
* signal source
Vs 1 0 sin (0V 1V 1Hz)
Rs 1 2 100k
* stage 1
Ri1 2 0 1Meg
E1 3 0 2 0 10
R1 3 4 1k
* stage 2
Ri2 4 0 100k
E2 5 0 4 0 100
R2 5 6 1k
* stage 3
Ri3 6 0 10k
E3 7 0 6 0 1
R3 7 8 10
* output load
Rl 8 0 100

** Analysis Requests **
* compute transient response from t=0 to 5s in time steps of
* 10ms with an internal time-step no greater than 10ms.
.TRAN 10ms 5s 0s 10ms

** Output Requests **
* graphical post-processor
.PROBE
.end
```

Figure 1.14 Spice input deck for circuit shown in Fig. 1.13.

post-processor facility of PSpice, to compute the various signal gains. This requires that we enter the .Probe command into the Spice input file as shown in Fig. 1.14.

On completion of Spice, the results of this analysis can be viewed using the Probe facility of PSpice. In Fig. 1.15(a) we plot both the input voltage (v(1)) and output voltage (v(8)) as a function of time. The range of the y-axis was set equal to the peak-to-peak value of the signal that is shown. Because the two-signals are in-phase, the voltage gain V(8)/V(1) is simply the ratio of the peak values, which, in this case, turns out to be 743.8 V/V.

Similarly, the waveforms of the input current (i_i=i(Ri1)) and the load current (i_o=i(Rl)) are shown in Fig. 1.15(b). The resulting current gain provided by this network is then calculated to be 8.18 MA/A.

As a final calculation, Probe was used to compute and display the instantaneous power delivered to the amplifier stage (v(2)*i(Ri1)) and to its load (v(8)*i(Rl)). These waveforms

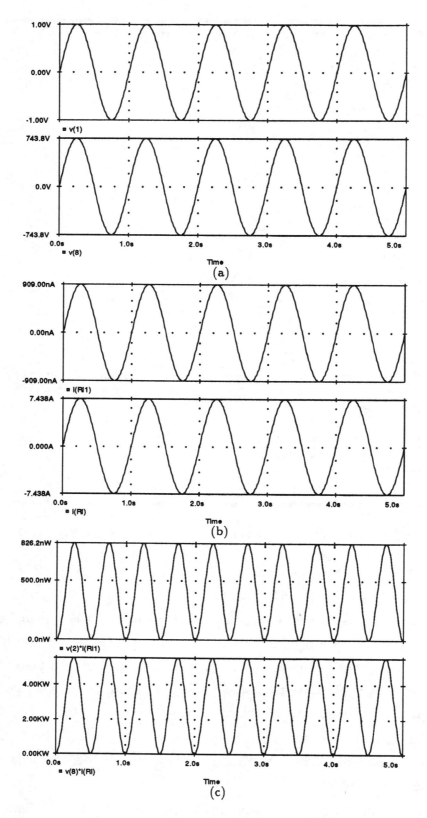

Figure 1.15 Various transient results for the circuit of Fig. 1.13: (a) input and output voltage signals; (b) input and output current signals; (c) power delivered to amplifier and load.

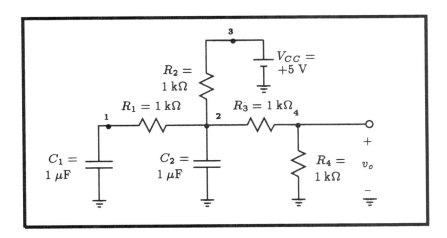

Figure 1.16 An RC network for investigating the different ways in which Spice sets the initial conditions prior to the start of a transient analysis.

are shown in Fig. 1.15(c). The power gain is then found to be 6.69 GW/W or 98.3 dB.

Comparing these results with those performed by Sedra and Smith using hand analysis we find that we are in exact agreement, as expected.

1.4.3 Example 3: Setting Circuit Initial Conditions During A Transient Analysis

There are three different ways in which the initial conditions of a circuit can be set at the start of a transient analysis. In the following we shall demonstrate these three ways on the simple RC circuit shown in Fig. 1.16.

In the first case, let us begin our transient analysis with the DC operating point of the circuit establishing the initial conditions for the circuit. The Spice input file for this circuit is shown in Fig. 1.17. The transient analysis request seen there simply commands Spice to compute the behavior of the RC circuit over a 10 ms interval using a 500 μs step interval. For the purpose of comparison, we shall request that Spice plot the voltages across each capacitor as well as the voltage that appears at the output terminals. In this way, we can observe the initial conditions established by Spice at the start of the transient analysis and the effect that they have on the output.

On completion of Spice, the results of the analysis are shown plotted in Fig. 1.18. The top graph displays the voltage appearing across capacitor C_1, the middle graph displays the voltage appearing across capacitor C_2, and the bottom graph displays the voltage appearing at the output. As is clearly evident in all three cases, no change in the output voltage is taking place. This suggests that the initial conditions found by Spice were also the final time values. To illustrate that these initial conditions correspond to the DC operating point solution, we list below the results of an .OP analysis:

1 Introduction To Spice

```
Investigating Initial Conditions Established By Spice

** Circuit Description **
Vcc 3 0 DC +5V
R1 1 2 1k
R2 3 2 1k
R3 2 4 1k
R4 4 0 1k
C1 1 0 1uF
C2 2 0 1uF

** Analysis Requests **
.Tran 500us 10ms 0ms 500us

** Output Requests **
.PLOT TRAN V(1) V(2) V(4)
.probe
.end
```

Figure 1.17 Spice input deck for circuit shown in Fig. 1.16. No explicit initial conditions are indicated.

```
****    SMALL SIGNAL BIAS SOLUTION       TEMPERATURE =   27.000 DEG C
*****************************************************************************

 NODE   VOLTAGE     NODE   VOLTAGE     NODE   VOLTAGE     NODE   VOLTAGE

(   1)    3.3333  (   2)    3.3333  (   3)    5.0000  (   4)    1.6667
```

Let us consider setting the initial voltage across capacitor C_1 at +1 V and observe the effect that it has on the circuit operation. To do this, we simply modify the element statement for C_1 according to

C1 1 0 1uF IC=+1V

and specify that Spice use this initial condition by attaching on the end of the .TRAN statement the flag "UIC" according to

.TRAN 500us 10ms 0ms 500us UIC.

The results of this analysis are shown in Fig. 1.19. Unlike the previous case, the voltages in the circuit are now changing with time. At time $t=0$, we see that the voltage across capacitor C_1 is +1 V, as expected. The voltage across C_2 at this time is 0 V by default (since it was not specified in the Spice deck), and as a result, the output voltage is initially zero. As time progresses, we see that these three voltages converge to values identical to those found in the previous case (ie. 3.33 V, 3.33 V and 1.667 V, respectively).

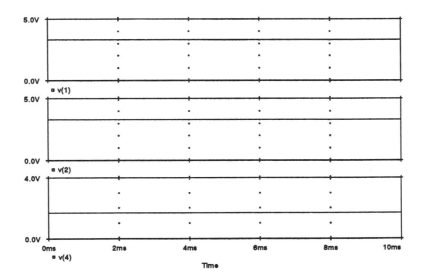

Figure 1.18 Voltage waveforms associated with the RC circuit shown in Fig. 1.16. No initial conditions were explicitly given, instead the DC operating point solution is used as the circuit initial conditions. The top graph displays the voltage appearing across capacitor C_1, the middle graph displays the voltage appearing across capacitor C_2, and the bottom graph displays the voltage appearing at the output.

Another way of setting the initial conditions is with the .IC command line. This method is essentially a combination of the two previous methods. The specific node voltages can be explicitly set and the remaining nodes will take on values that result from the DC operating point analysis (with the initial value of appropriate nodes taken into account) instead of defaulting to zero.

For example, let us set the voltage at node 1 by using the following .IC command line:

```
.IC V(1)=+1V
```

Furthermore, we shall remove any previous reference to the initial conditions, that is, change the element statement for C_1 back to its original form,

```
C1 1 0 1uF
```

and remove the UIC flag on the .TRAN statement so that it re-appears in the Spice deck as follows:

```
.TRAN 500us 10ms 0ms 500us.
```

The results of this analysis are shown in Fig. 1.20. Here we see that at time $t = 0$ the

1 Introduction To Spice

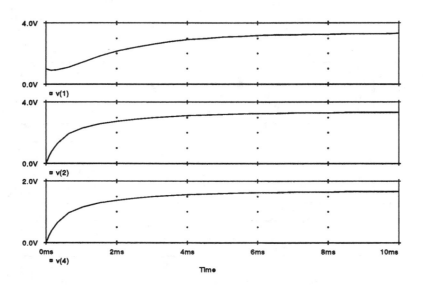

Figure 1.19 Voltage waveforms associated with the RC circuit shown in Fig. 1.16 when the voltage across capacitor C_1 is initially set to +1 V using IC=+1V on the element statement of this capacitor.

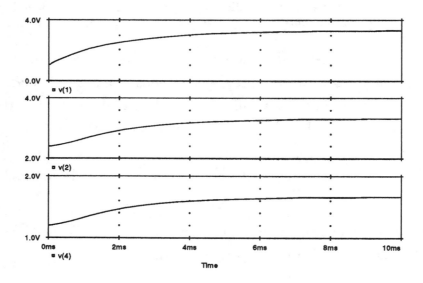

Figure 1.20 Voltage waveforms associated with the RC circuit shown in Fig. 1.16 when the voltage at node 1 is initially set to +1 V using an .IC command.

voltage at node 1 begins at +1 V as specified. In contrast to the previous case, the voltage at the second node does not begin at 0 V but instead begins with a value of 2.4 V. Likewise, the output voltage is no longer zero but instead begins at a voltage level of 1.20 V. The final

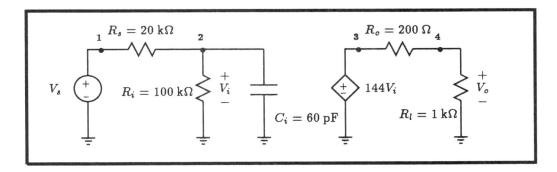

Figure 1.21 A frequency-dependent voltage amplifier with signal input and load.

values settle to the same values found previously in the other two cases.

1.4.4 Example 4: Frequency Response Of A Linear Amplifier

The final example of this chapter is to demonstrate how Spice is used to compute the frequency response of a linear amplifier. Consider the small-signal equivalent circuit of a one-stage amplifier shown in Fig. 1.21. The corresponding Spice input file used to describe this circuit is listed in Fig. 1.22. The input to the circuit is a 1 V AC voltage source whose frequency will be varied between 1 Hz and 100 MHz logarithmically with 5 points-per-decade, as is indicated by the .AC analysis command. By selecting a 1 V input level, the output voltage level will also correspond to the voltage transfer function V_o/V_s since $V_s = 1$.

The frequency response behavior of this amplifier was calculated by Spice and the magnitude and phase of the output voltage V_o was plotted using Probe. The plot seen on the screen of the computer monitor is shown in Fig. 1.23. It consists of two graphs: the top graph indicates the magnitude response and the bottom graph indicates the phase response of the amplifier. As a rough estimate of the 3 dB bandwidth of this amplifier, we see from this graph that it ranges somewhere around 100 kHz. A better estimate of the 3 dB bandwidth is obtained using the cursor facility of Probe and found to be 158.5 kHz.

1.5 Spice Tips

- Spice is an acronym for *Simulation Program with Integrated Circuit Emphasis*. It was originally developed for large main-frame computers.
- PSpice is a PC-version of Spice and a student version is distributed freely by MicroSim Corporation.
- Circuits are designed by people not computers; Spice can only verify circuit operation not design them.

```
Frequency Response Behavior Of A Voltage Amplifier

** Circuit Description **
* signal source
Vs 1 0 AC 1V
Rs 1 2 20k
* frequency-dependent amplifier
Ri 2 0 100k
Ci 2 0 60p
Eamp 3 0 2 0 144
Ro 3 4 200
* load
Rl 4 0 1k

** Analysis Requests **
* compute AC frequency response from 1 Hz to 100 MHz
*    using 5 frequency steps per decade.
.AC DEC 5 1 100Meg

** Output Requests **
* print the magnitude  and phase of the output voltage
*    as a function of frequency
.PRINT AC Vm(4) Vp(4)
.PROBE
.end
```

Figure 1.22 Spice input deck for circuit shown in Fig. 1.21.

- A Spice input file consists of three main parts: circuit description, analysis requests and output requests.
- The first line in a Spice input file must be a title statement and the last line must be an .end statement.
- A circuit is described to Spice by a sequence of element statements describing how each element is connected to the rest of the circuit and specifying its value.
- Each element type has a unique first-letter representation, eg. R for resistor, C for capacitor.
- Spice performs three main analyses: nonlinear DC analysis, transient analysis and small-signal AC analysis.
- Spice can perform other analysis as special cases of the three main analysis types. One example shown was the DC Sweep command used to compute DC transfer characteristics.
- The results of circuit simulation are placed in an output file in either tabular or graphical form.

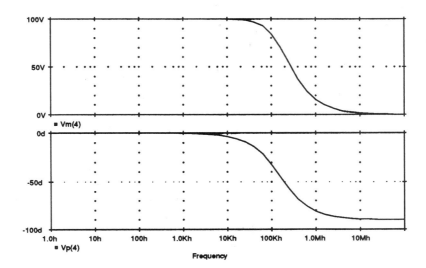

Figure 1.23 The magnitude and phase response behavior of the amplifier circuit shown in Fig. 1.21.

- PSpice is equipped with a post-processor called Probe which allows for interactive graphical display of simulation results.
- Probe can perform many powerful mathematical functions such as differentiation and integration on any network variable created during circuit simulation.

1.6 Bibliography

A. Vladimirescu, K. Zhang, A. R. Newton, D. O. Pederson, and A. Sangiovanni-Vincentelli, "SPICE Version 2G6 User's Guide," Dept. of Electrical Engineering and Computer Sciences, University of California, Berkeley, CA, 1981.

Staff, *PSpice Users' Manual*, MicroSim Corporation, Irvine, California, Jan. 1991.

1.7 Problems

1.1 For each of the circuits shown in Fig. P1.1 compute the corresponding node voltages using Spice.

1.2 For each of the circuits shown in Fig. P1.2 compute the corresponding node voltages and branch currents indicated using Spice.

1.3 Using the simple circuit arrangement shown in Fig. P1.3, generate a voltage waveform across the 1 Ω resistor having the following described form:

1 Introduction To Spice

(i) $10 \cdot \sin(2\pi \cdot 60 \cdot t)$

(ii) $1 + 0.5 \cdot \sin(2\pi \cdot 1000 \cdot t)$

(iii) $1 + 1 \cdot e^{-0.005t} \sin(2\pi \cdot 1000 \cdot t)$

(iv) $v_I = \begin{cases} 1 & \text{for } t < 1 \text{ ms} \\ 1 + 5 \cdot \sin(2\pi \cdot 60 \cdot t) & \text{for } t \geq 1 \text{ ms} \end{cases}$

Verify your results using Spice by plotting the voltage waveform that appears across the 1 Ω resistor for at least 6 cycles of its waveform. Use a time step that samples at least 20 points on one cycle of the waveform.

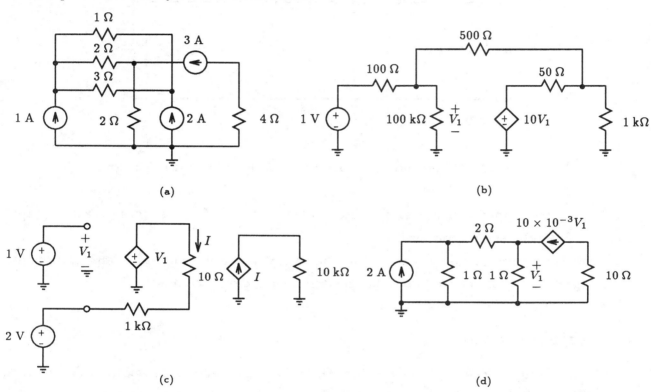

Fig. P1.1

1.4 Using the simple circuit arrangement shown in Fig. P1.3, generate a voltage waveform across the 1 Ω resistor having the following described form:

(i) 10 V peak-to-peak symmetrical square-wave at a frequency of 1 kHz. Let the rise and fall times be 0.1% of the total period of the waveform.

(ii) 8 V peak-to-peak asymmetrical square-wave at a frequency of 100 kHz having a DC offset of 2 V. Assign the rise and fall times of this waveform to be 0.1% of the total period of this waveform.

(iii) 10 V peak-to-peak asymmetrical square-wave at a frequency of 5 kHz having a DC offset of 2 V. Let the rise and fall times be 0.1% of the total period of the waveform.

Verify your results using Spice by plotting the voltage waveform that appears across the

1 Ω resistor for at least 6 cycles of its waveform. Use a time step that samples at least 20 points on one cycle of the waveform.

1.5 Replace the voltage source seen in Fig. P1.3 by a current source. Generate a current into the 1 Ω resistor using the PULSE source statement of Spice such that it has a triangular shape with an amplitude of 2 mA and a period of 2 ms. The average value of the waveform is zero. Plot this current for at least 6 cycles of its waveform using 10 points per period. *Hint: Spice will not accept a pulse width of zero, so use a value that is at most 0.1% of the pulse period.*

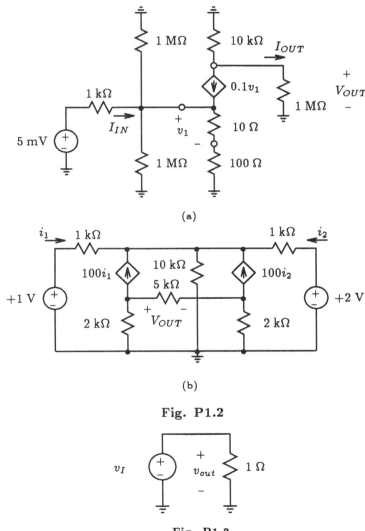

Fig. P1.2

Fig. P1.3

1.6 Using the PULSE source statement of Spice, together with the circuit setup shown in Fig. P1.3, generate the saw-tooth voltage waveform shown in Fig. P1.6. Verify your results by plotting the voltage across the 1 Ω resistor for at least 6 cycles of its waveform.

1.7 Using the PWL source statement of Spice, together with the circuit setup shown in Fig.

P1.3, generate the voltage waveform shown in Fig. P1.7. Verify your results by plotting the voltage across the 1 Ω resistor for the full duration of this waveform. What voltage appears across the 1 Ω resistor if the simulation time extends beyond 50 ms?

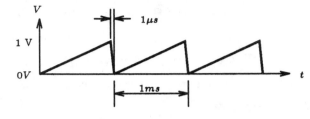

Fig. P1.6

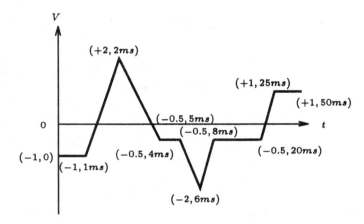

Fig. P1.7

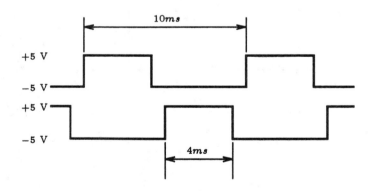

Fig. P1.8

1.8 Using two voltage sources with appropriate loads, generate two signals that are non-overlapping complementary square-waves, such as those shown in Fig. P1.8. Verify your results by plotting the voltage across each load resistor for at least 10 cycles of each waveform.

1.9 Using the simple circuit arrangement shown in Fig. P1.3, generate a 0 to 1 V step signal across the 1 Ω resistor having a rise-time of no more than 1 μs. Verify your results using

Spice by plotting the voltage waveform that appears across the 1 Ω resistor for at least 1 ms using a 50 μs time step.

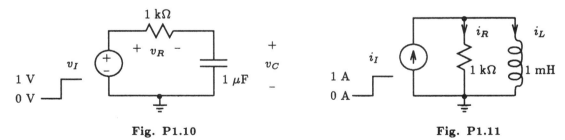

Fig. P1.10   **Fig. P1.11**

1.10 For the first-order RC circuit shown in Fig. P1.10, simulate the behavior of this circuit with Spice subject to a 0 to 1 V step input having a rise time of no more than 10 ns. Plot the voltage waveform that appears across the 1 kΩ resistor and the 1 μF capacitor. Verify that the voltage across the capacitor changes by 63% of its final value in a time of one time-constant.

1.11 For the first-order RL circuit shown in Fig. P1.11, simulate the behavior of this circuit with Spice subject to a 0 to 1 A step input having a rise time of no more than 10 ns. Plot the current that flows in both the 1 kΩ resistor and the 1 mH inductor. Verify that the current that flows in the inductor changes by 37% of its final value in a time of one time-constant. *Hint: One way of monitoring the current through a resistor or inductor is to connect a zero-valued voltage source in series with that element.*

1.12 Repeat Problem P1.10 with the 1 μF capacitor initially charged to 0.5 V.

1.13 Repeat Problem P1.11 with the 1 mH inductor initially conducting a current of 2 A.

1.14 For the first-order RC circuit shown in Fig. P1.10 subject to a 1 V peak sine-wave input signal of 1 kHz frequency, simulate the behavior of this circuit with Spice. Plot the voltage waveform that appears across the 1 μF capacitor for at least 6 cycles of the input signal. Use a time step that acquires at least 20 points per period.

1.15 Repeat Problem P1.14 with a 1 V peak symmetrical square-wave input of 10 kHz frequency. How would you describe the voltage waveform that appears across the capacitor?

1.16 For the first-order RL circuit shown in Fig. P1.11 subject to a 1 A peak-to-peak triangular input signal of 1 kHz frequency, simulate the behavior of this circuit with Spice. Plot the current waveform that flows through the 1 kΩ resistor for at least 6 cycles of the input signal. Use a time step that acquires at least 20 points per period. How would you describe this current waveform?

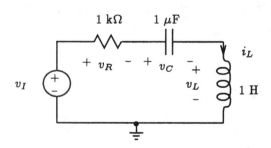

Fig. P1.17

1.17 For the second-order RLC circuit shown in Fig. P1.17 subject to a 1 V step input, simulate the transient behavior of the circuit and plot the voltage waveform that appears across each element for about 40 ms. Use a time step of no more than 100 μs.

1.18 Repeat Problem 1.17 with value of the resistor decreased by a factor of 10. How do the waveforms compare with that in Problem 1.17.

1.19 Repeat Problem 1.17 with value of the resistor decreased by a factor of 100. How do the waveforms compare with that in Problem 1.17.

1.20 Compute the frequency response behavior of the RC circuit shown in Fig. P1.10 using Spice for a 1 V AC input signal. Plot both the magnitude and phase behavior of the voltage across the resistor and the voltage across the capacitor over a frequency range of 0.1 Hz to 10 MHz. Use 20 points-per-decade in your plot.

1.21 Compute the frequency response behavior of the RL circuit shown in Fig. P1.11 using Spice for a 1 A AC input signal. Plot both the magnitude and phase behavior of the current through the resistor and inductor over a frequency range of 1 mHz to 1 MHz. Use 10 points-per-decade in your plot.

1.22 Compute the frequency response behavior of the RLC circuit shown in Fig. P1.17 using Spice for a 1 V AC input signal. Plot both the magnitude and phase behavior of the voltage across the resistor, inductor and capacitor over the frequency range of 1 Hz to 1 kHz. Use 10 points-per-octave in your plot.

1.23 Compute the frequency response behavior of the RLC circuit shown in Fig. P1.17 with R having values of 10, 100 and 1 kΩ. Plot the magnitude and phase response of each case and compare them. Select an appropriate frequency range and number of points that best illustrate your results.

Chapter 2

Operational Amplifiers

Having just been introduced to several aspects of circuit simulation using Spice, we are now ready to re-enforce our understanding of linear circuits constructed with Operational Amplifiers. We shall begin our investigation by developing a simple voltage-controlled voltage-source (VCVS) representation of the op amp from which to study various types of op amp circuits. Progressively, we shall increase the complexity of the op amp model in order to capture more of the true-life behavior and the effect that this behavior has on closed-loop circuit operation. Several new Spice concepts will be discussed here; largely in order to describe the nonlinear circuit behavior of an op amp to Spice.

2.1 Modeling An Ideal Op Amp With Spice

An ideal op amp as shown in Fig. 2.1 may be modeled as a voltage-controlled voltage source with infinite voltage gain (ie. $A \to \infty$). The input resistance is very high, infinite in fact, and the output resistance is considered to be zero since the output node is driven directly by a voltage source. Moreover, the voltage gain is assumed to be independent of frequency. At a first glance, a Spice model for the ideal op amp may seem to be trivial — a one line VCVS Spice statement. Unfortunately, Spice has no concept of infinity, hence, the infinite voltage gain can not be specified as a Spice value. Instead, we must compromise the accuracy of our ideal model by specifying a large, but finite, voltage gain value. Normally, a value of 10^6 V/V is sufficient without any significant deviation from the ideal. Under this gain condition, we shall consider the op amp as pseudo-ideal.

2 Operational Amplifiers

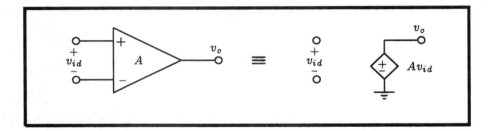

Figure 2.1 Equivalent circuit of the ideal op amp ($A \to \infty$).

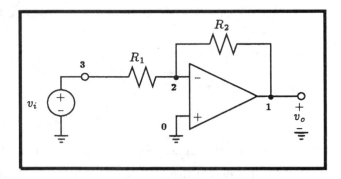

Figure 2.2 The inverting amplifier circuit.

2.2 Analyzing The Behavior Of Ideal Op Amp Circuits

We have now come to a point where we can use Spice to analyze the behavior of various types of op amp circuits, and thus develop a better understanding of these circuits.

2.2.1 Inverting Amplifier

Consider the inverting op amp circuit shown in Fig. 2.2 which consists of one ideal op amp and two resistors R_1 and R_2. We would like to determine the DC transfer function of this circuit when R_1 and R_2 assume values of 1 kΩ and 10 kΩ, respectively.

To perform this calculation using Spice we shall make use of the Transfer Function (.TF) command briefly discussed in the last chapter. The syntax of this command was not discussed there, rather we shall present it here and use it to analyze the op amp circuit shown in Fig. 2.2.

The transfer-function analysis command of Spice computes the DC small-signal gain from the input of a circuit driven by some signal source to some pre-specified network variable. In addition, this command will also calculate the input resistance of the circuit as seen by the input source, and the output resistance seen looking back into the circuit from the port formed by the output variable and ground. Alternatively, this command can be viewed as calculating the Thevenin or Norton equivalent circuit of the network from the point-of-view

2.2 Analyzing The Behavior Of Ideal Op Amp Circuits

Analysis Requests	Spice Command
Small Signal Transfer Function	.TF *output_variable input_source_name*

Table 2.1 Small-signal transfer function analysis request.

of the input and output ports.

A general description of the syntax of the Transfer Function analysis command (.TF) is given in Table 2.1. The different fields of this command should be self-evident from the discussion above. The command line begins with the keyword .TF followed by the output variable, either a voltage at a node or a current through a voltage source, and the name of the input signal source for which the output will be referenced to. The results of the .TF command are directly sent to the Spice output file in much the same way as that performed previously by the .OP command. No .PRINT or .PLOT statement is required in the Spice input file in order to view the results of the .TF command.

Returning to the circuit shown in Fig. 2.2, one can create the Spice input file shown in Fig. 2.3. Here the op amp is modeled as a VCVS with a voltage gain of 10^6 V/V. A one-volt DC signal is applied to the input of the circuit and a .TF analysis request was included to compute the DC small-signal voltage gain.

The results of the Transfer Function calculations, as produced by Spice, are listed below together with the small signal bias solution:

```
****    SMALL SIGNAL BIAS SOLUTION        TEMPERATURE =   27.000 DEG C

NODE    VOLTAGE     NODE    VOLTAGE     NODE    VOLTAGE     NODE    VOLTAGE

(   1)   -9.9999  (   2) 10.00E-06  (   3)    1.0000

****    SMALL-SIGNAL CHARACTERISTICS

    V(1)/Vi = -1.000E+01

    INPUT RESISTANCE AT Vi =  1.000E+03

    OUTPUT RESISTANCE AT V(1) = 0.000E+00
```

Clearly, the voltage gain from the DC input source to the op amp output of −10 as calculated by Spice agrees with the expected voltage gain determined by the ratio $-R_2/R_1$. In addition, the input and output resistances are listed. Combining this information with that of the gain calculation allows us to represent the op amp circuit of Fig. 2.2 with the equivalent circuit model shown in Fig. 2.4.

2 Operational Amplifiers

```
Inverting Amplifier Configuration
** Circuit Description **
* signal source
Vi 3 0 DC 1v
* inverting amplifier circuit description
R1 3 2 1k
R2 2 1 10k
Eopamp 1 0 0 2 1e6
** Analysis Requests **
.TF V(1) Vi
** Output Requests **
* none required
.end
```

Figure 2.3 Spice input deck for calculating the small-signal characteristics of the circuit shown in Fig. 2.2.

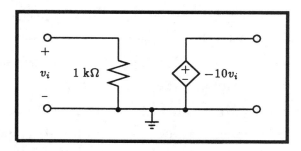

Figure 2.4 Equivalent circuit model of the inverting amplifier configuration of Fig. 2.2 as calculated by Spice.

Evident in the "SMALL SIGNAL BIAS SOLUTION" is that the negative terminal of the op amp (node 2) is not exactly at ground potential. This error is caused by the finite DC gain used to model the terminal behavior of the ideal op amp and would be expected to decrease with increasing the op amp DC gain. In most practical cases, an error of this magnitude is considered insignificant and need not be a cause for alarm.

2.2.2 The Miller Integrator

Our next example illustrates another important application of the inverting op amp configuration. Consider replacing R_2 in the inverting amplifier circuit shown in Fig. 2.2 with a capacitor C_2. The resulting configuration is known as the inverting or Miller integrator circuit and is shown in Fig. 2.5. Here we wish to determine the transient response of a Miller integrator with $R_1 = 1$ kΩ and $C_2 = 10$ μF subject to a 1-volt step-input. On completion of this, we would also like to determine its AC frequency response.

The Spice input file used to calculate the transient response of the Miller Integrator is

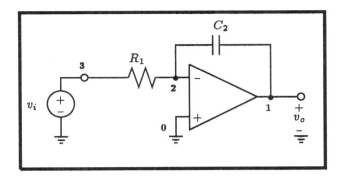

Figure 2.5 The Miller Integrator.

```
The Miller Integrator
** Circuit Description **
* signal source
Vi 3 0 PWL (0 0V 1ms 0V 1.001ms 1V 10ms 1V)
* components of the Miller integrator
R1 2 3 1k
C2 2 1 10uF
Eopamp 1 0 0 2 1e6
** Analysis Requests **
.TRAN 100us 5ms 0ms 100us
** Output Requests **
.PRINT TRAN V(3) V(1)
.probe
.end
```

Figure 2.6 Spice input deck for computing the step response of the circuit shown in Fig. 2.5.

given in Fig. 2.6. At this point of the text most statements contained in this file should be self explanatory; however, we need to clarify the description provided for realizing the step input. An ideal step function does not exist in Spice, so we need to approximate the step with a series of piece-wise linear segments. The pulse is held low at 0 V for 1 ms and then made to rise to 1 V with a rise-time of 1 μs, and then held at 1 V for 9 ms. If the rise time of this pulse was made equal to zero then we would have realized a step function exactly; unfortunately, Spice will not accept a waveform having a rise-time of zero. Alternatively, we could decrease the rise time of the step input and more closely approximate the step function; however, this only increases the time to complete a simulation. For this particular example, a rise-time of 1 μs was found to be sufficient.

The results of the transient simulation are shown in Fig. 2.7. The top curve represents the input step signal and the bottom curve represents the output of the integrator. Clearly, the output is the time-integral of the input, ie. the integral of a step function is a ramp

2 Operational Amplifiers

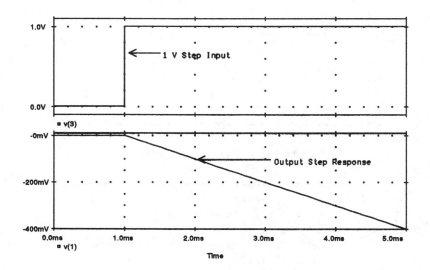

Figure 2.7 Step response of the Miller integrator circuit shown in Fig. 2.5 when $R_1 = 1$ kΩ and $C_2 = 10$ μF.

function. The rate at which the ramp output decreases is -400 mV/4 ms or -100 V/s. The magnitude of this rate can easily be shown to be equal to $V_I/C_2 R_1$ where V_I is the magnitude of the input step. It should then be obvious why this circuit is called an integrator.

The AC frequency response of this circuit can be easily computed using Spice by simply changing the input source in the Spice input file of Fig. 2.6 from a time-varying PWL voltage source to an AC voltage source with the following syntax:

```
Vi 3 0 AC 1V.
```

In addition, we must replace the .TRAN statement by an .AC statement specifying the range of frequencies that we are interested in. For this particular case, we are interested in a fairly broad frequency range of 1 Hz to 1 kHz, thus we decided to use a log sweep of the input frequency using the following .AC analysis statement:

```
.AC DEC 5 1Hz 1kHz.
```

The AC frequency response of the Miller integrator as calculated by Spice is displayed in Fig. 2.8. The magnitude of the output node voltage (V(1)) is large at low frequencies and rolls-off at a rate of -20 dB for each decade increase in frequency. Using the Probe post-processor available with PSpice, we found that the frequency at which the magnitude of the output voltage crosses the 0 dB level is 15.9 Hz. This corresponds exactly with the

2.2 Analyzing The Behavior Of Ideal Op Amp Circuits

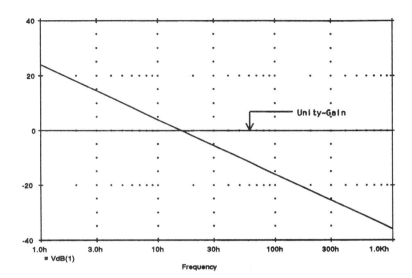

Figure 2.8 Frequency response of the Miller integrator circuit shown in Fig. 2.5 when $R_1 = 1$ kΩ and $C_2 = 10$ μF.

result of substituting the circuit parameters into the expression $1/(2\pi R_1 C_2)$.

2.2.3 A Damped Miller Integrator

The DC instability of a Miller integrator is a result of the very high (ideally infinite) DC gain that the Miller integrator has. The DC gain can be made finite by connecting a feedback resistor across integrating capacitor C_2. Unfortunately, however, this makes the integrator nonideal, known as a *damped integrator*. To obtain near ideal response over a large frequency range the feedback resistor should be as large as possible.

In the following we shall investigate the behavior of a damped integrator for two different-sized feedback resistors. Specifically, we shall compare the step response of a damped integrator with feedback resistors of 1 MΩ and 100 kΩ, to that of an ideal integrator. The Spice input file for each case is quite similar to that seen in Fig. 2.6 for the Miller integrator except for the following changes. The amplitude of the input step is reduced to 1 mV from the original 1 V level. This is to keep the output signal level within practical limits (ie. between typical power supply levels). The element statement for this step input signal would then appear as follows:

```
Vi 3 0 PWL ( 0 0V 1ms 0V 1.001ms 1mV 2s 1mV ).
```

Secondly, the element statement for the feedback resistor is added to the Spice deck. For the case of 1 MΩ feedback resistor, we would add the following statement:

2 Operational Amplifiers

$$\text{R2 1 2 1Meg}$$

and in the case of the 100 kΩ feedback resistor we would add:

$$\text{R2 1 2 100k.}$$

A separate Spice deck is created for each case and concatenated together into one file. This will enable us to graphically compare the final results on a single graph using the Probe post-processing facility available in PSpice.

As the final change, the transient analysis command statement is changed to read as follows:

$$\text{.TRAN 100ms 2s 0s 100ms.}$$

To make direct comparisons with the ideal-integrator step-response, we shall also add the Spice deck for the Miller integrator shown in Fig. 2.6 (with the appropriate changes made) to the file containing the two Spice decks of damped integrators. For reference, we list the complete file containing the three Spice decks in Fig. 2.9. Notice that no space separates the end of one Spice deck, denoted by .end, and the start of the next one. If this is not adhered to, then the file will probably be rejected by Spice.[†]

Submitting this input file to Spice for processing, results in the three step responses shown in Fig. 2.10; one result is for the ideal integrator and the other two are for the different damped integrators. Up to about 0.1 s, all three step responses are almost identical. After this, the integrator that is damped with 100 kΩ begins to deviate from the other two. Moreover, as the time progresses, the curve corresponding to this one begins to settle towards −100 mV. In the case of the 1 MΩ damped integrator, similar behavior is also observed but with a different time scale. The step response of this integrator begins to significantly deviate from the ideal after about 1 s. Although not shown, if we were to have the transient response calculation of Spice run longer, we would see the step response of this integrator settle to a −1 V level.

In either of the two damped integrator cases, we see that the step response deviates significantly from the ideal situation in about one-tenth the time-constant formed by the integrating capacitor C_2 and the damping resistor R_2 (ie. $C_2R_2/10$). We can therefore conclude that the output of a damped integrator behaves much like an ideal integrator for

---

[†] Spice version 2G6 and later have a built-in command called .ALTER that allows the user to specify changes to the circuit without having to re-type the entire file as we do here. Unfortunately, the student version of PSpice does not have this command or one that accomplishes the same thing, so we have opted to re-create a new Spice deck for each circuit change and concatenate them together into one file for processing. In this way, we can view the results together using Probe.

```
The Miller integrator

** Circuit Description **
* signal sources
Vi 3 0 PWL (0 0V 1ms 0V 1.001ms 1mV 2s 1mV)
* components of the Miller integrator
R1 2 3 1k
C2 2 1 10uF
Eopamp 1 0 0 2 1e6
** Analysis Requests **
.TRAN 100ms 2s 0ms 100ms
** Output Requests
.PLOT TRAN V(1)
.probe
.end
The Damped Miller integrator (R=1M)

** Circuit Description **
* signal sources
Vi 3 0 PWL (0 0V 1ms 0V 1.001ms 1mV 2s 1mV)
* components of the Miller integrator
R1 2 3 1k
R2 1 2 1Meg
C2 2 1 10uF
Eopamp 1 0 0 2 1e6
** Analysis Requests **
.TRAN 100ms 2s 0ms 100ms
** Output Requests
.PLOT TRAN V(1)
.probe
.end
The Damped Miller integrator (R=100k)

** Circuit Description **
* signal sources
Vi 3 0 PWL (0 0V 1ms 0V 1.001ms 1mV 2s 1mV)
* components of the Miller integrator
R1 2 3 1k
R2 1 2 100k
C2 2 1 10uF
Eopamp 1 0 0 2 1e6
** Analysis Requests **
.TRAN 100ms 2s 0ms 100ms
** Output Requests
.PLOT TRAN V(1)
.probe
.end
```

Figure 2.9 The complete Spice deck for computing the step response of the two damped integrator circuits and one ideal integrator circuit. It consists of three separate Spice decks concatenated into one file.

2 Operational Amplifiers

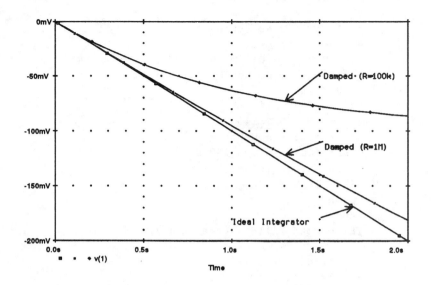

Figure 2.10 Comparing the 1 mV step response of two differently damped integrator circuits with that of an ideal Miller integrator circuit.

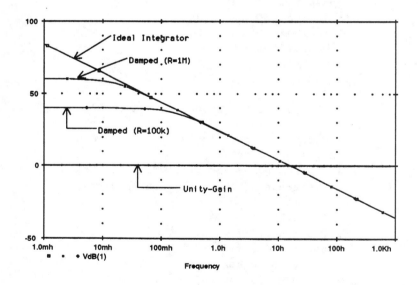

Figure 2.11 Comparing the magnitude response behavior of two damped integrator circuits with that of an ideal Miller integrator.

times less than one-tenth the time constant formed by the feedback capacitor and resistor.

It is also interesting to observe the magnitude response behavior of the two damped integrators as a function of frequency and compare it to that obtained for the Miller integrator. Consider adding the following AC source statement and analysis command to each Spice

2.2 Analyzing The Behavior Of Ideal Op Amp Circuits

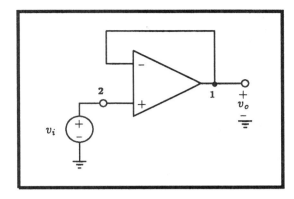

Figure 2.12 The unity-gain buffer.

input deck described in Fig. 2.9:

```
Vi 3 0 AC 1V
.AC DEC 5 1mHz 1kHz
```

On completion of the AC analysis, we obtain the results shown in Fig. 2.11. In contrast to the results of the ideal integrator, the two damped integrators have a magnitude response that consists of two parts: A low frequency component that is essentially independent of frequency and a second component that rolls off linearly with frequency at a rate of −20 dB/dec. The frequency point that divides the two regions is approximately the reciprocal of the time-constant formed by the feedback resistor and capacitor (ie. $\frac{1}{2\pi C_2 R_2}$).

For frequencies about ten-times larger than the corresponding break or 3 dB frequency of the damped integrator, both the ideal and the damped integrator have essentially identical frequency response behavior. Thus, for input signal frequencies larger than 10 times the 3 dB frequency of the damped integrator, the response of the damped integrator closely approximates that of the ideal Miller integrator.

2.2.4 The Unity-Gain Buffer

This next example will repeat the same DC transfer function analysis performed above for the inverting amplifier on a unity-gain buffer. Although, the circuit itself will not prove very challenging for Spice, it will be used to highlight an apparent problem with Spice and the technique used to alleviate it.

Consider the unity-gain buffer shown in Fig. 2.12. The output of the op amp is feed directly back to it's negative terminal and the input signal generator v_i is connected to the positive terminal. The Spice input file for this circuit is listed in Fig. 2.13 and, on submission to Spice, the following results are found in the output file (excluding the input circuit description):

2 Operational Amplifiers

```
Unity-Gain Buffer
** Circuit Description **
* signal source
Vi 2 0 DC 1V
* op-amp in unity-gain configuration
Eopamp 1 0 2 1 1e6
** Analysis Requests **
.TF V(1) Vi
** Output Requests **
* none required
.end
```

Figure 2.13 Spice input deck for calculating the DC small-signal gain of the circuit shown in Fig. 2.12. Spice rejects this file because of the lack of two connections at nodes 1 and 2.

```
ERROR: Less than 2 connections at node    2
ERROR: Less than 2 connections at node    1
```

Obviously, something went wrong. Spice is complaining about having less than 2 connections at nodes 1 and 2. This is a topological restriction imposed by Spice in order to guarantee that the solution calculated is a unique one. Prior to the start of any simulation, Spice performs a topological check during the equation formulation phase to ensure that this restriction is not violated. If it is, then Spice reports in the output file the nodes that violate this condition.

On counting the number of connections made at nodes 1 and 2, there appears to be exactly two connections. So why then does Spice complain? The answer lies in the fact that Spice does not consider the input port of a voltage-controlled dependent source as an element. To eliminate this problem and have Spice get on with the circuit simulation, one must "fool" Spice into thinking that it has at least 2 connections at nodes 1 and 2. A very elegant solution is to connect zero-valued current sources between node 1 and ground, and node 2 and ground. In this way each current source acts as an open circuit and has no effect on the operation of the circuit. Alternatively, one could connect high-valued resistors between each node in question and ground. If the resistors added are large enough relative to other resistors in the circuit, then their presence will be insignificant.

For the above example we added two zero-valued current sources. One between node 1 and ground, and the other between node 2 and ground. The revised Spice input file including these zero-valued current sources is listed in Fig. 2.14 and the small-signal characteristics, as calculated by Spice, are shown below:

2.2 Analyzing The Behavior Of Ideal Op Amp Circuits

```
Unity-Gain Buffer
** Circuit Description **
* signal source
Vi 2 0 DC 1V
* op-amp in unity-gain configuration
Eopamp 1 0 2 1 1e6
Iopen1 1 0 0A    ; redundant current-sources to eliminate
Iopen2 2 0 0A    ; less-than-two connection at nodes 1 and 2
** Analysis Requests **
.TF V(1) Vi
** Output Requests **
* none required
.end
```

Figure 2.14 Revised Spice input deck for the unity-gain buffer shown in Fig. 2.12. Also included is two zero-valued current sources to rid the circuit of its less than 2 connections problem at nodes 1 and 2.

```
****      SMALL SIGNAL BIAS SOLUTION       TEMPERATURE =   27.000 DEG C

   NODE   VOLTAGE      NODE   VOLTAGE      NODE   VOLTAGE      NODE   VOLTAGE

(    1)    1.0000  (    2)    1.0000

****      SMALL-SIGNAL CHARACTERISTICS

        V(1)/Vi =   1.000E+00

        INPUT RESISTANCE AT Vi =    1.000E+12

        OUTPUT RESISTANCE AT V(1) =    0.000E+00
```

Notice that the input resistance of the unity-gain buffer is not infinite as it should be, but rather 10^{12} Ω. This is an artifact of the algorithm used by Spice to calculate the DC bias solution of a network. For all practical purposes, in a circuit such as the one shown in Fig. 2.12, one can consider a resistance value of 10^{12} Ω as being equivalent to a value of infinity.

2.2.5 Instrumentation Amplifier

Our next example is a two-stage instrumentation amplifier shown in Fig. 2.15 consisting of three op amps. Such an amplifier is usually employed as the front-end of an instrument that measures a differential signal between the amplifier input terminals (20 mV in Fig. 2.15). We would like to investigate the effect of a 60 Hz common-mode signal on such a measurement, as shown in Fig. 2.15. This example involves more op amps than previous examples; however, it does not pose any special difficulty for Spice. Any difficulty will usually

2 Operational Amplifiers

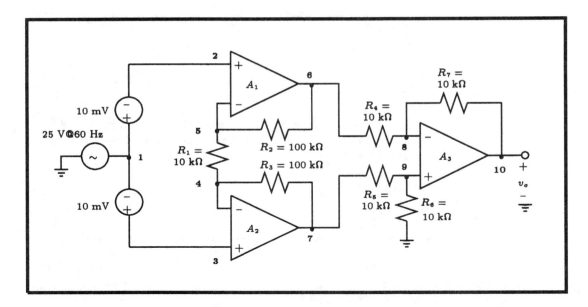

Figure 2.15 A two-stage three op amp instrumentation amplifier.

be experienced by the user in the amount of effort that's required to type the correct circuit description into a computer file. In general, as more information has to be typed there exists a greater chance of entering erroneous data.

To simplify matters, and reduce the chances of error, a provision has been made in Spice for defining *subcircuits*. A subcircuit is considered separate and isolated from the main circuitry except that it is connected to the main circuit through specific nodes. Such a subcircuit can, of course, be used repeatedly in the same main circuit. This concept is analogous to the subroutine concept found in most computer programming languages such as FORTRAN. Experience has shown that Spice input files are easier to read, and simpler to debug, when a large circuit is described using subcircuits. In addition, once a subcircuit is created for a specific building block, it can be re-used by other circuits constructed from the same building blocks. Thus, subcircuits provide a convenient way of creating a library of basic building blocks for future use.

The format and syntax of a subcircuit definition is displayed in Fig. 2.16. It begins by a .SUBCKT statement followed by a unique name and a list of the internal nodes of the subcircuit that will be allowed to connect to the main circuit. Subsequent statements are used to describe the subcircuit to Spice in exactly the same way as that performed for the main circuit. The nodes of the subcircuit are local to the subcircuit and can have the same numbers as those used in the main circuit or any other subcircuit. One exception is the ground node (node 0) which is common to all circuits. Similarly, the names of the elements making up the subcircuit are also local to the subcircuit and can be the same as those used

2.2 Analyzing The Behavior Of Ideal Op Amp Circuits

> .SUBCKT subcircuit_name list_of_nodes
> Circuit Description
> Power Supplies / Signal Sources
> Element Descriptions
> Model Statements
> .ENDS [subcircuit_name]

Figure 2.16 Subcircuit format and syntax.

> Xname   node_connections_to_subcircuit   subcircuit_name

Figure 2.17 Accessing a subcircuit by the main circuit.

for other elements in the main circuit. To conclude the definition of the subcircuit, the final statement must be an .ENDS statement. Notice that this includes an "S" appended on its end to distinguish it from an end to the Spice input file. The name of the subcircuit may also be included on the same .ENDS statement to clearly mark the end of the subcircuit.

Once a subcircuit has been defined it can be incorporated into the main circuit in much the same way as a circuit element is described to Spice. The beginning of the statement is a unique alphanumeric name prefixed with the letter "X" followed by a list of the nodes of the subcircuit that connect it to the main circuit. The final field must then specify the name of the subcircuit that is being referenced. Obviously, there must be a one-to-one correspondence between the list of nodes given on this statement and the ones listed on the .SUBCKT statement. For quick reference, the syntax of this statement is displayed in Fig. 2.17.

Now returning to the instrumentation amplifier displayed in Fig. 2.15, we see that it consists of three identical op amps A_1, A_2 and A_3. These can be conveniently described to Spice using the subcircuit facility. Consider the op amp equivalent circuit displayed in Fig. 2.1 and the following corresponding subcircuit description:

```
.subckt ideal_opamp 1 2 3
* connections:       | | |
*              output | |
*             +ve input |
*              -ve input
Eopamp 1 0 2 3 1e6
Iopen1 2 0 0A      ; redundant connection made at +ve input terminal
```

2 Operational Amplifiers

```
        Iopen2 3 0 0A        ; redundant connection made at -ve input terminal
    .ends ideal_opamp
```

One element statement is used to specify the VCVS and the other two are used to make redundant circuit connections at the floating nodes of the VCVS. Notice also that we added a set of comment statements clarifying the list of nodes that interface the subcircuit and the main circuit. One should always include such a description in every subcircuit for easy reference.

Each op amp of the instrumentation amplifier can now be described to Spice using the following subcircuit calls:

```
        Xop_A1   6 2 5  ideal_opamp
        Xop_A2   7 3 4  ideal_opamp
        Xop_A3  10 9 8  ideal_opamp
```

The complete Spice input file including the op amp subcircuit for the instrumentation amplifier is on display in Fig. 2.18. In order to determine whether the 60 Hz common-mode signal is affecting the DC measurement being performed by the instrumentation amplifier, a transient analysis is to be performed by Spice.

On completion of Spice, we observe (Fig. 2.19) that the voltage waveform appearing at the output of the instrumentation amplifier (node 10) is a constant DC signal of 420 mV, even though the voltage appearing at the inputs of the instrumentation amplifier (nodes 2 and 3) are time-varying voltage waveforms riding on the very small DC signal of 10 mV. Comparing the DC level found at the output with that predicted by the equation derived by Sedra and Smith in Example 2.5 of Section 2.6 for the instrumentation amplifier and repeated here below with the constraint $R_2 = R_3$, $R_4 = R_5$ and $R_6 = R_7$:

$$v_o = -\frac{R_6}{R_4}\left(1 + \frac{2R_2}{R_1}\right)(v_1 - v_2),$$

one finds that we are in perfect agreement when the appropriate component values are substituted. There is no evidence of the 60 Hz common-mode signal at the output. This highlights an important feature of the instrumentation amplifier; its ability to *reject* common-mode signals while amplifying much smaller difference signals.

In the above simulation we assumed that the appropriate pairs of resistors (R_4, R_5, R_6 and R_7) were perfectly matched. In practise, this is rarely the case and, as a result, the lack of matching will affect the common-mode rejection capability of the instrumentation amplifier. To illustrate this, consider altering one of the resistors of the difference amplifier, say R_5, by 1% of its nominal value. In other words we shall repeat the above simulation with

2.3 Nonideal Op Amp Performance

```
Instrumentation Amplifier

* op-amp subcircuit
.subckt ideal_opamp 1 2 3
* connections:        | | |
*               output | |
*             +ve input |
*              -ve input
Eopamp 1 0 2 3 1e6
Iopen1 2 0 0A       ; redundant connection made at +ve input terminal
Iopen2 3 0 0A       ; redundant connection made at -ve input terminal
.ends ideal_opamp

** Main Circuit **
* signal sources
Vcm 1 0 SIN (0 25V 60Hz)
Vdc1 1 2 DC 10mV
Vdc2 3 1 DC 10mV
* instrumentation amplifier
Xop_A1 6 2 5 ideal_opamp
Xop_A2 7 3 4 ideal_opamp
Xop_A3 10 9 8 ideal_opamp
R1 5 4 10k
R2 5 6 100k
R3 4 7 100k
R4 6 8 10k
R5 7 9 10k
R6 9 0 10k
R7 8 10 10k
** Analysis Requests **
.TRAN 0.1ms 66.68ms 0 0.1ms
** Output Requests **
.PRINT TRAN V(2) V(3) V(10)
.probe
.end
```

Figure 2.18 Spice input deck for calculating the transient response of the instrumentation amplifier shown in Fig. 2.15.

R_5 having a value of 10.1 kΩ. On doing so, one finds at the output of the instrumentation amplifier a 250 mV peak-to-peak 60 Hz frequency component riding on a DC level of 420 mV. The actual simulation result for the output waveform is displayed in Fig. 2.20. Clearly, the common-mode signal is now present at the output. This illustrates the importance of having closely matched resistors in the difference amplifier portion of the instrumentation amplifier.

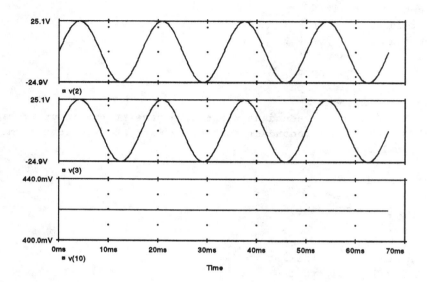

Figure 2.19 Input and output signals of the instrumentation amplifier. Notice that no AC signal is appearing at the output (V(10)) even though it is present at the two input nodes (V(2) and V(3)).

2.3 Nonideal Op Amp Performance

In previous sections we performed several Spice simulations assuming that the op amps used had a very large open-loop gain and that the gain was independent of frequency. In the remainder of this chapter we shall incorporate more of the nonidealities of practical monolithic op amps into our simulations; thus obtaining results that are more realistic of closed-loop op amp behavior. Rather than simulate detailed op amp circuitry at the transistor level, which, of course, is a very feasible approach provided one has the knowledge of the internal circuit (Chapter 10), in this section we shall develop an equivalent circuit representation for the op amp that relies only on the information found on the manufacturer's data sheet.

Op amp nonidealities can be incorporated into the op amp circuit model in various ways. One method involves constructing a circuit whose terminal behavior provides a close resemblance to actual op amp terminal behavior. Obviously, one tries to find a simple circuit that captures the op amp nonideal behavior. Alternatively, with newer versions of Spice, specifically PSpice, terminal behavior can be specified using a mathematical expression in either functional or piecewise-linear form. This new approach to circuit modeling has come to be known as *analog behavior modeling* and forms a very elegant and powerful means of

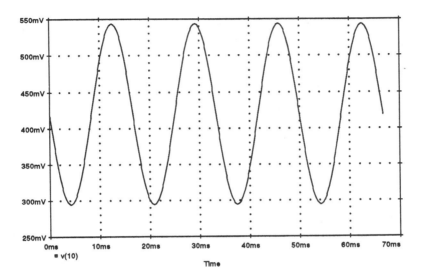

Figure 2.20 The output voltage signal of the instrumentation amplifier when there is a 1% mismatch between R_5 and R_6 in the difference amplifier of Fig. 2.15.

specifying nonlinear circuit behavior. We shall utilize this approach below to investigate the effect of large-signal properties of an op amp on the closed-loop response of op amp circuits. For the small-signal performance we shall use a lumped circuit model (an equivalent circuit) to represent the frequency response of the op amp.

2.3.1 Small-Signal Frequency Response Of Op Amp Circuits

The differential small-signal open-loop gain of an internally-compensated op amp can be mathematically described as

$$A(s) = \frac{A_0}{1 + s/\omega_b} \qquad (2.1)$$

where A_0 denotes the DC gain and ω_b is the 3-dB break frequency. Typically, A_0 is very large, on the order of 10^6 V/V for modern bipolar op amps such as the 741 op amp and ω_b typically ranges between 1 and 100 rad/sec. The single-capacitor circuit shown in Fig. 2.21 has infinite input resistance and zero output resistance, much like the ideal op amp, and it can be shown that it has the following single-pole transfer function:

$$\frac{V_o}{V_{id}}(s) = \frac{G_m R_1}{1 + sR_1 C_1} \qquad (2.2)$$

2 Operational Amplifiers

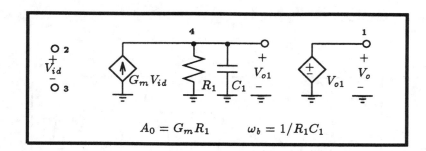

Figure 2.21 A one-pole circuit representation of the small-signal open-loop frequency response of an internally compensated op amp.

Clearly, if we let $G_m R_1 = A_0$ and $R_1 C_1 = 1/\omega_b$, then the circuit shown in Fig. 2.21 can be used to model the small-signal frequency response of the op amp in Spice. As an example, consider typical frequency response parameters for the 741 op amp: it has a DC gain of 2.52×10^5 V/V and a 3 dB frequency of 4 Hz. Using the above equations we can write two equations in terms of three unknowns. Thus, we have at our disposal a single degree of freedom which we can exercise to obtain the other two circuit parameters. That is, if we let $C = 30$ pF then we can solve for $G_m = 0.190$ mA/V and $R_1 = 1.323 \times 10^9$ Ω.

It is imperative that one check that the op amp model behaves as expected otherwise false conclusions would later be drawn. We shall perform this check in conjunction with our next example where we investigate the effects that the limited op amp gain and bandwidth have on the closed-loop gain of an inverting amplifier.

Consider calculating the frequency response of the inverting amplifier shown in Fig. 2.2 for nominal gains of -1, -10, -100 and -1000 using the one-pole op amp model calculated above. Furthermore, we will want to contrast the frequency response obtained in these four closed-loop cases with the open-loop response of the op amp using Probe. This is easily done by simply concatenating the Spice input decks into one file and submitting this larger file to Spice as if it were a single job. The results of all the analyses would then be found in one Spice output file or, for our purposes here, be accessible by Probe to graphically display and compare. Rather than list the entire contents of this input file because of its excessive length, we list only the Spice description for the inverting amplifier having a gain of -1 and the subcircuit used to represent the op amp, in Fig. 2.22. The Spice description for the other amplifiers would be identical to this one, except that R_2 would change to reflect the gain required in each case. Also included in this concatenated file is a Spice description for computing the open-loop frequency response of the op amp (ie. from the circuit of Fig. 2.2 with R_1 shorted and R_2 removed).

The frequency response behavior of the inverting amplifier under different gain settings is displayed in Fig. 2.23 together with the op amp open loop frequency response. One

```
Inverting Amplifier With Gain -1

* op-amp subcircuit
.subckt small_signal_opamp 1 2 3
* connections:           | | |
*                 output | |
*              +ve input |
*              -ve input
Ginput 0 4 2 3 0.19m
Iopen1 2 0 0A      ; redundant connection made at +ve input terminal
Iopen2 3 0 0A      ; redundant connection made at -ve input terminal
R1 4 0 1.323G
C1 4 0 30p
Eoutput 1 0 4 0 1
.ends small_signal_opamp

** Main Circuit **
* signal source
Vi 3 0 AC 1V 0Degrees
Xopamp 1 0 2 small_signal_opamp
R1 3 2 1k
R2 2 1 1k
** Analysis Requests **
.AC DEC 5 0.1Hz 100MegHz
** Output Requests **
.PRINT AC V(3) V(1)
.probe
.end
```

Figure 2.22 The Spice input deck for investigating the small-signal frequency response behavior of the inverting amplifier shown in Fig. 2.2 having a gain of −1. Other Spice decks can be appended to the end of this one to enable one to compare results with Probe.

sees clearly the effect of increasing amplifier gain on its bandwidth. Moreover, the gain and bandwidth are seen not to exceed those values of the open-loop frequency response.

2.3.2 Modeling The Large-Signal Behavior Of Op Amps[†]

The above subcircuit model for the op amp is limited to circuit situations where the output voltage of the op amp is small. In this section we shall elaborate on the op amp subcircuit so that it can be used to model op amp behavior when subjected to both small

---

[†] Study of this section relies heavily on the analog behavior modeling feature of PSpice. Those readers who do not have access to a version of Spice equipped with this feature can still perform the op amp simulations presented here provided the op amp macromodel present in Section 12.1 is substituted for the op amp subcircuit developed in this section; albeit with the appropriate modifications to the subcircuit call.

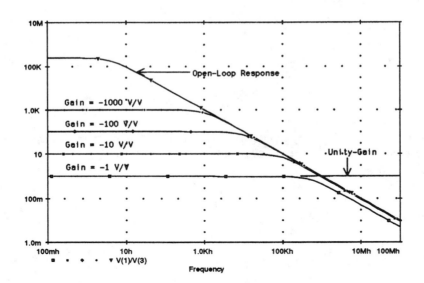

Figure 2.23 Frequency response of an inverting amplifier having nominal closed-loop gains of −1, −10, −100 and −1000. Also shown is the open-loop frequency response of the op amp used in the inverting amplifier ($A_0 = 2.52 \times 10^5$ V/V and $f_b = 4$ Hz).

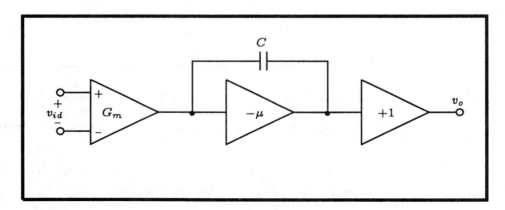

Figure 2.24 Block diagram of the internal structure of an internally-compensated op amp.

and large input signals. To accomplish this, consider that the internal structure of an actual op amp consists of three parts as shown in Fig. 2.24 in contrast to the single stage model put forth above in the last subsection. The front-end stage consists of a differential-input transconductance amplifier, followed by a high gain voltage amplifier with gain $-\mu$, together with a feedback compensation capacitor C, and the final stage is simply an output unity-gain buffer providing a low output resistance. By modeling the terminal behavior of each stage of the op amp, accommodating both linear and nonlinear behavior, realistic op amp

2.3 Nonideal Op Amp Performance

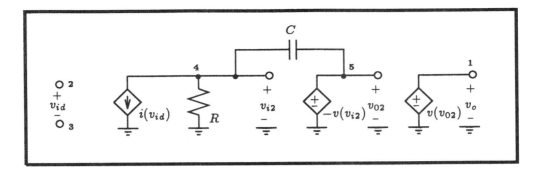

Figure 2.25 Equivalent circuit representation of the internal behavior of an internally-compensated op amp. In the small-signal region of the op amp $i(v_{id}) = G_m v_{id}$, $v(v_{i2}) = \mu v_{i2}$, and $v(v_{o2}) = v_{o2}$.

characteristics can be captured by Spice without simulating detailed internal circuitry.

To describe the behavior of these three stages to Spice, we make use of the equivalent circuit shown in Fig. 2.25. Here the gain of each dependent sources is expressed as an unspecified function of the controlling signal. The gain of these stages are written this way in order to convey to the reader that both the linear and nonlinear behavior of each of the internal stages is to be captured by these dependent sources. Spice has provisions for specifying nonlinear dependent sources; however, the functional description must be expressible in polynomial form. Unfortunately, this makes specifying arbitrary functions difficult. Instead, newer versions of Spice, specifically PSpice, allow users to specify the control function as a mathematical expression or in piecewise-linear form. We shall elaborate on this further below.

Limited Current-Output Of The Front-End Transconductance Stage

The first stage of an op amp is a circuit that converts a differential input voltage signal into a corresponding current. The operation of this stage is largely linear with transconductance G_m; however, the maximum output current that it can source or sink is limited to a value I_{max}. Conversely, the transconductance stage will behave linearly provided the input voltage levels are restricted to lie within the range $-I_{max}/G_m$ to I_{max}/G_m. To help the reader visualize these transfer characteristics we display them in Fig. 2.26. Now to try and capture this behavior with a set of polynomials would be very difficult, if not, impossible. Instead, it can be very easily incorporated into the circuit representation using the analog behavior modeling capability of PSpice.

The analog behavior model feature of PSpice is a set of extensions made to the VCVS and VCCS statements. These extensions allow the user to specify the controlled signal (voltage or current) in terms of the controlling variable (again, voltage or current) as either

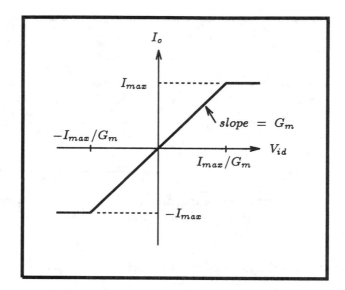

Figure 2.26 Transfer characteristic of the input transconductance amplifier of an op amp.

a mathematical expression or in piecewise-linear form. For example, the level of a VCCS with its output terminals connected between node 2 and ground can be made dependent on the square of the input voltage V(1) by simply using the following PSpice description,

```
Gsquare 2 0 VALUE = { V(1)*V(1) }.
```

The first three fields of this VCCS statement is like before; a unique name beginning with the letter G, followed by the nodes that the current source output is connected to. The subsequent fields are the extensions that we made reference to above. The keyword VALUE indicates to PSpice that this VCCS has a functional description defined by the field enclosed by the braces {} found after the equal sign. In this particular case, the functional description is the product of the voltage at node 1 with itself. In general, the functional description consists of an arithmetic expression in terms of the network variables and arbitrary constants. The arithmetic operators allowable are +, −, * and /. Parenthesis may also be used to simplify the notation. In addition, functions can also be included in each expression; however this is beyond the scope of this text. Interested readers can consult the *PSpice Users' Manual* for more details.

Another extension to the PSpice voltage-controlled source statement is one that allows the transfer function characteristics of the dependent source to be described in terms of a table of values. For example, the transfer characteristics of the input transconductance amplifier of the op amp, shown in Fig. 2.26, is simply described using the following PSpice statement:

```
Ginput 2 0 TABLE { V(1) } = (-Imax/Gm,-Imax) (+Imax/Gm,+Imax)
```

The first three fields of this statement should be obvious at this point from the above discussion. The keyword TABLE indicates that the controlled source has a tabular description, with the field between braces specifying the control variable V(1). The value of the controlled source is specified by the table of ordered pairs on the right hand side of the expression. The output values corresponding to input values that fall between specified points are computed by linearly interpolating between them. Thus, one can view the table of ordered pairs as points interconnected by straightlines. Thus, in the above case, a straightline having a slope of G_m interconnects the points $(-I_{max}/G_m, -I_{max})$ and $(+I_{max}/G_m, +I_{max})$. The output value corresponding to an input that falls outside the limits of the table is considered to be equal to the output value corresponding to the smallest (or largest) specified input, thus forming the two saturating limits of the amplifier.

Output Saturation

Op amps behave linearly over a limited range of output voltages, usually bounded by the voltage levels of the power supplies. In a similar manner to the input transconductance stage above, we can specify some piecewise-linear function for the voltage gain of either the middle-stage voltage amplifier or the output buffer, or both. An example of this will be given in the Spice subcircuit description listed below in the subsection: A PSpice Large-Signal Op Amp Model.

Frequency Response

Within the linear region of the equivalent circuit devised for the op amp in Fig. 2.25 (ie. $i(v_{id}) = G_m v_{id}$, $v(v_{i2}) = \mu v_{i2}$, and $v(v_{o2}) = v_{o2}$) one can show that the input-output transfer function is given by

$$\frac{V_o}{V_{id}} = \frac{\mu G_m R}{1 + sC(1+\mu)R}. \tag{2.3}$$

One can then draw a comparison between this equation and the one-pole model of the frequency response for the op amp given in Eq. (2.1) and deduce the following: $A_0 = \mu G_m R$ and $w_b = 1/C(1+\mu)R$. Multiplying these together results in the gain-bandwidth product ω_t given by

$$\omega_t = \frac{G_m}{C} \frac{\mu R}{(1+\mu)R}. \tag{2.4}$$

Since usually $\mu \gg 1$,

2 Operational Amplifiers

$$\omega_t \approx \frac{G_m}{C}. \tag{2.5}$$

A PSpice Large-Signal Op Amp Subcircuit

Combining the nonlinear effects described above, together with the op amp limited frequency response behavior, we can create a subcircuit description for the op amp that is valid under both large and small signal conditions. Of course, we shall make use of the op amp equivalent circuit shown in Fig. 2.25.

Consider an op amp characterized by a DC gain of 2.52×10^5 V/V, a unity-gain frequency of 1 MHz, and a slew-rate of 0.633 V/μs. Furthermore, we shall assume that the internal compensation capacitor C is 30 pF and that the op amp output stage saturates at ± 10 V. Now, it has been shown by Sedra and Smith that the slew-rate of an op amp is related to the transconductance stage current limit I_{max} and the capacitor C according to

$$SR = \frac{I_{max}}{C}. \tag{2.6}$$

Thus, for this particular op amp example, I_{max} is limited to 19 μA. Likewise, the transconductance of the first stage is simply deduced from Eqn. (2.5) above to be 0.19 mA/V. The remaining two parameters μ and R are now left to be determined. Unfortunately, they can not be uniquely determined from the information provided. We shall assume $R = 2.5 \times 10^6$ Ω and derive μ from $A_0 = \mu G_m R$ to get $\mu = 529$ V/V.

A subcircuit capable of capturing both the small- and large-signal behavior of the op amp described above is listed below (Refer to Fig. 2.25):

```
.subckt large_signal_opamp 1 2 3
* connections:             | | |
*                    output | |
*                    +ve input |
*                    -ve input
R 4 0 2.5Meg
C 4 5 30p
Ginput 4 0 Table {V(2)-V(3)} = (-0.1V,-19uA) (+0.1V,19uA)
Emiddle 5 0 4 0 -529
Eoutput 1 0 Table {V(5)} = (-10V,-10V) (10V,10V)
.ends large_signal_opamp
```

2.4 The Effects Of Op Amp Large-Signal Nonidealities On Closed-Loop Behavior

```
DC Transfer Characteristics Of An Inverting Amplifier With Gain -10

* op-amp subcircuit
.subckt large_signal_opamp 1 2 3
* connections:            | | |
*                  output | |
*               +ve input |
*                -ve input
R 4 0 2.5Meg
C 4 5 30p
Ginput 4 0 Table {V(2)-V(3)} = (-0.1V,-19uA) (0.1V,19uA)
Emiddle 5 0 4 0 -529
Eoutput 1 0 Table {V(5)} = (-10V,-10V) (10V,10V)
.ends large_signal_opamp

** Main Circuit **
* signal source
Vi 3 0 DC 1V
Xopamp 1 0 2 large_signal_opamp
R1 3 2 1k
R2 2 1 10k
** Analysis Requests **
.DC Vi -15V +15V 100mV
** Output Requests **
.PLOT DC V(1)
.probe
.end
```

Figure 2.27 Spice input deck for calculating the DC transfer characteristic of an inverting amplifier containing a nonideal op amp.

2.4 The Effects Of Op Amp Large-Signal Nonidealities On Closed-Loop Behavior

Now that we have created a subcircuit for the op amp that accounts for several of its nonidealities, let us explore some of the idiosyncrasies of the op amp in various closed-loop configurations.

2.4.1 DC Transfer Characteristic Of An Inverting Amplifier

Consider the inverting amplifier circuit first shown in Fig. 2.2 with resistors $R_1 = 1$ kΩ and $R_2 = 10$ kΩ. Let us calculate the DC transfer characteristic of this circuit using PSpice assuming that the op amp is nonideal and modeled as described above in the last section for large signal operation. The PSpice input file is given in Fig. 2.27.

The results of the DC sweep calculations are displayed in Fig. 2.28. As can be seen from this graph, input signals of magnitude less than 1 V will experience a signal gain of

2 Operational Amplifiers

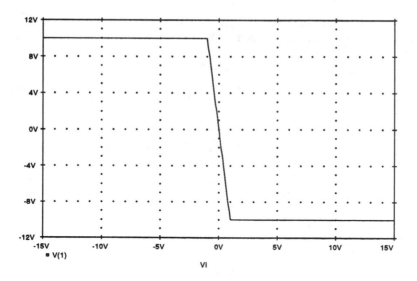

Figure 2.28 DC transfer characteristic of the inverting amplifier as calculated by Spice.

−10 without distortion as inferred from the slope of the line in this region. Whereas, signal amplitudes exceeding this limit will not be amplified, but instead the output will be held at a constant voltage of ±10 V depending on the sign of the input signal. Thus, linear circuit operation is limited to input signals less than a volt in magnitude.

To see how these transfer characteristics manifest themselves in the time-domain consider applying a sinewave of 400 mV peak amplitude at a 1 kHz frequency to the input of the amplifier and then repeat the same analysis using an input signal that has a signal amplitude larger than the 1 volt limit. In this particular case, we shall use a signal level of 1.1 V peak. We can use the same PSpice input deck as was just used above for calculating the amplifier DC transfer characteristic by replacing the DC source statement for the first case by the following

```
Vi 3 0 SIN (0 400mV 1kHz),
```

and by changing the 400 mV parameter to 1.1 V for this new case. Also necessary is that a transient analysis command replace the DC sweep command statement such as the following

```
.TRAN 10u 5ms 0s 10u.
```

The results of these two transient analyses are shown in Fig. 2.29. Clearly, when the input level exceeds the 1 V limit, the output signal becomes clipped. Whereas, the other

2.4 The Effects Of Op Amp Large-Signal Nonidealities On Closed-Loop Behavior

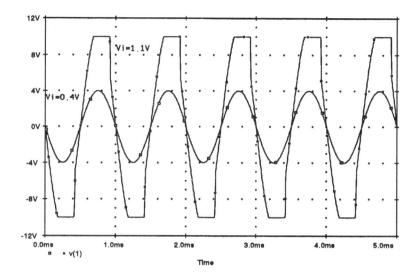

Figure 2.29 Evidence of output voltage clipping when the input signal level is too high.

signal is amplified without any distortion.

2.4.2 Slew-Rate Limiting

Another important phenomenon caused by internal amplifier saturation effects is that of slew-rate limiting. This effect plays an important role in determining the high-frequency operation of op amp circuits. To obtain a better understanding of this effect, let us simulate a commonly used experimental set-up for characterizing op amp slewing. This set-up consists of an op amp in a unity-gain configuration and a generator supplying a voltage step as an input signal. The reader first encountered this configuration in Fig. 2.12 of Section 2.2.3, albeit the signal source was a sinewave generator instead of a of PWL generator. We shall use exactly the same large-signal op amp subcircuit developed in the last section; however, we shall add additional terminals to the op amp subcircuit so that we may monitor the output current of the first stage. As we shall see, op amp slew-rate behavior is fully explained by the behavior of this current.

For the first part of our simulation, consider applying a very small step input of 1 mV and observe the transient response at the output of the amplifier and the current that flows between the first and second stages. The PSpice input file is listed below in Fig. 2.30 and the results are displayed in Fig. 2.31(a). The top curve of Fig. 2.31(a) displays both the step input voltage signal and the corresponding output response, the curve below it represents

75

```
              Investigating Op-Amp Slew-Rate Limiting

* op-amp subcircuit
.subckt large_signal_opamp 1 2 3 6 4
* connections:            | | | | |
*                  output | | | \ /
*               +ve input | |
*               -ve input |
*     current monitor of 1st stage
Iopen1 2 0 0A   ; redundant connection made at +ve input terminal
Iopen2 3 0 0A   ; redundant connection made at -ve input terminal
R 4 0 2.5Meg
C 4 5 30p
Ginput 6 0 Table {V(2)-V(3)} = (-0.1V,-19uA) (0.1V,19uA)
Emiddle 5 0 4 0 -529
Eoutput 1 0 Table {V(5)} = (-10V,-10V) (10V,10V)
.ends large_signal_opamp

** Main Circuit **
* signal source
Vi 2 0 PWL (0,0V 1us,0V 1.01us,1mV 1s,1mV )
Xopamp 1 2 1 4 5 large_signal_opamp
Vmonitor 4 5 0

** Analysis Requests **
.TRAN 10ns 5us 0s 10ns
** Output Requests **
.PLOT TRAN V(2) V(1) I(Vmonitor)
.probe
.end
```

Figure 2.30 Spice input deck for investigating op amp slew-rate limiting.

the voltage appearing between the input terminals of the amplifier, and the bottom curve represents the output current of the first-stage. As is evident, the output voltage signal increases towards its final state in an exponential manner. Similarly, the voltage appearing between the input terminals of amplifier and the output current of the first-stage follows a similar exponential pattern. Moreover, we see that these two signals are proportional to one another; albeit with a negative sign. These results are expected of an op amp whose dynamic behavior is modeled as a single-time-constant network.

Conversely, if a 1 V step input is applied to the input terminal, then instead of an exponential increase in the output voltage, the output voltage ramps up at a constant rate as shown in the top graph of Fig. 2.31(b). This rate, of course, is the slew-rate of the op amp at 0.633 V/μs. To help understand this effect, refer to the voltage and current waveforms shown in the two graphs below the top one. We see from the graph of the current waveform (bottom-most graph) that, for the most part, it is saturated at a constant level of -19 μA, which, of course, is the maximum current deliverable by the first stage ie. I_{max}. It is not before the voltage between the input terminals of the op amp (middle graph) goes between $\pm I_{max}/G_m$ or ± 100 mV does the transconductance stage enter its linear region and that the current generated by this stage begins to decrease at an exponential rate.

In terms of sinusoidal inputs, slew-rate limiting manifests itself in the output as a distorted sinewave. Consider applying a 100 kHz sinewave of 1.5 V amplitude to the input of the unity-gain buffer described above in the PSpice input file listed in Fig. 2.30. This requires that one change the step input statement to a sinewave input using the following source statement:

```
Vi 2 0 SIN (0 1.5V 100kHz).
```

The results of the PSpice simulation are illustrated in Fig. 2.32. In addition to the input and output voltage waveforms, and the evidence of distortion in the output signal, the cause of this distortion should be clear from the waveform of the current signal delivered by the front-end transconductance stage. Under small-signal conditions, this current waveform should be sinusoidal like the input, but, obviously, the front-end stage is being push beyond its linear capability.

2.5 Other Op Amp Nonidealities

A practical op amp deviates from its ideal behavior in many other ways than those discussed previously. Some of these additional nonidealities would include common-mode signal gain, finite input impedances, nonzero output impedance and, DC bias and offset signals. In this section we shall demonstrate how to incorporate these nonidealities into circuit simulation using Spice.

2.5.1 Common-Mode Gain

Accounting for op amp common-mode gain is accomplished by simply adding a dependent voltage source in series with the positive input terminal of the op amp, as shown in Fig. 2.33. The op amp shown in Fig. 2.33 can be represented either by an ideal op amp model or one that captures more of its true-life small- and large-signal behavior.

As an example of common-mode signal error, we shall return to the instrumentation amplifier example discussed in section 2.2.4 and assume that the op amps have a finite common-mode rejection ratio (CMRR). Recall that CMRR is defined as the ratio of the op

2 Operational Amplifiers

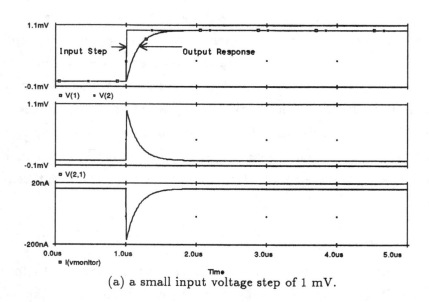

(a) a small input voltage step of 1 mV.

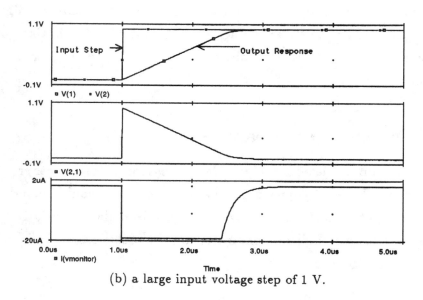

(b) a large input voltage step of 1 V.

Figure 2.31 Input and output waveforms of the unity-gain amplifier when both a small and large voltage step input is applied. The middle curve of each graph is the voltage between the two input terminals of the op amp. Also shown in the lower curve of each graph is the current supplied by the front-end transconductance stage.

amp differential gain to the corresponding common-mode gain. The rest of the circuit will be assumed ideal and perfectly matched. Thus, under ideal op amp conditions, common-mode signals appearing at the input of the instrumentation amplifiers should not appear at its output. However, some common-mode signal will appear at the output and we wish to

2.5 Other Op Amp Nonidealities

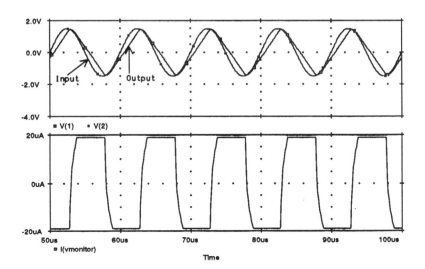

Figure 2.32 The upper two curves are the input and output waveforms of the unity-gain amplifier subjected to a 100 kHz sinusoidal input signal of 1.5 V amplitude. The lower waveform is of the current being supplied by the front-end transconductance stage.

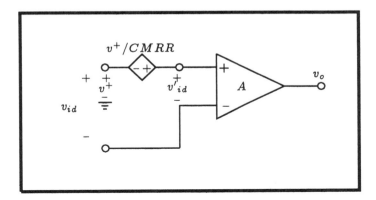

Figure 2.33 Accounting for the common-mode signal gain of an op amp circuit by attaching a dependent voltage source in series with the positive terminal of the op amp.

determine the magnitude of this signal when the op amps have an assumed CMRR of 80 dB and a finite DC gain of 120 dB. The Spice input deck for this particular example is identical to the one used in section 2.2.4, and listed in Fig. 2.18, except that the op amp subcircuit *ideal_opamp* is replaced with the following one:

2 Operational Amplifiers

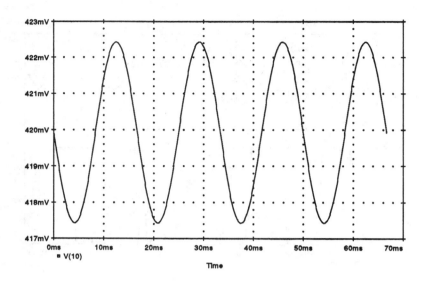

Figure 2.34 Common-mode signal feeding through from the input of the instrumentation amplifier to its output because of the finite CMRR of the op amp.

```
.subckt common_mode_opamp 1 2 3
* connections:              | | |
*                     output | |
*                    +ve input |
*                     -ve input
Iopen1 2 0 0A
Iopen2 3 0 0A
Iopen3 4 0 0A
Eerror 2 4 4 0 100u  ; Gain = 1/CMRR
Eamp 1 0 4 3 1e6
.ends common_mode_opamp
```

The Spice results are shown in Fig. 2.34 and show that a 2.5 mV 60 Hz signal component appears at the output. Thus, this instrumentation amplifier has a common-mode voltage gain (A_{CM}) of 2.5 mV / 25 V, or 10^{-4} V/V. Combining this with the result found previously for the differential gain (A_d) of 420 mV / 20 mV, or 21 V/V, we can compute the overall CMRR for this instrumentation amplifier to be 21 V/V $\big/$ 10^{-4} V/V or 106 dB.

2.5.2 Input And Output Resistances

Input and output impedances can be added to the op amp subcircuit in a straightforward way. We encourage the reader to try and incorporate input and output resistors for the various op amp models presented above. On completion, calculate the input and output impedance of some closed-loop op amp circuit as a function of frequency. The results should prove interesting.

Also note that many of the redundant connections made by the zero-valued current sources in these subcircuits will no longer be needed once the input and output resistances are added. They can therefore be removed to simplify the subcircuit.

2.5.3 DC Problems

Consider the equivalent circuit for the op amp shown in Fig. 2.35 which is separated into two parts: the first part consists of a set of input-referred DC signal sources representing the offsets and bias currents of the op amp, and the second part consists of an op amp circuit free of any DC offsets or bias currents. The two current sources labeled I_B represent the average value of the DC current flowing into the op amp input terminals. The polarity of these sources will depend on the front-end nature of the amplifier, ie. positive for front-end *npn* transistors and negative for a *pnp* transistorized front-end. The current source I_{OS} represents the difference between the actual currents that flow into the op amp input terminals. The voltage source V_{OS} represents the input-referred offset voltage. A subcircuit taking into account these bias and offset signals is listed below:

```
.subckt dc_opamp 1 2 3
* connections:   | | |
*         output | |
*          +ve input |
*           -ve input
IB1    4 0 DC 200nA
IB2    3 0 DC 200nA
IOS/2  3 4 DC 5nA
VOS    4 2 DC 5mV
Xdc_free_opamp 1 4 3 ideal_opamp
.ends dc_opamp
```

Here we have specified that the op amp will have an input-referred offset voltage of +5 mV, an input bias current of 200 nA and an offset current of 5 nA. Notice that we are calling the previously defined op amp subcircuit named *ideal_opamp* inside this new subcircuit. This form of nesting subcircuit calls is valid provided each subcircuit is found within the Spice input file. But note that a subcircuit cannot be made to call itself.

Given the above model accounting for DC offset effects, let us consider the effect that these offsets have on integrator behavior. Returning to the Spice input file for the Miller integrator shown in Fig. 2.5, we have modified it to include the above described op amp model, as shown in Fig. 2.36. A transient analysis request is included in this file. An

2 Operational Amplifiers

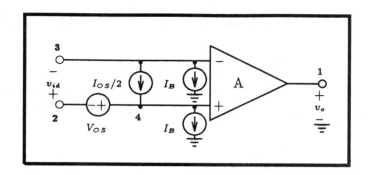

Figure 2.35 Modeling the effect of op amp DC offsets. Here we have included a DC voltage source in series with the positive input terminal of the op amp to account for its voltage offset, two equal current sources connected to the input terminals of the ideal op amp to account for its input bias currents, and a third current source to account for the input offset current.

additional UIC (use initial conditions) flag is included on this analysis request in order to ensure that the transient analysis begins with all nodes in the integrator circuit at 0 V. Without UIC, the transient analysis would begin with the nodes set at their DC operating point which turns out to be the final conditions of the transient analysis.[†] The input to the integrator is set to zero in order for us to observe separately the effect of op amp offsets.

The results of the transient analysis are shown in Fig. 2.37. We see here that even though the input to the integrator is zero, its output ramps upwards towards the positive power supply; clearly, an undesirable situation. In practise, one usually finds that eventually the op amp output voltage reaches the positive saturation level of the op amp. We would not see this here since our op amp does not model output voltage saturation effects. But, of course, this could easily be included by replacing the ideal op amp model with one that accounts for op amp large-signal behavior, such as the one described in Section 2.3.2.

A practical method commonly used to eliminate this run away effect is to connect a resistor across the integrator capacitor. Consider adding a 100 kΩ resistor across C_2. This requires that we add the following Spice statement to the Spice deck listed in Fig. 2.36:

```
R2 1 2 100k.
```

If we run the revised Spice file and observe the voltage that appears at the integrator output, one would find the voltage signal shown in Fig. 2.38. Here we do not see the output voltage increase without bound, but rather, exponentially settle to a constant output signal of about 526 mV. This is the net output offset voltage of the integrator. This value can

---

[†] When calculating the transient behavior of a circuit that contains no time-varying voltage or current sources, a UIC command should always be included on the .TRAN statement. In this way, the DC operating point of the circuit will not be used as the initial condition for the circuit.

```
The Effect Of DC Offsets On A Miller Integrator

* op-amp subcircuits

.subckt ideal_opamp 1 2 3
* connections:      | | |
*            output | |
*            +ve input |
*              -ve input
Iopen1 2 0 0A
Iopen2 3 0 0A
Eoutput 1 0 2 3 1e6
.ends ideal_opamp

.subckt dc_opamp 1 2 3
* connections:      | | |
*            output | |
*            +ve input |
*              -ve input
IB1    4 0 DC 200nA
IB2    3 0 DC 200nA
IOS/2  3 4 DC 5nA
VOS    4 2 DC 5mV
Xdc_free_opamp 1 4 3 ideal_opamp
.ends dc_opamp

** Main Circuit **

* inverting amplifier
Vi 3 0 DC 0
Xopamp 1 0 2 dc_opamp
R1 3 2 1k
C2 2 1 10uF
** Analysis Requests **
.TRAN 500ms 10s 0s 500ms UIC
** Output Requests **
.PLOT TRAN V(1)
.probe
.end
```

Figure 2.36 A Spice input file for calculating the effect of op amp DC offsets on integrator output when the input is set to zero. Since no initial conditions are explicitly indicated, Spice assumes that all nodes are initially at 0 V.

easily be calculated by hand analysis.

2.6 Spice Tips

- An ideal op amp can be modeled as a voltage-controlled voltage source with a large DC

2 Operational Amplifiers

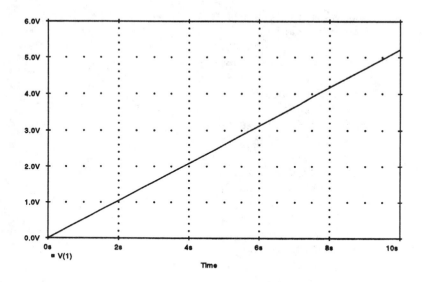

Figure 2.37 The effect of op amp DC offsets on the Miller integrator output when the input is grounded.

gain of at least 10^6 V/V.

- Nodes that do not have a DC path to ground are considered as floating nodes. Spice will not run with floating nodes. To circumvent this problem, connect large resistors between each floating node and ground.

- At least 2 or more connections must be made at each node in a circuit in order for Spice to run. Situations often arise when using controlled sources that this condition is violated. One can get around this problem by either connecting a large resistor between the node in question and ground, or by connecting a zero-valued current source between the node in question and ground. The latter method has the advantage that it does not disturb the operation of the network in any way.

- To simplify the writing of Spice input files, subcircuits of basic building blocks of the main circuit can be used to separate different portions of the circuit into smaller, more manageable, circuit blocks.

- When concatenating several Spice decks together into one file there can not be any blank lines that separate the end of one Spice deck (denoted by an .end statement) from the start of another Spice deck.

- Newer versions of Spice have the capability of describing the terminal behavior of a dependent source in either functional or piecewise-linear form. This is known as analog

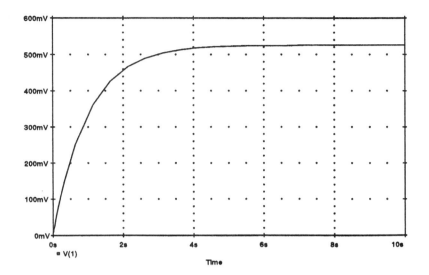

Figure 2.38 The effect of op amp DC offsets on the Miller integrator output when the input is grounded but the feedback capacitor is shunted with a 100 kΩ damping resistor.

behavior modeling and provides a very elegant means of describing circuit behavior to Spice.

- The initial conditions of a transient analysis can come from three sources: (1) DC operating point, (2) all node voltages and branch currents assumed 0, or (3) predefined initial conditions. One should be aware of the initial conditions used by Spice during a transient analysis so that the results are meaningful.

- As a general rule, when performing a transient analysis of a circuit that does not contain any time-varying sources, a UIC (use initial conditions) command should be included on a .TRAN statement.

2.7 Bibliography

Staff, *PSpice Users' Manual*, MicroSim Corporation, Irvine, California, Jan. 1991.

2.8 Problems

2.1 Assuming a pseudo-ideal op amp model for each op amp (ie. DC gain of 10^6), determine, with the aid of Spice, the voltage gain v_o/v_i and input and output resistance of each of

2 Operational Amplifiers

the circuits in Fig. P2.1.

2.2 Design an inverting op amp circuit for which the gain is -4 V/V and the total resistance used is 100 kΩ. Verify your design using Spice.

2.3 Repeat Problem 2.1 for the case the op amps have a finite gain $A = 1000$.

2.4 A Miller integrator incorporates an ideal op amp, a resistor R of 100 kΩ and a capacitor C of 0.1 μF. Using the AC analysis capability of Spice, together with the ideal op amp represented by a high-gain VCVS, determine the following:

(a) At what frequency are the input and output signals equal in amplitude?

(b) At this frequency how does the phase of the output sine-wave relate to that of the input?

(c) If the frequency is lowered by a factor of 10 from that found in (a), by what factor does the output voltage change, and in what direction (smaller or larger)?

(d) What is the phase relation between the input and output in situation (c)?

Confirm each of these situations by applying a 1 V peak sine-wave at the appropriate frequency and compare the voltage waveform appearing at the output using the transient analysis of Spice.

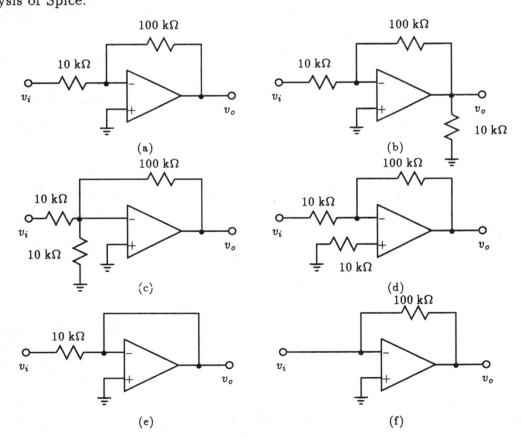

Fig. P2.1

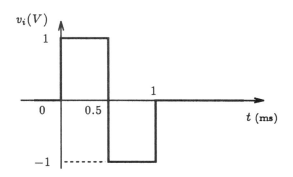

Fig. P2.5

2.5 A Miller integrator whose input and output voltages are initially zero and whose time constant is 1 ms is driven by the signal shown in Fig. P2.5. Using Spice compute the output transient waveform that results. Repeat the above simulation with the input levels increased to ± 2 V and the time constant raised to 2 ms. How does this output compare to that when the time constant is 1 ms?

2.6 Consider a Miller integrator having a time constant of 1 ms, and whose output is initially zero, when fed with a string of pulses of 10-μs duration and a 1 V amplitude rising from 0 V (see Fig. P2.6). Use Spice to to obtain a plot of the output voltage waveform. How many pulses are required for an output voltage change of 1 V?

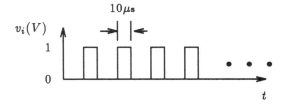

Fig. P2.6

2.7 In order to limit the low-frequency gain of a Miller integrator, a resistor is often shunted across the integrating capacitor. Consider the case when the input resistor is 100 kΩ, the capacitor is 0.1 μF, and the shunt resistor is 10 MΩ.

(a) Use the AC analysis capability of Spice to compute the Bode plot of the magnitude response of the resulting circuit and contrast it with that of an ideal integrator (that is, without the shunt resistor). At what frequency does the circuit begin to behave less as an integrator and more as an amplifier?

(b) Use the transient analysis capability of Spice to obtain a plot of the output voltage waveform resulting when an input pulse of 0.1 V height and 1 ms duration is applied. Consider the cases without and with the shunt resistor.

2.8 A differentiator utilizes a pseudo-ideal op amp, a 10 kΩ resistor, and a 0.01 μF capacitor.

Using the AC analysis command of Spice, determine the frequency f_o at which its input and output sine-wave signals have equal magnitude. What is the output signal for a 1 V peak-to-peak sine-wave input with frequency equal to $10 f_o$?

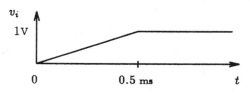

Fig. P2.9

2.9 An op amp differentiator with a 1 ms time constant is driven by the rate-controlled step shown in Fig. P2.9. Initializing the output voltage at 0 V, compute the voltage time waveform that appears at the output using Spice over a time interval of at least 5 ms.

2.10 A weighted summer circuit using a pseudo-ideal op amp has three inputs using 100 kΩ resistors and a feedback resistor of 50 kΩ. A signal v_1 is connected to two of the inputs, while a signal v_2 is connected to the third. With the aid of Spice, determine the output voltage v_o if $v_1 = 3$ V and $v_2 = -3$ V.

2.11 Design an op amp circuit to provide an output $v_o = -[3v_1 + v_2/2]$. Choose relatively low values of resistors but ones for which the input current (from each input signal source) does not exceed 0.1 mA for 2 V input signals. Verify all attributes of your design using Spice.

2.12 In an instrumentation system, there is a need to take the difference between two signals, one being, $v_1 = 3\sin(2\pi \times 60t) + 0.01\sin(2\pi \times 1000t)$ volts, and another, $v_2 = 3\sin(2\pi \times 60t) - 0.01\sin(2\pi \times 1000t)$ volts. Design a difference amplifier that meets the above requirements using two op amps. In addition, amplify the resulting difference by a factor of 10. Verify your design using Spice by simulating the transient behavior of your circuit. Plot both the input and output signals. Model each op amp with a high-gain VCVS.

2.13 It is required to connect a 10 V source with a source resistance of 100 kΩ to a 1 kΩ load. Find the voltage that will appear across the load if:

(a) the source is connected directly across the load.

(b) an op amp unity-gain buffer is inserted between the source and load.

In each case, find the load current and the current supplied by the source. Also, monitor the current supplied by the op amp. Assume that the op amp is pseudo-ideal with a DC gain of 10^6.

2.14 Consider the instrumentation amplifier of Fig. 2.15 with a common-mode input voltage of +5 V (dc) and a differential input signal of 10 mV peak, 1 kHz sine-wave. Let $R_1 = 1$ kΩ, $R_2 = 0.5$ MΩ, and $R_3 = R_4 = 10$ kΩ. With the aid of Spice, plot the voltage waveform at each node in the circuit for at least 2 periods of the input signal.

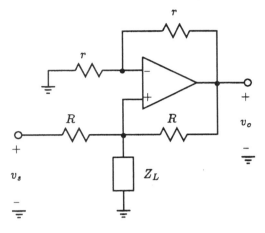

Fig. P2.15

2.15 For the negative impedance converter circuit of Fig. P2.15 with $R = 1$ kΩ and $v_s = 1$ V, use Spice to determine the voltages across the load and at the output of the op amp, for load resistances of 0 Ω, 100 Ω, 1 kΩ and 2 kΩ.

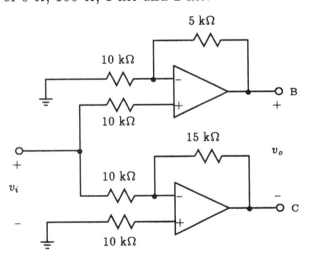

Fig. P2.16

2.16 The circuit shown in Fig. P2.16 is intended to supply current to floating loads while making greatest possible use of the available power supplies. With a 1 V peak-to-peak, 1 Hz sine-wave applied to its input, plot the voltage waveform appearing at nodes B and C. Also, plot v_o. What is the voltage gain v_o/v_i?

2.17 Measurements performed on an internally compensated op amp shows that at low frequencies its gain is 4.2×10^4 V/V and has a 3 dB frequency located at 100 Hz. Create a small-signal equivalent circuit model of this op amp and verify using the AC frequency analysis of Spice that it satisfies the above measurements.

2 Operational Amplifiers

2.18 A noninverting amplifier with a nominal gain of +20 V/V employs an op amp having a DC gain of 10^4 and a unity-gain frequency of 10^6 Hz. Model this behavior using an equivalent circuit, and with the aid of Spice, plot the magnitude response of the closed-loop amplifier and determine its 3 dB frequency f_{3dB}. What is the gain at $0.1 f_{3dB}$ and at $10 f_{3dB}$?

2.19 Consider a unity-gain follower utilizing an internally compensated op amp with $f_t = 1$ MHz. Assume that the low frequency gain is 10^6 V/V. Using Spice, determine the 3 dB frequency of the follower by plotting the magnitude response of the follower using the AC analysis command. At approximately what frequency is the gain of the follower 1% below its low frequency magnitude. What is the corresponding phase shift at this frequency? If the input to the follower is a 1 V step, determine the 10% to 90% rise time of the output voltage using the transient analysis command of Spice.

2.20 Consider an inverting summer with two inputs V_1 and V_2 and with $V_o = -(V_1 + V_2)$. With the aid of Spice, determine the 3 dB frequency of each of the gain functions V_o/V_1 and V_o/V_2 assuming that the small-signal behavior of the op amp is modeled after the 741 op amp. How do they compare with that predicted by theory? (*Hint: In each case, the other input to the summer can be set to zero.*)

2.21 A Miller integrator uses a 100 kΩ input resistor and a feedback capacitor of 0.1 μF. Using Spice, compare the magnitude and phase response of this Miller integrator assuming that the op amp is: a) pseudo-ideal, and b) internally compensated with a dc gain of 10^5 V/V and a 3 dB frequency of 10 Hz. For comparison purposes, its easiest to concatenate the two Spice files together and submit them together as one file to Spice. In this way, the results of the two situations can be directly compared. What is the "excess phase" that the nonideal integrator has at the unity-gain frequency of the (pseudo-)ideal integrator? Is the excess phase of the lag or lead type?

2.22 A particular family of op amps has an internal structure that can be modeled as that in Fig. 2.25. Four separate designs are being considered that have the following small-signal parameters:

G_m (mA/V)	R (MΩ)	μ (V/V)	C (pF)
0.2	2	500	30
0.02	5	10^4	1
20	1	10^3	30
2	0.5	700	20

Create a Spice subcircuit that captures the above four circuit descriptions and then use it to plot the input-output voltage gain as a function of frequency. What is the DC gain, 3 dB and unity-gain frequencies for each op amp?

2.23 Consider that the large-signal behavior of an internally compensated op amp can be modeled as that shown in Fig. 2.25. Assume that the resistor R has a 1 MΩ value, capacitor C is 30 pF and the three control sources are described by the following mathematical expressions:

$$i(v_{id}) = 10^{-3} v_{id} \quad \text{provided } |v_{id}| \leq 1 \text{ V},$$
$$v(v_{i2}) = 500 v_{i2} \quad \text{for all } v_{i2},$$
$$v(v_{o2}) = v_{o2} \quad \text{provided } |v_{o2}| \leq 12 \text{ V}.$$

Create a Spice (PSpice) subcircuit that captures the above behavior and verify that it indeed satisfies the given descriptions. Next, connect the op amp in a unity-gain configuration and apply a 5 V step input. Compute the expected output transient response and determine the positive-going slew-rate for this amplifier. How does this value compare with that predicted by theory?

2.24 For the large-signal op amp described above in Problem 2.23, use Spice to compute the magnitude response for this amplifier as a function of frequency between 0.1 Hz and 100 MHz. What is the corresponding DC gain, 3 dB and unity-gain frequencies?

2.25 Consider an inverting amplifier configuration having a gain of -10. Using the large-signal op amp model described in Section 2.3, confirm that the highest frequency of a 15 V peak-to-peak sine-wave that passes undistorted through the amplifier is 13.5 kHz. Do this by having Spice (PSpice) compute the transient behavior of the amplifier subject to the 1.5 V peak-to-peak input sine-wave of 13.4 kHz and compare this result to that when the frequency of the input signal is increased to 13.6 kHz.

2.26 To demonstrate the many trade-offs that a designer faces when designing with op amps, consider investigating the limitations imposed on a noninverting amplifier having a nominal gain of 10 with an op amp that has a unity-gain bandwidth (f_t) of 2 MHz, a slew rate (SR) of 1 V/μs and an output saturation voltage (V_{omax}) of 10 V. Model the op amp using the large-signal macromodel described in Section 2.3. Assume a sine-wave input with peak amplitude V_i.

(a) If $V_i = 0.5$ V, what is the maximum frequency of the input signal that can be applied to this amplifier before the output signal shows visible distortions?

(b) If the frequency of the input signal is 20 kHz, what is the maximum value of V_i before the output distorts?

(c) If $V_i = 50$ mV, what is the useful frequency range of operation?

(d) If $f = 5$ kHz, what is the useful input voltage range?

2.27 An op amp with a DC gain of 10^6 V/V and a CMRR of only 40 dB is used in a noninverting configuration with a closed-loop gain of 2. Model the behavior of this op amp using a Spice subcircuit consisting of several VCVSs. Plot the voltage waveform that appears at the output of the amplifier when an input signal wave of 1 kHz and 10 V peak-to-peak is applied to this amplifier configuration.

2.28 A particular op amp, for which $A_0 = 10^4$, $R_{iCM} = 10$ MΩ, and $R_{id} = 10$ kΩ, is connected in the noninverting configuration with a closed-loop gain of 10 (ideally). Model the op amp with a VCVS, then, with the aid of Spice, determine the input resistance seen by the source for low frequencies?

2.29 An op amp for which $R_{iCM} = 50$ MΩ, $R_{id} = 10$ kΩ, $A_0 = 10^4$, and $f_t = 10^6$ Hz is used to design a noninverting amplifier with a nominal closed-loop gain of 10. Model the op amp with a single pole small-signal equivalent circuit model and create a Spice subcircuit for it before arranging the resistor feedback around it. With the aid of the AC analysis capability in Spice, apply a 1 V AC voltage signal to the amplifier input and plot the magnitude of the admittance seen by this source over a frequency interval of 0.1 Hz to 10^7 Hz.

2.30 An inverting amplifier for which $R_1 = 10$ kΩ and $R_2 = 100$ kΩ is constructed with an op amp whose open-loop output resistance is 1 kΩ, whose dc gain is 10^4, and whose 3 dB frequency is 100 Hz. Evaluate the magnitude of the output impedance of the closed loop amplifier over a frequency interval of 0.01 Hz to 10^7 Hz. Select a logarithmic sweep of the input signal frequency.

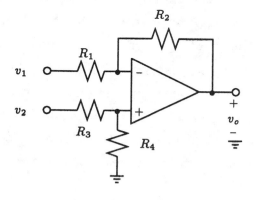

Fig. P2.32

2.31 A noninverting amplifier with a gain of 100 uses an op amp having an input offset voltage of ± 2 mV. Assume that the rest of the op amp can be modeled as a high-gain VCVS. If a 10 mV peak sine-wave of 1 Hz frequency is applied to the input of the amplifier, observe the voltage waveform that appears at the amplifier output using Spice.

2.32 Consider the differential amplifier circuit in Fig. P2.32. Let $R_1 = R_3 = 10$ kΩ and $R_2 = R_4 = 1$ MΩ. If the op amp has $V_{OS} = 3$ mV, $I_B = 0.2$ μA, and $I_{OS} = 50$ nA, determine, with the help of Spice, the dc offset voltage that appears at the amplifiers output. Assume that the rest of the op amp is pseudo-ideal (ie. modeled as a high-gain VCVS).

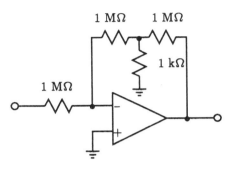

Fig. P2.33

2.33 The circuit shown in Fig. P2.33 uses an op amp that has a ± 5 V offset. Assume that the op amp is pseudo-ideal except for the dc offset. Determine the output offset voltage using Spice. What does the output offset become with the input ac coupled through a 1 μF capacitor? If, instead, the 1 kΩ resistor is capacitively coupled to ground, what does the output offset become?

2.34 An op amp is connected in a closed-loop with gain of $+100$ utilizing a feedback resistor of 1 MΩ. With the aid of Spice, answer the following:

(a) If the input bias current is 100 nA and everything else about the op amp is assumed ideal, what output voltage results with the input grounded?

(b) If the input offset voltage is ± 1, and the input bias current as in (a), what is the largest possible output that can be observed with the input grounded?

(c) If bias current compensation is used, what is the value of the required resistor? Verify that this indeed reduces the output offset voltage.

Chapter 3　　　　　　　　　　　　　　　　　　　　Diodes

The most fundamental nonlinear element of electronic circuits is the semiconductor junction diode. Because of its importance in both discrete and integrated circuits, the creators of Spice have made provisions for a built-in model of the semiconductor diode. In this chapter we shall outline how to access the diode model, and how to alter the model parameters to suit particular applications. Finally, using Spice, we shall investigate the behavior of some commonly used diode circuits, and compare the results with those obtained through the approximate hand analysis methods presented in the text by Sedra and Smith.

3.1   Describing Diodes To Spice

A *pn* junction diode is described to Spice using two statements. One statement is necessary to specify the diode type and the manner in which it is connected to the rest of the network, and the other statement is required to specify the values of the parameters of the built-in model for the diode type specified by the first statement. In the following we shall describe these two statements.

3.1.1   Diode Element Description

The presence of a *pn* junction diode in a circuit is described to Spice through the Spice input file using an element statement beginning with the letter D. If more than one diode exists in a circuit then a unique name must be attached to D to uniquely identify each diode. This is then followed by the two nodes that the anode and cathode of the diode are connected to. Subsequently, on the same line, the name of the model that will be used to characterize this particular diode is given. The name of this model must correspond to the name given on a model statement containing the values of the model parameters. Lastly, one has the

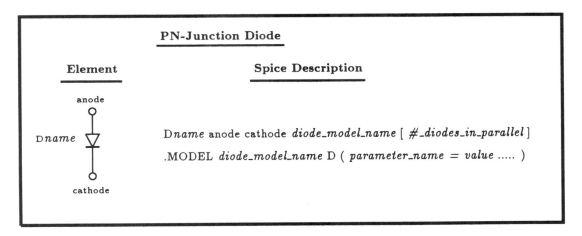

Figure 3.1 Spice element description for the *pn* junction diode. Also listed is the general form of the associated diode model statement. A partial listing of the parameter values applicable to the *pn* junction diode is given in Table 3.1.

option of specifying the number of diodes that are considered to be connected in parallel. This acts as a convenient way of scaling the cross-sectional area of the device.

For quick reference, we depict in Fig. 3.1 the syntax of the Spice statements pertaining to the *pn* junction diode. We shall discuss the model statement more fully next.

3.1.2 Diode Model Description

As is evident from Fig. 3.1, the model statement for the *pn* junction begins with the keyword .MODEL and is followed by the name of the model used by a diode element statement, the letter D to indicate that it is a diode model, and a list of the values of the model parameters (enclosed between brackets). There are quite a few parameters associated with the *pn* junction diode model used by Spice, and their individual meanings are rather involved. Instead of trying to describe the meaning of every parameter, we shall only consider here the parameters of the Spice diode model that are relevant to our introductory study of diodes in this chapter.

The large-signal model used for the semiconductor junction diode in Spice is shown in Fig. 3.2. The DC characteristic of the diode is modeled by the nonlinear current source i_D which depends on v_D according to the following equation:

$$i_D = I_S(e^{v_D/nV_T} - 1). \tag{3.1}$$

Here I_S, n and V_T are device parameters: I_S is referred to as the saturation current, n the emission coefficient, and V_T as the thermal voltage. Both I_S and n are related to the physical make-up of the diode, in contrast to the thermal voltage V_T which depends on the

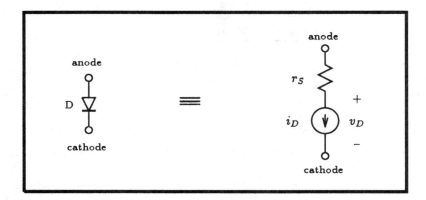

Figure 3.2 The Spice large-signal pn junction diode model.

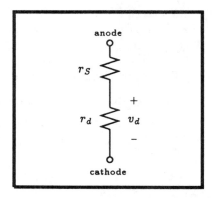

Figure 3.3 A small-signal equivalent circuit used by Spice to represent a semiconductor diode.

temperature of the device and two physical constants, ie.

$$V_T = \frac{kT}{q}. \tag{3.2}$$

Here k is Boltzmann's constant ($k = 1.381 \times 10^{-23}$ V C/°K), T is the absolute temperature in degrees Kelvin, and q is the charge of an electron ($q = 1.602 \times 10^{-19}$). In many approximate analyses, V_T is assumed to equal 25 mV at a room temperature of 290 °K. However, Spice does not approximate this quantity, and instead uses the exact values of k, T and q to determine V_T.

Finally, the series resistance r_S in the diode model shown in Fig. 3.2 simply models the lump resistance of the silicon on both sides of the semiconductor junction. The value of r_S may range from 10 to 100 Ω in low-power diodes.

Under small-signal conditions, Spice adopts the diode equivalent circuit shown in Fig. 3.3. Here r_d is the incremental resistance of the diode around its quiescent operating point and is expressed in terms of the DC bias current I_D as follows:

Symbol	Spice Name	Model Parameter	Units	Default
I_S	Is	Saturation current	Amps	1×10^{-14}
r_S	Rs	Ohmic resistance	Ω	0
n	n	Emission coefficient		1
V_{ZK}	BV	Reverse-bias breakdown voltage	Volts	∞
I_{ZK}	IBV	Reverse-bias breakdown current	Amps	1×10^{-10}

Table 3.1 A partial listing of the Spice parameters for a static pn junction diode model.

$$r_d = \left(\left. \frac{\partial i_D}{\partial v_D} \right|_{OP} \right)^{-1} = \frac{nV_T}{I_D}. \tag{3.3}$$

Spice does not use the name r_d for the small-signal resistance of the diode, instead it refers to this resistance in the Spice output file as REQ.

Under large reverse-bias conditions (ie. $v_D << 0$), the operation of the semiconductor diode is dominated by physical effects other than those which give rise to Eqn. (3.1). More specifically, under large reverse-bias conditions, the semiconductor diode enters its breakdown region and begins to conduct a large reverse current. In Spice, the dependence of this reverse current on reverse voltage is modeled as an exponential function – much like that of the forward bias region. In fact, for voltages below a commonly specified reverse-bias voltage $-V_{ZK}$ (breakdown voltage) the i-v characteristic is a near vertical straight line. Correspondingly, at this breakdown voltage $-V_{ZK}$, the diode is said to conduct a reverse current of $-I_{ZK}$. Thus, V_{ZK} and I_{ZK} specify the start of the breakdown region of the semiconductor diode. Note that these two quantities are positive numbers.

A partial listing of the parameters associated with the Spice pn junction diode model under static conditions is given in Table 3.1. The first column lists the symbol used to denote each parameter as described in this chapter. For the most part, the symbol of each of these parameters corresponds with those used in Chapter 3 of Sedra and Smith. The next column lists the corresponding name of the parameter used by Spice; this is the only name that can be used for this parameter in the list of model parameters on a .MODEL statement. Also listed are the associated default values which the parameter assumes if a particular value is not specified on the .MODEL statement. To specify a parameter value one simply writes, for example, Is=1e-13, n=2, BV=50, IBV=1e-12, etc.

3 Diodes

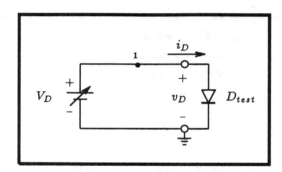

Figure 3.4 Simple circuit arrangement for determining the i-v characteristic of a 1 mA diode.

3.2 Spice As A Curve Tracer

Whenever a model of a device is created, whether it be a model of an op-amp, diode or some other electronic device, one should make certain the expected terminal characteristics are actually captured by the model. In the laboratory, a curve tracer is an instrument that is commonly used to measure the DC terminal characteristics of semiconductor devices. Using Spice we can emulate the behavior of the curve-tracer and therefore determine whether the DC model chosen to represent the diode is suitable.

A curve-tracer is an instrument designed to measure the i-v characteristics of semiconductor devices over a wide range of voltage and current levels. A typical curve-tracer instrument contains a set of variable voltage and current sources that are capable of varying their level beginning at some lower value and proceeding to some maximum value in discrete steps. Concurrently, the current supplied to — or the voltage that develops across — the externally attached device is measured and recorded. On completion, the results are displayed on the screen of a cathode ray tube (CRT) as a graph. To emulate this behavior using Spice we make use of the DC sweep command in conjunction with some DC source.

For example, in Fig. 3.4 we display a voltage source (V_D) with a single diode D_{test} as its load. We shall assume that the diode is a "1 mA diode" (meaning that it conducts a current of 1 mA at a forward bias voltage of 0.7 V) and that its voltage drop changes by 0.1 V for every decade change in current. Thus, using the diode current equation listed in Eqn. (3.1), one can show that this diode is characterized by $I_S = 100$ pA and $n = 1.679$. To describe this particular 1 mA diode to Spice, we use the following model statement:

```
.model 1mA_diode D ( Is=100pA n=1.679 ).
```

Now, to compute and display the forward-bias i-v characteristics of this 1 mA diode we shall sweep V_D in Fig. 3.4 from 0 V to 800 mV in 10 mV steps using the following DC sweep command

3.2 Spice As A Curve Tracer

```
Spice As A Curve Tracer: Diode I-V Characteristics

** Circuit Description **
VD 1 0 DC 700mV
Dtest 1 0 1mA_diode
* diode model statement
.model 1mA_diode D (Is=100pA n=1.679)
** Analysis Requests **
* vary diode voltage and measure diode anode current
.DC VD 0V 800mV 10mV
** Output Requests **
.plot DC I(VD)
.probe
.end
```

Figure 3.5 Spice input deck for determining diode forward-bias characteristics.

.DC VD 0V 800mV 10mV

and then plot the diode current as a function of this voltage. Although the diode current is not directly accessible by Spice[†] it is equal to the current supplied by the voltage source V_D which is readily accessible by Spice. We shall therefore plot the current supplied by V_D. Recall that according to Spice conventions, the current supplied by a voltage source is negative, thus the current plotted by Spice will be opposite to that which flows through the diode. Fortunately, the plot routine of Probe enables one to plot –I(VD).

The Spice input deck for this particular example is listed in Fig. 3.5 and the resulting i-v characteristic for this particular diode is shown in Fig. 3.6 in two different forms. The top curve displays the diode i-v characteristic on a linear scale, whereas the bottom curve is on a semi-logarithmic scale.

The reverse-bias characteristics of a particular diode are computed in exactly the same manner as that used to compute the forward-bias diode characteristics. In fact, one can combine both the forward- and reverse-bias characteristics onto one i-v plot. Consider comparing the forward- and reverse-bias characteristics of the 1 mA diode used above (the breakdown region of which was not specified, ie. left for Spice to use default values) to a similar one that has a breakdown region defined by $V_{ZK} = 10$ V and $I_{ZK} = 1$ nA. The Spice model statement for this new diode would be as follows:

.model bkdwn_diode D (Is=100pA n=1.679 BV=10V IBV=1nA).

[†] PSpice does allow the user access to this current, as outlined in Chapter 1. In some examples of this text we will make use of this extra feature of PSpice as a matter of convenience.

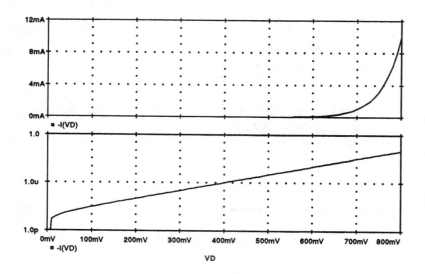

Figure 3.6 Forward-bias characteristics of a 1 mA diode with an emission coefficient of 1.679; upper curve: linear scale, lower curve: semilogarithmic.

The resulting i-v characteristics for these two diode are displayed in Fig. 3.7. Clearly, both these diodes have identical forward-bias characteristics but very different reverse-bias behavior.

3.2.1 Extracting The Small-Signal Diode Parameters

If an operating point (.OP) analysis command is included within the Spice input file, then the small-signal parameters of each diode within the circuit described to Spice will be evaluated by Spice and listed in the output file. For example, the small-signal parameters of the 1 mA diode used in the above example biased at 700 mV, would be found in the Spice output file as follows:

```
****      OPERATING POINT INFORMATION      TEMPERATURE =    27.000 DEG C

**** DIODES

NAME         Dtest
MODEL        1mA_diode
ID           1.00E-03
VD           7.00E-01
REQ          4.34E+01
CAP          0.00E+00
```

Included in this list of operating point information is the name of the diode assigned by the user and the name of the model used to characterize the diode. This is followed by the DC

3.2 Spice As A Curve Tracer

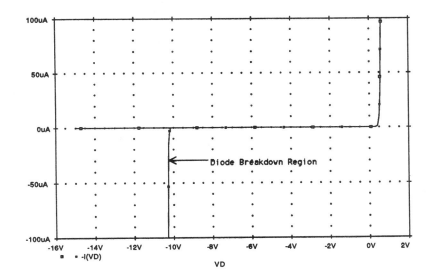

Figure 3.7 Comparing the *i-v* characteristic of a 1 mA diode with a breakdown region specified by $V_{ZK} = 10$ V and $I_{ZK} = 1$ nA, and one that has no breakdown region specified.

bias point information, ID and VD, and then, the incremental resistance of the diode REQ. At the bottom of this list is the small-signal capacitance CAP associated with this diode. We shall defer discussion of this parameter until Chapter 7.

At this point it would be highly instructive to check that the small-signal resistance of the diode computed by Spice in the list of operating point information agrees with that computed by the simple formula $r_d = nV_T/I_D$. Consider, for this particular diode, $n = 1.679$, $V_T = 25.8$ mV and $I_D = 1$ mA. Thus, on substituting these values into the expression for r_d, we get 43.47 Ω, almost in perfect agreement with the value computed by Spice (43.4 Ω).

One will notice in the above list of operating point information that the temperature at which the analysis has been performed is also indicated. In this case, the operating point analysis was performed at a room temperature of 27°C. The temperature at which the circuit is simulated by Spice can be changed by the user. The next example will demonstrate how this is accomplished.

3.2.2 Temperature Effects

To investigate the effect of temperature variation on diode behavior we simply repeat the above curve-tracer analysis at different temperatures. To accomplish this, one just adds a .TEMP statement with the temperature (in degrees Celsius) at which the analysis should

Analysis Requests	Spice Command
Temperature Analysis	.TEMP *temperature_list*

Table 3.2 The general syntax of the Spice command for setting the temperature of a circuit.

be performed. If more than one temperature is listed, then the analysis will be repeated for each temperature. A general description of the syntax of this command is provided in Table 3.2. For the example above, let us compute the diode i-v characteristics for temperatures of 0°, 27° and 125° C. The statement that should be added to the Spice deck shown in Fig. 3.5 is:

.TEMP 0 27 125.

The Spice job is then re-run and the results are shown in Fig. 3.8. The results displayed in this graph were restricted to lie within a 0 to 1 mA current range by adjusting the scale of the y-axis. This was deemed necessary in order to best illustrate the diode characteristics for all three temperatures on one graph. It should be evident form Fig. 3.8 that as the temperature increases the i-v curve for the diode shifts to the left. Close scrutiny using the cursor facility of Probe (found in PSpice) reveals that for a constant current of 0.4 mA the forward diode voltage decreases by about 1.7 mV for every degree C increase in temperature.

3.3 Approximating Ideal Diode Behavior

The ideal diode concept often simplifies the analysis of electronic circuits containing diodes by assuming that the forward bias voltage across a conducting diode is zero regardless of the level of current flowing through it, and conversely, when in the blocking state, it prevents current from flowing through it regardless of the level of reverse bias voltage appearing across it. An ideal model for a diode does not exist in Spice for the simple reason that an ideal diode does not exist in practise. However, there are numerous occasions when an ideal diode is useful to have; especially when attempting to represent an arbitrary nonlinear function.

Consider the equation for the static diode current given in Eqn. (3.1) and repeated below for convenience,

$$i_D = I_S(e^{v_D/nV_T} - 1).$$

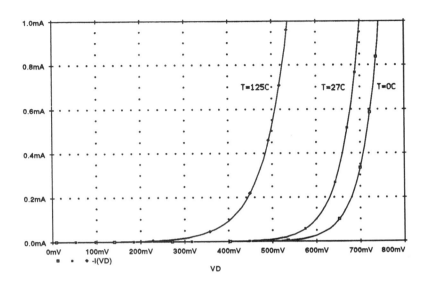

Figure 3.8 Illustrating the temperature dependence of the forward bias characteristics of a 1 mA diode.

There are two parameters associated with this equation that are under our direct control: the saturation current I_S and the emission coefficient n. Our problem is to adjust one, or both, of these parameters, such that (a) v_D tends towards zero when the diode is considered on, and (b) i_D tends towards zero when it is considered off. On examination of Eqn. (3.1) we see that condition (b) is satisfied when I_S is reduced and/or n is increased (recall that v_D is negative). Conversely, from the equation depicting the diode voltage (a simple rearrangement of Eqn. (3.1)),

$$v_D = nV_T \ln(\frac{i_D}{I_S} + 1), \qquad (3.4)$$

we see that alterations will tend to increase v_D under forward bias condition instead of reducing it. This suggests that conditions (a) and (b) can not both be met simultaneously by adjusting either I_S or n. We notice from Eqn. (3.1), however, that under reverse bias conditions that regardless of the value of n, i_D will never exceed I_S — a near-zero value. Thus, a good approximation to ideal diode behavior is obtained by making n small because this will reduce the diode forward voltage drop and maintain low reverse-bias current. Experience has shown that a value for n between 0.01 and 0.001 works well. Values smaller than 0.001 usually result in DC convergence problems. Fig. 3.9 illustrates the i-v characteristics of a simple diode characterized by the following model statement:

```
.model ideal_diode D (Is=100pA n=0.01).
```

3 Diodes

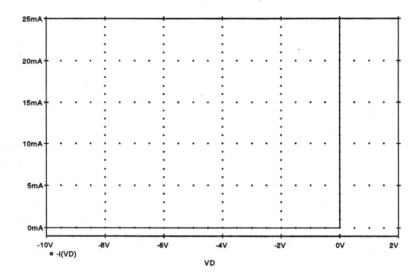

Figure 3.9 i - v characteristics of a near-ideal diode.

For all intensive purposes, the curve shown in Fig. 3.9 is an excellent representation of ideal diode behavior.

3.4 Voltage Regulation Using A String Of Diodes

Connecting one or more diodes in series with a resistor and a power supply provides a simple means of creating a relatively constant voltage, somewhat independent of fluctuations in the power supply level. One example of this is demonstrated in Fig. 3.10(a) where three diodes and a 1 kΩ resistor are connected in series with a +10 V DC voltage source. Since the forward voltage drop of each diode remains almost constant at approximately +0.7 V for a wide range of diode currents, the voltage that appears at the output of this regulator circuit is about +2.1 V. With the aid of Spice, we would like to investigate the effect of the fluctuations in the +10 V supply on the output voltage. We shall assume that the fluctuations are caused by a 60 Hz, 1 V-peak sinusoidal signal riding on the +10 V DC voltage level. This arrangement is illustrated in part (b) of Fig. 3.10. This same example was analyzed by hand in Example 3.7 of Sedra and Smith. Once the Spice results are obtained we shall compare them with those obtained by hand.

In Fig. 3.11 we list the Spice input file for this circuit. The three diodes are assumed to have model parameters: $I_S = 831.5$ pA and $n = 2$. A transient analysis is requested to

3.4 Voltage Regulation Using A String Of Diodes

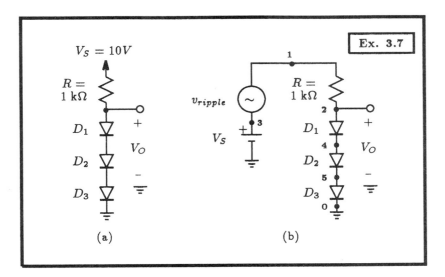

Figure 3.10 (a) A 2.1 V voltage regulator circuit consisting of three diodes in series. (b) Representing the power supply fluctuation with a 60 Hz, 1 V-peak sinusoidal signal superimposed on a constant DC voltage of +10 V.

```
A Three Diode String Voltage Regulator Circuit

** Circuit Description **
* DC supply + AC ripple
Vs 3 0 DC +10V
Vripple 1 3 sin (0 1V 60Hz)
* diode circuit
R 1 2 1k
D1 2 4 diode
D2 4 5 diode
D3 5 0 diode
* diode model statement
.model diode D (Is=831.5pA n=2)
** Analysis Requests **
.OP
.TRAN 0.5ms 100ms 0ms 0.5ms
** Output Requests **
.PLOT TRAN V(2)
.PROBE
.end
```

Figure 3.11 The Spice input file for computing the time-varying output voltage of the voltage regulator circuit shown in Fig. 3.10.

compute the behavior of this regulator circuit over six periods of the 60 Hz input signal. On average, 33.3 points-per-period of the input signal (ie. a sampling interval of 0.5 ms) are to be collected. This will provide enough points so that a smooth graph of the output waveform can be plotted. Also included in this Spice deck is a request for an operating point (.OP)

3 Diodes

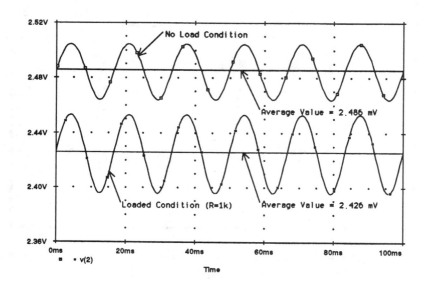

Figure 3.12 The output voltage signals from the three-diode string voltage regulator shown in Fig. 3.10 assuming that the power supply voltage varies sinusoidally with an amplitude of 1 V. The bottom waveform is the output of the regulator with a 1 kΩ load and the one above it is with no load.

analysis.

Before we submit this input file to Spice for processing we shall concatenate on the end of it another Spice deck. This Spice deck will be identical to that seen in Fig. 3.11 except that the output of the voltage regulator circuit is loaded with a 1 kΩ resistor. The element statement required to be added to the Spice deck would be as follows:

```
Rload 2 0 1k.
```

On completion of the two Spice jobs, we plot the output waveforms from the voltage regulator circuit in Fig. 3.12. The bottom curve is the output voltage waveform of the regulator circuit having a 1 kΩ load, and the curve above it is the output voltage signal of the regulator circuit under no-load conditions. As is evident, the output voltage waveform from the unloaded regulator circuit is sinusoidal, having the same frequency as the power supply fluctuations (60 Hz). It has an average value of 2.486 V, slightly different from that estimated above (2.1 V) assuming the voltage drop of each diode is 0.7 V. The peak-to-peak excursion of this output voltage signal is found using the cursor facility of Probe to be 40.7 mV. This is obviously much less than the peak-to-peak excursions of the power supply fluctuations of 2 V. In the case of the loaded regulated circuit, we see that the output voltage signal is riding on a DC level of 2.426 V, a decrease in the DC level of 60 mV below that of

3.4 Voltage Regulation Using A String Of Diodes

the unloaded regulator circuit. The peak-to-peak amplitude of the ripple superimposed on the output voltage is found to have increased to 57.4 mV.

According to small-signal analysis (see Example 3.7 of Sedra and Smith), the expected output voltage ripple of the unloaded diode regulator circuit subject to a power supply fluctuations v_{ac} is simply given by the following expression:

$$v_o = \frac{3r_d}{3r_d + R} v_{ac}, \qquad (3.5)$$

where r_d is the incremental resistance of each diode. The incremental resistance of each diode as calculated by Spice is found in the Spice output file amongst the DC operating point information. For the unloaded regulator case, these are shown below:

```
**** DIODES

NAME       D1         D2         D3
MODEL      diode      diode      diode
ID         7.51E-03   7.51E-03   7.51E-03
VD         8.29E-01   8.29E-01   8.29E-01
REQ        6.88E+00   6.88E+00   6.88E+00
```

As is evident, each diode has an incremental resistance (r_d) of 6.88 Ω. Interestingly enough, this is quite close to the value predicted by simple hand analysis at 6.3 Ω; a result obtained by assuming a 0.7 volt-drop across each diode leading to a diode current of 7.9 mA. Substituting, $r_d = 6.88$ Ω into Eqn. (3.5) above, and assuming a 2 V peak-to-peak input signal, results in an estimated output peak-to-peak voltage signal of 40.4 mV: very close to the value of 40.7 mV found by Spice.

In the case of the 1 kΩ loaded regulator circuit, hand analysis also leads to a similar confirmation. Small-signal analysis of the three-diode regulator circuit with a 1 kΩ load leads to an output voltage given by the expression:

$$v_o = \frac{3r_d \| R_{load}}{(3r_d \| R_{load}) + R} v_{ac}, \qquad (3.6)$$

Substituting the appropriate circuit parameters, together with the incremental resistance of each diode as calculated by Spice, and shown below,

```
**** DIODES

NAME       D1         D2         D3
MODEL      diode      diode      diode
ID         5.15E-03   5.15E-03   5.15E-03
VD         8.09E-01   8.09E-01   8.09E-01
REQ        1.01E+01   1.01E+01   1.01E+01
```

we find

3 Diodes

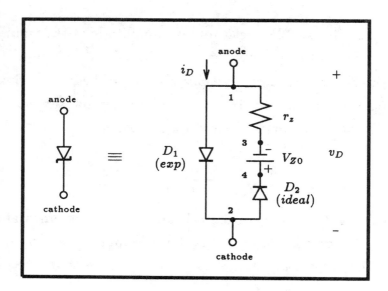

Figure 3.13 A circuit model of a Zener diode.

$$v_o = 28.56 \times 10^{-3} v_{ac}.$$

It is interesting to note that the addition of the load resistance has acted to increase the incremental resistance of each diode. This is a result of the 1 kΩ load drawing a current away from the diode string (the diode current has been reduced by 2.36 mA). For a 2 V peak-to-peak power supply variation, the expected time-varying output voltage signal is then 57.1 mV, in close agreement to the value of 57.4 mV calculated by Spice.

3.5 Zener Diode Modeling

Although the start of the diode breakdown region defined by V_{ZK} and I_{ZK} can be specified on a diode model statement using the Spice parameters BV and IBV, no control is provided for the user to specify the characteristics of the breakdown region, ie. slope of the i-v curve in the breakdown region.[†] It is common for zener diode manufacturers to specify the shape of the breakdown region by specifying the inverse of the slope of the almost-linear i-v curve at some operating point (V_Z, I_Z) inside the breakdown region. This inverse-slope parameter has dimensions of resistance and is known as the dynamic resistance of the zener, denoted as r_z.

To model a zener diode the equivalent circuit shown in Fig. 3.13 is sometimes used. When $v_D > -V_{Z0}$, ideal diode D_2 is consider cut-off and the terminal characteristics of the

---

† PSpice has since modified the built-in model for a diode so that this region of diode operation could be specified on the same model statement. However, this facility is unique to PSpice and will therefore not be used in this text.

3.5 Zener Diode Modeling

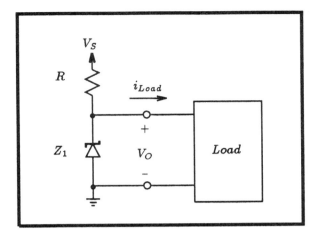

Figure 3.14 A simple voltage regulator circuit with load using a single zener diode.

zener diode are determined solely by diode D_1. Conversely, when $v_D \leq -V_{Z0}$, ideal diode D_2 turns on, whereby a voltage of $v_D + V_{Z0}$ appears across resistor r_z. The resulting current that flows through this resistance will be much greater than the reverse-bias leakage current that flows through diode D_1, and therefore, the current that dominates the breakdown region of the zener diode is given by the following equation,

$$i_D \approx \frac{v_D + V_{Z0}}{r_z}. \tag{3.7}$$

The value of V_{Z0} is not specified directly by the zener diode manufacturer but can be derived from the operating point information at which the dynamic resistance is obtained. It is found from the expression

$$V_{Z0} = V_Z - r_z I_Z. \tag{3.8}$$

The following example will illustrate a common application of a zener diode.

Voltage Regulation Using A Zener Diode

Rather than using a string of diodes to create a simple voltage regulator circuit, a single zener diode can be used in its place, as shown in Fig. 3.14. In Example 3.9 of Sedra and Smith, one such regulator circuit was designed for an output voltage of approximately 7.5 V, assuming that the raw supply voltage fluctuates between 15 and 25 V and that the load current can vary between 0 and 15 mA. The zener diode available has a voltage drop of $V_Z = 7.5$ V at a current of 20 mA, and its r_z equals 10 Ω. The current limiting resistor R in series with the zener diode was chosen at 383 Ω so that the minimum current through the diode never drops below 5 mA. Based on this design, both the line and load regulation

3 Diodes

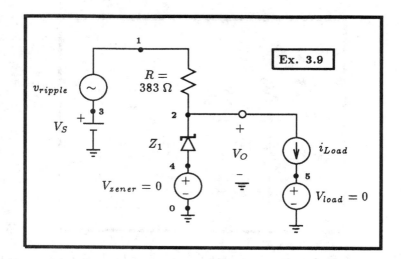

Figure 3.15 Circuit setup used to investigate the line and load regulation of the simple zener diode voltage regulator circuit of Fig. 3.14. A zero-valued voltage source is placed in series with each of the zener diode and the current source i_{Load} to monitor their respective currents. They play no part in the circuit operation.

were found to be 25.4 mV/V and –9.7 mV/mA, respectively. Using Spice, together with the model described above for the zener diode, we would like to confirm that the design requirements are indeed met. Further, we would like to check the line and load regulation directly from simulation results.

To carry out this investigation, we use the circuit setup shown in Fig. 3.15. Here the raw power supply level is modeled with two sources: a DC voltage source V_S to model the average value of the power supply level, and a time-varying voltage source v_{ripple} to model the fluctuations of the power supply. Specifically, the level of the DC source is set at 20 V, and the fluctuations are modeled as a sinusoidal signal having a peak amplitude of 5 V. The frequency of this source is arbitrarily selected to be 60 Hz. To mimic possible load current fluctuations, a single current source is connected across the output terminals of the voltage regulator. We shall begin our first simulation with the value of this current source set to zero in order to first determine the behavior of the regulator under no-load conditions. A zero-valued voltage source is connected in series with the zener diode to monitor the current through it, and another is also placed in series with the current source i_{Load} for the same reason.

The model for the zener diode is the equivalent circuit shown in Fig. 3.13. The value of r_z is simply that specified in the problem at 10 Ω. The value of V_{Z0} is determined from Eqn. (3.8) and the data supplied for the zener diode (given above), from which we find $V_{Z0} = 7.3$ V. Diode D_1 will be modeled as a 1 mA diode with an emission coefficient of 1.679. The ideal diode, D_2, will be model with $I_S = 100$ pA and $n = 0.01$. The subcircuit describing this particular zener diode would then appear as follows:

```
.subckt zener_diode    1 2
* connections:         | |
*                  anode |
*                        cathode
Dforward 1 2 1mA_diode
Dreverse 2 4 ideal_diode
Vz0 4 3 DC 7.3V
Rz 1 3 10
* diode model statement
.model   1mA_diode   D (Is=100pA n=1.679 )
.model   ideal_diode D (Is=100pA n=0.01 )
.ends zener_diode
```

The complete Spice input file describing the circuit shown in Fig. 3.15 is listed in Fig. 3.16. A transient analysis is requested so that 6 periods of the output voltage signal can be observed. The results of the Spice analysis are shown in Fig. 3.17. Both the power supply voltage $V_S + v_{ripple}$ and the voltage across the zener diode are shown. The top graph displays the supply voltage and the bottom graph displays the corresponding zener diode voltage waveform. As expected the voltage fluctuation of the power supply is sinusoidal having a peak-to-peak amplitude of 10 V riding on a DC level of 20 V. The output voltage from the regulator circuit is also sinusoidal having a peak-to-peak amplitude of 254.7 mV and riding on a DC level of 7.628 V. The precise value of these two levels were determined using the cursor facility of Probe. One can also notice that both the input and the output voltage waveforms are in phase. Thus, we can conclude that a line voltage change of +10 V gives rise to an output voltage change of +254.7 mV. Thus, the line regulation, given by

$$\text{Line regulation} = \frac{\Delta V_O}{\Delta V_S},$$

is calculated to be +25.47 mV/V. This seems to agree almost exactly with the value determine by the simple expression for line regulation given by $r_z/(r_z + R)$ derived in Chapter 3 of Sedra and Smith (ie. 25.4 mV/V). This should not be too surprising here given that the circuit has been operating entirely in its linear region.

We can perform a similar analysis to that above, but this time, with the load current varying between 0 and 15 mA. In this way we can determine the load regulation by observing the output voltage waveform. The power supply voltage will be assumed constant at +20 V. We shall assume that the load current is triangular with a minimum value of 0 mA and a maximum value of 15 mA. The frequency of this signal will be made arbitrarily equal to 30 Hz, corresponding to a period of 33.33 ms. This signal will correctly model the minimum and maximum fluctuations of the load current. Consider revising the statement for the output

3 Diodes

```
Zener Diode Voltage Regulator Circuit (No Load)

* zener diode subcircuit
.subckt zener_diode 1 2
* connections:         | |
*                anode |
*              cathode
Dforward 1 2 1mA_diode
Dreverse 2 4 ideal_diode
Vz0 4 3 DC 7.3V
Rz 1 3 10
* diode model statement
.model 1mA_diode D (Is=100pA n=1.679 )
.model ideal_diode D (Is=100pA n=0.01 )
.ends zener_diode

** Main Circuit **
* power supply
Vs 3 0 DC +20V
Vripple 1 3 sin ( 0V 5V 60Hz )
* zener diode voltage regulator circuit
R    1 2 383
XD1 4 2 zener_diode
Vzener 4 0 0
* simulated load condition
Iload 2 5 0A
Vload 5 0 0
** Analysis Requests **
.OP
.TRAN 0.5ms 100ms 0ms 0.5ms
** Output Requests **
.PLOT TRAN V(1) V(2)
.PROBE
.end
```

Figure 3.16 The Spice input file for computing the time-varying no-load output voltage of the zener diode voltage regulator circuit shown in Fig. 3.15.

current source i_{Load} seen in the Spice deck of Fig. 3.16 according to the following:

```
Iload 2 5 PULSE ( 0mA 15mA 0 16.66ms 16.66ms 1us 33.33ms).
```

Here a triangular waveform is emulated using the source PULSE statement. The rise time and fall time are set equal to one-half the period of the triangular waveform of 33.33 ms. The pulse width is assigned a very small value of 1 us because Spice will not accept a zero value for the pulse width.

The amplitude of the ripple voltage superimposed on the DC supply voltage should be set to 0 V in order to eliminate its presence during this analysis. Therefore, the revised Spice

3.5 Zener Diode Modeling

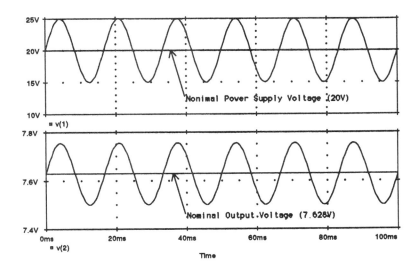

Figure 3.17 Several waveforms associated with the zener diode regulator circuit of Fig. 3.14 under no-load conditions. The top graph displays the voltage generated by the power supply and the bottom graph displays the corresponding output voltage from the regulator circuit.

statement for this source is changed to the following:

```
Vripple 1 3 sin ( 0V 0V 60Hz ).
```

Submitting the revised Spice job for processing, results in the output voltage waveform V_O for the regulator circuit shown in the bottom graph of Fig. 3.18. The waveform shown in the top graph depicts the load current i_{Load}. Here the load current is triangular, as it should be, with a 15 mA peak-to-peak amplitude. The corresponding output voltage signal is also triangular of the same frequency. The peak-to-peak amplitude of this signal was found using the cursor facility of Probe to be 146.33 mV. For an average load current of 7.5 mA, the output voltage corresponds to 7.555 V. It is important to notice that the phase of the output voltage waveform is opposite to that of the load current. This suggests that for a change in the load current of +15 mA, the output voltage changes by −146.33 mV, thus suggesting that the load regulation, expressed as

$$\text{Load regulation} = \frac{\Delta V_O}{\Delta I_{load}},$$

would be −146.33 mV / 15 mA, or −9.7 mV/mA. This agrees exactly with the value determine by the simply expression for load regulation given by $-r_z \| R$ derived in Chapter 3 of Sedra and Smith.

113

3 Diodes

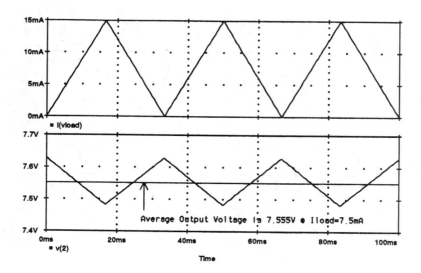

Figure 3.18 The output voltage of the zener diode regulator circuit of Fig. 3.14 when the load current varies between 0 and 15 mA. The top graph displays the current drawn by the load, and the bottom graph displays the corresponding output voltage from the regulator circuit.

As a final check on this design, let us investigate the minimum current that flows through the zener diode. Observe that the zener diode current is minimum when the power supply voltage is at its minimum and the load current is at its maximum. In keeping with our earlier approach, we shall maintain the load current as a triangular wave varying linearly between 0 and 15 mA. The ripple voltage v_{ripple} simulating fluctuations in the power supply voltage will be set to a constant −5 V level. This requires that the Spice statement for this source be changed to the following:

```
Vripple 1 3 DC -5V.
```

Re-running the Spice job with the revised source statement results in the three waveforms shown in Fig. 3.19. The top graph displays the constant +15 V level associated with the power supply and the graph below it displays the load current waveform. The lowest-most graph displays the current waveform associated with the zener diode. As is evident, the current through this diode varies between 5 mA and 20 mA.

For all practical purposes, based on the Spice results above, we can conclude that all aspects of the design did indeed meet the required conditions.

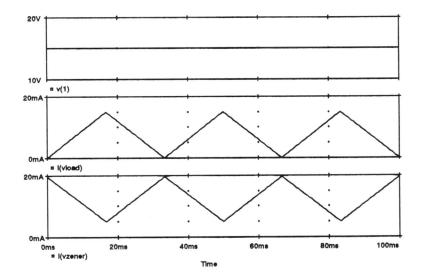

Figure 3.19 Observing a worst-case situation: The top-most graph displays the minimum power supply voltage, the middle graph displays the time-varying load current, and the bottom graph displays the current flowing through the zener diode.

3.6 Rectifier Circuits

One of the most important applications of semiconductor diodes is in the design of rectifier circuits. In the following, with the aid of Spice, we shall investigate two common types of rectifier circuits: the half-wave and the full-wave rectifier. In the case of the full-wave rectifier, we shall consider it in conjunction with a peak rectifier circuit. Subsequently, we shall combine this full-wave rectifier circuit with a zener diode voltage regulator circuit to form a complete power supply circuit.

3.6.1 Half-Wave Rectifier

A half-wave rectifier circuit is shown in Fig. 3.20. It consists of a transformer with a 14:1 turns ratio, a single diode D_1 of the commercial type 1N4148, and a load resistance R_{load} of 1 kΩ. The source resistance of 0.5 Ω of the AC line is also included in this circuit. The purpose of the transformer is to step down the main household AC power supply voltage of 120 V-rms to a 12 V-peak level. Spice does not make provision for an ideal transformer, probably for a good reason; one does not exist in practice. Instead, Spice allows coupled inductors to be described having a coefficient of coupling k less than one. Two inductors, say for example, L_P and L_S, which share a common magnetic path and have a coefficient of coupling k very close to unity, say 0.999, would be a reasonably good model of many practical

3 Diodes

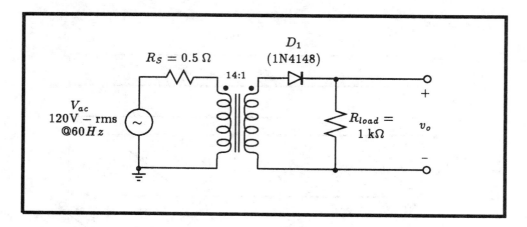

Figure 3.20 Half-wave rectifier circuit using a transformer with a 14:1 turns ratio to step down the line voltage of 120 V-rms to 12 V-peak.

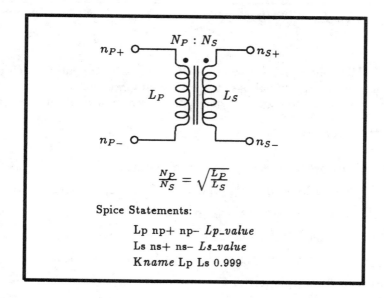

Figure 3.21 The general syntax of the Spice statements used to describe a (nonideal) transformer. The transformer turns ratio N_P:N_S is determined by the appropriate selection of primary and secondary inductor values, L_P and L_S, respectively.

transformers. The turns ratio N_P/N_S of such a transformer is given by the square-root of the ratio of the primary to secondary inductance, ie. $\frac{N_P}{N_S} = \sqrt{\frac{L_P}{L_S}}$.

To describe such a transformer to Spice, three element statements are required: One statement for each inductor, and a statement that describes the coefficient of coupling between the two inductors. Inductor coupling is described to Spice using a new statement that begins with the letter K. If more than two inductors share a common magnetic path, then a unique name is attached to K to uniquely identify each coefficient of coupling. This is then followed by the names of the two inductors that are magnetically coupled together. These

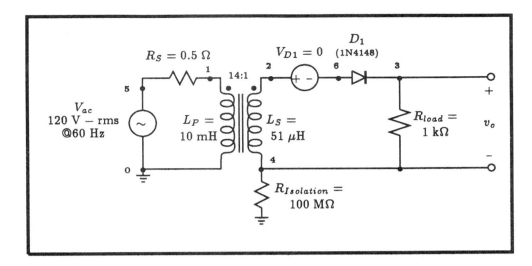

Figure 3.22 Preparing the half-wave rectifier circuit shown in Fig. 3.20 for Spice analysis: A large-valued isolation resistor (100 MΩ) is placed between the secondary side of the transformer and ground. This provides a DC path between the secondary side of the transformer and the common reference node (0). Also added is a zero-valued voltage source in series with the rectifier diode. This will allow indirect access to the diode current.

names must correspond with the names of two inductors described in the present Spice deck. Subsequently, the final field of this statement describes the coefficient of coupling k, which can take on a value between 0 and 1. Since Spice does not accept a value of k equal to unity, we shall always in this book use $k = 0.999$. The transformer "dot" convention is adhered to in Spice. Observe that the "dot" of each transformer is located at the positive node $(n+)$ of each inductor. Extension to three or more coupled inductors should be self-evident. For a transformer, Fig.3.21 illustrates the three Spice statements necessary to describe it to Spice.

Returning to the half-wave rectifier circuit of Fig. 3.20, we can create a Spice description of this circuit. We shall assume that the inductance of the primary side of the transformer is 10 mH, and the inductance of secondary side is 51 μH. This will provide an effective transformer turns ratio of 14:1. Continuing, the alert student will quickly realize that the circuit on the secondary side of the transformer has no DC path to ground and will therefore be rejected by Spice. To circumvent this situation, we simply add a large resistor between ground and one point on the secondary side. The value of this resistor should be chosen such that it does not significantly interfere with the operation of the circuit. Fig. 3.22 illustrates the addition of a 100 MΩ between ground and node 4 of the rectifier. Also shown in this figure is a zero-valued voltage source placed in series with the rectifier diode D_1. This will enable us to monitor the current flowing through the diode. The resulting Spice deck for this modified circuit is seen listed in Fig. 3.23. A transient analysis is requested to compute the voltage appearing across the load resistance, the voltage appearing across the primary-

```
Half-Wave Rectifier Circuit

** Circuit Description **

* ac line voltage
Vac 5 0 sin (0 169V 60Hz)
Rs 5 1 0.5
* transformer section
Lp 1 0 10mH
Ls 2 4 51uH
Kxfrmr Lp Ls 0.999
* isolation resistor (allows secondary side to pseudo-float)
Risolation 4 0 100Meg
* diode current monitor
VD1 2 6 0
* rectifier circuit
D1 6 3 D1N4148
Rload 3 4 1kOhm
* diode model statement
.model D1N4148 D (Is=0.1pA Rs=16 CJO=2p Tt=12n Bv=100 Ibv=0.1p)

** Analysis Requests **
.TRAN 0.5ms 100ms 0ms 0.5ms
** Output Requests **
.plot TRAN V(3,4) V(2,4) V(1)
.plot TRAN V(6,3)
.plot TRAN I(VD1)
.probe
.end
```

Figure 3.23 The Spice input file for calculating the transient behavior of the half-wave rectifier circuit shown in Fig. 3.20.

and secondary-side of the transformer, and finally, the AC line voltage. The Spice model of the commercial diode, 1N4148, was obtained from a library of Spice models for various electronic components included in PSpice[†].

The results of the Spice analysis are shown in Fig. 3.24. The top graph displays the voltage waveform of the AC line voltage (V_{ac}) and the voltage appearing across the primary-side of the transformer. Here we see that the voltage across the transformer experiences a short transient effect, quickly settling into its steady-state with the transformer voltage slightly lagging behind the line voltage. The bottom graph displays the rectified voltage appearing across the load resistance and the voltage appearing across the secondary-side of the transformer. A blown-up view of a half period of the rectified output voltage and the

---

† The user can find these devices in a file named NOM.LIB in the same directory that the PSpice program is located in.

3.6 Rectifier Circuits

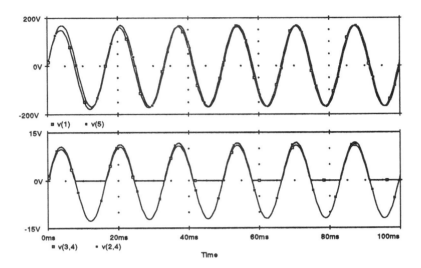

Figure 3.24 Various voltage waveforms associated with the half-wave rectifier circuit shown in Fig. 3.22. The top graph displays both the AC line voltage and the voltage appearing across the primary-side of the transformer. The bottom graph displays the voltage appearing across the load resistor and the voltage appearing across the secondary-side of the transformer.

transformer secondary-side voltage is shown in Fig. 3.25.

An important consideration in the design of rectifier circuits is the diode current-handling capability, determined by the largest current that it has to conduct, and the peak inverse voltage (PIV) that the diode must be able to withstand without breakdown. In Fig. 3.26 we display both the voltage across the diode and the current that it conducts. We see that the PIV of this particular rectifier circuit is 12 V. Because the diode has not broken down, we can assume that the breakdown voltage of the 1N4148 commercial diode is larger than 12 V. In fact, data sheets of the 1N4148 diode indicate that its breakdown voltage is in the vicinity of 100 V. The maximum current that the diode has to conduct is seen to be about 11 mA. Using the cursor facility of Probe, we find that it is 11.1 mA. The data sheets of the 1N4148 indicate that this diode can handle a peak current of no more than 100 mA, thus our rectifier design is well within the limits of the 1N4148.

3.6.2 Full-Wave Peak Rectifier

Fig. 3.27(a) displays a circuit for a full-wave peak rectifier. It consists of a full-wave rectifier — diodes D_1 and D_2 and a center-tapped transformer — and a filter capacitor C

3 Diodes

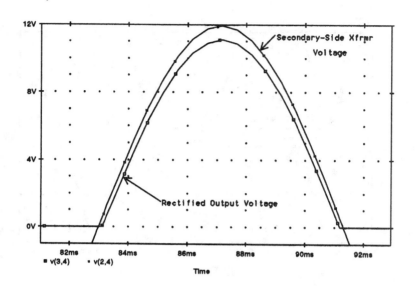

Figure 3.25 Zooming-in on a half cycle of the voltage waveform appearing across the load resistor and comparing it to the voltage developed across the secondary-side of the transformer.

to smooth the voltage that appears across the load resistor R. Also shown is the resistance of the input voltage source of 0.5 Ω. The transformer is center-tapped with each coil on the secondary side having a turns ratio of 14:1 with respect to the primary coil. In Fig. 3.27(b) we display the circuit that we shall actually describe to Spice for analysis. An isolation resistance $R_{Isolation}$ has been added to the circuit on the secondary-side of the transformer in order to provide a DC path to ground. Also added are two zero-valued voltage sources, one in series with each diode. This will enable us to view the current that flows through each diode.

In the following we shall analyze the rectifier circuit of Fig. 3.27 with Spice assuming that the peak rectifier has a load resistance of 1 kΩ and a smoothing capacitor of 50 μF. The two rectifier diodes will be assumed to be modeled after the commercial 1N4148. The Spice deck describing this circuit can be seen listed in Fig. 3.28. The center-tapped transformer is described by three inductor statements and three corresponding coupling statements specifying the coefficients of coupling:

```
* transformer section with center-tap
Lp    2 0 10mH
Ls1   3 4 51uH
Ls2   4 5 51uH
K12 Lp  Ls1 0.999
```

3.6 Rectifier Circuits

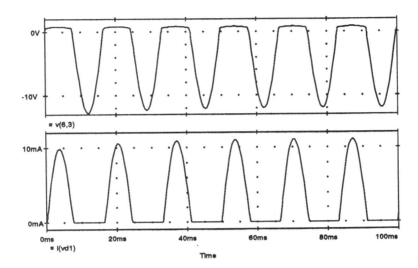

Figure 3.26 The voltage and current waveform associated with diode D_1. The peak inverse voltage (PIV) is seen to be 12 V and the maximum diode current is 11.1 mA.

```
K13  Lp   Ls2  0.999
K23  Ls1  Ls2  0.999
```

To obtain a transformer with a primary to secondary turns ratio of 14:1 for each coil on the secondary side, we have assumed that each coil of the secondary-side has an inductance of 51 μH. This, then, implies that the inductance of the primary must be 10 mH according to the relationship: $\frac{N_P}{N_S} = \sqrt{\frac{L_P}{L_S}}$. Further, we have also assumed a rather ideal coefficient of coupling between each coil at a value near unity (0.999).

As our first analysis, we shall plot the voltage waveform that appears across the load resistor and the voltage waveform that appears across one coil on the secondary-side of the transformer. This analysis will be performed over six periods of the input line voltage. Subsequently, we shall investigate the current that flows through each diode of the rectifier.

Fig. 3.29 displays the results of this analysis. As is evident, the voltage appearing across the load resistor initially ramps up from 0 V to a steady-state value that ripples somewhere between 10 and 11 V. Using the cursor facility of Probe, we are able to determine more precisely the minimum and maximum values of this output voltage waveform to be 9.53 and 10.8 V, respectively. Thus, the average voltage appearing across the load resistor is 10.17 V having a peak-to-peak ripple voltage of 1.27 V. Also shown in the graph of Fig. 3.29 is the voltage waveform appearing across the secondary-side coil L_{S1} of the transformer (see

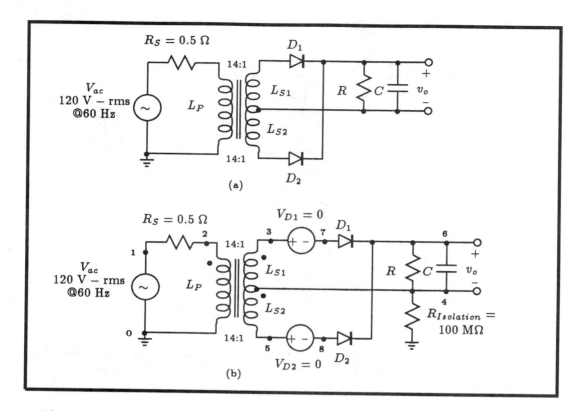

Figure 3.27 (a) A full-wave peak rectifier circuit. (b) Actual circuit set-up simulated by Spice.

Fig. 3.27). Here we see that it settles into a sinusoidal with a peak value of 12 V. Thus, confirming that the transformer circuit is operating correctly and converting the AC line voltage of 120 V-rms to a 12 V-peak level.

According to a hand analysis, together with the assumption that the forward voltage drop across each diode is 0.8 V, the expected ripple associated with this circuit using the equation $V_r = \frac{V_P}{2fCR}$ developed by Sedra and Smith in Section 3.6 of their text is 1.86 V. This appears to be slightly larger than that observed from the above simulations. As we shall soon discover, the reason for the discrepancy is largely due to the large series resistance r_S of 16 Ω associated with each diode (see the diode model for the 1N4148).

It is interesting to point out that the average load voltage ($V_L = V_P - V_r/2$) agrees quite closely with the value obtained from the above simulation at 10.26 V.

To further investigate the behavior of the full-wave peak rectifier circuit shown in Fig. 3.27, we have plotted the waveforms of the current that flows through each diode in Fig. 3.30. The top graph displays the current that flows through diode D_1 and the graph below it displays the current waveform associated with diode D_2. As is evident from the top graph, diode D_1 conducts a rather large initial current pulse having a peak value of about 165 mA, and seems to have reached steady-state behavior by the forth current pulse. The current

3.6 Rectifier Circuits

```
Full-Wave Peak Rectifier Circuit

** Circuit Description **

* ac line voltage
Vac  1 0 sin(0 169V 60Hz)
Rs  1 2 0.5
* transformer section with center-tap
Lp   2 0 10mH
Ls1  3 4 51uH
Ls2  4 5 51uH
K12 Lp  Ls1 0.999
K13 Lp  Ls2 0.999
K23 Ls1 Ls2 0.999
* isolation resistor
Risolation 4 0 100Meg
* monitor diode current
VD1 3 7 0
VD2 5 8 0
* full-wave peak rectifier circuit
D1 7 6 D1N4148
D2 8 6 D1N4148
C 6 4 50u
R 6 4 1k
* diode model statement
.model D1N4148 D (Is=0.1pA Rs=16 CJO=2p Tt=12n Bv=100 Ibv=0.1p)

** Analysis Requests **
.TRAN 0.1ms 100ms 0ms 0.1ms
** Output Requests **
.plot TRAN V(6,4) V(3,4)
.plot TRAN I(VD1) I(VD2)
.probe
.end
```

Figure 3.28 The Spice input file for calculating the transient response of a full-wave peak rectifier circuit.

waveform associated with diode D_2 does not reveal such a dramatic transient, instead it appears to have reached steady-state by the third current pulse.

To see more closely a single steady-state pulse of the current that flows through diode D_2, we expanded the horizontal scale of the bottom graph shown in Fig. 3.30 between 93 and 97 ms, and plotted an expanded view of the current waveform shown there in Fig. 3.31. Here we find that the current pulse has a peak value of 57.7 mA, and extends between 93.95 ms and 96.37 ms for a conduction period of 2.42 ms. We also notice that the shape of this current pulse is rather gaussian, unlike the right-angled triangular current pulse that results when the diode is assumed ideal.

3 Diodes

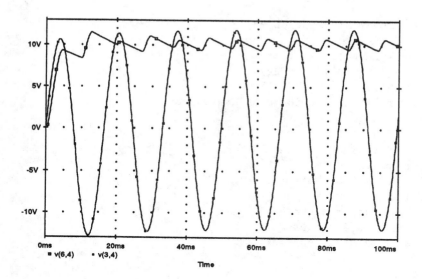

Figure 3.29 Voltage waveforms associated with the peak-rectifier circuit shown in Fig. 3.27. One waveform is the voltage that appears across the load resistance R and the other waveform is the voltage that appears across L_{S1} on the secondary-side of the transformer.

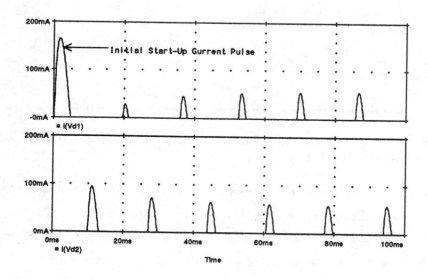

Figure 3.30 The instantaneous current flowing through each diode of the rectifier.

According to the simple theory developed for full-wave rectifier circuits in Section 3.6 of Sedra and Smith, the expected conduction period Δt of each diode is given by

124

3.6 Rectifier Circuits

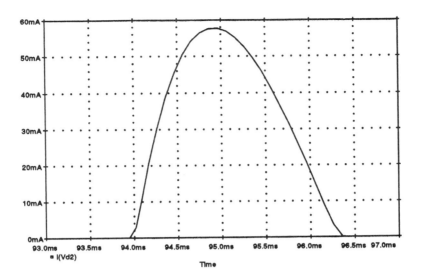

Figure 3.31 A close-up view of a single current pulse in steady-state flowing through diode D_2.

$$\Delta t = \frac{1}{\pi f} \sqrt{\frac{V_r}{2V_P}}. \tag{3.9}$$

and the peak diode current $i_{D_{MAX}}$ (in steady-state) is given by

$$i_{D_{MAX}} = I_L \left(1 + 2\pi \sqrt{\frac{V_P}{2V_r}} \right). \tag{3.10}$$

Substituting the appropriate numerical values we estimate the conduction period of either diode in steady-state to be 1.53 ms. Similarly, the peak diode current is expected to be 122.1 mA. Comparing these two results with those found through simulation (2.42 ms and 57.7 mA), we find that they differ significantly.

On investigation we discovered that the major reason for the discrepancies between theory and the Spice simulation is the presence of the nonzero bulk diode resistance r_S. In the mathematical development of the formulae that describe the full-wave rectifier circuit, the series resistance associated with a practical diode was not accounted for (mainly to keep the mathematical description simple). If we repeat the above simulation with the r_S term in the model statement set to zero we find that our simulation results correlate much more closely with those predicted by the above equations.

To illustrate this, we re-simulated the full-wave peak rectifier circuit shown in Fig. 3.27(b) with the following diode model for the 1N4148:

3 Diodes

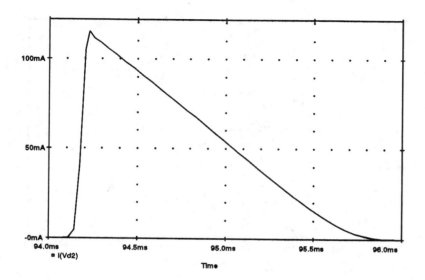

Figure 3.32 A close-up view of a single current pulse in steady-state flowing through diode D_2 when the bulk diode resistance r_S is set to zero.

```
.model D1N4148 D (Is=0.1pA Rs=0 CJO=2p Tt=12n Bv=100 Ibv=0.1p)
```

Here the Spice model parameter Rs was re-assigned a value of 0. On completion of the Spice job, we plotted the voltage across the load resistor and found a waveform that is similar to that seen previously in Fig. 3.29 and is therefore not shown here. The output voltage had an average value of 10.57 V with a peak-to-peak voltage ripple of 1.46 V.

The steady-state current pulse associated with either diode D_1 or D_2 is quite different from that seen previously. As an example we display in Fig. 3.32 a single current pulse in steady-state flowing through diode D_2. Here the shape of the current pulse is very much a right-angle triangle having a peak value of 115.2 mA and a conduction period of 1.60 ms. When these results are compared to those results predicted by Eqns. (3.9) and (3.10), (ie. 122.1 mA and 1.53 ms), we see that we are in much better agreement.

To conclude this section we present in Table 3.3 a comparison of various parameters of the full-wave peak rectifier circuit found from the above two Spice analysis with corresponding parameter values computed using the formulas derived in Section 3.6 of Sedra and Smith. These formulas are also provided in this table for easy reference. Included in this table is the average current that flows through each diode while it is forward conducting. Spice provides no direct way of accessing this current value from the simulated results. Instead, the area under a single current pulse was computed by using the numerical integration facility of

Parameter	Expression	Hand	Spice $Rs = 16\Omega$	$Rs = 0\Omega$
Ripple Voltage	$V_r = \frac{V_P}{2fCR}$	1.86 V	1.27 V	1.46 V
Average Load Voltage	$V_L = V_P - V_r/2$	10.26 V	10.17 V	10.57 V
Average Load Current	$I_L = \frac{V_P - V_r/2}{R}$	10.26 mA	10.17 mA	10.57 mA
Conduction Period	$\Delta t = \frac{1}{\pi f}\sqrt{\frac{V_r}{2V_P}}$	1.53 ms	2.42 ms	1.60 ms
Maximum Diode Current	$i_{D_{MAX}} = I_L\left(1 + 2\pi\sqrt{\frac{V_P}{2V_r}}\right)$	122.1 mA	57.7 mA	115.2 mA
Average Diode Current ‡	$i_{D_{AV}} = I_L\left(1 + \pi\sqrt{\frac{V_P}{2V_r}}\right)$	66.2 mA	35.5 mA	56.2 mA

‡ The numerical integration facitlity of Probe was used to compute the area under a single current pulse. This value was then divided by the conduction period Δt to obtain the average current.

Table 3.3 Various parameters of the full-wave peak rectifier circuit shown in Fig. 3.27 as computed by hand and the two Spice analyses. Here f represents the frequency of the input AC line voltage applied to the primary-side of the transformer.

Probe, then it was divided by the time that the diode is conducting current (ie. conduction period) to obtain the average current flowing through the diode.

3.6.3 A Voltage Regulated Power Supply

To complete this section on rectifier circuits we shall analyze a commonly used power supply configuration with Spice. Consider the circuit shown in Fig. 3.33. It can be thought of as consisting of three parts: a full-wave peak rectifier, a zener diode voltage regulator and the load. The peak rectifier circuit acts to supply a relatively stable DC voltage to the zener regulator which, in turn, reduces any voltage fluctuation (ripple) that appears on it. In addition, the voltage regulator acts to maintain a constant voltage across the load for a wide range of load currents. Resistor $R_{Isolation}$ is used for Spice simulation purposes and plays no role in the circuit function.

Let us consider using the circuit configuration shown in Fig. 3.33 to design a 5 V power supply for an application that requires a maximum load current of 20 mA. The 120 V-rms AC household voltage is stepped down to a 12 V-peak level using a center-tapped transformer with each coil on the secondary side having a turns ratio of 14:1 with respect to the primary coil. Further, we have at our disposal a zener diode that has $V_Z = 5.1$ V at a current of 20 mA and has a dynamic resistance $r_z = 10\ \Omega$. We also know that the minimum zener diode current must be limited to 5 mA if we are to maintain the diode in its breakdown region.

3 Diodes

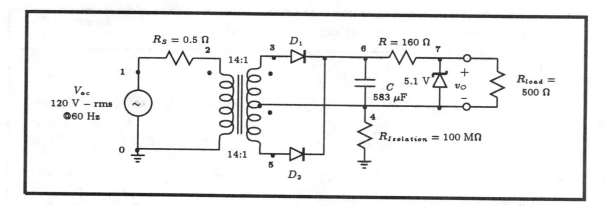

Figure 3.33 A 5 V regulated power supply.

Assuming that the input voltage to the voltage regulator circuit ranges between 9 and 12 V, we choose the current limiting resistor R from the expression (derived in Section 3.5 of Sedra and Smith):

$$R = \frac{V_{Smin} - V_{Z0} - r_z I_{Zmin}}{I_{Zmin} + I_{Lmax}}. \tag{3.11}$$

Thus, we obtain $R = 160\ \Omega$. As a point of reference, under worst-case conditions, the expected minimum output voltage is about 4.95 V as calculated from the expression, $V_{Omin} = V_{Z0} + r_z I_{Zmin}$.

The size of the smoothing capacitor is to be determined so that the voltage applied to the regulator circuit does not go below 9 V. Assuming that the peak voltage appearing across the secondary-side of the center-tapped transformer is 12 V, then the worst-case ripple voltage must be limited to no more than 3 V. We shall limit the ripple voltage to a more conservative 1 V level in case the peak voltage level changes. We can then estimate the size of the capacitor we require by using the formula for the ripple voltage provided in Table 3.3 for a full-wave peak rectifier. Substituting $R = 160\ \Omega$, $V_r = 1$ V, $V_P = 11.2$ V (accounting for a 0.8 V diode drop) and $f = 60$ Hz, we get $C = 583\ \mu$F. This capacitance may seem large but is typical of the size of capacitor used in power supplies.

To investigate whether our design meets the required specifications, we shall simulate the power supply circuit shown in Fig. 3.33 with an initial load resistance of 500 Ω. This load should draw an average current of no more than 10 mA – well within the maximum load current condition. The Spice deck for this circuit is seen listed in Fig. 3.34. The transformer is represented by a primary inductance of 10 mH, and the two coils on the secondary-side are each assigned a value of 51 μH. The two rectifier diodes are assumed to be modeled after the commercial diode type 1N4148.

The first analysis that we shall perform with Spice is to determine whether the output

3.6 Rectifier Circuits

```
A Regulated Power Supply

* zener diode subcircuit
.subckt zener_diode 1 2
* connections:        | |
*                anode |
*                cathode
Dforward 1 2 1mA_diode
Dreverse 2 4 ideal_diode
Vz0 4 3 DC 4.9V
Rz 1 3 10
* diode model statements
.model 1mA_diode D   (Is=100pA n=1.679 )
.model ideal_diode D (Is=100pA n=0.01 )
.ends zener_diode

** Main Circuit **
* ac line voltage
Vac 1 0 sin(0 169V 60Hz)
Rs 1 2 0.5
* transformer section with center-tap
Lp   2 0 10mH
Ls1  3 4 51uH
Ls2  4 5 51uH
K12 Lp  Ls1 0.999
K13 Lp  Ls2 0.999
K23 Ls1 Ls2 0.999
* isolation resistor
Risolation 4 0 100Meg
* full-wave peak rectifier circuit
D1 3 6 D1N4148
D2 5 6 D1N4148
C 6 4 583uF
R 6 7 160
* zener diode
XZ1 4 7 zener_diode
* load
Rload 7 4 500
* diode model statement
.model D1N4148 D (Is=0.1pA Rs=16 CJO=2p Tt=12n Bv=100 Ibv=0.1p)

** Analysis Requests **
.OPTIONS ITL5=0
.TRAN 0.5ms 200ms 0ms 0.5ms UIC
** Output Requests **
.plot TRAN V(7,4) V(6,4)
.probe
.end
```

Figure 3.34 The Spice input file for calculating the time-varying output voltage of the 5 V regulated power supply shown in Fig. 3.33. Changing the ITL5 internal parameter of Spice through an .OPTIONS command resets the transient analysis total iteration limit. Setting ITL5=0 increases this limit to infinity.

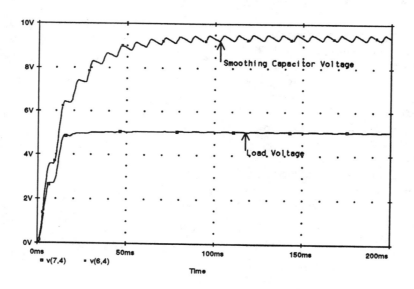

Figure 3.35 The voltage across the smoothing capacitor C of the peak rectifier, and the output voltage across the 500 Ω load resistance.

voltage is nominally 5 V. This can be determined by observing the voltage across the load resistance. Due to the presence of the large smoothing capacitor C, a long charge-up time will be necessary before the power supply circuit reaches steady-state. Therefore, a request for a long transient analysis is necessary to observe steady-state behavior. Here we have selected that the transient be computed over a 200 ms interval. Moreover, because of this long time interval, many iterations of the transient analysis algorithm will be performed by Spice. If the number of iterations performed by Spice exceeds 5000, Spice will stop the analysis. To allow more than 5000 iteration, we can reset this limit by reassigning a new value to the Spice internal parameter ITL5. This is accomplished by including an .OPTIONS command line in the Spice deck with the new value of ITL5, ie.

```
.OPTIONS ITL5=0.
```

Note that Spice recognizes that ITL5=0 really means ITL5=∞, or in other words, ignore this test and allow as many iterations as is necessary. In almost all cases, the number of iterations required to complete a given analysis is not know *a priori*, thus it is simplest to just use ITL5=0.

On completion of the Spice analysis, the voltage waveform that appears across the 500 Ω load resistor is shown in Fig. 3.35. Also shown is the voltage that appears across the smoothing capacitor C. As we can see, the voltage across the capacitor has an average value of about 9.5 V and a peak-to-peak ripple of 0.25 V. In contrast, the voltage across the load

3.6 Rectifier Circuits

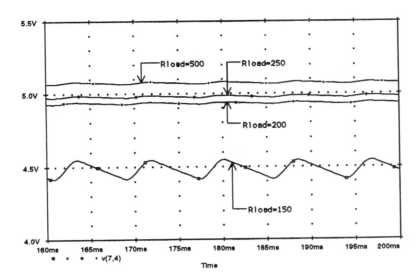

Figure 3.36 The output voltage waveform from the 5 V power supply for load resistances of 150, 200, 250 and 500 Ω. The voltage regulation is lost at a load resistance of 150 Ω.

resistor is quite close to 5 V. Using the cursor facility of Probe, we find that the load voltage ripples slightly, between 5.065 and 5.080 V, a ripple voltage of only 15 mV. We therefore see that the above power-supply design is operating quite close to the nominal design, providing an output voltage of 5 V at a load current of about 10 mA.

To see the effect of larger current demands on the power supply, consider reducing the load resistance. In order to compare the effect of different loads, we shall re-simulate the circuit with load resistances of: 150, 200 and 250 Ω. Assuming that these load resistances do not significantly affect the output voltage, they would correspond to a load current of: 33.3 mA, 25 mA and 20 mA, respectively. Using the same Spice deck as shown in Fig. 3.34 with only the value of the load resistance altered, we concatenated three similar files, together with the original one having a load of 500 Ω, into one file for processing. As a result of the analysis, we display a view of the output voltage over the time interval 160 to 200 ms. As is clearly evident, for load resistances greater than and including 200 Ω, the output voltage is maintained very near the 5 V level with very little ripple visible. However, for a load resistance of 150 Ω, we see that the output voltage level has dropped down to an average value of about 4.5 V. Also, we see that the ripple voltage associated with this signal has increased significantly. This suggests that the output voltage is no longer being regulated. This is because the zener diode has been starved of its current and has turned off.

We conclude that the power supply circuit shown in Fig. 3.33 will provide a constant 5

3 Diodes

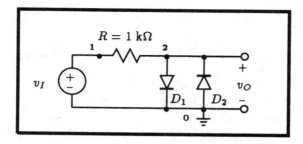

Figure 3.37 A back-to-back diode limiter circuit.

```
A Diode Limiter Circuit

** Circuit Description **

Vi 1 0 DC 0V
R 1 2 1k
D1 2 0 D1N4148
D2 0 2 D1N4148
* diode model statement
.model D1N4148 D (Is=0.1pA Rs=16 CJO=2p Tt=12n Bv=100 Ibv=0.1p)

** Analysis Requests **
* sweep the input voltage level from -5 V to +5 V in 100 mV increments
.DC Vi -5V 5V 100mV
** Output Requests **
.PLOT DC V(2)
.probe
.end
```

Figure 3.38 The Spice input file for calculating the DC transfer characteristic of the back-to-back diode limiter circuit of Fig. 3.37.

V output level for load currents at least as large as 25 mA.

3.7 Limiting And Clamping Circuits

In the following we shall simulate the circuit operation of several commonly used diode circuits. This will include the analysis of a back-to-back diode limiter circuit, a DC restorer circuit and a voltage doubler circuit.

A Diode Limiter Circuit

In Fig. 3.37 we present a simple back-to-back diode limiter circuit constructed with two diodes of the 1N4148 type. Using Spice we would like to observe the transfer characteristic of such a circuit. The Spice deck for this circuit is shown in Fig. 3.38. Here we are sweeping

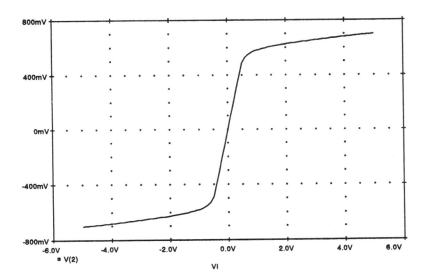

Figure 3.39 DC transfer characteristic of the back-to-back diode limiter shown in Fig. 3.37.

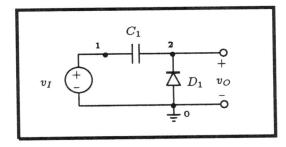

Figure 3.40 A DC restorer circuit.

the input DC source v_I between -1 and $+1$ V.

The results of this analysis are shown in Fig. 3.39. Here we see that the transfer characteristic exhibits rather soft limiting with the linear region ranging between -0.5 V and $+0.5$ V. The slope in the linear region is found to be unity.

A DC Restorer Circuit

In Fig. 3.40 we present a DC restorer or a clamped capacitor circuit. Using Spice we would like to observe the transient behavior of this circuit with a square-wave input having a 10 V peak-to-peak amplitude, a $+2$ V DC offset, and a 1 kHz frequency. The diode will be assumed to be of the 1N4148 type and the capacitor has a 1 μF value. The Spice deck for this particular example is shown in Fig. 3.41. The square-wave input is described by the following source PULSE statement:

3 Diodes

```
A DC Restorer Circuit

** Circuit Description **
Vi 1 0 PULSE ( -3 7 0s 10us 10us 0.490ms 1ms )
C1 1 2 1u
D1 0 2 D1N4148
* diode model statement
.model D1N4148 D (Is=0.1pA Rs=16 CJO=2p Tt=12n Bv=100 Ibv=0.1p)

** Analysis Requests **
.TRAN 100u 10m 0m 100u
** Output Requests **
.PLOT TRAN V(1) V(2)
.probe
.end
```

Figure 3.41 The Spice input file for calculating the time-varying output voltage from the DC restorer circuit shown in Fig. 3.40.

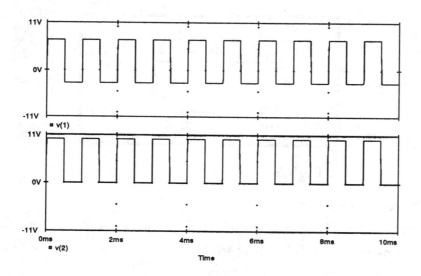

Figure 3.42 The input and output voltage waveforms of the associated with the DC restorer circuit of Fig. 3.40.

Vi 1 0 PULSE (-3 7 0s 10us 10us 0.490ms 1ms).

It goes between −3 and 7 V with 0 s delay, a rise and a fall time of 10 μs, and has a period of 1 ms.

The results of the Spice transient analysis are shown in Fig. 3.42. The top graph displays

3.7 Limiting And Clamping Circuits

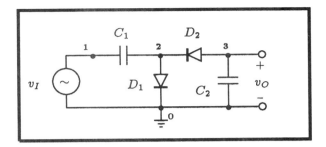

Figure 3.43 A voltage doubler circuit.

```
A Voltage Doubler Circuit

** Circuit Description **
Vi 1 0 sin ( 0 10V 1kHz )
C1 1 2 1u
C2 3 0 1u
D1 2 0 D1N4148
D2 3 2 D1N4148
* diode model statement
.model D1N4148 D (Is=0.1pA Rs=16 CJO=2p Tt=12n Bv=100 Ibv=0.1p)

** Analysis Requests **
.TRAN 100u 10m 0m 100u
** Output Requests **
.PLOT TRAN V(1) V(2) V(3)
.probe
.end
```

Figure 3.44 The Spice input file for calculating the transient response of the voltage doubler circuit shown in Fig. 3.43.

the input 10 V square-wave signal and the bottom graph shows the corresponding signal that appears at the output. We see that it is also a 10 V square-wave but its DC level has now changed to a 5 V level (one-half the peak-to-peak value).

Voltage Doubler Circuit

In Fig. 3.43 we show a voltage doubler circuit. Using Spice we would like to observe the transient behavior of this circuit for an input sine-wave signal having a 10 V amplitude and a 1 kHz frequency. The diodes will be assumed to be of the 1N4148 type and the two capacitors are of the same value at 1 μF. The Spice deck for this particular example is shown in Fig. 3.44.

The results of the Spice transient analysis are shown in Fig. 3.45. The top graph displays the input 10 V-peak sine-wave, the middle graph displays the voltage that appears

3 Diodes

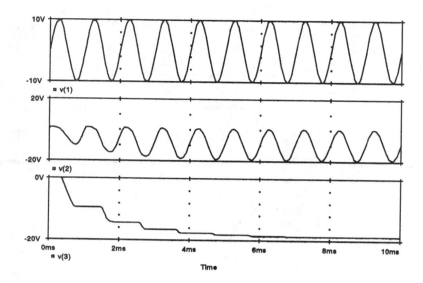

Figure 3.45 Various voltage waveforms of the voltage doubler circuit shown in Fig. 3.43. The top graph displays the input sine-wave voltage signal, the middle graph displays the voltage across diode D_1, and the bottom graph displays voltage that appears at the output.

across diode D_1 and the bottom graph shows the voltage that appears at the output of the doubler circuit. Looking at the output voltage waveform in the bottom graph we see that it experiences a transient that lasts for about 7 ms and then settles to a constant level of -18.8 V. The magnitude of this signal is approximately twice the peak value of the input sine-wave signal. The voltage across diode D_1 settles into a 10 V-peak sine-wave signal with a -10 V DC offset.

3.8 Spice Tips

- In some circuit situations, a DC path does not exist between a node in the circuit and the ground reference node 0. This causes a problem for Spice. The problem can be circumvented by placing a large resistor between this node and ground. The value of this resistor should be large enough so as not to interfere with the circuit operation.
- Spice can emulate the behavior of a laboratory curve-tracer through the application of the DC sweep command.
- The effect of temperature on a circuit can be investigated using Spice.
- An ideal diode can be very well approximated by setting the emission coefficient n of the diode model between 0.01 and 0.001.

- A zener diode can not be properly represented by the diode model statement found in Spice. Instead, a zener diode is represented by a subcircuit consisting of several diodes, a voltage source and a resistor – See Fig. 3.13.

- An ideal transformer is not represented in Spice; instead it is closely approximated by two inductors with a coefficient of mutual coupling very close to one (ie. $k = 0.999$).

- Commercial vendors of electronic components are making available Spice models of their devices and circuits. This should greatly improve the range of circuits that can be simulated by Spice.

- Provided with the student version of PSpice is a series of device models for several commercial parts. These are found in a file named NOM.LIB located in the same directory that the PSpice program is located in.

- When performing long transient runs, to ensure that Spice does not stop after 5000 iterations of the transient analysis, the parameter ITL5 can reset this limit to some other value. Since, in most circuit simulations, one does not know *a priori* the number of iterations the transient analysis will require, it is best to set this limit to infinity. This is accomplished by setting ITL5=0.

3.9 Problems

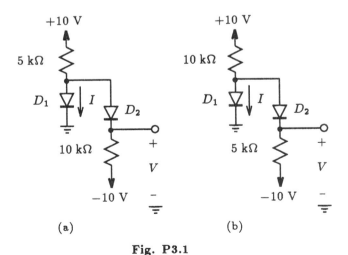

Fig. P3.1

3.1 Assuming that the diodes in the circuits of Fig. P3.1 are modeled with parameters: $I_S = 10^{-14}$ and $n = 2$. Determine, with the aid of Spice, values of the labeled voltages and currents. Repeat with diode parameters: $I_S = 10^{-12}$ and $n = 1$.

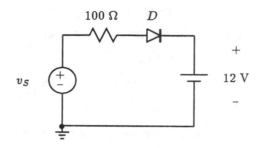

Fig. P3.2

3.2 Consider the battery charger circuit shown in Fig. P3.2. If v_S is a 60 Hz sinusoid with 24 V-peak amplitude, find the fraction of each cycle during which the diode conducts. Also find the peak value of the diode current and the maximum reverse-bias voltage that appears across the diode. Assume that the diode has Spice model parameters: $I_S = 10^{-12}$ and $n = 1.6$.

3.3 If the sinusoidal source of Problem 3.2 is replaced by a square-wave of the same frequency and amplitude, for what fraction of the cycle does the diode conduct?

3.4 Repeat Problem 3.2 for a symmetrical triangular wave of 24 V amplitude.

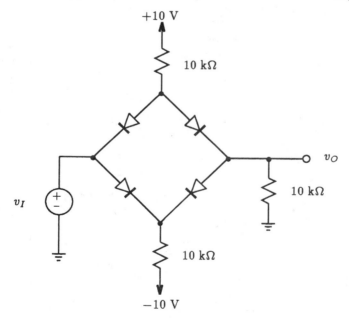

Fig. P3.5

3.5 Simulate the circuit shown in Fig. P3.5 with Spice and determine the DC transfer characteristic. Assume that each diode is of the 1N4148 type. See Fig. 3.23 for the Spice model parameters for this diode.

3.6 Use the Spice operating point command (.OP) to determine what the incremental resistance is for 10 diodes of the 1N4148 type connected in parallel and fed with a DC current source of 10 mA. See Section 3.6 for the model parameters of the 1N4148.

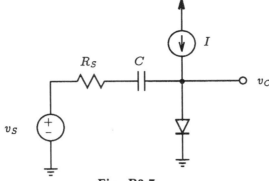

Fig. P3.7

3.7 For the circuit shown in Fig. P3.7 with $R_s = 1$ kΩ, $C = 1.0$ μF, and I having several values of 1 mA, 0.1 mA, and 1 μA, verify that the signal component in the output voltage for each case is given by:

$$v_o = v_s \frac{nV_T}{nV_T + IR_s}.$$

Using Spice, simulate the transient behavior of this circuit with an input 100 Hz sinewave signal of 1 mV-peak amplitude. Assume that the diode has model parameters $I_S = 10$ fA and $n = 2$.

3.8 A voltage regulator consisting of two diodes in series fed with a constant current source is used as a replacement for a single carbon-zinc (battery) cell of nominal voltage 1.5 V. The regulator load current varies from 2 to 7 mA. Compare this regulator circuit for three different current source levels of 5, 10 and 15 mA as the load current varies over its full range. What is the change in the output voltage for each case. Assume that the diodes have a 0.7 V drop at a 1 mA current and $n = 2$.

3.9 A zener shunt regulator of the type shown in Fig. 3.14 has been designed to provide a regulated voltage of about 10 V. The zener diode is of the type 1N4740 which is specified to have a 10 V drop at a test current of 25 mA. At this current its $r_z = 7$ Ω. The raw supply available has a nominal value of 20 V but can vary by as much as $\pm 25\%$. The regulator is required to supply a load current of 0 to 20 mA. Assuming that the minimum zener current is to be 5 mA, the resistance R was determine to be 200 Ω. Using Spice, simulate the behavior of this circuit and:

(a) Find the load regulation. By what percentage does V_O change from the no-load to full-load condition?

(b) Find the line regulation. What is the change in V_O expressed as a percentage, corresponding to the $\pm 25\%$ change in V_S.

(c) What is the maximum current that the zener diode has to conduct? What is maximum zener power dissipation?

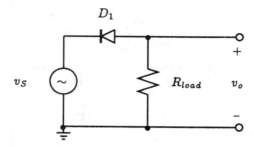

Fig. P3.10

3.10 For the half-wave rectifier circuit of Fig. P3.10. Let v_S be a sinusoid with a 10 V-peak amplitude, and let $R = 1$ kΩ. Assume that diode has diode parameters $I_S = 100$ pA and $n = 1.679$. Using Spice:

(a) Calculate and plot the transfer characteristic.

(b) Calculate and plot the output voltage waveform assuming that the input signal frequency is 1 kHz. Estimate the average value of this output signal.

(c) Determine the peak current in the diode.

(d) Determine the PIV of the diode.

3.11 Consider a half-wave rectifier circuit with a triangular input of 20 V peak-to-peak amplitude and zero average. The resistance in series with the diode is 1 kΩ. Assume that the diode has model parameters $I_S = 100$ pA and $n = 1.679$. Calculate the output voltage and estimate the average value of this output signal.

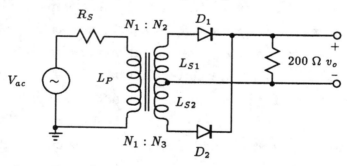

Fig. P3.12

3.12 It is required to design a full-wave rectifier circuit using the circuit of Fig. P3.12 to provide an average output voltage of:

(a) 10 V.

(b) 100 V.

In each case find the required turns ratio of the transformer assuming $L_P = 10$ mH and check your results using Spice. Assume that the diodes are of the 1N4148 type, the ac line voltage is 120 V-rms and has a source resistance of 0.5 Ω.

3.9 Problems

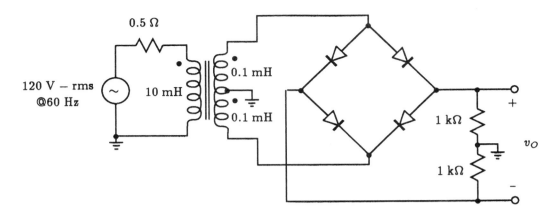

Fig. P3.13

3.13 The circuit in Fig. P3.13 implements a complementary-output rectifier. Simulate the behavior of this circuit using Spice and plot the voltage that appears across the two output terminals. Assume that the diodes have model parameters $I_S = 100$ fA and $n = 2$. What is the PIV of each diode?

3.14 Design a half-wave peak rectifier circuit that provides an average DC output voltage of 15 V on which a maximum ± 1 V ripple is allowed. The rectifier feeds a load of 150 Ω and the rectifier is fed from the line voltage (120 V-rms at 60 Hz) through a transformer. The diodes are assumed of the type 1N4148. Simulate the operation of your design using Spice and verify that it indeed satisfies the design requirements.

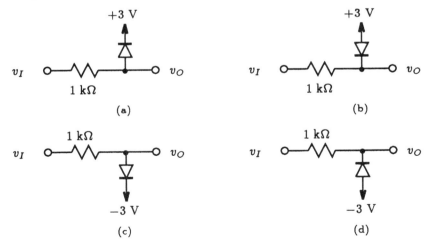

Fig. P3.15

3.15 Using Spice, plot the transfer characteristic v_O versus v_I for the four limiter circuits shown in Fig. P3.15. Assume that the diodes have model parameters $I_S = 100$ fA and $n = 1.6$.

3 Diodes

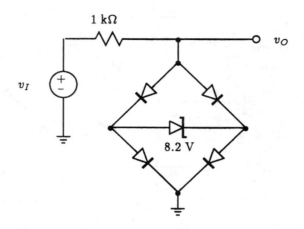

Fig. P3.16

3.16 Using Spice, plot the transfer characteristic v_O versus v_I of the circuit in Fig. P3.16 for $-20\text{ V} \le v_I \le +20\text{ V}$. Assume that the diodes have model parameters $I_S = 100$ fA and $n = 1.6$, and that the zener diode has a reverse-bias voltage drop of 8.2 V at a current of 10 mA and $r_z = 20\text{ }\Omega$.

3.17 For the circuits in Fig. P3.17, each utilizing diodes of the 1N4148 type, plot the output waveform of the circuit for a 10 V-peak amplitude input square-wave having a frequency of 1 kHz.

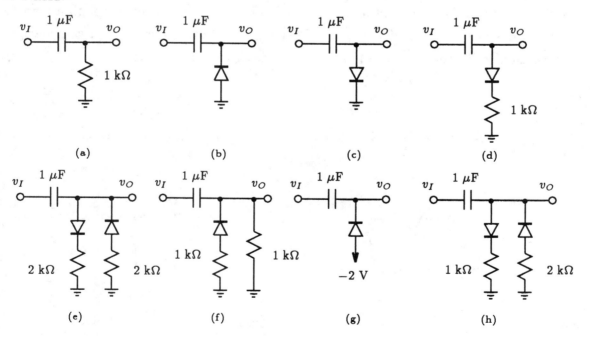

Fig. P3.17

Chapter 4

Bipolar Junction Transistors (BJTs)

In the previous chapter we outlined how the semiconductor diode is described to Spice using a diode element and model statement. Further, we illustrated how the terminal characteristics of a diode are modeled within Spice and how the user can alter the parameters of this model to more closely characterize specific diode behavior. In this chapter on the bipolar junction transistor (BJT) we shall proceed in a similar fashion, first outlining the two statements that are required to describe a BJT to Spice. This will then be followed by a brief description of the model used to represent the BJT within Spice. On completion of this, we shall use Spice to investigate the low-frequency behavior of various types of electronic circuits containing BJTs. The types of circuits that will be simulated will range from single *npn* and *pnp* transistor amplifiers to multiple-transistor amplifier circuits, as well as circuits that utilize the transistor as an on-off switch.

4.1 Describing BJTs To Spice

Two statements are required to described any particular semiconductor device to Spice. One statement is necessary to describe the nature of the semiconductor device and the manner in which it is connected to the rest of the network, and the other statement is required

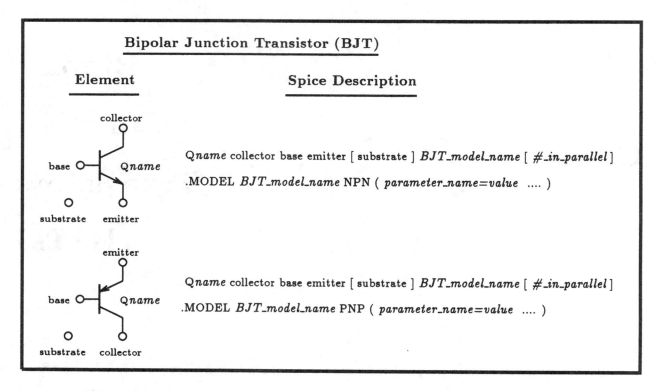

Figure 4.1 Spice element description for the *npn* and *pnp* BJT. Also listed is the general form of the associated BJT model statement. A partial listing of the parameter values applicable to either the *npn* or *pnp* BJT given in Table 4.1.

to describe the parameters of the built-in model of the semiconductor device described by the first statement. In the following we shall describe these two statements as they apply to the BJT.

4.1.1 BJT Element Description

The presence of a BJT in a circuit is described to Spice through the Spice input file using an element statement beginning with the letter Q. If more than one transistor exists in a circuit then a unique name must be attached to Q to uniquely identify each transistor. This is then followed by a list of the three nodes to which the collector, base, and emitter of the BJT are connected to. One can also include the node that the substrate is connected to if it is an integrated transistor. Subsequently, on the same line, the name of the model that will be used to characterize this particular BJT is given. The name of this model must correspond to the name given on a model statement containing the parameter values that characterize this transistor to Spice. Lastly, one has the option of specifying the number of BJTs that are considered to be connected in parallel.

For quick reference we depict the syntax for the Spice statement describing the BJT

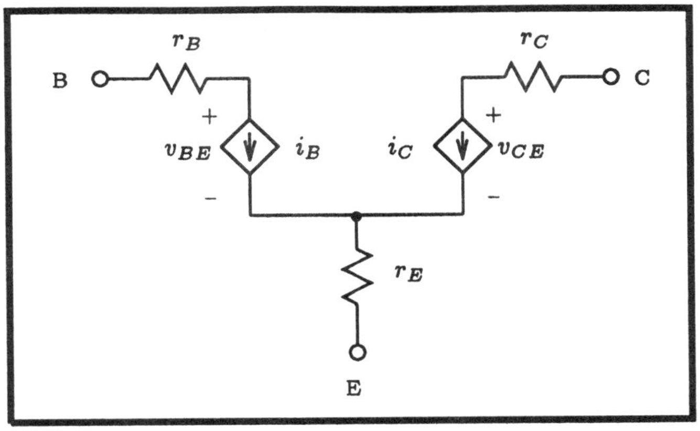

4.1 Describing BJTs To Spice

Figure 4.2 The Spice large-signal BJT model for DC analysis.

in Fig. 4.1. Also shown is the syntax for the model statement (.MODEL) that must be present in any Spice input file that makes reference to the built-in BJT model of Spice. This statement defines the terminal characteristics of the BJT by specifying the values of particular parameters of the BJT model. We shall discuss the model statement more fully next.

4.1.2 BJT Model Description

As is evident from Fig. 4.1, the model statement for either the *npn* or *pnp* transistor begins with the keyword .MODEL and is followed by the name of the model used by a BJT element statement, the nature of the BJT (ie. *npn* or *pnp*), and a list of the parameters characterizing the terminal behavior of the BJT, enclosed between brackets. The number of parameters associated with the Spice model of the BJT is rather large (40 in total), and their individual meanings are rather involved. Instead of trying to describe the meaning of each parameter of the BJT model, we shall simply outline the parameters of the Spice BJT model that are relevant to the discussion contained within this chapter.

The Spice BJT model is illustrated schematically in Fig. 4.2. The ohmic resistances of the base, collector, and emitter regions of the BJT are lumped into three linear resistances r_B, r_C and r_E, respectively. The DC characteristics of the intrinsic BJT are determined by the nonlinear dependent current sources i_B and i_C. The exact functional descriptions of these two currents – as adopted by Spice – are rather complex and will not be given here. Interested readers can consult [Nagel, 1975] for more details. For a transistor operated in its active mode, a first-order representation of these two currents can be described by the following two equations

4 Bipolar Junction Transistors (BJTs)

Symbol	Spice Name	Model Parameter	Units	Default
I_S	Is	Saturation current	Amps	1×10^{-16}
β_F	Bf	Forward current gain		100
V_{AF}	VAf	Forward Early voltage	Volts	∞
r_B	Rb	Base ohmic resistance	Ω	0
r_C	Rc	Collector ohmic resistance	Ω	0
r_E	Re	Emitter ohmic resistance	Ω	0

Table 4.1 A partial listing of the Spice parameters for a static BJT model.

$$i_B = \frac{I_S}{\beta_F} e^{v_{BE}/V_T} \tag{4.1}$$

and

$$i_C = I_S e^{v_{BE}/V_T} \left(1 + \frac{v_{CE}}{V_{AF}}\right). \tag{4.2}$$

Here I_S is the saturation current (similar to the diode saturation current) and V_T is the thermal voltage. The constant β_F is the forward common-emitter current gain. In Sedra and Smith, this constant is designated simply as β. Spice attaches the subscript F to distinguish β_F from another current gain β_R which represents the common-emitter current gain of the same transistor when operated in the reverse mode (that is, with the emitter and collector interchanged). Finally, V_{AF} is the forward early voltage (denoted V_A in Sedra and Smith).

A partial listing of the parameters associated with the Spice BJT model under static conditions is given in Table 4.1. Also listed are the associated default values; if the value of a particular parameter is not specified on the .MODEL statement, the parameter assumes its default value. To specify a parameter value one simply writes, for example, Is=1e-14, Bf=100, etc.

4.1.3 Verifying NPN Transistor Circuit Operation

As the first example of this chapter, consider verifying the *npn* transistor circuit designed in Example 4.1 of Sedra and Smith and repeated here in Fig. 4.3. This particular transistor circuit was designed to have a collector current of 2 mA and a collector voltage of +5 V.

For the purpose of design, the *npn* transistor was assumed to have a $\beta_F = 100$ and to exhibit a v_{BE} of 0.7 V at $i_C = 1$ mA. The first condition can be directly specified on the

4.1 Describing BJTs To Spice

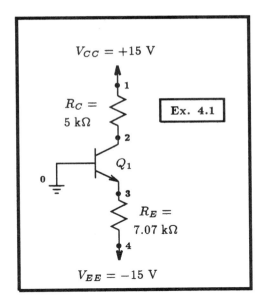

Figure 4.3 Transistor circuit design created by Sedra and Smith in Example 4.1 of their text. Spice is used to calculate the DC operating point of this circuit.

BJT model statement using $\beta_F = 100$, however, the latter condition needs to be translated into a BJT parameter. From Eqn. (4.2) we can write

$$1 \times 10^{-3} = I_S e^{0.7/25.89 \times 10^{-3}} \left(1 + \frac{v_{CE}}{V_{AF}}\right) \quad (4.3).$$

Now, to make matters simpler, we shall assume that $V_{AF} = \infty$, thus reducing the above equation to

$$1 \times 10^{-3} = I_S e^{0.7/25.89 \times 10^{-3}}. \quad (4.4)$$

We can then solve to obtain $I_S = 1.8104 \times 10^{-15}$ A. Our model statement for this particular transistor is then described to Spice using the following statement:

```
.model npn_transistor npn (Is=1.8104e-15 Bf=100) .
```

Notice that we did not specify the value of V_{AF} in the list of parameters, rather, we are relying on the default value assigned to V_{AF}. Of course, the same could have also been done for the parameter β_F.

Assuming the nodes of the circuit are labeled as shown in Fig. 4.3, the corresponding Spice input file is listed in Fig. 4.4. An operating point analysis command (.OP) is included in this file to tell Spice to calculate the DC operating point of this circuit. Submitting this file to Spice, results in the following DC operating point information:

4 Bipolar Junction Transistors (BJTs)

```
Example 4.1: Verifying Transistor Circuit Design

** Circuit Description **
Vcc 1 0 DC +15V
Vee 4 0 DC -15V
Q1 2 0 3 npn_transistor
Rc 1 2 5k
Re 3 4 7.07k
* transistor model statement
.model npn_transistor npn (Is=1.8104e-15 Bf=100)
** Analysis Requests **
* calculate DC bias point information
.OP
** Output Requests **
* none required
.end
```

Figure 4.4 The Spice input file for calculating the collector current and voltage of the transistor circuit shown in Fig. 4.3.

```
****     SMALL SIGNAL BIAS SOLUTION      TEMPERATURE =   27.000 DEG C
*************************************************************************

NODE   VOLTAGE     NODE   VOLTAGE     NODE   VOLTAGE     NODE   VOLTAGE

(  1)   15.0000  (  2)    4.9990  (  3)   -.7172  (  4)  -15.0000

        VOLTAGE SOURCE CURRENTS
        NAME         CURRENT

        Vcc         -2.000E-03
        Vee          2.020E-03

        TOTAL POWER DISSIPATION   6.03E-02  WATTS
```

As is evident, the collector of the transistor (node 2) is at 4.9990 V and the collector current I_C, as inferred from the current supplied by the voltage source V_{CC} is 2.000 mA. (The negative sign is a result of the convention used by Spice). Although, the collector voltage is not exactly at +5 V, the deviation from this value is extremely small (1 mV). If one were to back-track to find why this error occurred, one would find that the value of $V_T = 25$ mV assumed during the design phase is different than the value of $V_T = 25.89$ mV that Spice used, assuming a circuit temperature of 27°C. Repeating the design procedure outlined in Example 4.1 with the exact value of $V_T = 25.89$ mV would result in an emitter resistance of $R_E = 7.0703$ kΩ and a collector voltage much closer to +5 V.

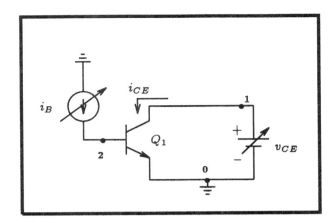

Figure 4.5 Spice curve-tracer arrangement for calculating the i_C - v_{CE} characteristics of a BJT.

4.2 Using Spice As A Curve Tracer

A typical curve-tracer arrangement for measuring the $i_C - v_{CE}$ characteristics of a transistor is illustrated in Fig. 4.5. Here two independent sources v_{CE} and i_B are varied, and the collector current of the transistor is calculated. The collector current would then be plotted as a function of v_{CE} and i_B. For example, consider plotting the forward $i_C - v_{CE}$ characteristics of an *npn* transistor characterized by $I_S = 1.8104 \times 10^{-15}$ A, $\beta_F = 100$, and a forward Early voltage $V_{AF} = 35$ V, for a base current of 10 μA. Using the circuit displayed in Fig. 4.5 as a guide, we can create the Spice input file shown in Fig. 4.6 to accomplish this task. We make use of the DC Sweep command available in Spice to vary the collector-emitter voltage of transistor Q_1 from 0 V to +10 V in 100 mV steps. The resulting $i_C - v_{CE}$ characteristic as calculated by Spice is displayed in Fig. 4.7.

Typically, one also wants to vary the base current i_B while at the same time varying v_{CE} of the transistor. This can be accomplished with Spice by augmenting the DC Sweep command with another source name and the range of values it should be stepped through. For example, the DC Sweep command required to sweep v_{CE} from 0 V to +10 V in 100 mV steps while at the same time sweeping i_B from 1 μA to 10 μ A in 1 μA steps is simply the following:

```
.DC Vce 0V +10V 100mV Ib 1u 10u 1u.
```

In essence, this command tells Spice to perform the voltage sweep for each value specified by the current sweep. (For those familiar with computer programming, this command should remind them of a set of programming loops: the inner sweep being nested within the outer sweep.)

Revising the Spice input deck with this augmented DC Sweep command and re-submitting

4 Bipolar Junction Transistors (BJTs)

```
Spice As A Curve Tracer:  BJT I-V Characteristics

** Circuit Description **
Vce 1 0 DC 0V
Ib 0 2 DC 10uA
* device under test
Q1 1 2 0 npn_transistor
* transistor model statement
.model npn_transistor NPN (Is=1.8104e-15A Bf=100 VAf=35V)
** Analysis Requests **
* vary Vce from 0V to 10V in steps of 100mV
.DC Vce 0V +10V 100mV
** Output Requests **
.plot DC I(Vce)
.probe
.end
```

Figure 4.6 The Spice input file for calculating the collector current of the transistor circuit shown in Fig. 4.5 for a given base current and collector-emitter voltage.

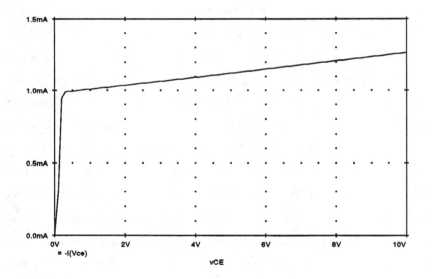

Figure 4.7 The i_C - v_{CE} curve at a base current of 10 μA for a transistor characterized by $I_S = 1.8104 \times 10^{-15}$ A, $\beta_F = 100$, and $V_{AF} = 35$ V.

this job to Spice, the $i_C - v_{CE}$ characteristics displayed in Fig. 4.8 are obtained.

Clearly, other arrangements of the two independent sources are possible, thus allowing one to investigate other characteristics of the transistor. We encourage the reader to investigate some of them using the approach just outlined.

4.3 Spice Analysis Of Transistor Circuits At DC

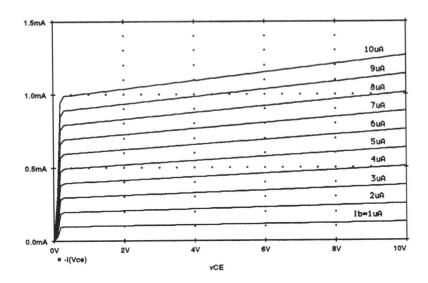

Figure 4.8 A family of i_C - v_{CE} curves for a base current varied between 1 μA and 10 μA in steps of 1 μA for a transistor characterized by $I_S = 1.8104 \times 10^{-15}$ A, $\beta_F = 100$, and $V_{AF} = 35$ V.

4.3 Spice Analysis Of Transistor Circuits At DC

We are now ready to investigate the DC operating point of several simple transistor circuits using Spice. Throughout this section we shall assume that the transistor is characterized by a $\beta_F = 100$, exhibits a v_{BE} of 0.7 V at $i_C = 1$ mA, and that its Early voltage is infinite. The primary goal of this section is to determine, with the aid of Spice, the mode of operation that a transistor is working in.

4.3.1 Transistor Modes Of Operation

Depending on the bias condition imposed across the emitter–base junction (EBJ) and the collector–base junction (CBJ), different modes of operation of the BJT are obtained, as shown in Table 4.2. In the following we shall look at several transistor circuits and use Spice to determine the mode of operation of each.

Active Region

Consider the circuit shown in Fig. 4.9. This same circuit was analyzed by Sedra and Smith using hand analysis in Example 4.2 of their text and transistor Q_1 was shown to be operating in the active region. We shall repeat this example using Spice to calculate the DC operating point and show that, indeed, transistor Q_1 is operating in the active mode.

4 Bipolar Junction Transistors (BJTs)

Mode	EBJ	CBJ
Cutoff	Reverse	Reverse
Active	Forward	Reverse
Saturation	Forward	Forward

Table 4.2 BJT modes of operation.

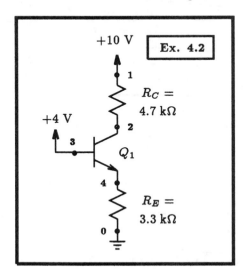

Figure 4.9 A simple *npn* transistor circuit (Example 4.2 of Sedra and Smith).

A Spice description of this circuit is listed in Fig. 4.10 and a DC operating point analysis request is included. The results of the analysis are listed below.

```
****     SMALL SIGNAL BIAS SOLUTION       TEMPERATURE =   27.000 DEG C
*************************************************************************

 NODE   VOLTAGE     NODE   VOLTAGE     NODE   VOLTAGE     NODE   VOLTAGE

(   1)   10.0000  (   2)    5.3452  (   3)    4.0000  (   4)    3.3009

****     OPERATING POINT INFORMATION      TEMPERATURE =   27.000 DEG C
*************************************************************************

**** BIPOLAR JUNCTION TRANSISTORS

NAME         Q1
MODEL        npn_transistor
IB           9.90E-06
IC           9.90E-04
VBE          6.99E-01
VBC          -1.35E+00
VCE          2.04E+00
```

Included amongst the list of node voltages is a partial summary of the DC operating point conditions associated with the transistor. This includes the currents I_B and I_C, and the voltages V_{BE}, V_{BC}, and V_{CE}. Other information is included in this list of operating point

4.3 Spice Analysis Of Transistor Circuits At DC

```
Example 4.2: NPN Transistor Operated In Active Mode

** Circuit Description **
Vcc 1 0 DC +10V
Vbb 3 0 DC +4V
Q1 2 3 4 npn_transistor
Rc 1 2 4.7k
Re 4 0 3.3k
* transistor model statement
.model npn_transistor npn (Is=1.8104e-15 Bf=100)
** Analysis Requests **
* calculate DC bias point information
.OP
** Output Requests **
* none required
.end
```

Figure 4.10 The Spice input file for calculating the DC operating point of the circuit shown in Fig. 4.9.

information but is not shown here. Also, we do not show the small-signal parameters associated with the hybrid-pi model of the transistor, deferring discussion of this topic to a later stage.

What one should notice here, amongst this operating point information, is that the base-emitter junction is forward biased with $V_{BE} = 0.699$ V and the collector-base junction is reversed biased with $V_{BC} = -1.35$ V; thus confirming that transistor Q_1 is indeed operating in its active region. One could have deduced this same information from the DC node voltages, but for multiple transistor circuits this can be a tedious endeavor. Also, the collector and base currents agree reasonably well with the values calculated by hand analysis.

Saturation Region

If we increase the base voltage of the circuit of Fig. 4.11 from +4 V to +6 V, transistor Q_1 leaves the active region and moves into the saturation region. To see this, change the value of Vbb in the Spice input file shown in Fig. 4.10 to +6 V and re-run the Spice job. The results are:

```
****    SMALL SIGNAL BIAS SOLUTION      TEMPERATURE =   27.000 DEG C
*********************************************************************

NODE    VOLTAGE     NODE    VOLTAGE     NODE    VOLTAGE     NODE    VOLTAGE

(   1)  10.0000  (   2)   5.3147  (   3)   6.0000  (   4)   5.2808

****    OPERATING POINT INFORMATION     TEMPERATURE =   27.000 DEG C
*********************************************************************
```

4 Bipolar Junction Transistors (BJTs)

```
****  BIPOLAR JUNCTION TRANSISTORS

NAME        Q1
MODEL       npn_transistor
IB          6.03E-04
IC          9.97E-04
VBE         7.19E-01
VBC         6.85E-01
VCE         3.39E-02
```

As is evident, both V_{BE} and V_{BC} are forward biased, suggesting that Q_1 is operating in its saturation region. The saturation region of operation will be studied further in Sections 4.4 and 4.5.

Cutoff Region

Finally, if we reduce the base voltage to zero volts, then the transistor becomes cutoff. Altering the Spice input deck to reflect this (ie. setting Vbb=0) and re-running the Spice job results in the following:

```
****    SMALL SIGNAL BIAS SOLUTION       TEMPERATURE =   27.000 DEG C
***************************************************************************

  NODE   VOLTAGE    NODE   VOLTAGE    NODE   VOLTAGE    NODE   VOLTAGE

(   1)   10.0000  (   2)   10.0000  (   3)    0.0000  (   4)  33.01E-09

****    OPERATING POINT INFORMATION      TEMPERATURE =   27.000 DEG C
***************************************************************************

****  BIPOLAR JUNCTION TRANSISTORS

NAME        Q1
MODEL       npn_transistor
IB          -1.00E-11
IC           2.00E-11
VBE         -3.30E-08
VBC         -1.00E+01
VCE          1.00E+01
```

Here, both V_{BE} and V_{BC} are reversed biased, indicating that Q_1 is now cutoff. Notice, however, that even under cutoff conditions the transistor is still conducting a base and collector current (and, of course, an emitter current). These currents are mainly leakage currents associated with the two reverse biased junctions of the transistor.

4.3.2 Computing DC Bias Of A PNP Transistor Circuit

The above circuit examples consisted of *npn* transistors only. In this next example we shall calculate the DC operating point of a circuit containing a *pnp* transistor. As far as Spice is concerned, *pnp* transistor circuits are no more complicated than *npn* transistor circuits.

4.3 Spice Analysis Of Transistor Circuits At DC

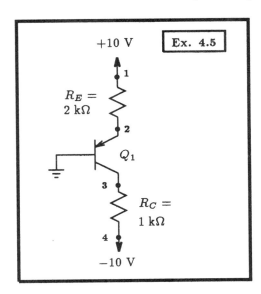

Figure 4.11 A simple *pnp* transistor circuit (Example 4.5 of Sedra and Smith).

```
Example 4.5: PNP Transistor DC Operating Point Calculation

** Circuit Description **
Vcc 1 0 DC +10V
Vee 4 0 DC -10V
Q1 3 0 2 pnp_transistor
Re 1 2 2k
Rc 3 4 1k
* transistor model statement
.model pnp_transistor pnp (Is=1.8104e-15 Bf=100)
** Analysis Requests **
* calculate DC bias point information
.OP
** Output Requests **
* none required
.end
```

Figure 4.12 The Spice input file for calculating the DC operating point of the circuit shown in Fig. 4.11.

Consider the circuit shown in Fig. 4.11. This example corresponds to Example 4.5 of Sedra and Smith. The Spice input file for this circuit is listed in Fig. 4.12 and the results of the DC analysis are shown below:

```
****    SMALL SIGNAL BIAS SOLUTION      TEMPERATURE =   27.000 DEG C
***********************************************************************

    NODE    VOLTAGE     NODE    VOLTAGE     NODE    VOLTAGE     NODE    VOLTAGE

    (  1)   10.0000   (   2)     .7387   (   3)   -5.4152   (   4)  -10.0000

        VOLTAGE SOURCE CURRENTS
```

4 Bipolar Junction Transistors (BJTs)

```
           NAME           CURRENT

           Vcc            -4.631E-03
           Vee             4.585E-03

           TOTAL POWER DISSIPATION   9.22E-02  WATTS

   ****       OPERATING POINT INFORMATION     TEMPERATURE =   27.000 DEG C
   *************************************************************************

   **** BIPOLAR JUNCTION TRANSISTORS

           NAME           Q1
           MODEL          pnp_transistor
           IB             -4.58E-05
           IC             -4.58E-03
           VBE            -7.39E-01
           VBC             5.42E+00
           VCE            -6.15E+00
```

From the above results, it is obvious that the transistor is operating in the active mode. Notice that the polarity of the transistor currents are opposite to those of an *npn* transistor operating in its active region (see the previous subsection). Further, one should also note that Spice defines a positive collector current of a *pnp* transistor as the current flowing into the collector. This is opposite to that which Sedra and Smith uses. One should be careful of this. When one compares the above Spice results with those generated by a simple hand calculation, say, for example, as was performed in Example 4.5 of Sedra and Smith, we find we are in very good agreement; about a 0.4% difference.

In the above analysis, the Early effect was neglected. In practise this is never the case. In the following, let us repeat the above analysis assuming the *pnp* transistor has an Early voltage V_{AF} of 100 V. We shall then compare the resulting transistor collector current with that obtained previously when the Early effect was ignored.

Consider replacing the transistor model statement for the *pnp* transistor given in the Spice deck of Fig. 4.12 with the following one:

```
.model pnp_transistor pnp (Is=1.8104e-15 Bf=100 Vaf=100V).
```

Re-running the Spice job, we find in the Spice output file the following results:

```
   ****       SMALL SIGNAL BIAS SOLUTION      TEMPERATURE =   27.000 DEG C
   *************************************************************************

   NODE   VOLTAGE     NODE   VOLTAGE     NODE   VOLTAGE     NODE   VOLTAGE

   (  1)  10.0000  (   2)     .7373  (   3)   -5.4122  (   4)  -10.0000

           VOLTAGE SOURCE CURRENTS
           NAME           CURRENT

           Vcc            -4.631E-03
           Vee             4.588E-03
```

4.3 Spice Analysis Of Transistor Circuits At DC

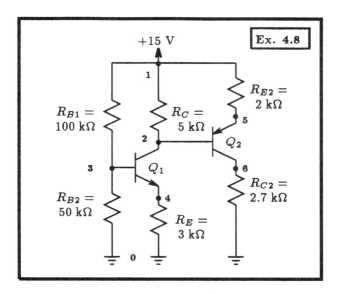

Figure 4.13 A multiple transistor circuit (Example 4.8 of Sedra and Smith).

```
****    OPERATING POINT INFORMATION         TEMPERATURE =   27.000 DEG C
*******************************************************************************

**** BIPOLAR JUNCTION TRANSISTORS

     NAME         Q1
     MODEL        pnp_transistor
     IB           -4.35E-05
     IC           -4.59E-03
     VBE          -7.37E-01
     VBC          5.41E+00
     VCE          -6.15E+00
```

We see from the above results that the collector current is now 4.59 mA. This is in contrast to the previous situation when the Early effect was ignored and a Spice analysis revealed a collector current of 4.58 mA. In the latter case, a similar result would also be obtained by a simple hand calculation, as was noted above. Thus, if we compare the result obtained by Spice with the Early effect included to that obtained by a simple hand calculation, we see a difference of about 0.2%. For most practical engineering situations, a 5% to 10% accuracy obtained from a simple hand analysis is usually quite acceptable. But, more importantly, to account for the Early effect during hand analysis would complicate the analysis significantly that it would probably mask any insight. Thus, if greater accuracy is thought necessary then one should make use of a detailed Spice analysis with the transistor Early effect included.

4.3.3 DC Operating Point Of A Multiple Transistor Circuit

Consider the multiple transistor circuit shown in Fig. 4.13. This same circuit was ana-

4 Bipolar Junction Transistors (BJTs)

Example 4.8: Multiple Transistor Circuit Bias Point Calculation

```
** Circuit Description **
Vcc 1 0 DC +15V
* npn stage
Q1 2 3 4 npn_transistor
Rc 1 2 5k
Rb1 1 3 100k
Rb2 3 0 50k
Re 4 0 3k
* pnp stage
Q2 6 2 5 pnp_transistor
Re2 1 5 2k
Rc2 6 0 2.7k
* transistor model statement
.model npn_transistor npn (Is=1.8104e-15 Bf=100)
.model pnp_transistor pnp (Is=1.8104e-15 Bf=100)
** Analysis Requests **
* calculate DC bias point information
.OP
** Output Requests **
* none required
.end
```

Figure 4.14 The Spice input file for calculating the DC operating point of the multiple transistor circuit shown in Fig. 4.13.

lyzed by hand for its node voltages and branch currents by Sedra and Smith, and designated as Example 4.8 in their text. We shall compute the DC operating point of this circuit using Spice. The Spice input file is listed in Fig. 4.14 with an operating point analysis command included. The results generated by Spice are listed below:

```
****    SMALL SIGNAL BIAS SOLUTION      TEMPERATURE =   27.000 DEG C
*************************************************************************

NODE    VOLTAGE     NODE    VOLTAGE     NODE    VOLTAGE     NODE    VOLTAGE

(   1)  15.0000  (   2)   8.7526  (   3)   4.5744  (   4)   3.8688
(   5)   9.4779  (   6)   7.3810


    VOLTAGE SOURCE CURRENTS
    NAME            CURRENT

    Vcc            -4.115E-03

    TOTAL POWER DISSIPATION   6.17E-02  WATTS

****    OPERATING POINT INFORMATION     TEMPERATURE =   27.000 DEG C
*************************************************************************

**** BIPOLAR JUNCTION TRANSISTORS

NAME            Q1                      Q2
MODEL           npn_transistor          pnp_transistor
```

4.3 Spice Analysis Of Transistor Circuits At DC

```
IB       1.28E-05           -2.73E-05
IC       1.28E-03           -2.73E-03
VBE      7.06E-01           -7.25E-01
VBC     -4.18E+00            1.37E+00
VCE      4.88E+00           -2.10E+00
```

Comparing these results with those calculated by hand in Sedra and Smith, we see that the results generated by hand analysis (eg. $I_{C_1} = 1.28$ mA and $I_{C_2} = 2.75$ mA) are quite close to the values generated by Spice. Moreover, we see that both transistors are operating in the active mode.

In the following we repeat the analysis with each transistor having an Early voltage of 100 V. This requires that we modify the two transistor model statement given in the Spice deck of Fig. 4.14 to the following:

```
.model npn_transistor npn (Is=1.8104e-15 Bf=100 Vaf=100V)
.model pnp_transistor pnp (Is=1.8104e-15 Bf=100 Vaf=100V).
```

On doing so, and re-submitting the revised Spice deck for processing, we obtain the following results:

```
****    SMALL SIGNAL BIAS SOLUTION        TEMPERATURE =   27.000 DEG C
*******************************************************************************

 NODE   VOLTAGE     NODE   VOLTAGE     NODE   VOLTAGE     NODE   VOLTAGE

(   1)   15.0000 (    2)    8.7227 (    3)    4.5894 (    4)    3.8847
(   5)    9.4478 (    6)    7.4222

        VOLTAGE SOURCE CURRENTS
        NAME         CURRENT

        Vcc         -4.136E-03

        TOTAL POWER DISSIPATION   6.20E-02  WATTS

****    OPERATING POINT INFORMATION       TEMPERATURE =   27.000 DEG C
*******************************************************************************

**** BIPOLAR JUNCTION TRANSISTORS

NAME     Q1                 Q2
MODEL    npn_transistor     pnp_transistor
IB       1.23E-05           -2.71E-05
IC       1.28E-03           -2.75E-03
VBE      7.05E-01           -7.25E-01
VBC     -4.13E+00            1.30E+00
VCE      4.84E+00           -2.03E+00
```

Here we see that the collector current of Q_1 remains at 1.28 mA, but the collector current of Q_2 has increased to a level of 2.75 mA, a 0.02 mA increase. This example, like the previous one, demonstrates the effect that transistor Early voltage has on the DC bias levels in a circuit and shows that it is usually small. Thus, not accounting for it during a

159

4 Bipolar Junction Transistors (BJTs)

hand analysis seems reasonable.

4.4 BJT Transistor Amplifiers

Transistors operated in their active region find important applications as linear amplifiers. Design and analysis of linear amplifiers is facilitated by the use of small-signal models for the BJT. In the following we discuss the small-signal model used by Spice.

4.4.1 BJT Small-Signal Model

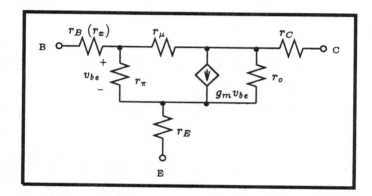

Figure 4.15 The small-signal BJT Spice model under static conditions.

Under static and small-signal conditions, the linearized BJT Spice model is the familiar hybrid-pi model shown in Fig. 4.15. Both the *npn* and *pnp* transistor have the same hybrid-pi model, so we make no distinction between the two. The transconductance g_m is related to the DC bias current I_C according to the following

$$g_m = \frac{\partial i_C}{\partial v_{BE}}\bigg|_{OP} + \frac{\partial i_C}{\partial v_{BC}}\bigg|_{OP} + \frac{\partial i_B}{\partial v_{BC}}\bigg|_{OP} \approx \frac{I_C}{V_T}. \qquad (4.5)$$

Similarly, the input and output resistances are also related to the transistors DC operating point, according to the following

$$r_\pi = \left(\frac{\partial i_B}{\partial v_{BE}}\bigg|_{OP}\right)^{-1} \approx \frac{V_T}{I_B} \qquad (4.6)$$

$$r_o = \left(\frac{\partial i_C}{\partial v_{BC}}\bigg|_{OP} + \frac{\partial i_B}{\partial v_{BC}}\bigg|_{OP}\right)^{-1} \approx \frac{V_A}{I_C} \qquad (4.7)$$

and r_μ is approximated as

$$r_\mu = \left(\left.\frac{\partial i_B}{\partial v_{BC}}\right|_{OP}\right)^{-1} \approx 10\frac{V_A}{I_C}\beta_F. \tag{4.8}$$

Also included in the hybrid-model are the ohmic resistances of the three junctions; r_C, r_B (r_x), and r_E. Note that r_E is **not** the r_e used in the text by Sedra and Smith.

The small-signal parameters of a transistor are usually computed by Spice prior to most analyses. In many types of analysis, the small-signal model of the BJT is paramount to the analysis. Spice will list the small-signal model parameters of all transistors in a given circuit when an .OP command is included in the Spice input file. Because of the importance of the base resistance r_B on small-signal operation, Spice will also list this value amongst the small-signal parameters in the output file. It will be denoted by Rx.

As an example of the small-signal parameters that are listed in the Spice output file as a result of an .OP analysis command, we show below the operating point information for transistors Q_1 and Q_2 in our previous example given in subsection 4.3.3:

```
**** BIPOLAR JUNCTION TRANSISTORS

NAME        Q1                  Q2
MODEL       npn_transistor      pnp_transistor
IB          1.23E-05            -2.71E-05
IC          1.28E-03            -2.75E-03
VBE         7.05E-01            -7.25E-01
VBC         -4.13E+00           1.30E+00
VCE         4.84E+00            -2.03E+00
BETADC      1.04E+02            1.01E+02
GM          4.96E-02            1.06E-01
RPI         2.10E+03            9.53E+02
RX          0.00E+00            0.00E+00
RO          8.12E+04            3.69E+04
CBE         0.00E+00            0.00E+00
CBC         0.00E+00            0.00E+00
CBX         0.00E+00            0.00E+00
CJS         0.00E+00            0.00E+00
BETAAC      1.04E+02            1.01E+02
FT          7.89E+17            1.69E+18
```

Here we see a list containing the DC operating point and small-signal model parameters for both the *npn* and *pnp* transistors in the circuit of Fig. 4.13. Besides the parameters that have already been explained, this list also contains other parameters, such as capacitances CBE, CBC, CBX and CJS. These capacitances model dynamic effects of the transistor and will be further explained in Chapter 7. Likewise, BETAAC, and FT, are parameters that are derived from the dynamic small-signal model of the transistor and will also be explained in Chapter 7.

4 Bipolar Junction Transistors (BJTs)

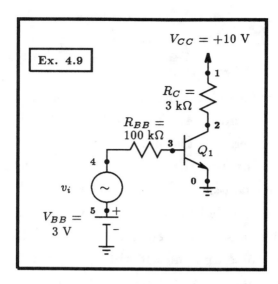

Figure 4.16 A simple common-emitter voltage amplifier (Example 4.9 of Sedra and Smith).

4.4.2 Single-Stage Voltage-Amplifier Circuits

Consider the voltage amplifier circuit shown in Fig. 4.16. This circuit was presented and analyzed for its voltage gain by Sedra and Smith in Example 4.9 of their text. Here we shall use Spice to calculate the voltage gain of this circuit assuming that the transistor has a $\beta_F = 100$ and $I_S = 1.8104 \times 10^{-15}$ A. The other parameters of the BJT model will take on their default values.

In Fig. 4.17 we list the Spice input file for this circuit. A transfer function command (.TF) is included to calculate the voltage gain v_o/v_i where v_o is taken at node 2. In addition, we are also requesting an operating point calculation (.OP) so that we may cross-check the results found using the .TF command with those computed by hand using the equations derived by Sedra and Smith in Example 4.9 of their text.

The results of the .TF command are as follows:

```
****    SMALL-SIGNAL CHARACTERISTICS

        V(2)/Vi = -2.966E+00

        INPUT RESISTANCE AT Vi =  1.011E+05

        OUTPUT RESISTANCE AT V(2) =  3.000E+03
```

Hence, the voltage gain of this circuit is −2.966 V/V. Also, the input and output resistance of this circuit are 101.1 kΩ and 3 kΩ, respectively. Referring back to the circuit shown in Fig. 4.16, these resistance levels seem to be in line with what one would expect.

We can collaborate the voltage gain calculation found using the .TF command by per-

4.4 BJT Transistor Amplifiers

Example 4.9: Calculating Voltage Gain Of A Transistor Amplifier

```
** Circuit Description **
* dc supplies
Vcc 1 0 DC +10V
Vbb 5 0 DC +3V
* small-signal input
Vi 4 5 DC 1mV
* amplifier circuit
Q1 2 3 0 npn_transistor
Rc 1 2 3k
Rbb 4 3 100k
* transistor model statement
.model npn_transistor npn (Is=1.8104e-15 Bf=100)
** Analysis Requests **
* calculate small-signal transfer function: Vo/Vi
.TF V(2) Vi
* we shall also calculate the small-signal parameters of the transistor
.OP
** Output Requests **
* none required
.end
```

Figure 4.17 The Spice input file for calculating the small-signal voltage gain of the circuit shown in Fig. 4.16.

forming the same gain calculation using the small-signal equivalent circuit representation of the amplifier circuit shown in Fig. 4.16. This was performed by Sedra and Smith, and they found that the voltage gain, in terms of the small-signal parameters of the transistor, is given by

$$\frac{v_o}{v_i} = -g_m R_C \frac{r_\pi}{r_\pi + R_{BB}}. \tag{4.9}$$

In this derivation the output resistance of the transistor was assumed infinite. The small-signal parameters of the hybrid-pi model for this particular transistor as calculated by Spice are found amongst the transistor operating point information in the Spice output file. A partial listing of the operating point information found in the output file for transistor Q_1 is given below:

```
**** BIPOLAR JUNCTION TRANSISTORS

NAME      Q1
MODEL     npn_transistor
IB        2.28E-05
IC        2.28E-03
VBE       7.21E-01
VBC       -2.44E+00
VCE       3.16E+00
BETADC    1.00E+02
```

4 Bipolar Junction Transistors (BJTs)

Example 4.10: Time-Domain Waveforms Of A Transistor Amplifier

```
** Circuit Description **
* dc supplies
Vcc 1 0 DC +10V
Vbb 5 0 DC +3V
* small-signal input
Vi 4 5 PULSE (-1.1V 1.1V 0 0.5ms 0.5ms 1us 1ms)
* amplifier circuit
Q1 2 3 0 npn_transistor
Rc 1 2 3k
Rbb 4 3 100k
* transistor model statement
.model npn_transistor npn (Is=1.8104e-15 Bf=100)
** Analysis Requests **
.TRAN 10us 4ms 0ms 10us
** Output Requests **
.PLOT TRAN V(2)
.probe
.end
```

Figure 4.18 The Spice input file for calculating the collector voltage of the circuit shown in Fig. 4.16 subject to a triangular waveform input.

```
        GM      8.82E-02
        RPI     1.13E+03
        RX      0.00E+00
        RO      1.00E+12
```

From the above list, we see that $g_m = 88.2$ mA/V and $r_\pi = 1.13$ kΩ. Substituting these values, together with $R_C = 3$ kΩ and $R_{BB} = 100$ kΩ, into Eqn. (4.9), one finds the voltage gain to be -2.96 V/V. Comparing this result with that computed using the .TF command of Spice, as expected, one finds almost exact agreement. The difference is due the number of significant digits used in the two calculations. In principle, these two calculations should be in perfect agreement because the .TF command calculates the voltage gain in the exact same way.

To visualize the operation of this amplifier, consider simulating the circuit shown in Fig. 4.16 with a time-varying input. More specifically, we shall consider that v_i is time-varying with a triangular waveform. We shall consider that the input signal has a period of 1 ms and an input amplitude of 1.1 V. Moreover, we shall repeat the same example with the input amplitude reduced to 0.8 V. We shall then want to compare the results.

To emulate a triangular waveform, we make use of the following PULSE statement:

```
Vi 4 5 PULSE (-1.1V 1.1V 0 0.5ms 0.5ms 1us 1ms)
```

4.4 BJT Transistor Amplifiers

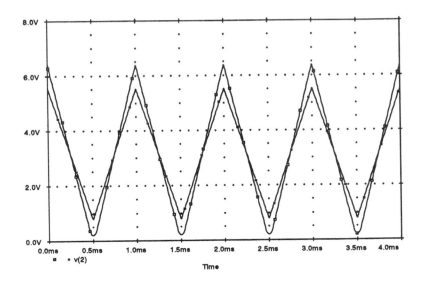

Figure 4.19 The output voltage waveform of the amplifier shown in Fig. 4.16 for two different inputs levels of 0.8 V and 1.1 V. The larger output signal is seen to be clipped on its lower peaks.

The rise and fall time is set to equal half the period of the triangle waveform of 1 ms. Mathematically, the pulsewidth field should be zero, however, due to a Spice technicality, the pulsewidth can not be specified as being equal to zero. To circumvent this problem, we set the pulsewidth to a small, but, nonzero value – in this case 1 μs. Furthermore, we added a .TRAN statement to calculate the time response of the circuit over a 4 ms interval. The resulting Spice input file for the first example is listed in Fig. 4.18. The same file can be used for the second case by simply reducing the peak level of the input signal from 1.1 V to 0.8 V. For comparison purposes, we shall concatenate the two files together before submitting the job to Spice.

The voltage waveforms at the collector of the amplifier for both input levels, as calculated by Spice, are displayed in Fig. 4.19. Both signals are riding on a DC component of 3.2 V, corresponding to the DC bias point at the collector. One will also notice that the larger output waveform is clipped on its lower peak. This can be traced back to the transistor being forced into its cutoff region when the input signal level exceeds a 0.82 V level. The other waveform does not appear to be clipped, confirming that the transistor stays out of its cutoff region.

Another Example:

As another example of a transistor being used to form a linear amplifier, consider the

4 Bipolar Junction Transistors (BJTs)

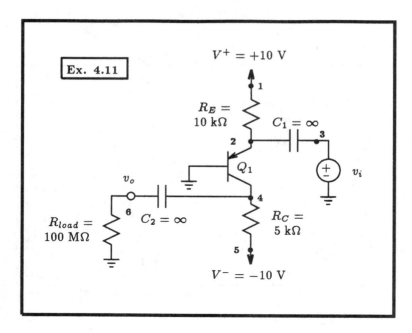

Figure 4.20 A *pnp* transistor amplifier circuit (Example 4.11 of Sedra and Smith).

circuit shown in Fig. 4.20. In this particular case, a *pnp* transistor is the central component of the design. This circuit was analyzed by hand by Sedra and Smith in Example 4.11 of their text. From their analysis they found that the amplifier had a noninverting voltage gain of 183.3 V/V. They reasoned that, because the input signal appears directly across the base-emitter junction of the *pnp* transistor, the input signal amplitude must remain below 10 mV to justify small-signal operation and obtain this voltage gain. In the following we shall simulate the amplifier circuit shown in Fig. 4.20 subject to a triangular waveform input signal having a peak value of 10 mV and compare it to one with twice the amplitude.

The Spice input file for the *pnp* transistor amplifier circuit shown in Fig. 4.20 subject to a 10 mV triangular input signal is listed below in Fig. 4.21. The infinite-valued capacitors, C_1 and C_2, are represented in the Spice input file by large 100 μF capacitors that are initially charged to the DC bias voltage that will eventually appear across them. These are usually found by first determining the DC operating point of the circuit using an .OP command with the input set to zero. This "trick" speeds up the circuit simulation by starting the circuit off near its steady-state operation. If these initial conditions were not specified, then our transient simulation time would be excessively long. This stems from the fact that these large coupling capacitors form large time-constants and need a long time to charge to their steady-state values. Even then, we are still required to allot some simulation time for the circuit to reach its true steady-state operation. This is the reason for delaying the output on the .TRAN statement until 10 ms (or 10 periods of the input signal) after its start. Also

4.4 BJT Transistor Amplifiers

Example 4.11: Time-Domain Waveforms For A PNP Transistor Amplifier

```
** Circuit Description **
* dc supplies
V+ 1 0 DC +10V
V- 5 0 DC -10V
* small-signal triangular-wave input
Vi 3 0 PULSE (-10mV 10mV 0 0.5ms 0.5ms 1us 1ms)
* amplifier circuit
Q1 4 0 2 pnp_transistor
Rc 4 5 5k
Re 1 2 10k
* dc blockers - very large capacitors
C1 3 2 100uF IC=-0.6972V
C2 4 6 100uF IC=-5.3946V
* output node (very large resistance as not to load the amplifier)
Ro 6 0 100Meg
* transistor model statement
.model pnp_transistor pnp (Is=1.8104e-15 Bf=100)
** Analysis Requests **
* calculate transient response using initial conditions
.TRAN 100us 14ms 10ms 100us UIC
** Output Requests **
.PLOT TRAN V(4) V(3)
.probe
.end
```

Figure 4.21 The Spice input file for calculating the input and output signal waveforms of the circuit shown in Fig. 4.20.

note that we must specify on the .TRAN statement that we want Spice to use the initial conditions established for each capacitor by specifying the field flag UIC.

Submitting this file, concatenated together with another file having a triangular input signal of 20 mV amplitude, to Spice results in the input and output waveforms shown in Fig. 4.22. The larger output signal in the lower trace due to the 20 mV input signal clearly deviates from an ideal triangle waveform; it looks more parabolic than linear. The smaller output signal caused by the 10 mV input signal though much more linear, exhibits some deviation from the ideal, thus suggesting that the small-signal linear range of operation is actually less than 10 mV.

We also see from Fig. 4.22 that for the 10 mV peak input signal, the corresponding output signal has a peak to peak excursion of 3.5 V, corresponding to a signal gain of 175 V/V. This result is, therefore, in good agreement with the small-signal gain that was calculated by hand (183.3 V/V) in Example 4.11 of Sedra and Smith.

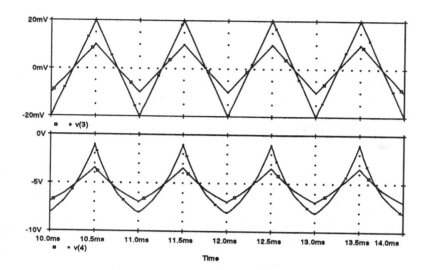

Figure 4.22 Input and output waveforms of the *pnp* amplifier shown in Fig. 4.20. The top graph display the two input triangular waveforms and the bottom graph displays the output response to the two input signals. Distortion is evident in both output waveforms.

4.5 DC Bias Sensitivity Analysis

Sensitivity To Component Variations

Bias networks are used in transistor amplifier design to establish the proper DC operating point of each transistor. In order to maintain consistent operation, the DC operating point of each transistor must be held constant. This is normally achieved by a bias network that maintains the emitter current of each transistor relatively constant under potential circuit variations, eg. variations in transistor β, etc.

Provisions have been made in Spice for investigating the DC stability of circuit behavior in the face of component variations through a DC sensitivity analysis. When invoked, through a .SENS analysis command placed in the Spice input file, Spice will compute the derivatives, or what is also known as the sensitivities, of selected variables of the circuit to most of the components of the circuit. This includes the sensitivities to many of the model parameters of the BJT, such as β and I_S. If diodes are present in the circuit, the sensitivities to the diode model parameters will also be computed and listed. Unfortunately, Spice does not compute the sensitivities to the FET model parameters.

The general description of the syntax of the Sensitivity Analysis command (.SENS) is provided in Table 4.3. The command line begins with the keyword .SENS followed by a list

4.5 DC Bias Sensitivity Analysis

Analysis Requests	Spice Command
DC Sensitivity	.SENS *output_variables*

Table 4.3 General form of the DC sensitivity analysis request.

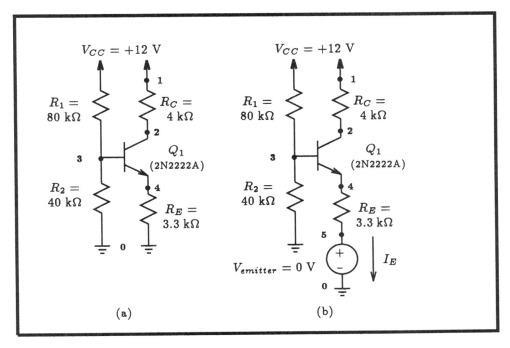

Figure 4.23 (a) An amplifier biased using a single power supply resistive network arrangement. (b) Monitoring the emitter current of the amplifier circuit by including a zero-valued voltage source ($V_{emitter}$) in series with the emitter of the transistor.

of the variables whose derivatives will be computed with respect to the different components of the circuit. The results of the .SENS command are sent directly to the Spice output file in much the same way as that performed by the operating point or transfer function analysis commands seen previously. No .PRINT or .PLOT statement is required in the Spice input file. It should be noted that Spice first computes the DC operating point of the circuit and then computes the appropriate derivatives. This next example will demonstrate the results of a Spice sensitivity analysis.

Consider the BJT amplifier shown in Fig. 4.23(a) which is biased using a single power supply resistive network arrangement. Let us compute the sensitivities of the emitter current of Q_1 using Spice assuming that the BJT is modeled after the widely available commercial transistor 2N2222A. This requires that we place a zero-valued voltage source $V_{emitter}$ in the

4 Bipolar Junction Transistors (BJTs)

Investigating The Sensitivity Of the Emitter Current To Amplifier Components

```
** Circuit Description **
* power supply
Vcc 1 0 DC +12V
* amplifier circuit
Q1 2 3 4 Q2N2222A
Rc 1 2 4k
R1 1 3 80k
R2 3 0 40k
Re 4 5 3.3k
Vemitter 5 0 0
* transistor model statement for the 2N2222A
.model Q2N2222A NPN (Is=14.34f Xti=3 Eg=1.11 Vaf=74.03 Bf=255.9 Ne=1.307
+                    Ise=14.34f Ikf=.2847 Xtb=1.5 Br=6.092 Nc=2 Isc=0 Ikr=0 Rc=1
+                    Cjc=7.306p Mjc=.3416 Vjc=.75 Fc=.5 Cje=22.01p Mje=.377 Vje=.75
+                    Tr=46.91n Tf=411.1p Itf=.6 Vtf=1.7 Xtf=3 Rb=10)
** Analysis Requests **
.SENS I(Vemitter)
** Output Requests **
* none required
.end
```

Figure 4.24 The Spice input file for calculating the DC sensitivities of the emitter current of the BJT amplifier shown in Fig. 4.23.

emitter branch of the amplifier circuit to monitor the emitter current, as is shown in Fig. 4.23(b). The model parameters for the 2N2222A were derived from the information contained on the manufacturer's data sheets for the 2N2222A. The details of converting manufacturer's data on supplied transistor parts to a Spice model is beyond the scope of this text. We note, however, that, software is now available from various development houses that greatly simplify this task, eg. MicroSim Corporation, the suppliers of PSpice. An interesting paper [Malik,1990] describes a step-by-step procedure for determining numerical parameter values for the Spice model of a BJT from manufacturer's measured data.

The Spice input file describing this circuit is provided in Fig. 4.24. Here the sensitivity analysis command,

.SENS I(Vemitter)

will request that Spice compute the sensitivity of the emitter current to various components in the circuit. The results of this analysis are then found in the output file, together with some of the DC bias solution, as follows:

4.5 DC Bias Sensitivity Analysis

```
 ****     SMALL SIGNAL BIAS SOLUTION       TEMPERATURE =  27.000 DEG C
 *******************************************************************************

    NODE   VOLTAGE     NODE   VOLTAGE     NODE   VOLTAGE     NODE   VOLTAGE

 (   1)   12.0000  (   2)    8.1567   (   3)    3.8346   (   4)    3.1912
 (   5)    0.0000

       VOLTAGE SOURCE CURRENTS
       NAME           CURRENT

       Vcc           -1.063E-03
       Vemitter       9.670E-04

       TOTAL POWER DISSIPATION   1.28E-02  WATTS

 ****     DC SENSITIVITY ANALYSIS           TEMPERATURE =  27.000 DEG C
 *******************************************************************************

 DC SENSITIVITIES OF OUTPUT I(Vemitter)

            ELEMENT        ELEMENT         ELEMENT         NORMALIZED
            NAME           VALUE           SENSITIVITY     SENSITIVITY
                                           (AMPS/UNIT)     (AMPS/PERCENT)

            Rc             4.000E+03       -6.110E-10      -2.444E-08
            R1             8.000E+04       -9.727E-09      -7.782E-06
            R2             4.000E+04        1.827E-08       7.309E-06
            Re             3.300E+03       -2.771E-07      -9.144E-06
            Vcc            1.200E+01        9.594E-05       1.151E-05
            Vemitter       0.000E+00       -2.865E-04       0.000E+00
 Q1
            RB             1.000E+01       -1.773E-09      -1.773E-10
            RC             1.000E+00       -6.110E-10      -6.110E-12
            RE             0.000E+00        0.000E+00       0.000E+00
            BF             2.559E+02        1.053E-07       2.695E-07
            ISE            1.434E-14       -1.395E+09      -2.001E-07
            BR             6.092E+00       -2.946E-15      -1.795E-16
            ISC            0.000E+00        0.000E+00       0.000E+00
            IS             1.434E-14        1.584E+09       2.272E-07
            NE             1.307E+00        2.913E-04       3.808E-06
            NC             2.000E+00        0.000E+00       0.000E+00
            IKF            2.847E-01        5.546E-07       1.579E-09
            IKR            0.000E+00        0.000E+00       0.000E+00
            VAF            7.403E+01       -3.712E-08      -2.748E-08
            VAR            0.000E+00        0.000E+00       0.000E+00
```

The above list of sensitivities consists of two parts: the top portion of the table lists the sensitivities to the DC sources and passive components of the circuit and the bottom part of the table lists the sensitivities to the parameters of the models of the active components in the circuit. In this particular case, there is only one active component in the circuit, Q_1.

In the above table, the first column indicates the element that the sensitivity of the emitter current I_E is taken with respect to. The second column indicates the nominal value of that element as it appears in the Spice input file. The third column indicates the sensitivity quantity $\frac{\partial I_E}{\partial x}$ where x is the corresponding element appearing in the leftmost column. In general, the units of this sensitivity quantity are the units of the output variable specified on the .SENS statement divided by the units of the element x. For instance, in the case of the 3.3 kΩ emitter resistor R_E, denoted by Re in the above sensitivity list as the fourth element from the top, the sensitivity quantity $\frac{\partial I_E}{\partial R_E}$ is expressed in A/Ω and has a value of -2.771×10^{-7} A/Ω. The final column that appears on the right, is a normalized sensitivity measure. It simply expresses the sensitivity in more convenient units of A/%. Mathematically, it is written as $\frac{\partial I_E}{(\partial R_E/R_E)_\%}$ where $(\partial R_E/R_E)_\%$ is the relative change of R_E expressed in per-cent. For R_E, the normalized sensitivity measure is -9.144×10^{-6} A/%.

From the above sensitivity analysis we can obtain estimates of how stable the emitter current I_E is in the face of component variations. For example, if the emitter resistor R_E changes by 10% we can expect that the emitter current will approximately change according to the relation

$$\Delta I_E \approx \frac{\partial I_E}{(\partial R_E/R_E)_\%} \times (\Delta R_E/R_E)_\% \qquad (4.10)$$

Thus substituting the appropriate numerical values, we find that a 10% change in R_E results in a 91.4 μA change in the emitter current of Q_1. This is about 9.5% of its quiescent value of 967.0 μA.

Similarly, we can estimate the effect a 20% variation in transistor β_F has on the emitter current of Q_1. Consider from the above Spice generated sensitivities we see that $\frac{\partial I_E}{(\partial \beta_F/\beta_F)_\%} = 2.695 \times 10^{-7}$ A/%. Thus, a 20% change in β_F gives rise to a 5.39 μA change in the emitter current of Q_1. This is about 0.56% of the quiescent value.

It should be obvious that this same approach can be used for any of the components of the circuit in Fig. 4.23 whose sensitivities are given in the above table. One can also investigate the effect that different component changes has on the emitter current by making use of the mathematical concept of a total derivative. An example of this is provide in section 5.6 as it applies to a MOSFET amplifier circuit.

Sensitivity To Temperature Variations

To investigate the bias stability of an amplifier subject to temperature changes requires a different approach than that just seen using the sensitivity analysis (.SENS) command of Spice. For such cases, we make use of an extension to the DC sweep command which will allow the user to sweep the temperature of the circuit over a specified range while repeatedly performing a DC analysis. The syntax of this command is identical to any other DC sweep command with the keyword *TEMP* replacing the field marked by *source_name*. The range

4.5 DC Bias Sensitivity Analysis

Investigating Emitter Current Dependence On Temperature

```
** Circuit Description **
* power supply
Vcc 1 0 DC +12V
* amplifier circuit
Q1 2 3 4 Q2N2222A
* resistive biasing network accounting for temperature dependence
Rc 1 2 4k TC=1200u
R1 1 3 80k TC=1200u
R2 3 0 40k TC=1200u
Re 4 5 3.3k TC=1200u
Vemitter 5 0 0
* transistor model statement for the 2N2222A
.model Q2N2222A NPN (Is=14.34f Xti=3 Eg=1.11 Vaf=74.03 Bf=255.9 Ne=1.307
+                    Ise=14.34f Ikf=.2847 Xtb=1.5 Br=6.092 Nc=2 Isc=0 Ikr=0 Rc=1
+                    Cjc=7.306p Mjc=.3416 Vjc=.75 Fc=.5 Cje=22.01p Mje=.377 Vje=.75
+                    Tr=46.91n Tf=411.1p Itf=.6 Vtf=1.7 Xtf=3 Rb=10)
** Analysis Requests **
* vary temperature of circuit beginning at 0C to 125C in temperature steps of 25C
.DC TEMP 0 125 25
** Output Requests **
.PLOT DC I(Vemitter)
.probe
.end
```

Figure 4.25 The Spice input file for investigating the dependence of the emitter current on temperature in the amplifier circuit shown in Fig. 4.23.

of temperature values is specified in exactly the same way as for a voltage or current source. For this particular example, we will monitor the emitter current over a temperature range beginning at 0°C and ending at 125°C in steps of 25°C. The exact syntax of this DC sweep command can be seen in the Spice input file for this example in Fig. 4.25.

The built-in model for the BJT accounts for variations in temperature. The same can not be said for the resistors. In order to account for changes in resistance due to temperature variations, one must indicate on each resistor statement in the Spice input file its temperature dependence using the two temperature coefficients TC_1 and TC_2. The formula used by Spice to calculate the the value of a resistor at a temperature (Temp) other than 27°C, is given by

$$R(Temp) = R(27°) \left[1 + TC_1(Temp - 27°) + TC_2(Temp - 27°)^2 \right]. \quad (4.11)$$

For our purposes here, we shall consider only the linear dependence of resistance on temperature, and therefore, consider $TC_2 = 0$. We shall assume for this example, that $TC_1 = 1200$ $ppm/°C$, where ppm denotes parts-per-million or a multiplication factor of 10^{-6}. To specify the dependence of a resistor on temperature in the Spice input file, one

4 Bipolar Junction Transistors (BJTs)

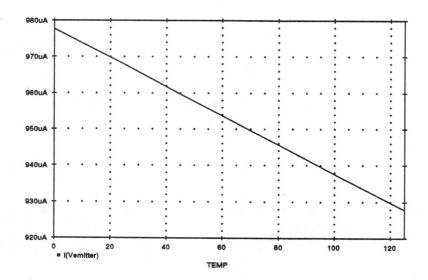

Figure 4.26 Temperature variation of the emitter current in the amplifier circuit shown in Fig. 4.23(a).

simply appends to the resistor statement, after the field indicating the nominal value of the resistor, TC=TC_1. In this particular case, we will write TC=1200u. This is also indicated in the Spice input file listed in Fig. 4.25.

The results of the emitter current dependence on temperature are summarized in Fig. 4.26. For a 125°C change in circuit temperature, the emitter current changed by no more than 50 μA. This is sometimes expressed as a temperature coefficient in units of $ppm/°C$ as the ratio of the change in emitter current divided by the emitter current at the nominal temperature of 27°C to the change in circuit temperature creating this current change. In this case, the emitter bias current has a temperature coefficient of -413 $ppm/°C$.

4.6 Basic Single-Stage BJT Amplifier Configurations

In this section we shall investigate the three basic types of amplifier configurations shown in Fig. 4.27: the common-emitter (CE), common-base (CB) and common-collector (CC). The primary goal of this section is to determine the small-signal parameters of these amplifiers, such as input resistance R_i, output resistance R_o, voltage gain A_v and current gain A_i, using the AC analysis facility of Spice. These results will then be compared to those predicted by the formulae derived by hand in Sedra and Smith using small-signal analysis. The transistor used in each amplifier of Fig. 4.27 will be assumed modeled after the commercial 2N2222A *npn* transistor.

4.6 Basic Single-Stage BJT Amplifier Configurations

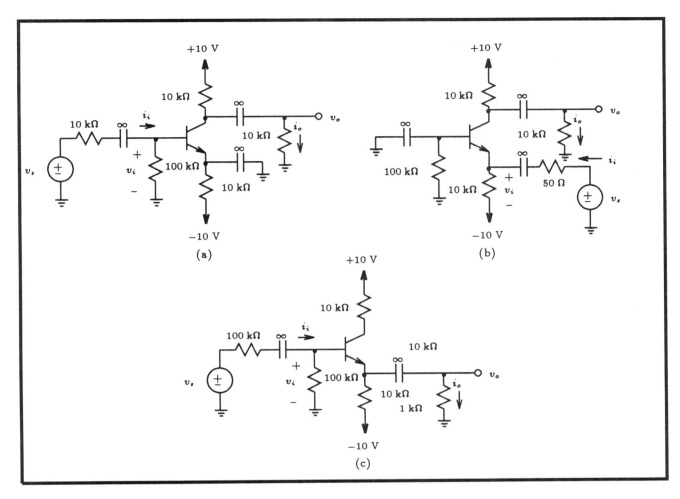

Figure 4.27 Basic single-stage BJT amplifier configurations: (a) common-emitter amplifier, (b) common-base amplifier and (c) common-collector amplifier.

The Common-Emitter Amplifier

As the first example of this section, let us compute the small-signal parameters of the common-emitter (CE) amplifier shown in Fig. 4.27(a). To obtain all four parameters, ie. A_v, A_i, R_i and R_o, we will have to run two separate Spice analyses; one for computing the input current and the output voltage for a known voltage applied to the input of the amplifier, and the other for computing the current supplied by a voltage source connected to the output terminal of the amplifier when the input voltage source is set to zero. These two situations are depicted in Fig. 4.28. In order to monitor the current flowing through the load resistor we placed a zero-valued voltage source in series with it, as shown in Fig. 4.28(a). The input current to the amplifier can be determine by monitoring the current supplied by the the input voltage source v_s. The four parameters of the amplifier can then be computed from these results according to the following:

4 Bipolar Junction Transistors (BJTs)

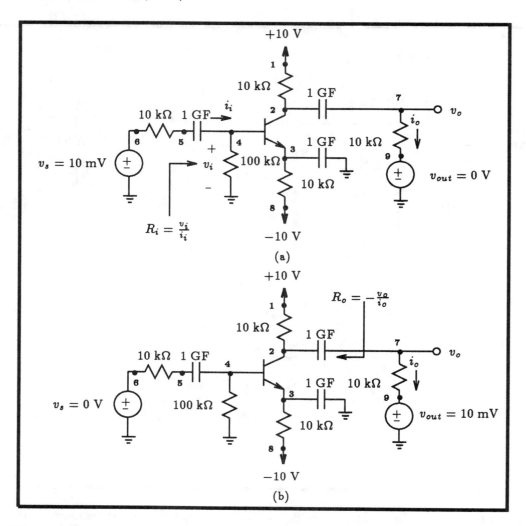

Figure 4.28 Circuit setups used to determine the amplifier's small-signal parameters: (a) circuit arrangement for computing i_i, v_o and i_o (b) circuit arrangement for computing i_o due to voltage source connected to output terminal of amplifier.

$$A_v = \frac{v_o}{v_s} \tag{4.12}$$

$$A_i = \frac{i_o}{i_i} \tag{4.13}$$

$$R_i = \frac{v_i}{i_i} \tag{4.14}$$

$$R_o = -\frac{v_o}{i_o} \tag{4.15}$$

For the first case depicted in Fig. 4.28(a), consider applying a 10 mV AC signal to the input of the amplifier. A DC input voltage signal would not be useful here since the input

4.6 Basic Single-Stage BJT Amplifier Configurations

Common-Emitter Amplifier Stage

```
** Circuit Description **
* power supplies
Vcc 1 0 DC +10V
Vee 8 0 DC -10V
* input signal
Vs 6 0 AC 10mV
Rs 5 6 10k
* amplifier
C1 4 5 1GF
Rb 4 0 100k
Q1 2 4 3 Q2N2222A
Rc 1 2 10k
Re 3 8 10k
C2 2 7 1GF
C3 3 0 1GF
* load + ammeter
Rl 7 9 10k
Vout 9 0 0
* transistor model statement for the 2N2222A
.model Q2N2222A NPN (Is=14.34f Xti=3 Eg=1.11 Vaf=74.03 Bf=255.9 Ne=1.307
+                    Ise=14.34f Ikf=.2847 Xtb=1.5 Br=6.092 Nc=2 Isc=0 Ikr=0 Rc=1
+                    Cjc=7.306p Mjc=.3416 Vjc=.75 Fc=.5 Cje=22.01p Mje=.377 Vje=.75
+                    Tr=46.91n Tf=411.1p Itf=.6 Vtf=1.7 Xtf=3 Rb=10)
** Analysis Requests **
* calculate DC bias point information
.OP
.AC LIN 1 1kHz 1kHz
** Output Requests **
*   voltage gain Av=Vo/Vs
.PRINT AC Vm(6) Vp(6) Vm(7) Vp(7)
*   current gain Ai=Io/Ii
.PRINT AC Im(Vs) Ip(Vs) Im(Vout) Ip(Vout)
*   input resistance Ri=Vi/Ii
.PRINT AC Vm(4) Vp(4) Im(Vs) Ip(Vs)
.end
```

Figure 4.29 The Spice input file for calculating the input current and the output voltage of the common-emitter amplifier shown in Fig. 4.28(a) subject to a 10 mV AC input signal.

source to the amplifier is AC-coupled. Thus, the frequency of the input signal should be selected to be in the midband frequency range of the amplifier. With the choice of decoupling and by-pass capacitors selected here (each selected very large at 1 GF),[†] an input frequency

---

[†] A 1 GF capacitor is impossible to buy! We use such a large capacitance in the simulation in order to eliminate the effect of the coupling capacitor on the performance of the circuit. The freedom to do that is a definite advantage of circuit simulation as compared to experimentation with actual electronic

4 Bipolar Junction Transistors (BJTs)

of 1 kHz is sufficiently midband. The Spice input file describing this circuit setup is provided in Fig. 4.29. Both a DC and an AC analysis requests are specified. The results of the AC analysis will be used to indirectly calculate the small-signal parameters of the amplifier, as mentioned above. The results of the DC analysis will provide us with the small-signal parameters of the transistor from which we can use the formulae derived in Sedra and Smith to predict the small-signal parameters of the amplifier and compare them to those computed by the indirect AC analysis approach.

Submitting the Spice input file for processing, we find on completion that the magnitude and phase of the input and output voltage of the amplifier is as follows:

```
****        AC ANALYSIS                        TEMPERATURE =   27.000 DEG C

    FREQ        VM(6)       VP(6)       VM(7)       VP(7)

    1.000E+03   1.000E-02   0.000E+00   5.068E-01   1.799E+02
```

Thus, we find that this amplifier has a midband voltage gain A_v of -50.68 V/V. Likewise, the input and output currents associated with this amplifier are found to be:

```
****        AC ANALYSIS                        TEMPERATURE =   27.000 DEG C

    FREQ        IM(Vs)      IP(Vs)      IM(Vout)    IP(Vout)

    1.000E+03   6.803E-07  -1.796E+02   5.068E-05   1.799E+02
```

These two current results indicate that the midband current gain A_i of this amplifier is -74.49 A/A. Finally, the input resistance of the CE amplifier can be computed from the following results,

```
****        AC ANALYSIS                        TEMPERATURE =   27.000 DEG C

    FREQ        VM(4)       VP(4)       IM(Vs)      IP(Vs)

    1.000E+03   3.198E-03  -9.300E-01   6.803E-07  -1.796E+02
```

Thus, the midband input resistance R_i of this amplifier is 4.7 kΩ.

Repeating this same process but with the input voltage source level set to 0 V and the level of the voltage source in series with the load resistance increased to 10 mV AC, we can determine the output resistance of this amplifier. This requires that we modify the statements for these two AC sources in the Spice deck seen listed in Fig. 4.29 to read as follows:

```
Vs   6  0  AC  0V
```

---

components.

4.6 Basic Single-Stage BJT Amplifier Configurations

Vout 9 0 AC 10mV

Further, to access the resulting current supplied by the output voltage source V_{out} and the voltage developed across the output terminal of the CE amplifier, we include the following PRINT statement in the Spice deck:

.PRINT AC Vm(7) Vp(7) Im(Vout) Ip(Vout)

Re-running the Spice job, we find the following results in the output file:

```
****      AC ANALYSIS                  TEMPERATURE =   27.000 DEG C

    FREQ       VM(7)       VP(7)       IM(Vout)    IP(Vout)

  1.000E+03   4.728E-03   0.000E+00   5.272E-07   1.800E+02
```

From these results we compute the output resistance R_o to be 8.97 kΩ.

To compare the above results with those predicted by hand analysis, we list below the expressions for A_v, A_i, R_i and R_o in terms of the small-signal model parameters of the transistor, as derived in section 4.11 of Sedra and Smith:

$$R_i = R_B \| r_\pi \tag{4.16}$$

$$R_o = R_C \| r_o \tag{4.17}$$

$$A_v \simeq -\frac{\beta(R_C \| R_L \| r_o)}{r_\pi + R_S} \tag{4.18}$$

$$A_i \simeq -\frac{\beta(R_C \| R_L \| r_o)}{R_L} \tag{4.19}$$

Here β is the small-signal transistor current gain[†] defined in terms of g_m and r_π according to:

$$\beta = g_m r_\pi \tag{4.20}$$

With the inclusion of the .DC operating point command in the above mentioned Spice decks we find that the parameters of the hybrid-pi model for the 2N2222A transistor of the common-emitter amplifier are as follows:

[†] Spice computes two different β's and denotes them β_{ac} and β_{dc}. In general they are not equal to one another. β_{ac} refers to the small-signal current gain equal to $g_m r_\pi$ and β_{dc} is the DC current gain equal to I_C/I_B. Moreover, β_{dc} depends on the collector current level I_C and its peak value corresponds with that specified on the BJT model statement using parameter Bf.

4 Bipolar Junction Transistors (BJTs)

Parameter	Formula	Hand Analysis	Spice	% Error
R_i	$R_B \| r_\pi$	4.69 kΩ	4.70 kΩ	-0.21%
R_o	$R_C \| r_o$	8.97 kΩ	8.97 kΩ	0%
A_v	$-\frac{\beta(R_C\|R_L\|r_o)}{r_\pi + R_S}$	-52.3 V/V	-50.68 V/V	3.2%
A_i	$-\frac{\beta(R_C\|R_L\|r_o)}{R_L}$	-78.0 A/A	-74.49 A/A	4.7%

Table 4.4 Comparing the small-signal amplifier parameters of the common-emitter amplifier shown in Fig. 4.27(a) as calculated by hand analysis and that indirectly computed using Spice.

```
**** BIPOLAR JUNCTION TRANSISTORS

NAME       Q1
MODEL      Q2N2222A
IB         5.84E-06
IC         8.72E-04
VBE        6.42E-01
VBC        -1.87E+00
VCE        2.51E+00
BETADC     1.49E+02
GM         3.36E-02
RPI        4.92E+03
RX         1.00E+01
RO         8.71E+04
```

Substituting the appropriate parameter value, together with values for the different circuit components, we find: $A_v = -52.3$ V/V, $A_i = -78.0$ A/A, $R_i = 4.69$ kΩ and $R_o = 8.97$ kΩ. To compare these with those computed indirectly by Spice above, we list in Table 4.4 a comparison of the results of these two approaches. The first and second column of this table list the small-signal parameters of the amplifier and its formula in terms of the small-signal transistor model parameters. The third and fourth columns indicate the value predicted by hand analysis and that obtained indirectly with Spice. The fifth column indicates the relative difference between the two results as a means of indicating how accurate the hand analysis is. As is evident from the results listed in this table, the results compare quite well.

To demonstrate the usefulness of small-signal analysis, let us apply a small-amplitude sinusoidal signal to the input of the amplifier and observe the time-varying signal that appears at the output of this amplifier. Say for the sake of illustration, we apply a 10 mV sinusoidal of 1 kHz frequency to the input of the amplifier. The Spice deck for this particular situation is provided in Fig. 4.30. A transient analysis command is specified to compute the output waveform over three cycles of the input signal beginning at a time of 5 ms. This is to ensure that the circuit has had sufficient time to reach steady-state. The dc-coupling and bypass

4.6 Basic Single-Stage BJT Amplifier Configurations

Common-Emitter Amplifier Stage With Sine-Wave Input

```
** Circuit Description **
* power supplies
Vcc 1 0 DC +10V
Vee 8 0 DC -10V
* input signal
Vs 6 0 SIN ( 0V 10mV 1kHz )
Rs 5 6 10k
* amplifier
C1 4 5 10uF
Rb 4 0 100k
Q1 2 4 3 Q2N2222A
Rc 1 2 10k
Re 3 8 10k
C2 2 7 10uF
C3 3 0 10uF
* load + ammeter
Rl 7 9 10k
Vout 9 0 0
* transistor model statement for the 2N2222A
.model Q2N2222A NPN (Is=14.34f Xti=3 Eg=1.11 Vaf=74.03 Bf=255.9 Ne=1.307
+                    Ise=14.34f Ikf=.2847 Xtb=1.5 Br=6.092 Nc=2 Isc=0 Ikr=0 Rc=1
+                    Cjc=7.306p Mjc=.3416 Vjc=.75 Fc=.5 Cje=22.01p Mje=.377 Vje=.75
+                    Tr=46.91n Tf=411.1p Itf=.6 Vtf=1.7 Xtf=3 Rb=10)
** Analysis Requests **
.OP
.TRAN 50us 8ms 5ms 50us
** Output Requests **
.Plot TRAN V(7) V(6)
.probe
.end
```

Figure 4.30 The Spice input file for calculating the time-varying output signal of the common-emitter amplifier subject to a 10 mV, 1 kHz sinusoidal input signal. Unlike the previous AC analysis, which is a small-signal analysis, a transient analysis performs a large signal analysis even when the input level is small.

capacitors have been assigned more practical values of 10 μF each, unlike those used in the previous AC analysis. In this way, the time to reach steady-state is drastically reduced. One should note that by the very nature of a Spice transient analysis, the analysis performed here is a large-signal analysis and not a small-signal analysis even though the input signal level is small.

The results of the transient analysis are shown in Fig. 4.31. The top graph displays the 10 mV peak, 1 kHz input sinusoidal signal and the bottom graph displays the corresponding output signal. With the aid of Probe, we found that the amplitude of the output signal is 500 mV. Thus, the voltage gain of this amplifier is seen to be about −50 V/V, which agrees

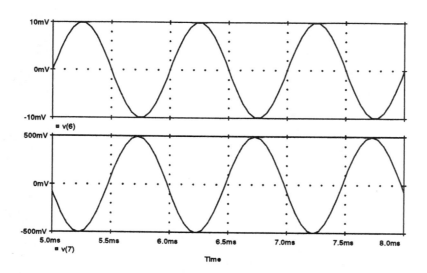

Figure 4.31 The input and output time-varying voltage waveforms associated with the common-emitter amplifier shown in Fig. 4.27(a).

with that predicted by the two analyses above.

The Common-Base Amplifier

Another important amplifier configuration is the common-base (CB) amplifier shown in Fig. 4.27(b). Here the base of the transistor is AC grounded, the input signal source is connected to the emitter, and the load is connected to the collector. In the following we shall repeat the analysis method carried out on the common-emitter amplifier using Spice and compare the results with those estimated by the small-signal formulas derived by hand analysis.

The Spice input file for this common-base amplifier is shown in Fig. 4.32. Like the previous case of the common-emitter amplifier, a zero-valued voltage source is placed in series with the 10 kΩ load resistor. This voltage source is used to monitor the current supplied to the load resistor. It will also be used to determine the output resistance of the amplifier. Both a DC and an AC analysis are specified. The DC analysis will provide us with the small-signal model parameters of the transistor. The AC analysis will enable us to determine the current supplied to the amplifier by the input voltage source, and the voltage and current supplied to the load resistor.

Submitting the Spice input file for processing, we find on completion that the magnitude and phase of the input and output voltage of the amplifier are as follows:

4.6 Basic Single-Stage BJT Amplifier Configurations

```
Common-Base Amplifier Stage

** Circuit Description **
* power supplies
Vcc 1 0 DC +10V
Vee 8 0 DC -10V
* input signal
Vs 6 0 AC 10mV
Rs 5 6 50
* amplifier
C1 4 0 1GF
Rb 4 0 100k
Q1 2 4 3 Q2N2222A
Rc 1 2 10k
Re 3 8 10k
C2 2 7 1GF
C3 3 5 1GF
* load + ammeter
Rl 7 9 10k
Vout 9 0 0
* transistor model statement for the 2N2222A
.model Q2N2222A NPN (Is=14.34f Xti=3 Eg=1.11 Vaf=74.03 Bf=255.9 Ne=1.307
+                    Ise=14.34f Ikf=.2847 Xtb=1.5 Br=6.092 Nc=2 Isc=0 Ikr=0 Rc=1
+                    Cjc=7.306p Mjc=.3416 Vjc=.75 Fc=.5 Cje=22.01p Mje=.377 Vje=.75
+                    Tr=46.91n Tf=411.1p Itf=.6 Vtf=1.7 Xtf=3 Rb=10)
** Analysis Requests **
* calculate DC bias point information
.OP
.AC LIN 1 1kHz 1kHz
** Output Requests **
*   voltage gain Av=Vo/Vs
.PRINT AC Vm(6) Vp(6) Vm(7) Vp(7)
*   current gain Ai=Io/Ii
.PRINT AC Im(Vs) Ip(Vs) Im(Vout) Ip(Vout)
* input resistance Ri=Vi/Ii
.PRINT AC Vm(5) Vp(5) Im(Vs) Ip(Vs)
.end
```

Figure 4.32 The Spice input file for calculating the input current and the output voltage of the common-base amplifier shown in Fig. 4.27(b) subject to a 10 mV AC input signal.

```
****        AC ANALYSIS                      TEMPERATURE =   27.000 DEG C

    FREQ       VM(6)      VP(6)      VM(7)      VP(7)

  1.000E+03  1.000E-02  0.000E+00  6.096E-01  -6.826E-05
```

Therefore we determine that this common-base amplifier has a midband voltage gain A_v of +60.96 V/V. Likewise, the input and output current associated with this amplifier are found

4 Bipolar Junction Transistors (BJTs)

to be:

```
****    AC ANALYSIS                    TEMPERATURE =   27.000 DEG C

   FREQ        IM(Vs)      IP(Vs)     IM(Vout)    IP(Vout)

  1.000E+03   1.231E-04   1.800E+02   6.096E-05   -6.826E-05
```

Thus the input current to the amplifier is 123.1 μA and the current supplied to the load is 60.96 μA, giving a current gain 0.495 A/A. Finally, the input resistance of the CB amplifier can be computed from the following results,

```
****    AC ANALYSIS                    TEMPERATURE =   27.000 DEG C

   FREQ        VM(5)       VP(5)       IM(Vs)      IP(Vs)

  1.000E+03   3.845E-03   1.819E-03   1.231E-04   1.800E+02
```

Thus, the midband input resistance R_i is 31.2 Ω.

Repeating this same process but with the input voltage source level set to 0 V and the level of the voltage source in series with the load resistance increased to 10 mV AC, we find that the output resistance R_o of this amplifier is 9.58 kΩ. This was found from another Spice run of a modified version of the Spice deck shown in Fig. 4.32 where the two AC input and output voltage sources were modified to read as follows:

```
Vs    6 0 AC 0V
Vout  9 0 AC 10mV
```

and the following PRINT statement was added to the Spice deck:

```
.PRINT AC Vm(7) Vp(7) Im(Vout) Ip(Vout)
```

The results of this analysis are then found below:

```
****    AC ANALYSIS                    TEMPERATURE =   27.000 DEG C

   FREQ        VM(7)       VP(7)      IM(Vout)    IP(Vout)

  1.000E+03   4.894E-03   0.000E+00   5.106E-07   1.800E+02
```

As a means of comparison we summarizes in Table 4.5 the small-signal formulae pertinent to the common-base amplifier derived by hand in Sedra and Smith, together with their values as calculated by substituting the relevant transistor small-signal model parameters into each equation. The small-signal model parameters for the 2N2222A transistor are identical to those found in the common-emitter case in the previous subsection. This is not too surprising given that the transistor of the common-base amplifier is biased in exactly the same manner as in the common-emitter amplifier. This was also confirmed by the DC analysis performed

Parameter	Formula	Hand Analysis	Spice	% Error
R_i	r_e	29.8 Ω	31.2 Ω	−4.5%
R_o	R_C	10 kΩ	9.58 kΩ	4.4%
A_v	$\frac{R_C \| R_L}{r_e + R_S}$	62.6 V/V	60.9 V/V	2.7%
A_i	$\alpha \frac{R_C \| R_L}{R_L}$	0.5 A/A	0.49 A/A	1.0%

Table 4.5 Comparing the small-signal parameters of the common-base amplifier shown in Fig. 4.27(b) as calculated by hand analysis to those indirectly computed using Spice.

Parameter	Formula	Hand Analysis	Spice	% Error
R_i	$R_B \| (\beta+1)\left[r_e + R_E \| r_o \| R_L\right]$	60.7 kΩ	58.2 kΩ	4.2%
R_o	$R_E \| \left[r_e + (R_S \| R_B)/(\beta+1)\right]$	320.4 kΩ	354.1 Ω	−9.5%
A_v	$\frac{R_i}{R_i + R_S}$	0.37 V/V	0.36 V/V	3.4%
A_i	$\frac{R_i}{R_L}$	60.7 A/A	56.2 A/A	8.0%

Table 4.6 Comparing the small-signal parameters of the common-collector amplifier shown in Fig. 4.27(c) as calculated by hand analysis to those indirectly computed using Spice.

by Spice on the common-base amplifier. A fourth column is also included in this table listing the small-signal parameters computed from the Spice analysis above. As is evident from the fifth column, indicating the relative error between the two methods of estimating the amplifier small-signal parameters, we see that they are in good agreement.

The Common-Collector Amplifier

This same analysis can be repeated for the common-collector (CC) amplifier configuration shown in Fig. 4.27(c). Rather than list much of the same as that seen previously, we simply summarize the results in Table 4.6. As before, the Spice results are in good agreement with those obtained with hand analysis.

4.7 The Transistor As a Switch

As the final example of this chapter, consider the simple transistor switching circuit shown in Fig. 4.33. This particular circuit was designed by Sedra and Smith in Example 4.14 of their text such that transistor Q_1 is in saturation with an overdrive factor larger

4 Bipolar Junction Transistors (BJTs)

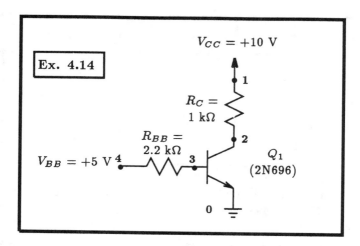

Figure 4.33 A simple transistor switching circuit (Example 4.14 of Sedra and Smith).

than 10 when the transistor has a minimum β of 50. With the aid of Spice, we would like to determine the overdrive factor of the transistor in this circuit when the transistor is assumed to be the commercial 2N696 *npn* device and the base driving voltage V_{BB} equals 5 V.

The overdrive factor (OD) of a saturated transistor is defined as the ratio of the base current I_B divided by the minimum base current that will force the transistor into saturation, denoted as I_{Bsat}. Thus,

$$OD = I_B/I_{Bsat}. \qquad (4.21)$$

Correspondingly, at the minimum base current condition, the transistor is beginning to conduct a collector current I_{Csat}. Moreover, as the device is driven further into saturation the collector current remains fairly constant at I_{Csat}. Thus, the point at which the collector current begins to saturates denotes the edge of the saturation region. Using Spice we shall compute the input condition that gives rise to the transistor entering its saturation region in the circuit of Fig. 4.33. This can be accomplished by sweeping the base driving voltage V_{BB} between ground and 5 volts and observing both the collector current and the base current of the transistor. From this, we can then determine the base current I_{Bsat}. Moreover, at $V_{BB} = 5$ V, we can determine the base current I_B and thus the overdrive factor OD for this particular circuit.

The Spice input file for the switching circuit shown in Fig. 4.33 is listed in Fig. 4.34. A DC sweep analysis is requested that varies the input voltage V_{BB} between 0 and 5 volts in 0.1 volt increments. The BJT model parameters for the 2N696 *npn* transistor are also listed. These were derived from nominal device characteristics.

The results of the Spice analysis are shown plotted in Fig. 4.35. The upper trace displays

Example 4.14: Using The 2N696 As A Switch

```
** Circuit Description **
* power supplies
Vcc 1 0 DC +10V
* input signal
Vbb 4 0 DC +5V
* transistorized switch
Q1 2 3 0 Q2N696
Rc 1 2 1k
Rb 4 3 2.2k
* transistor model statement for the 2N696
.model Q2N696 NPN (Is=14.34f Xti=3 Eg=1.11 Vaf=74.03 Bf=65.62 Ne=1.208
+                  Ise=19.48f Ikf=.2385 Xtb=1.5 Br=9.715 Nc=2 Isc=0 Ikr=0 Rc=1
+                  Cjc=9.393p Mjc=.3416 Vjc=.75 Fc=.5 Cje=22.01p Mje=.377 Vje=.75
+                  Tr=58.98n Tf=408.8p Itf=.6 Vtf=1.7 Xtf=3 Rb=10)
** Analysis Requests **
.DC Vbb 0V 5V 0.1V
** Output Requests **
.PLOT DC I(Vcc) I(Vbb)
.probe
.end
```

Figure 4.34 The Spice input file for calculating the overdrive factor associated with the switching circuit shown in Fig. 4.33. The transistor is modeled after the commercial *npn* transistor 2N696.

the transistor collector current I_C as a function of the base driving voltage V_{BB}, and the lower trace displays the corresponding base current for the same input voltage. Using the cursor feature in Probe we find from the plot of collector current that the collector current begins to saturate at a current level of 9.79 mA when the base driving voltage V_{BB} exceeds 1.4 V. Correspondingly, from the trace of the transistor base current, we find that at an input voltage V_{BB} of 1.4 V, the base current is 313.6 μA. Hence, $I_{Bsat} = 313.6$ μA. Also, we find at an input voltage level of $V_{BB} = 5$ V, the transistor base current I_B is 1.932 mA. Therefore, using Eqn. (4.21), we compute the overdrive factor OD for transistor Q_1 to be 6.16.

4.8 Spice Tips

- A BJT is described to Spice using an element statement and a model statement.
- The element statement describes how the base, collector, and emitter of a transistor are connected to the rest of the network.
- The model statement contains a list of parameters describing the terminal characteristics of a BJT using the built-in model of Spice.

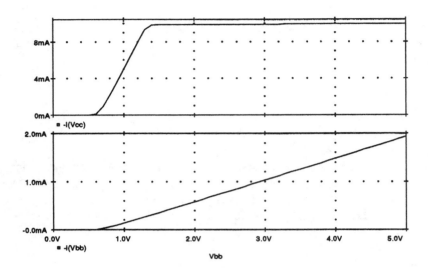

Figure 4.35 The collector and base current characteristics of the transistor in the circuit of Fig. 4.34 as functions of the base driving voltage V_{BB}. The top curve displays the collector current and the bottom trace displays the base current.

- There are 40 parameters associated with the built-in Spice model of the BJT. (We discussed only 6 of them, the remainder are described in the Appendix).
- A specific transistor mode of operation is deduced from the DC voltages that appears across its emitter-base junction and collector-base junction. This information can be found in the Spice output file as a result of an operating point (.OP) analysis.
- When an operating point (.OP) calculation is included in the Spice input file, a list of operating point information for each transistor is obtained. This information contains DC bias conditions for the transistor and the parameters associated with its small-signal model.
- Models of many commercial bipolar transistors are available from various semiconductor manufacturers. A library of transistor models for various commercial transistors are available with the PSpice program.
- Spice can be used to generate families of i-v curves for transistors, just like the laboratory curve-tracer instrument.
- A DC sweep command can be extended to include a sweep of temperature, allowing one to investigate the circuits dependence on temperature.
- The pulsewidth of a time-varying waveform generated by the PULSE source statement of Spice can not be made equal to zero. Instead, a value that is much smaller, at least several orders of magnitude less, than the period of the waveform is assigned to the

pulsewidth.

- The DC sensitivities of a circuit can be computed using the .SENS analysis command. This command will compute the sensitivities of a particular circuit variable to most of the components in the circuit. This includes the sensitivities of many of the model parameters of the BJT.
- If the effect of decoupling and by-pass capacitors is to be ignored then their values should be selected to be very large during the simulation (generally, a good value is 1 GF).

4.9 Bibliography

L. W. Nagel, *SPICE2 - A computer program to simulate semiconductor circuits*, Memorandum no. ERL-M520, May 1975, Electronic Research Laboratory, University of California, Berkeley.

N. R. Malik, Determining Spice parameter values for BJT's, *IEEE Transaction on Education*, vol. 33, No. 4, pp. 366-368, Nov. 1990.

4.10 Problems

4.1 Consider the case of a transistor whose base is connected to ground, the collector is connected to a 10 V dc source through a 1 kΩ resistor, and a 5 mA current source is connected to the emitter with the polarity so that current is drawn out of the emitter terminal. If $\beta_F = 100$ and $I_S = 10^{-14}$ A, find the voltages at the emitter and the collector using Spice. Further, determine the base current.

4.2 Using the Spice model parameters for the commercial 2N2222A *npn* BJT given in Section 4.5, determine its leakage current I_{CBO} at room temperature (ie. 27°C). If the temperature of the device is raised to 75°C, what is the new I_{CBO}? *Hint: Connect the base to ground, the collector to +5 V, and do not connect the emitter terminal. The current supplied by the +5 V voltage source is I_{CBO}.*

4.3 Consider a *pnp* transistor having $I_S = 10^{-13}$ A and $\beta_F = 40$. If the emitter is connected to ground, the base is connected to a current source that pulls out of the base terminal a current of 10 μA, and the collector is connected to a negative supply of −10 V via a 10 kΩ resistor, find the base voltage, the collector voltage, and the emitter current using the operating point (.OP) command of Spice.

4.4 Using Spice as a curve tracer, plot the i_C - v_{CE} characteristics of an *npn* transistor having $I_S = 10^{-15}$ A and $V_A = 100$ V. Provide curves for $v_{BE} = 0.65, 0.70, 0.72, 0.73$ and 0.74 volts. Show the characteristics for v_{CE} up to 15 V.

4 Bipolar Junction Transistors (BJTs)

4.5 For a BJT having an Early voltage of 200 V, what is its output resistance at 1 mA as calculated by Spice? at 100 μA?

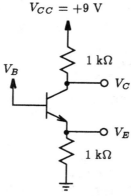

Fig. P4.6

4.6 The transistor in the circuit of Fig. P4.6 has a very high β_F (assume at least 10^6 in the Spice file). The other parameters of the transistor Spice model can assume default values. Find V_E and V_C for V_B equal to (a) +3 V, (b) +1 V, and (c) 0 V. What is the transistor mode of operation in each case?

4.7 For the circuit in Fig. P4.6, with the aid of Spice and V_B set equal to +2 V, find all node voltages for (a) β_F very high ($> 10^6$), and (b) $\beta_F = 99$.

4.8 In the circuit of Fig. 4.16, the input signal v_i is described by $0.004\sin(\omega t)$ volts and the transistor is assumed modeled after the 2N2222A type. In addition, V_{BB} is reduced from 3 V to 1 V. Using Spice, plot the base and collector current of Q_1 as a function of time for at least one period of the input signal. Likewise, plot the collector voltage of Q_1. What is the voltage gain of this amplifier?

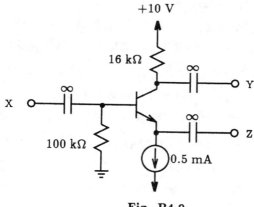

Fig. P4.9

4.9 The transistor shown in the circuit of Fig. P4.9 has $\beta_F = 100$ and $V_A = 80$ V. Use Spice to answer the following questions, but also compare your results with those obtained through a hand analysis:

(a) Find the dc voltages at the base, emitter, and collector using Spice. Represent the infinite-valued capacitors with a very large, but finite, value.

(b) Find g_m, r_π, and r_o.

(c) If terminal Z is connected to ground, X to signal source V_s having a 10 kΩ source resistance, what is the voltage gain v_y/v_s. Since the capacitor connected to node Y is left floating, either connected a large load to node Y or remove this capacitor and take the measurement directly from the collector of the transistor.

(d) If terminal X is connected to ground, Z to an input signal source v_s having a 200 Ω source resistance, and Y to a load resistance of 10 kΩ, find the voltage gain v_y/v_s.

(e) If terminal Y is connected to ground, X to an input signal source v_s having a 100 kΩ source resistance, and Z to a load resistance of 1 kΩ, find the voltage gain v_z/v_s.

4.10 With the aid of Spice, verify the design of a version of the circuit in Fig. 4.23 that uses ±9.5 V supplies for operation at 10 mA (for high β_F), such that the total variation in I_E is less than 5% for β_F as low as 50. To obtain the highest possible voltage gain, select the largest possible value for R_C; however, you must ensure that V_{CE} is never less than 2 V.

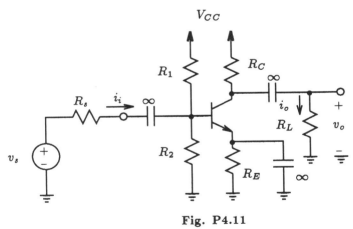

Fig. P4.11

4.11 For the common-emitter amplifier shown in Fig. P4.11, let $V_{CC} = 9$ V, $R_1 = 27$kΩ, $R_2 = 15$kΩ, $R_E = 1.2$kΩ, and $R_C = 2.2$kΩ. The transistor has $\beta_F = 100$ and $V_A = 100$ V. Using Spice, calculate the dc bias current I_E. If the amplifier operates between a source for which $R_s = 10$kΩ and a load of 2 kΩ, determine the values of R_i, G_m, R_o, A_v, and A_i.

4.12 Repeat Problem 4.11 with the transistor modeled after the commercial transistor 2N696. Use the Spice parameters for the 2N696 provided in Section 4.6.

4 Bipolar Junction Transistors (BJTs)

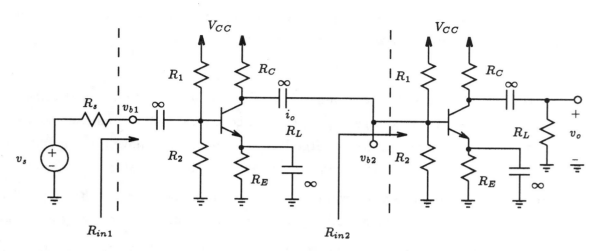

Fig. P4.13

4.13 The amplifier of Fig. P4.13 consists of two identical common-emitter amplifiers connected in cascade.

(a) For $V_{CC} = 15$ V, $R_1 = 100\text{k}\Omega$, $R_2 = 47\text{k}\Omega$, $R_E = 3.9\text{k}\Omega$, $R_C = 6.8\text{k}\Omega$, and $\beta_F = 100$, determine the dc collector current and collector voltage of each transistor using Spice. The amount of data that must be typed into the computer is reduced if a subcircuit is created for one of the stages and used twice to form the overall amplifier.

(b) Find R_{in1} and v_{b1}/v_s for $R_s = 5\text{k}\Omega$.

(c) Find R_{in2} and v_{b2}/v_{b1} for $R_s = 5\text{k}\Omega$.

(d) For $R_L = 2\text{k}\Omega$, find v_o/v_s.

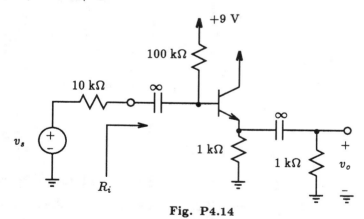

Fig. P4.14

4.14 For the emitter-follower circuit shown in Fig. P4.14 the BJT used is specified to have β_F values in the range of 20 to 200. For the two extreme values of β find:

(a) I_E, V_E and V_B.

(b) the input resistance R_i.

(c) the voltage gain v_o/v_s.

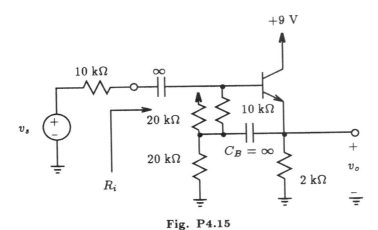

Fig. P4.15

4.15 For the bootstrapped follower circuit shown in Fig. P4.15 where the transistor is assumed to be of the 2N2222A type, compare the input and output resistance, and the voltage gain v_o/v_s, with and without the bootstrapping capacitor C_B in place.

4.16 One small-signal BJT model parameter that is used in the small-signal calculations of Spice but is not sent to the Spice output file when an .OP command is specified is r_μ. Devise a simple circuit arrangement for calculating this small-signal resistance with Spice.

Chapter 5

Field-Effect Transistors (FETs)

In this chapter we shall show how Spice is used to simulate circuits containing field-effect transistors (FETs). Spice has built-in models for two of the three FET types considered here, metal-oxide-semiconductor FETs (MOSFETs) and junction FETs (JFETs). In the case of metal-semiconductor FETs (MESFETs), we shall carry out our circuit simulations using the built-in model of PSpice. Various circuit examples involving the three types of FETs will be given.

5.1 Describing MOSFETs To Spice

MOSFETs are described to Spice using two statements; one statement describes the nature of the FET and its connections to the rest of the circuit, and the other specifies the values of the parameters of the built-in FET model. The following outlines the syntax of these two statements, including some details on the built-in "Level 1" MOSFET model of Spice.

5.1.1 MOSFET Element Description

The presence of a MOSFET in a circuit is described to Spice through the Spice input file using an element statement beginning with the letter M. If more than one MOSFET exists in a circuit then a unique name must be attached to M to uniquely identify each transistor.

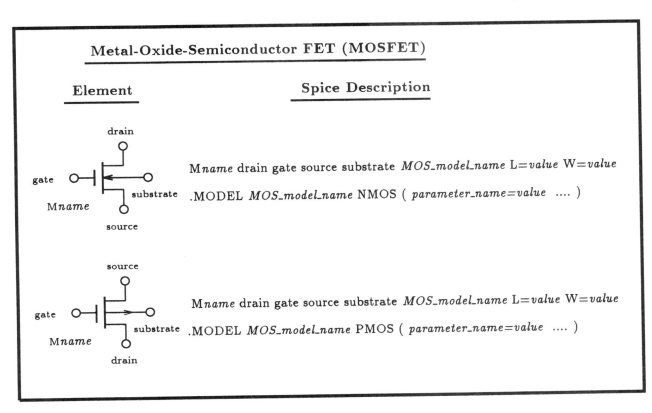

Figure 5.1 Spice element description for the NMOS and PMOS MOSFETs. Also listed is the general form of the associated MOSFET model statement. A partial listing of the parameter values applicable to either the NMOS or PMOS MOSFET is given in Table 5.1. Enhancement or depletion mode of operation is determined by the values assigned to these parameters.

This is then followed by a list of the nodes that the drain, gate, source, and substrate (body) of the MOSFET are connected to. Subsequently, on the same line, the name of the model that will be used to characterize a particular MOSFET is given. The name of this model must correspond to the name given on a model statement containing the parameter values that characterize this MOSFET to Spice. Finally, the length and width of the MOSFET are given. For quick reference, we depict in Fig. 5.1 the syntax for the Spice statement describing the MOSFET. Also listed is the syntax for the model statement (.MODEL) that must be present in any Spice input file that makes reference to the built-in MOSFET model of Spice. This statement specifies the terminal characteristics of the MOSFET by defining the values of particular parameters in the MOSFET model. Parameters of the model not specified are assigned default values by Spice. We shall briefly discuss the model statement next.

5 Field-Effect Transistors (FETs)

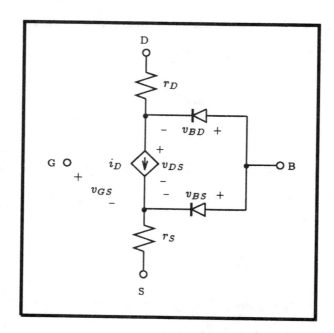

Figure 5.2 The general form of the Spice large-signal model for an n-channel MOSFET under static conditions.

5.1.2 MOSFET Model Description

As is evident from Fig. 5.1, the model statement for either the NMOS or PMOS transistor begins with the keyword .MODEL and is followed by the name of the model used by a MOSFET element statement, the nature of the MOSFET (ie. NMOS or PMOS), and a list giving the values of the model parameters (enclosed between brackets). The number of parameters associated with the Spice model of the MOSFET is large and their meaning complicated; besides, Spice has more than one large-signal model for the MOSFET. These are classified according to their levels of sophistication: Level 1, 2 and 3. The simplest MOSFET model being that described in Level 1. For the most part, MOSFET behavior described in Chapter 5 of Sedra and Smith is based on the Level 1 MOSFET model. The other two models are more complicated and their mathematical description will not be discussed here. Rather, we shall just mention their important differences. The Level 2 MOSFET model is a more complex version of the LEVEL 1 model which includes extensive second-order effects, largely dependent on the geometry of the MOSFET. The Level 3 MOSFET model of Spice is a semi-empirical model (having some model parameters that are not necessarily physically based), especially suited to short-channel MOSFETs (ie. $L \leq 5$ μm). For more details on these models, interested readers can consult reference [Vladimirescu, 1980]. Also, an advanced VLSI textbook by Geiger, Allen and Strader [Geiger, Allen and Strader, 1990] provides a good review of the three MOSFET models of Spice.

5.1 Describing MOSFETs To Spice

The general form of the DC Spice model for an n-channel MOSFET is illustrated schematically in Fig. 5.2. The bulk resistance of both the drain and source regions of the MOSFET are lumped into two linear resistances r_D and r_S, respectively. The DC characteristic of the intrinsic MOSFET is determined by the nonlinear dependent current source i_D, and the two diodes represent the two substrate junctions that define the channel region. A similar model exists for the p-channel device; the direction of the diodes, the current source and the polarities of the terminal voltages are all reversed. The mathematical relationship that describes the DC behavior of the dependent current source varies depending on which level of model is used. For the LEVEL 1 MOSFET model, the expression for drain current i_D, assuming that the drain is at a higher potential than the source, is described by the following:[†]

$$i_D = \begin{cases} 0 & \text{for } v_{GS} < V_t \\ K\left[2(v_{GS} - V_t)v_{DS} - v_{DS}^2\right](1 + \lambda v_{DS}) & \text{for } v_{GS} > V_t \text{ and } v_{DS} \leq v_{GS} - V_t \\ K(v_{GS} - V_t)^2(1 + \lambda v_{DS}) & \text{for } v_{GS} > V_t \text{ and } v_{DS} \geq v_{GS} - V_t \end{cases} \quad (5.1)$$

where the device constant K is related to process parameters and device geometry according to

$$K = \frac{1}{2}\mu C_{OX}\left(\frac{W}{L}\right) \quad (5.2)$$

and the threshold voltage V_t is given by

$$V_t = V_{t0} + \gamma\left[\sqrt{2\phi_f + V_{SB}} - \sqrt{2\phi_f}\right]. \quad (5.3)$$

Here we see that the drain current equations are determined by the eight parameters: W, L, μ, C_{OX}, V_{t0}, λ, γ and $2\phi_f$. Both W and L define the dimensions of the device. These two parameters are usually specified on the element statement of the MOSFET, although, if none is specified, Spice will assume that both W and L are 100 μm. Parameters μ and C_{OX} are process-related parameters that are multiplied together to form the process transconductance coefficient *kp*. It is *kp* that is usually specified in the parameter list of the model statement. The parameter V_{t0} is the zero-bias threshold voltage. V_{t0} is positive for enhancement-mode n-channel MOSFETs and depletion-mode p-channel MOSFETs. But, V_{t0} is negative for depletion-mode n-channel MOSFETs and enhancement-mode p-channel MOSFETs. The

---

[†] The expression for the drain current of the MOSFET in the triode region differs from that in Sedra and Smith in that it includes the factor $(1 + \lambda v_{DS})$. This ensures mathematical continuity between the triode and saturation regions.

5 Field-Effect Transistors (FETs)

Symbol	Spice Name	Model Parameter	Units	Default
	Level	Model type		1
μC_{OX}	kp	Transconductance coefficient	A/V^2	20μ
V_{t0}	Vto	Zero-bias threshold voltage	V	0
λ	lambda	Channel-length modulation	V^{-1}	0
γ	gamma	Body-effect parameter	V$^{1/2}$	0
$2\phi_f$	phi	Surface potential	V	0.6
r_D	Rd	Drain ohmic resistance	Ω	0
r_S	Rs	Source ohmic resistance	Ω	0

Table 5.1 A partial listing of the Spice parameters for the LEVEL 1 MOSFET model.

parameter λ is the channel-length modulation parameter and represents the influence that drain-source voltage has on the drain current i_D when the device is in saturation. In Spice, the sign of this parameter is always positive, regardless of the nature of the device type. This is unlike the convention adopted by Sedra and Smith. The last two parameters, γ and $2\phi_f$, are the body-effect parameter and the surface potential, respectively.

A partial listing of the parameters associated with the Spice MOSFET model under static conditions is given in Table 5.1. Also listed are default values which a parameter assumes if a value is not specified for it on the .MODEL statement. To specify a parameter value one simply writes, for example: level=1, kp=20u, Vto=1V, etc.

5.1.3 An Enhancement-Mode N-Channel MOSFET Circuit

In the following we shall calculate the DC conditions of a circuit containing a FET using Spice. Both the DC node voltages of the circuit and the DC operating point information of the FET will be determined.

Consider the circuit shown in Fig. 5.3, the particulars of this design were presented in Example 5.1 of Sedra and Smith. Here we would like to confirm that the FET is indeed biased at a current level of 0.4 mA and that the voltage appearing at the drain is +1 V. The NMOS transistor is assumed to have $V_t = 2$ V, $\mu_n C_{OX} = 20$ μA/V^2, $L = 10$ μm, and $W = 400$ μm. Furthermore, the channel-length modulation effect is assumed zero (ie. $\lambda = 0$). Assuming a level 1 MOSFET model, we can create the following Spice model statement for this FET using the above information as:

```
.model nmos_enhancement_mosfet nmos (kp=20u Vto=+2V lambda=0)
```

5.1 Describing MOSFETs To Spice

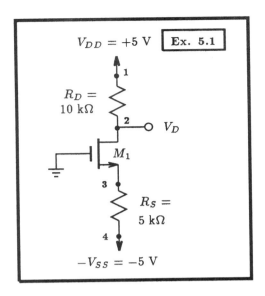

Figure 5.3 A MOSFET circuit example.

Example 5.1: Simple Enhancement-Mode MOSFET Circuit

```
** Circuit Description **
* dc supplies
Vdd 1 0 DC +5V
Vss 4 0 DC -5V
* MOSFET circuit
M1 2 0 3 3 nmos_enhancement_mosfet L=10u W=400u
Rd 1 2 10k
Rs 3 4 5k
* mosfet model statement (by default, level 1)
.model nmos_enhancement_mosfet nmos (kp=20u Vto=+2V lambda=0)
** Analysis Requests **
* calculate DC bias point
.OP
** Output Requests **
* none required
.end
```

Figure 5.4 Spice input file for computing the DC operating point of the circuit shown in Fig. 5.3.

Here, we have labeled the name of this model as: *nmos_enhancement_mosfet*. The meaning behind this name should be obvious. Transistor M_1, and its geometry, is then described to Spice using the following element statement:

M1 2 0 3 3 nmos_enhancement_mosfet L=10u W=400u

where we have assumed that the substrate (body) is connected to the source. Combining

5 Field-Effect Transistors (FETs)

these statements with the ones for the other devices of the circuit shown in Fig. 5.3, results in the Spice input file shown in Fig. 5.4. Submitting this file with an operating point (.OP) analysis request to Spice, results in the following DC circuit information:

```
****    SMALL SIGNAL BIAS SOLUTION      TEMPERATURE =   27.000 DEG C
*****************************************************************************

    NODE    VOLTAGE     NODE    VOLTAGE     NODE    VOLTAGE     NODE    VOLTAGE

    (  1)    5.0000   (  2)    1.0000   (  3)   -3.0000   (  4)   -5.0000

        VOLTAGE SOURCE CURRENTS
        NAME         CURRENT

        Vdd         -4.000E-04
        Vss          4.000E-04

        TOTAL POWER DISSIPATION   4.00E-03  WATTS

****    OPERATING POINT INFORMATION    TEMPERATURE =   27.000 DEG C
*****************************************************************************

**** MOSFETS

    NAME         M1
    MODEL        nmos_enhancement_mosfet
    ID           4.00E-04
    VGS          3.00E+00
    VDS          4.00E+00
    VBS          0.00E+00
    VTH          2.00E+00
    VDSAT        1.00E+00
```

As expected, these results confirm that the FET is biased at the intended current level of 0.4 mA, and also that the drain of the FET is at the correct voltage level of +1 V. As a further note, we see from the above results that M_1 is biased in its saturation region because $V_{DS} > V_{DS_{SAT}}$ where Spice uses the notation $V_{DS_{SAT}} = V_{GS} - V_t$.

5.1.4 Observing The MOSFET Current – Voltage Characteristics

The $i_D - v_{DS}$ characteristics of a MOSFET are easily obtained by sweeping the drain-to-source voltage through a range of DC voltages, all the mean while, the gate-to-source voltage is held constant at some voltage value. The drain current of the MOSFET is then monitored and plotted against the drain-source voltage. To demonstrate how one performs this operation on a MOSFET using Spice, consider the circuit shown in Fig. 5.5. Here two independent voltage sources, V_{GS} and V_{DS}, will be used to establish the different bias conditions on the enhancement-mode n-channel MOSFET whose source and body are connected together. It is assumed that the MOSFET has the same geometry and device parameters that were previously mentioned in the last subsection. Repeating them here, the NMOS transistor is characterized by $V_t = +2$ V, $\mu_n C_{OX} = 20$ μA/V^2, $L = 10$ μm, and $W = 400$ μm.

5.1 Describing MOSFETs To Spice

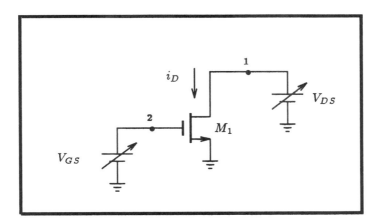

Figure 5.5 Spice curve-tracer arrangement for calculating the $i - v$ characteristics of a MOSFET. The $i_D - v_{DS}$ characteristic of the MOSFET is obtained by sweeping V_{DS} through a range of voltages while keeping V_{GS} constant at some value. The $i_D - v_{GS}$ characteristic of the MOSFET is obtained by sweeping V_{GS} through a range of values while V_{DS} is held constant.

```
Enhancement-Mode N-Channel MOSFET Id - Vds Characteristics

** Circuit Description **
* bias conditions
Vds 1 0 DC +10V      ; this value is arbitrary, we are going to sweep it
Vgs 2 0 DC +3V
* MOSFET under test
M1 1 2 0 0 nmos_enhancement_mosfet L=10u W=400u
* mosfet model statement (by default, level 1)
.model nmos_enhancement_mosfet nmos (kp=20u Vto=+2 lambda=0)
** Analysis Requests **
.DC Vds 0V 10V 100mV
** Output Requests **
.Plot DC I(Vds) V(1)
.Probe
.end
```

Figure 5.6 The Spice input file for computing the $i_D - v_{DS}$ characteristic of a MOSFET with device parameters: $V_t = +2$ V, $\mu_n C_{OX} = 20$ μA/V^2, $\lambda = 0$ V^{-1}, $L = 10$ μm, and $W = 400$ μm. The gate-source voltage V_{GS} is set equal to $+3$ V and the drain-source voltage is swept between 0 and $+10$ V in 100 mV increments.

The modulation factor will be assumed equal to zero, ie. $\lambda = 0$ V^{-1}. Let us consider setting $V_{GS} = +3$ V and sweep V_{DS} from 0 V to $+10$ V in steps of 100 mV. In this way, the MOSFET will be taken through both its triode and saturation regions of operation. The resulting drain current is then monitored by observing the current supplied by voltage source V_{DS}. The Spice input file representing this situation is listed in Fig. 5.6.

The results of this analysis are shown plotted in Fig. 5.7 as the curve marked by

5 Field-Effect Transistors (FETs)

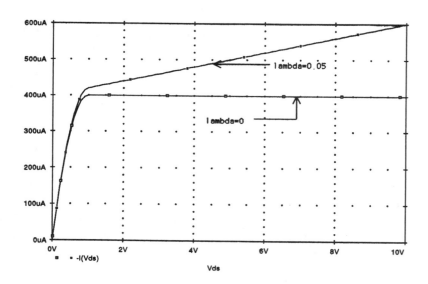

Figure 5.7 Comparing the $i_D - v_{DS}$ characteristics of a MOSFET with a channel-width modulation factor $\lambda = 0$ and $\lambda = 0.05$ V^{-1}. The gate-source voltage is held constant at +3 V.

lambda=0. Here the behavior of the MOSFET is as expected from Eqn. 5.1: For drain-source voltages less than $V_{DS_{SAT}}$, ($V_{DS_{SAT}} = V_{GS} - V_t = +1$ V), the MOSFET is in its triode region. For drain-source voltages above +1 V, the MOSFET saturates at a current level of 400 μA. The slope of the line drawn tangent to the output current in the saturation region is 0. Thus, the incremental drain-source resistance is infinite. It is interesting to compare the $i_{DS} - v_{DS}$ characteristic of a similar MOSFET having a nonzero channel-length modulation coefficient. For the case of an identical MOSFET but with $\lambda = 0.05$ V^{-1}, the $i_{DS} - v_{DS}$ curve is shown superimposed on the same graph as that in Fig. 5.7. This curve was obtained with results calculated with the same Spice deck seen previously in Fig. 5.6, except with the MOSFET model statement modified according to:

```
.model nmos_enhancement_mosfet nmos (kp=20u Vto=+2V lambda=0.05).
```

On comparison, we see that the effect of a nonzero channel-length modulation coefficient results in an output current that increases with increasing drain-source voltage at a rate of 20.314 μA/V. Thus, the corresponding incremental drain-source resistance is 49.23 kΩ. It is re-assuring that this increment resistance agrees quite closely with the value estimated by hand analysis; eg., for a drain current of approximately 400 μA, r_o which is given by $1/(\lambda I_D)$, is estimated to be 50 kΩ.

Another current – voltage characteristic that is used to describe the behavior of a MOS-

```
Enhancement-Mode N-Channel MOSFET Id - Vgs Characteristics

** Circuit Description **
* bias conditions
Vds 1 0 DC +5V
Vgs 2 0 DC +3V          ; this value is arbitrary, we are going to sweep it
* MOSFET under test
M1 1 2 0 0 nmos_enhancement_mosfet L=10u W=400u
* mosfet model statement (by default, level 1)
.model nmos_enhancement_mosfet nmos (kp=20u Vto=+2 lambda=0)
** Analysis Requests **
.DC Vgs 0V 10V 100mV
** Output Requests **
.Plot DC I(Vgs) V(2)
.Probe
.end
```

Figure 5.8 The Spice input file for computing the i_D - v_{GS} characteristic of a MOSFET with device parameters: $V_t = +2$ V, $\mu_n C_{OX} = 20$ μA/V^2, $\lambda = 0$ V^{-1}, $L = 10$ μm, and $W = 400$ μm. The drain-source voltage V_{DS} is set equal to +5 V and the gate-source voltage is swept between 0 and +5 V in 100 mV increments. The drain-source voltage is held constant at +5 V.

FET is: i_D versus v_{GS}. The circuit arrangement illustrated in Fig. 5.5 can also be used to obtain this current – voltage behavior. Instead of varying the drain-source voltage, this voltage is held constant and the gate-source voltage of the MOSFET is swept over a desired range. To demonstrate this, consider setting V_{DS} to +5 V and sweeping V_{GS} from 0 V to +5 V in increments of 100 mV. The Spice deck describing this is very similar to that used previously and is seen listed in Fig. 5.8. Here we have modeled the MOSFET like before with the channel-modulation coefficient set to zero. For comparison, we have created another Spice file that is identical to that seen in Fig. 5.8 except with the MOSFET having a channel-modulation coefficient $\lambda = 0.05$ V^{-1}. These two Spice decks are then concatenated together and submitted to Spice for analysis.

The results of these two Spice analyses are shown in Fig. 5.9. Here we have plotted the i_D - v_{GS} curves for the two cases of $\lambda = 0$ and $\lambda = 0.05$ V^{-1}. As is evident, the presence of a nonzero λ gives rise to a vertical shift in drain current. One interpretation of this is that for a given gate-source voltage, a MOSFET with non-zero channel-length modulation will draw more current from a circuit than one with zero channel-length modulation.

5.2 Spice Analysis Of MOSFET Circuits At DC

MOSFETs are classified as n-channel or p-channel devices depending on the material

5 Field-Effect Transistors (FETs)

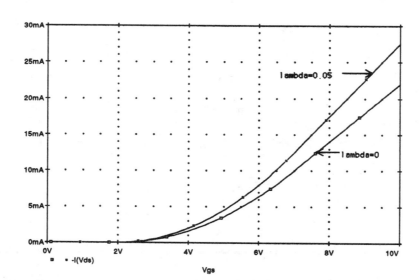

Figure 5.9 Comparing the $i_D - v_{GS}$ characteristics of a MOSFET with a channel-length modulation factor $\lambda = 0$ and $\lambda = 0.05$ V^{-1}. The drain-source voltage is held constant at +5 V.

used to form the channel. In addition, these devices are classified according to their mode of operation as enhancement or depletion type devices. As a result, the model statement characterizing these different FETs have subtle differences. In the following we shall highlight these differences as we analyze the DC operating point of several simple MOSFET circuits contained within Chapter 5 of Sedra and Smith. Our results can then be compared with those derived by hand analysis in Sedra and Smith.

5.2.1 An Enhancement-Mode P-Channel MOSFET Circuit

Consider the circuit shown in Fig. 5.10. Here we have a single transistor circuit containing a p-channel enhancement-mode MOSFET. The resistors have been chosen such that the FET is biased in its saturation region, has a bias current of 0.5 mA, and a drain voltage of +3 V. It is also assumed that the MOSFET has a threshold voltage of −1 V, a process transconductance parameter $\mu_p C_{OX}$ equal to 1 mA/V^2, and does not exhibit any channel-length modulation effect (ie. $\lambda = 0$). The details of this design can be found in Example 5.5 of Sedra and Smith.

This circuit would be described to Spice in the usual way; however because the dimensions of the MOSFET are not given we shall assume arbitrarily that the length and width of it are both equal to 10 μm. The model statement describing the characteristics of the p-channel

5.2 Spice Analysis Of MOSFET Circuits At DC

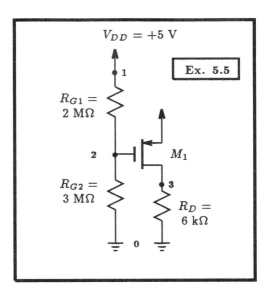

Figure 5.10 A transistor circuit designed in Example 5.5 of Sedra and Smith. Spice is used to calculate the DC operating point of this circuit.

```
Example 5.5: A Simple Enhancement-Mode PMOS Circuit (Rd=6k)

** Circuit Description **
* dc supplies
Vdd 1 0 DC +5V
* MOSFET circuit
M1 3 2 1 1 pmos_enhancement_mosfet L=10u W=10u
Rd 3 0 6k
Rg1 1 2 2Meg
Rg2 2 0 3Meg
* mosfet model statement (by default, level 1)
.model pmos_enhancement_mosfet pmos (kp=1m Vto=-1V lambda=0)
** Analysis Requests **
* calculate DC bias point
.OP
** Output Requests **
* none required
.end
```

Figure 5.11 Spice input file for calculating the DC operating point of the p-channel MOSFET circuit shown in Fig. 5.10.

MOSFET would appear as follows:

.model pmos_enhancement_mosfet pmos (kp=1m Vto=-1V lambda=0)

Here we have declared the MOSFET to be a PMOS device, and by assigning a negative threshold voltage, we are indicating to Spice that this particular PMOS device is of the enhancement-mode.

5 Field-Effect Transistors (FETs)

The Spice input deck for this particular example is listed in Fig. 5.11. We are requesting that Spice compute the DC operating point of this circuit. The results of this analysis are then found in the output file and some of its contents are on display below:

```
****    SMALL SIGNAL BIAS SOLUTION       TEMPERATURE =   27.000 DEG C
***********************************************************************

   NODE    VOLTAGE     NODE    VOLTAGE     NODE    VOLTAGE     NODE    VOLTAGE

   (  1)    5.0000   (  2)    3.0000   (  3)    3.0000


        VOLTAGE SOURCE CURRENTS
        NAME          CURRENT

        Vdd          -5.010E-04

        TOTAL POWER DISSIPATION   2.51E-03  WATTS

****    OPERATING POINT INFORMATION      TEMPERATURE =   27.000 DEG C
***********************************************************************

**** MOSFETS

NAME         M1
MODEL        pmos_enhancement_mosfet
ID           -5.00E-04
VGS          -2.00E+00
VDS          -2.00E+00
VBS           0.00E+00
VTH          -1.00E+00
VDSAT        -1.00E+00
```

As is evident from above, the p-channel MOSFET is biased at a current level of 0.5 mA and that its drain is at +3 V. The sign of I_D is negative because of the convention adopted by Spice; positive drain current flows into the drain terminal of a FET regardless of the device type. This is different than the convention used in Sedra and Smith. One should be aware of this convention difference. We also know that the device is biased in its saturation region because $V_{DS} < V_{DS_{SAT}}$. It is interesting to note that if we alter the value of R_D from 6 kΩ to 8 kΩ, and re-run the Spice input file, then we find that M_1 is biased on the edge of saturation (ie. $V_{DS} = V_{DS_{SAT}}$). This is evident from the results found in the Spice output file as follows:

```
****    SMALL SIGNAL BIAS SOLUTION       TEMPERATURE =   27.000 DEG C
***********************************************************************

   NODE    VOLTAGE     NODE    VOLTAGE     NODE    VOLTAGE     NODE    VOLTAGE

   (  1)    5.0000   (  2)    3.0000   (  3)    4.0000


        VOLTAGE SOURCE CURRENTS
        NAME          CURRENT

        Vdd          -5.010E-04

        TOTAL POWER DISSIPATION   2.51E-03  WATTS
```

5.2 Spice Analysis Of MOSFET Circuits At DC

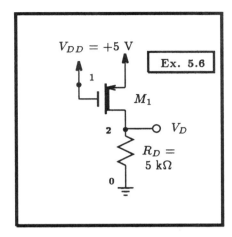

Figure 5.12 A depletion-mode p-channel MOSEFT circuit.

```
****        OPERATING POINT INFORMATION       TEMPERATURE =   27.000 DEG C
*************************************************************************

**** MOSFETS

NAME        M1
MODEL       pmos_enhancement_mosfet
ID          -5.00E-04
VGS         -2.00E+00
VDS         -1.00E+00
VBS          0.00E+00
VTH         -1.00E+00
VDSAT       -1.00E+00
```

Any increase in R_D above 8 kΩ will certainly cause M_1 to move out of the saturation region and into the triode region.

5.2.2 A Depletion-Mode P-Channel MOSFET Circuit

An example of a circuit incorporating a depletion-mode p-channel MOSFET is illustrated in Fig. 5.12. The depletion mode PMOS transistor is assumed to have $V_t = +1$ V, $\mu_p C_{OX} = 1$ mA/V^2 and $\lambda = 0$. With Spice, we would like to compute the drain current and the corresponding drain voltage. The Spice input file for this particular circuit is listed in Fig. 5.13. Take note of the model statement describing the depletion mode PMOS transistor, which we repeat below for convenience:

```
.model pmos_depletion_mosfet pmos (kp=1m Vto=+1V lambda=0)
```

In contrast to the enhancement mode PMOS transistor of the last example, the threshold voltage for a depletion mode PMOS transistor is made positive but everything else remains the same.

Submitting the input file to Spice, results in the following DC bias information:

5 Field-Effect Transistors (FETs)

Example 5.6: A Depletion-Mode PMOS Transistor Circuit

```
** Circuit Description **
* dc supplies
Vdd 1 0 DC +5V
* MOSFET circuit
M1 2 1 1 1 pmos_depletion_mosfet L=10u W=10u
Rd 2 0 5k
* mosfet model statement (by default, level 1)
.model pmos_depletion_mosfet pmos (kp=1m Vto=+1V lambda=0)
** Analysis Requests **
* calculate DC bias point
.OP
** Output Requests **
* none required
.end
```

Figure 5.13 Spice input file for calculating the DC operating point of the depletion-mode p-channel MOSFET circuit shown in Fig. 5.12.

```
****        SMALL SIGNAL BIAS SOLUTION       TEMPERATURE =   27.000 DEG C
***************************************************************************

   NODE   VOLTAGE     NODE   VOLTAGE     NODE   VOLTAGE     NODE   VOLTAGE

(    1)    5.0000  (    2)    2.5000


    VOLTAGE SOURCE CURRENTS
    NAME          CURRENT

    Vdd          -5.000E-04

    TOTAL POWER DISSIPATION   2.50E-03  WATTS

****        OPERATING POINT INFORMATION      TEMPERATURE =   27.000 DEG C
***************************************************************************

**** MOSFETS

NAME         M1
MODEL        pmos_depletion_mosfet
ID           -5.00E-04
VGS           0.00E+00
VDS          -2.50E+00
VBS           0.00E+00
VTH           1.00E+00
VDSAT        -1.00E+00
```

Thus the drain current and voltage of transistor M_1 are 0.5 mA and +2.5 V, respectively. Also, we see that the transistor is operating in the saturation region (ie. $V_{DS} < V_{DS_{SAT}}$). These results are identical to those computed by hand in Example 5.6 of Sedra and Smith.

In the above analysis, the effect of channel-length modulation was assumed zero. In

5.2 Spice Analysis Of MOSFET Circuits At DC

practise, this is not the case. In the following we repeat the above analysis assuming the transistor has a more realistic channel-length modulation coefficient of $\lambda = 0.02$ V^{-1}. We shall then compare the resulting transistor drain current with that obtained previously. This will give us some sense of how practical the assumption of neglecting the effect of channel-length modulation is when computing the DC bias point of a MOSFET circuit.

To carry out this task, we simply change the transistor model statement seen previously in Fig. 5.13 to the following:

```
.model pmos_depletion_mosfet pmos (kp=1m Vto=+1V lambda=0.02)
```

and re-submit the Spice deck for analysis. The results of the analysis are then found in the Spice output file. The pertinent details are shown below:

```
****    SMALL SIGNAL BIAS SOLUTION      TEMPERATURE =   27.000 DEG C
*******************************************************************************

    NODE    VOLTAGE     NODE    VOLTAGE     NODE    VOLTAGE     NODE    VOLTAGE

    (  1)    5.0000  (   2)    2.6190

        VOLTAGE SOURCE CURRENTS
        NAME            CURRENT

        Vdd            -5.238E-04

        TOTAL POWER DISSIPATION    2.62E-03  WATTS

****    OPERATING POINT INFORMATION     TEMPERATURE =   27.000 DEG C
*******************************************************************************

**** MOSFETS

NAME        M1
MODEL       pmos_depletion_mosfet
ID          -5.24E-04
VGS          0.00E+00
VDS         -2.38E+00
VBS          0.00E+00
VTH          1.00E+00
VDSAT       -1.00E+00
```

Thus, we see that the transistor drain current is now 524 μA as a result of the channel-modulation effect. This is about a 5% increase in the transistor drain current of 500 μA when no channel-length modulation effect was present. When performing "back-of-the-envelop" type calculations, neglecting the channel-length modulation effect seems to be quite reasonable. When more accuracy is required, one resorts to the use of Spice.

5.2.3 A Depletion-Mode N-Channel MOSFET Circuit

As the final example of this section, we shall look at a circuit containing a depletion-mode

5 Field-Effect Transistors (FETs)

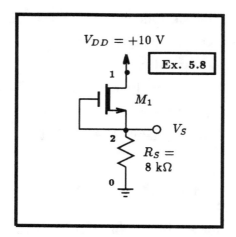

Figure 5.14 A depletion-mode n-channel MOSFET circuit.

Example 5.8: A Depletion-Mode NMOS Transistor Circuit

```
** Circuit Description **
* dc supplies
Vdd 1 0 DC +10V
* MOSFET circuit
M1 1 2 2 2 nmos_depletion_mosfet L=10u W=10u
Rd 2 0 100k
* mosfet model statement (by default, level 1)
.model nmos_depletion_mosfet nmos (kp=1m Vto=-1V lambda=0)
** Analysis Requests **
* calculate DC bias point
.OP
** Output Requests **
* none required
.end
```

Figure 5.15 The Spice input file for calculating the DC operating point of the depletion-mode n-channel MOSFET circuit shown in Fig. 5.14.

NMOS transistor. Consider the circuit shown in Fig. 5.14 where M_1 is assumed to have the following parameters: $V_t = -1$ V, $\mu_n C_{OX} = 1$ mA/V^2, and $\lambda = 0$. A Spice listing of this particular circuit is compiled in Fig. 5.15 and the results of an operating point analysis (.OP) are presented, in part, below:

```
****    SMALL SIGNAL BIAS SOLUTION        TEMPERATURE =   27.000 DEG C
************************************************************************

 NODE   VOLTAGE     NODE   VOLTAGE     NODE   VOLTAGE     NODE   VOLTAGE

(   1)   10.0000  (   2)    9.8956

    VOLTAGE SOURCE CURRENTS
    NAME         CURRENT
```

5.2 Spice Analysis Of MOSFET Circuits At DC

```
      Vdd         -9.896E-05

      TOTAL POWER DISSIPATION   9.90E-04  WATTS

 ****    OPERATING POINT INFORMATION      TEMPERATURE =   27.000 DEG C
 **********************************************************************

 **** MOSFETS

 NAME         M1
 MODEL        nmos_depletion_mosfet
 ID           9.90E-05
 VGS          0.00E+00
 VDS          1.04E-01
 VBS          0.00E+00
 VTH         -1.00E+00
 VDSAT        1.00E+00
```

Notice that in this case $V_{DS} < V_{DS_{SAT}}$, implying that the transistor is operating in the triode region.

In the following, we repeat the above analysis with the same MOSFET model used above with the addition that it has a channel-length modulation coefficient λ equal to 0.04 V^{-1}. On doing so, we find in the Spice output file the following DC operating information:

```
 ****    SMALL SIGNAL BIAS SOLUTION       TEMPERATURE =   27.000 DEG C
 **********************************************************************

 NODE   VOLTAGE    NODE   VOLTAGE    NODE   VOLTAGE    NODE   VOLTAGE

 (   1)  10.0000  (   2)   9.8960

      VOLTAGE SOURCE CURRENTS
      NAME            CURRENT

      Vdd         -9.896E-05

      TOTAL POWER DISSIPATION   9.90E-04  WATTS

 ****    OPERATING POINT INFORMATION      TEMPERATURE =   27.000 DEG C
 **********************************************************************

 **** MOSFETS

 NAME         M1
 MODEL        nmos_depletion_mosfet
 ID           9.90E-05
 VGS          0.00E+00
 VDS          1.04E-01
 VBS          0.00E+00
 VTH         -1.00E+00
 VDSAT        1.00E+00
```

Thus we see that only the voltage at the second (2) node has changed to 9.8960 V from the previous value of 9.8956 V. All other values listed appear to be the same as before; given the 3 or 4 digits of accuracy. Thus, the inclusion of the term $(1 + \lambda v_{DS})$ does not seem to affect the behavior of the circuit significantly when the device is operated in the triode region.

5 Field-Effect Transistors (FETs)

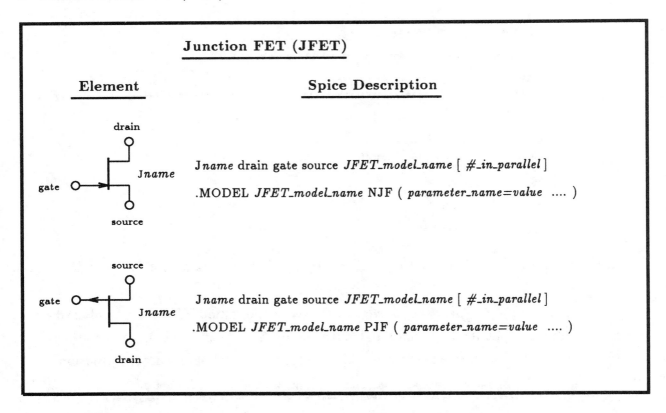

Figure 5.16 Spice element description for the n-channel and p-channel JFETs. Also listed is the general form of the JFET model statement. A partial listing of the parameters applicable to either an n-channel or p-channel JFET is given in Table 5.2.

5.3 Describing JFETs To Spice

Like MOSFETs, JFETs are described to Spice using an element statement and a model statement. The following outlines the syntax of these two statements, including the details of the built-in JFET model.

5.3.1 JFET Element Description

JFETs are describe to Spice using an element statement beginning with a unique name prefixed with the letter J. This is then followed by a list of the nodes that the drain, gate, and source of the JFET are connected to. The next field specifies the name of the model that characterizes its terminal behavior, and the final field of this statement is optional and its purpose is to allow one to scale the size of the device by specifying the number of JFETs connected in parallel. A summary of the element statement syntax for both the n-channel and p-channel JFETs is provided in Fig. 5.16. Included in this list is the syntax for the JFET model statement (.MODEL). This statement defines the terminal characteristics of the

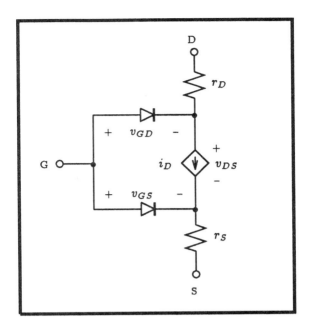

Figure 5.17 The Spice large-signal n-channel JFET model under static conditions.

JFET by specifying the values of parameters in the JFET model. Parameters not specified in the model statement are assigned default values by Spice. Details of this model will be discussed briefly next.

5.3.2 JFET Model Description

As is evident from Fig. 5.16, the model statement for either the n-channel or p-channel JFET transistor begins with the keyword .MODEL and is followed by the name of the model used by a JFET element statement, the nature of the JFET (ie. NJF or PJF), and a list of the JFET parameter values (enclosed between brackets). Specifically, the mathematical model of the JFET in Spice is very similar to that seen earlier for the Level 1 MOSFET model.

The general form of the DC Spice model for an n-channel JFET is illustrated schematically in Fig. 5.17. The bulk resistance of the drain and source regions of the JFET are lumped into two linear resistances r_D and r_S, respectively. The DC characteristic of the intrinsic JFET is determined by the nonlinear dependent current source i_D, and the two diodes represent the two substrate junctions that define the channel region. A similar model applies for the p-channel device; the direction of the diodes, the current source and the polarities of the terminal voltages are all reversed.

The expression for drain current i_D, assuming that the drain is at a higher potential than the source, is described by the following:

5 Field-Effect Transistors (FETs)

Symbol	Spice Name	Model Parameter	Units	Default
β	beta	Transconductance coefficient	A/V^2	100μ
V_t, V_P	Vto	Threshold voltage	V	-2.0
λ	lambda	Channel-length modulation	V^{-1}	0
r_D	Rd	Drain ohmic resistance	Ω	0
r_S	Rs	Source ohmic resistance	Ω	0

Table 5.2 A partial listing of the Spice parameters for the JFET model.

$$i_D = \begin{cases} 0 & \text{for } v_{GS} < V_t \\ \beta\left[2(v_{GS}-V_t)v_{DS} - v_{DS}^2\right](1+\lambda v_{DS}) & \text{for } v_{GS} > V_t \text{ and } v_{DS} \leq v_{GS} - V_t \\ \beta(v_{GS}-V_t)^2(1+\lambda v_{DS}) & \text{for } v_{GS} > V_t \text{ and } v_{DS} \geq v_{GS} - V_t \end{cases} \quad (5.4)$$

where the device parameters β and V_t are written in terms of I_{DSS} and V_P as

$$\beta = \frac{I_{DSS}}{V_P^2} \quad (5.5)$$

and

$$V_t = V_P. \quad (5.6)$$

From the above, we see that the JFET has 3 parameters that define its operation: I_{DSS}, V_P and λ. The parameter I_{DSS} is the drain current when $v_{GS} = 0$ V, and V_P corresponds to the pinch-off voltage of the channel. The parameter λ is the channel-length modulation parameter and represents the influence that the drain-source voltage has on the drain current i_D when the device is in pinch-off. The sign of this parameter is always positive, regardless of the nature of the device type. A note on notation: The transconductance coefficient β is denoted by the parameter K in Sedra and Smith.

A partial listing of the parameters associated with the Spice JFET model under static conditions is given in Table 5.2. Also listed are the default values which the parameter assume if no value is specified on the .MODEL statement. To specify a parameter value one simply writes, for example: beta=1m, Vto=-1V, lambda=0.01, etc.

5.3.3 An N-Channel JFET Example

To demonstrate how a circuit containing a JFET is described to Spice, consider the circuit shown in Fig. 5.18. Here the JFET is n-channel with parameters: $V_P = -4$ V,

5.3 Describing JFETs To Spice

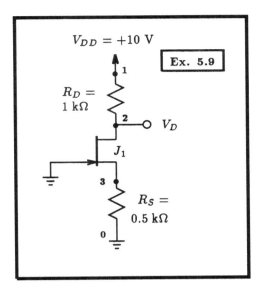

Figure 5.18 An n-channel JFET circuit example.

Example 5.9: A Simple N-Channel JFET Circuit

```
** Circuit Description **
* dc supplies
Vdd 1 0 DC +10V
* JFET circuit
J1 2 0 3  n_jfet
Rd 1 2 1k
Rs 3 0 0.5k
* n-channel jfet model statement
.model n_jfet NJF (beta=1m Vto=-4V lambda=0)
** Analysis Requests **
* calculate DC bias point
.OP
** Output Requests **
* none required
.end
```

Figure 5.19 The Spice input file for calculating the DC operating point of the JFET circuit shown in Fig. 5.18.

$I_{DSS} = 16$ mA and $\lambda = 0$. The model statement for this particular n-channel JFET that would appear in the Spice input file is as follows:

.model n_jfet NJF (beta=1m Vto=-4V lambda=0)

In this particular case, one had to convert the device parameters into terms that Spice is familiar with. This required that one compute β from I_{DSS} and V_P according to $\beta = I_{DSS}/V_P^2$. V_t is simply equal to V_P.

5 Field-Effect Transistors (FETs)

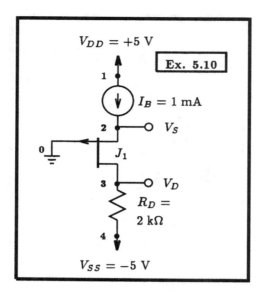

Figure 5.20 A p-channel JFET circuit.

In keeping with the discussion thus far, we shall compute the DC operating point of the circuit shown in Fig. 5.18. The Spice input file for this circuit, including the appropriate analysis request (ie. .OP command), is listed in Fig. 5.19. On completion of Spice, the following results are found in the output file:

```
****    SMALL SIGNAL BIAS SOLUTION     TEMPERATURE =   27.000 DEG C
******************************************************************************

  NODE    VOLTAGE      NODE    VOLTAGE     NODE    VOLTAGE     NODE    VOLTAGE

(    1)   10.0000   (    2)    6.0000   (    3)    2.0000

    VOLTAGE SOURCE CURRENTS
    NAME         CURRENT

    Vdd         -4.000E-03

    TOTAL POWER DISSIPATION    4.00E-02  WATTS

****    OPERATING POINT INFORMATION    TEMPERATURE =   27.000 DEG C
******************************************************************************

**** JFETS

NAME        J1
MODEL       n_jfet
ID          4.00E-03
VGS        -2.00E+00
VDS         4.00E+00
```

Here we see that the JFET is biased with a drain current of 4 mA and that the drain is at a voltage of 6 V.

Example 5.10: A Simple P-Channel JFET Circuit

```
** Circuit Description **
* dc supplies
Vdd  1 0 DC +5V
Vss  4 0 DC -5V
Ib   1 2 DC 1mA
* JFET circuit
J1 3 0 2 p_jfet
Rd 3 4 2k
* p-channel jfet model statement
.model p_jfet PJF (beta=1m Vto=-2V lambda=0)
** Analysis Requests **
* calculate DC bias point
.OP
** Output Requests **
* none required
.end
```

Figure 5.21 The Spice input file for calculating the DC operating point of the p-channel JFET circuit shown in Fig. 5.20.

5.3.4 A P-Channel JFET Example

This next example is along the same lines as the previous JFET example except that the JFET is p-channel. The p-channel JFET circuit that we will analyze for its DC bias conditions is on display in Fig. 5.20. Here the p-channel device has parameters: $V_P = +2$ V, $I_{DSS} = 4$ mA and $\lambda = 0$. The importance of this example is to highlight the fact that the threshold voltage (V_t) of the p-channel JFET is specified on the model statement as a negative value, even though V_P is positive. One must be careful of this for it is opposite to what is used in Sedra and Smith. The rationale for this is that the originators of Spice adopted the convention that a depletion-mode device, which the JFET is, would have $V_t < 0$ whether it be n-channel or p-channel.

The Spice input file describing the circuit shown in Fig. 5.20 is listed below in Fig. 5.21. On completion of Spice, the following results are found in the output file:

```
****   SMALL SIGNAL BIAS SOLUTION      TEMPERATURE =  27.000 DEG C
*******************************************************************************

 NODE   VOLTAGE    NODE   VOLTAGE    NODE   VOLTAGE    NODE   VOLTAGE

(   1)   5.0000  (   2)  -1.0000  (   3)  -3.0000  (   4)  -5.0000

    VOLTAGE SOURCE CURRENTS
    NAME         CURRENT

    Vdd         -1.000E-03
    Vss          1.000E-03
```

```
                TOTAL POWER DISSIPATION    1.00E-02  WATTS

        ****        OPERATING POINT INFORMATION      TEMPERATURE =   27.000 DEG C
        *************************************************************************

        **** JFETS

        NAME        J1
        MODEL       p_jfet
        ID          -1.00E-03
        VGS         1.00E+00
        VDS         -2.00E+00
```

As is evident, the transistor is biased at 1 mA, and the source and drain are at −1 V and −3 V, respectively.

To see how much the DC bias calculation changes with the inclusion of the transistor channel-length modulation effect, we repeat the above analysis with $\lambda = 0.04$ V^{-1}. Modifying the Spice deck shown listed in Fig. 5.21 to reflect this change, we then submit the revised Spice deck to Spice. The following results are then obtained:

```
        ****      SMALL SIGNAL BIAS SOLUTION        TEMPERATURE =   27.000 DEG C
        *************************************************************************

        NODE    VOLTAGE     NODE    VOLTAGE     NODE    VOLTAGE     NODE    VOLTAGE

        (  1)    5.0000   (  2)    -1.0371   (  3)   -3.0000   (  4)   -5.0000

            VOLTAGE SOURCE CURRENTS
            NAME         CURRENT

            Vdd         -1.000E-03
            Vss          1.000E-03

            TOTAL POWER DISSIPATION    1.00E-02  WATTS

        ****        OPERATING POINT INFORMATION      TEMPERATURE =   27.000 DEG C
        *************************************************************************

        **** JFETS

        NAME        J1
        MODEL       p_jfet
        ID          -1.00E-03
        VGS         1.04E+00
        VDS         -1.96E+00
```

Since the JFET is biased externally with a current source the effect of the channel-length modulation will manifest itself in the voltages that appear across the terminals of the device. For instance, the gate-source voltage increases from 1 V to 1.04 V. Likewise, the drain-source voltage experiences the same voltage change. In both cases, the resulting voltage change is small and this provides further justification that it is reasonable to neglect the presence of channel-length modulation when performing DC bias calculations by hand.

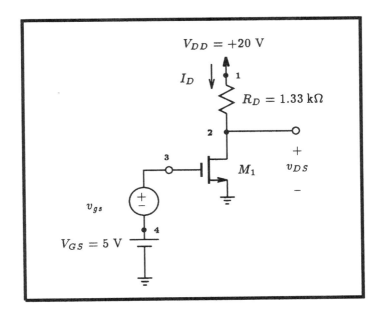

Figure 5.22 A fixed-bias n-channel MOSFET amplifier configuration.

5.4 FET Amplifier Circuits

Field-effect transistors are commonly employed in the design of linear amplifiers. Small-signal linear analysis is commonly employed as a means of estimating various attributes of amplifier behavior when subjected to small input signals. Examples of amplifier attributes would include input and output resistances, and current and voltage signal gain. Spice has a small-signal linear model of the MOSFET and another for the JFET. We shall describe these in this section. Subsequently, we shall illustrate the effectiveness of small-signal analysis as it applies to a common-source n-channel enhancement-mode MOSFET amplifier. This is accomplished by comparing the results calculated by the formulae presented in Sedra and Smith with those generated by Spice. But before these undertakings, we shall illustrate the importance of biasing a FET around an operating point that is well within the saturation region of the device.

5.4.1 Effect Of Bias Point On Amplifier Conditions

In this section we consider the effect that a misplaced DC operating point has on the large-signal operation of a MOSFET amplifier. For the enhancement-type n-channel MOSFET amplifier shown in Fig. 5.22 with a +5 V fixed-biasing scheme, the DC operating point of the MOSFET has been set at approximately $I_D = 9$ mA and $V_{DS} = 8$ V. This is a result of the MOSFET having an assumed threshold voltage V_t of +2 V, a conductance parameter $K = \frac{1}{2}\mu_n C_{OX}(\frac{W}{L}) = 1$ mA/V^2 and a channel-length modulation factor $\lambda = 0.01$ V^{-1}.

5 Field-Effect Transistors (FETs)

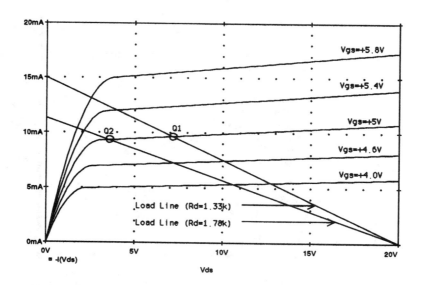

Figure 5.23 The i_D-v_{DS} characteristics of an n-channel MOSFET having model parameters $V_t = +2$ V, $\frac{1}{2}\mu_n C_{OX}(\frac{W}{L}) = 1$ mA/V^2 and $\lambda = 0.01$ V^{-1}. Load lines for the MOSFET amplifier shown in Fig. 5.22 for drain resistances of 1.33 kΩ and 1.78 kΩ. Q_1 and Q_2 indicate the DC operating point of each amplifier with a gate-source bias voltage of 5 V.

Figure 5.23 shows the $i_D - v_{DS}$ characteristics for this device obtained using the curve tracer method described in Section 5.1. Also shown superimposed on this graph is the 1.33 kΩ load line. We see that the resulting operating point Q_1 is located slightly to the left of the midway point between the triode and cut-off regions. If the drain resistance R_D is increase to 1.78 kΩ, then the DC operating point, Q_2, located at $I_D = 9.3$ mA and $V_{DS} = 3.6$ V, moves closer to the edge of the triode region.

To illustrate the effect that the location of the DC operating point has on the large-signal behavior of the amplifier, consider applying a 1 V peak-to-peak triangular waveform input signal of 1 kHz frequency in series with the 5 V DC biasing voltage as shown in Fig. 5.22. This will then be repeated for the same amplifier having a drain resistance of 1.78 kΩ, and the resulting output signals from each amplifier will be compared. The Spice input file describing the first situation is provided in Fig. 5.24. Here the triangular input is described using the piecewise linear source statement for a time duration of 3 ms. A transient analysis command is included to compute the behavior of this amplifier over the same time period.

The results of the two Spice analyses are shown in Fig. 5.25. The top graph in part (a) of this figure illustrates both the input and output waveforms for a drain resistance of 1.33 kΩ. As is evident the output signal is reasonably linear with a peak-to-peak level about

5.4 FET Amplifier Circuits

```
An Enhancement-Type NMOS Amplifier

** Circuit Description **
* dc supplies
Vdd 1 0 DC +20V
Vgs 4 0 DC +5V
* small-signal input
vi 3 4 PWL (0,0 0.25ms,+0.5V 0.75ms,-0.5V 1.25ms,+0.5V 1.75ms,-0.5V
+           2.25ms,+0.5V 2.75ms,-0.5V 3ms,0V)
* amplifier circuit
M1 2 3 0 0 nmos_enhancement_mosfet L=10u W=10u
Rd 1 2 1.33k
* mosfet model statement (by default, level 1)
.model nmos_enhancement_mosfet nmos (kp=2m Vto=+2V)
** Analysis Requests **
.OP
.TRAN 10us 3ms 0ms 10us
** Output Requests **
.PLOT TRAN V(2) V(3)
.probe
.end
```

Figure 5.24 The Spice input file for calculating the large-signal transient behavior of the MOSFET amplifier shown in Fig. 5.22 with a drain resistor of 1.33 kΩ.

8 times that of the input signal (more precisely, 7.53 times). This is in contrast with the case when the drain resistance is increased to 1.78 kΩ where we see in Fig. 5.25(b) that the output signal has become quite distorted in the lower portion of its waveform. Thus, this example helps to illustrate the importance of placing the DC operating point of an amplifier well within the saturation region of the MOSFET in order to maximize the output voltage swing of the amplifier while maintaining linear operation.

5.4.2 Small-Signal Model Of The FET

The linearized small-signal model for the MOSFET is shown in Fig. 5.26(a). It consists of two voltage-controlled current sources with transconductance g_m and g_{mb}. The MOSFET transconductance, g_m, is related to the DC bias current I_D and the device parameters (ignoring the channel-length modulation effect) as follows:

$$g_m = \left.\frac{\partial i_D}{\partial v_{GS}}\right|_{OP} \approx \sqrt{2\mu C_{OX}}\sqrt{W/L}\sqrt{I_D} \qquad (5.7)$$

Here $\left.\right|_{OP}$ indicates that the derivative is obtained at the DC operating point of the device. The next term, g_{mb}, is known as the body transconductance and is related to several MOSFET bias conditions and device parameters according to:

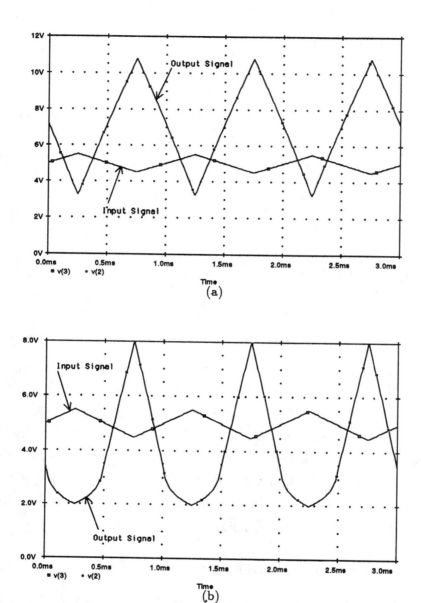

Figure 5.25 The input and output large-signal transient behavior of the MOSFET amplifier shown in Fig. 5.22 for a drain resistance of: (a) 1.33 kΩ (b) 1.78 kΩ.

$$g_{mb} = \left.\frac{\partial i_D}{\partial v_{BS}}\right|_{OP} \approx \frac{\gamma}{2\sqrt{2\phi_f + V_{SB}}} g_m \qquad (5.8)$$

Accounting for the presence of channel-length modulation, the output conductance g_{ds}, which is also equal to $1/r_o$, is given approximately by the expression

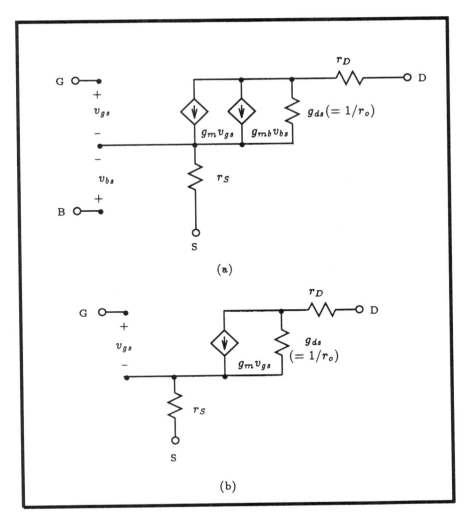

Figure 5.26 The static Spice small-signal model of: (a) MOSFET, (b) JFET.

$$g_{ds} = \frac{\partial i_D}{\partial v_{DS}}\bigg|_{OP} \approx \lambda I_D. \tag{5.9}$$

Finally, the resistances r_D and r_S represent the ohmic resistance of drain and source regions, respectively.

The small-signal linear model of a JFET is shown in Fig. 5.26(b). It consists of a single voltage-controlled current source having a transconductance g_m that is related to the DC bias current I_D and device parameters (ignoring the channel-length modulation effect) according to the following:

$$g_m = \frac{\partial i_D}{\partial v_{GS}}\bigg|_{OP} \approx \frac{2I_{DSS}}{|V_P|}\sqrt{\frac{I_D}{I_{DSS}}} \tag{5.10}$$

Accounting for the presence of channel-length modulation, the output conductance g_{ds},

5 Field-Effect Transistors (FETs)

which is also equal to $1/r_o$, is given approximately by the expression

$$g_{ds} = \left.\frac{\partial i_D}{\partial v_{DS}}\right|_{OP} \approx \lambda I_D. \tag{5.11}$$

Lastly, the resistances r_D and r_S represent the ohmic resistance of the drain and source regions, respectively.

Many of the analysis that is performed on a transistor circuit by Spice utilizes its linear small-signal equivalent circuit. The parameters of the small-signal model of each transistor in a given circuit, as computed by Spice, are available to the user through the operating point (.OP) command. To see this, consider a single NMOS transistor with its drain biased at +5 V, its gate biased at +3 V, and its source connected to ground. Furthermore, we shall assume that the substrate is biased at −5 V. The transistor is assumed to have the following parameters: $V_{t0} = +1$ V, $\mu_n C_{OX} = 20$ μA/V^2, $L = 10$ μm, and $W = 400$ μm. Furthermore, $\lambda = 0.05$ V^{-1} and $\gamma = 0.9$ V$^{1/2}$. The Spice input file for this circuit is provided below:

```
Small-Signal Model Of An N-Channel MOSFET

** Circuit Description **
* mosfet terminal bias
Vd 1 0 DC +5V
Vg 2 0 DC +3V
Vb 3 0 DC -5V
* mosfet under test
M1 1 2 0 3 nmos_enhancement_mosfet L=10u W=400u
* mosfet model statement (by default, level 1)
.model nmos_enhancement_mosfet nmos (kp=20u Vto=1V lambda=0.05 gamma=0.9)
** Analysis Requests **
.OP
** Output Requests **
* none required
.end
```

On completion of Spice, one finds in the Spice output file the following complete list of parameters pertaining to the small-signal MOSFET model computed by Spice:

```
****        OPERATING POINT INFORMATION        TEMPERATURE =   27.000 DEG C

**** MOSFETS

      NAME     M1
      MODEL    nmos_enhancement_mosfet
      ID       1.61E-04
      VGS      3.00E+00
      VDS      5.00E+00
      VBS      -5.00E+00
      VTH      2.43E+00
      VDSAT    5.67E-01
      GM       5.67E-04
```

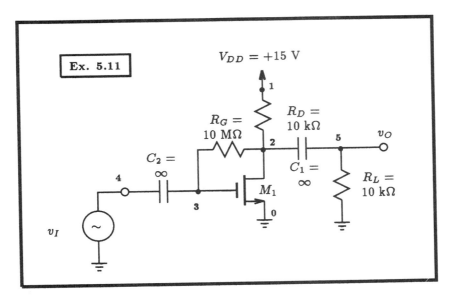

Figure 5.27 A common-source amplifier circuit.

```
GDS      6.44E-06
GMB      1.08E-04
CBD      0.00E+00
CBS      0.00E+00
CGSOV    0.00E+00
CGDOV    0.00E+00
CGBOV    0.00E+00
CGS      0.00E+00
CGD      0.00E+00
CGB      0.00E+00
```

Included in the above list of operating point information are: (a) DC bias conditions which include drain current and various terminal voltages; (b) device transconductances g_m and g_{mb}; (c) output conductance g_{ds}; and (d) device capacitances accounting for MOSFET dynamic effects. All the parameters in this list, except for the capacitances (these are all zero for the time being), have been discussed previously and therefore their meaning should be self-evident. A discussion of MOSFET capacitances will be deferred until Chapter 7.

Similar results extend to the JFET. We leave this to the reader to confirm.

5.4.3 A Basic FET Amplifier Circuit

Fig. 5.27 displays an enhancement-mode NMOS amplifier in which the input signal v_I is coupled to the gate of the MOSFET through a large capacitor, and the output signal at the drain is coupled to the load resistance R_L via another large capacitor. The MOSFET is assumed to have device parameters: $V_t = +1.5$ V, $\mu C_{OX} = 0.5$ mA/V^2 and $\lambda = 0.02$ V^{-1}. This same example was analyzed by hand in Example 5.11 of Sedra and Smith. There they found that this amplifier would have a voltage gain of -3.3 V/V and an input resistance of

5 Field-Effect Transistors (FETs)

```
Example 5.11: An Enhancement-Mode NMOS Amplifier

** Circuit Description **
* dc supplies
Vdd 1 0 DC +15V
* input AC signal
Vi 4 0 DC 0V
* amplifier circuit
M1 2 3 0 0 nmos_enhancement_mosfet L=10u W=10u
Rd 1 2 10k
Rg 2 3 10Meg
C1 2 5 1GF
C2 3 4 1GF
* load
Rl 5 0 10k
* mosfet model statement (by default, level 1)
.model nmos_enhancement_mosfet nmos (kp=0.25m Vto=+1.5V lambda=0.02)
** Analysis Requests **
.OP
** Output Requests **
* node required
.end
```

Figure 5.28 The Spice input file for calculating the DC operating point and small-signal model parameters of the amplifier circuit shown in Fig. 5.27.

2.33 MΩ. These results were obtained by performing a small-signal analysis of the amplifier circuit through a two step procedure. The first step was to obtain the DC bias conditions of the circuit, specifically the drain current, from which the small-signal model for the transistor is obtained. The second step was to analyze the linear small-signal equivalent circuit of the amplifier to obtain the voltage gain and the input resistance. In each step of the analysis some simplifying assumptions are made. For instance, during the DC bias calculation, the transistor channel-length modulation effect was ignored. In the second step, a simplifying assumption was made with regards to the current flowing through the feedback resistor R_G. We have seen in previous examples of this chapter that ignoring the channel-length modulation effect results in small variations in the drain current of a MOSFET, and thus, by extension, variations in the small-signal model parameters would also be small. In the following, with the aid of Spice, we would like to investigate the accuracy of these two steps as they apply to the common-source amplifier in Fig. 5.27 and demonstrate that these simplifying assumption are reasonable.

According to the hand analysis performed in Example 5.11 of Sedra and Smith (which ignores the channel-length modulation effect) the drain current of the MOSFET is found to be 1.06 mA. Thus, from Eqns. (5.7) and (5.11), the MOSFET transconductance g_m

Parameter	Hand Analysis	Spice	% Error
I_D	1.06 mA	1.07 mA	0.93 %
g_m	0.725 mA/V	0.762 mA/V	4.9 %
r_o	47 kΩ	50.8 kΩ	7.5 %

Table 5.3 Comparing the transistor drain current I_D and its corresponding small-signal model parameters g_m and r_o as computed by straightforward hand analysis and Spice.

equals 0.725 mA/V and the output resistance r_o is equal to 47 kΩ. To compare these results to those calculated by Spice, we have created the Spice input file listed in Fig. 5.28. The infinite-valued coupling capacitors are represented by 1 *giga-farads*. This will ensure that the capacitors behave as short circuits at the signal frequencies of interest. The results generated by Spice are as follows:

```
**** MOSFETS

NAME     M1
MODEL    nmos_enhancement_mosfet
ID       1.07E-03
VGS      4.31E+00
VDS      4.31E+00
VBS      0.00E+00
VTH      1.50E+00
VDSAT    2.81E+00
GM       7.62E-04
GDS      1.97E-05
GMB      0.00E+00
```

Here we see that the MOSFET is biased at a drain current of 1.07 mA, has a transconductance g_m equal to 0.762 mA/V and an output conductance of 19.7 μS, or an output resistance r_o of 50.8 kΩ. Comparing the hand calculated values of Sedra and Smith with those generated by Spice, we see that the hand calculated results are quite close, with at most, a 7.5% error. In Table 5.3 we list these two sets of results and also list the relative error between them.

Once the small-signal equivalent circuit of the amplifier is obtained, one proceeds to analyze the circuit using standard circuit analysis techniques to obtain pertinent amplifier parameters, such as voltage gain and input resistance. To gain insight into circuit behavior, closed-form expressions are usually derived from the equivalent circuit. In many cases, the expressions that result are complicated, large, and not very insightful. As a result, one makes simplifying assumptions based on practical considerations that lead to expressions that are simpler, but more insightful. For example, in Example 5.11 of Sedra and Smith, the linear small-signal equivalent circuit of the common-source amplifier shown in Fig. 5.27 was analyzed and the following expressions for its voltage gain and input resistance were

5 Field-Effect Transistors (FETs)

obtained after making several practical assumptions:

$$A_V = -g_m(R_D \| R_L \| r_o) \qquad (5.12)$$

and

$$R_{in} = \frac{R_G}{1 - A_V}. \qquad (5.13)$$

It is the simplicity of these two formulas that makes them useful in circuit design. The question then becomes: How accurate are they?

To verify their accuracy, we simply substitute the appropriate circuit parameters, together with the small-signal parameters of the MOSFET generated by Spice above, and evaluate. This is then compared with the results computed directly by Spice. For the first part, we find $A_V = -3.468$ V/V and $R_{in} = 2.238$ kΩ. With regard to the analysis we ask Spice to perform, one might be tempted to request a transfer function (.TF) analysis and directly obtain both the small-signal voltage gain and the amplifier input resistance. Unfortunately, the amplifier is AC coupled and intended to amplify signals containing frequencies other than DC. Since the .TF analysis calculates the small-signal input – output behavior of a circuit only at DC, in the situation at hand the results produced would not prove very useful. Instead, we shall apply a one-volt AC voltage signal to the input of the amplifier and compute the voltage appearing at the amplifier output using the .AC analysis command of Spice at a single midband frequency of, say, 1 Hz. A one-volt input level is usually chosen here because, in this way, the output voltage would be directly equal to the input – output transfer function. The input signal level of one-volt is not considered to be above the small-signal limit of the amplifier because the .AC analysis performed by Spice is performed directly with the small-signal equivalent circuit of the amplifier. Thus, any input level would work. The Spice statements necessary to invoke this analysis are as follows:

```
Vi 4 0 AC 1V
.AC LIN 1 1Hz 1Hz
```

In a similar vein, we can compute the input resistance of amplifier by determining the current that is supplied by the input voltage source and computing the input resistance as the ratio of the input voltage to input current. To obtain this information, we ask Spice to print both the magnitude and phase of the output voltage signal (node 5), followed by the magnitude and phase of the current supplied by Vi, using the following Spice statement:

```
.PRINT AC Vm(5) Vp(5) Im(Vi) Ip(Vi)
```

Recall that Spice uses complex variables to evaluate the AC small-signal response of a circuit,

5.4 FET Amplifier Circuits

Parameter	Hand Analysis		Spice (Direct Calculation)	% Error	
	Hand Estimate Of Small-Signal Model Parameters	Spice Small-Signal Model Parameters		Hand Estimate Of Small-Signal Model Parameters	Spice Small-Signal Model Parameters
A_V	-3.3 V/V	-3.468 V/V	-3.476 V/V	5.1 %	0.23 %
R_{in}	2.33 MΩ	2.238 MΩ	2.238 MΩ	-4.1 %	0 %

Table 5.4 Comparing the voltage gain and the input resistance of the amplifier shown in Fig. 5.27 as calculated by three different methods: (a) closed-form expression using hand estimates of the small-signal model parameters; (b) closed-form expression using Spice calculated small-signal model parameters; (c) direct calculation using Spice.

so one must indicate the form of the complex variable (ie. magnitude, real, imaginary, etc.) you want Spice to print.

The AC analysis results computed by Spice at 1 Hz would then be found in the output file as follows:

```
****    AC ANALYSIS                    TEMPERATURE =   27.000 DEG C

FREQ         VM(5)        VP(5)        IM(Vi)       IP(Vi)

1.000E+00    3.467E+00    -1.800E+02   4.467E-07    1.800E+02
```

The small-signal voltage gain of the amplifier (A_V) is therefore -3.476 V/V and the input resistance R_{in} is 2.238 MΩ (= $1/4.467 \times 10^{-7}$ MΩ). The phase of the input current IP(Vi) indicates that the sign of the current supplied by the input voltage source Vi is negative. This is consistent with the convention used by Spice; that is, current supplied by a source to a circuit is always negative.

Comparing these two sets of results, one set derived using hand analysis together with Spice generated small-signal model parameters (-3.468 V/V, 2.238 MΩ), and the other derived directly using Spice (-3.476 V/V, 2.238 MΩ), we see that these results are either identical to one another or quite close with a relative error of only 0.23%. We can therefore conclude that the simplifying assumptions used to derive the formula for small-signal voltage gain and input resistance in Sedra and Smith are very reasonable and introduce very little error. When the small-signal parameters computed by hand are used instead of the Spice generated model parameters, the accuracy of the results decrease but remain within practical limits (ie. $A_V = -3.3$ V/V and $R_{in} = 2.33$ MΩ). The relative error of the two calculations are in the vicinity of 5%. This error is largely due to the error incurred through the DC hand analysis which ignored the transistor channel-length modulation effect. A summary of

5 Field-Effect Transistors (FETs)

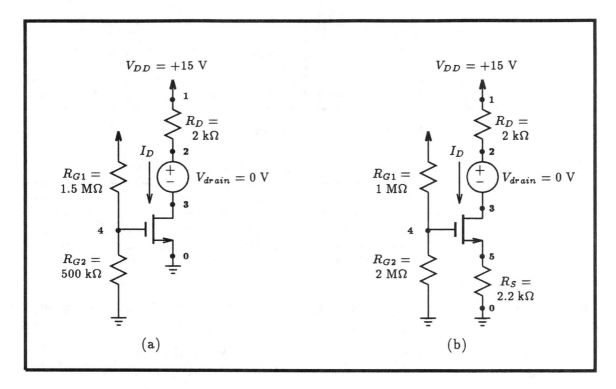

Figure 5.29 Two different MOSFET biasing arrangements: (a) fixed-biasing (b) biasing with source resistance feedback. The zero-valued voltage source in series with the drain terminal of each MOSFET is used to directly monitor the drain current.

the above discussion is provided in Table 5.4 for easy reference.

5.5 Investigating Bias Stability With Spice

To obtain a stable DC operating point in discrete transistor circuits, a biasing scheme utilizing some form of negative feedback is usually employed. This insures that the DC bias currents through the transistors in the circuit remain relatively constant under the influence of normal manufacturing or environmental variations.

To illustrate the effectiveness of incorporating negative feedback in the biasing network of a transistor amplifier, let us compare the DC sensitivities of the two amplifiers shown in Fig. 5.29 to the resistive biasing components. Each MOSFET will be assumed to have the following parameters: $V_t = +2$ V and $\mu_n C_{OX} = 2$ mA/V^2. For a fair comparison, each amplifier is biased at approximately the same current level of 3 mA. The n-channel enhancement MOSFET amplifier of Fig. 5.29(a) is simply biased by a voltage appearing at its gate terminal. Although, the MOSFET in the amplifier of Fig. 5.29(b) is also biased with a voltage at its gate through a voltage divider circuit, a resistor is included in the source lead of the MOSFET which provides a feedback action that acts to stabilize the drain current of

5.5 Investigating Bias Stability With Spice

```
Simple MOSFET Bias Network (No Feedback Mechanism)

** Circuit Description **
* dc supply
Vdd 1 0 DC +15V
* amplifier circuit
M1 3 4 0 0 nmos L=100u W=100u
Rg1 1 4 1.5Meg
Rg2 4 0 500k
Rd 1 2 2k
* drain current monitor
Vdrain 2 3 0
* mosfet model statement (by default, level 1)
.model nmos nmos (kp=2m Vto=+2V lambda=0)
** Analysis Requests **
.OP
.SENS I(Vdrain)
** Output Requests **
.end
```

Figure 5.30 The Spice input file for calculating the DC sensitivities of the drain current of the MOSFET amplifier shown in Fig. 5.29(a).

the MOSFET when subjected to change. What this means is, if one of the biasing elements undergoes a small change, then the resulting drain current of the MOSFET will experience less change than when no feedback action is present. To see this, we created the two Spice input files listed in Figs. 5.30 and 5.31 with the following sensitivity analysis request:

.SENS I(Vdrain)

This command will invoke Spice to compute the DC sensitivities of the drain current of each MOSFET as monitored by the zero-valued voltage source that appears in series with its drain terminal. The sensitivity analysis command of Spice was first introduced to the reader in Section 4.5 of this text.

The results of this analysis, as found in the Spice output file, are shown below for the amplifier circuit of Fig. 5.29(a). First we show the bias conditions of the amplifier, followed by a list of the circuit DC sensitivities.

```
****      OPERATING POINT INFORMATION      TEMPERATURE =   27.000 DEG C

**** MOSFETS

NAME      M1
MODEL     nmos
ID        3.06E-03
VGS       3.75E+00
VDS       8.88E+00
VBS       0.00E+00
VTH       2.00E+00
```

5 Field-Effect Transistors (FETs)

```
MOSFET Bias Network With Feedback

** Circuit Description **
* dc supply
Vdd 1 0 DC +15V
* amplifier circuit
M1 3 4 5 0 nmos L=100u W=100u
Rg1 1 4 1Meg
Rg2 4 0 2Meg
Rd 1 2 2k
Rs 5 0 2.2k
* drain current monitor
Vdrain 2 3 0
* mosfet model statement (by default, level 1)
.model nmos nmos (kp=2m Vto=+2V lambda=0)
** Analysis Requests **
.OP
.SENS I(Vdrain)
** Output Requests **
.end
```

Figure 5.31 The Spice input file for calculating the DC sensitivities of the drain current of the MOSFET amplifier shown in Fig. 5.29(b).

```
         VDSAT       1.75E+00

         ****    DC SENSITIVITY ANALYSIS       TEMPERATURE =   27.000 DEG C

         DC SENSITIVITIES OF OUTPUT I(Vdrain)

              ELEMENT      ELEMENT       ELEMENT       NORMALIZED
              NAME         VALUE         SENSITIVITY   SENSITIVITY
                                         (AMPS/UNIT)   (AMPS/PERCENT)

              Rg1          1.500E+06     -6.563E-09    -9.844E-05
              Rg2          5.000E+05      1.969E-08     9.844E-05
              Rd           2.000E+03     -3.063E-15    -6.125E-14
              Vdd          1.500E+01      8.750E-04     1.313E-04
              Vdrain       0.000E+00     -1.000E-12     0.000E+00
```

Likewise, the results of the sensitivity analysis for the amplifier circuit of Fig. 5.29(b), together with the DC operating point information for the MOSFET, are as follows:

```
         ****         OPERATING POINT INFORMATION    TEMPERATURE =   27.000 DEG C

         **** MOSFETS

              NAME        M1
              MODEL       nmos
              ID          2.87E-03
              VGS         3.69E+00
              VDS         2.96E+00
              VBS        -6.31E+00
              VTH         2.00E+00
              VDSAT       1.69E+00
```

5.5 Investigating Bias Stability With Spice

```
****    DC SENSITIVITY ANALYSIS         TEMPERATURE =    27.000 DEG C

DC SENSITIVITIES OF OUTPUT I(Vdrain)

        ELEMENT         ELEMENT         ELEMENT         NORMALIZED
        NAME            VALUE           SENSITIVITY     SENSITIVITY
                                        (AMPS/UNIT)     (AMPS/PERCENT)

        Rg1             1.000E+06       -1.336E-09      -1.336E-05
        Rg2             2.000E+06        6.679E-10       1.336E-05
        Rd              2.000E+03       -2.867E-15      -5.734E-14
        Rs              2.200E+03       -1.149E-06      -2.527E-05
        Vdd             1.500E+01        2.672E-04       4.008E-05
        Vdrain          0.000E+00       -1.000E-12       0.000E+00
```

Reviewing the above sensitivity results, we see that Spice has generated several columns of output. The first column indicates the element that the sensitivity of the drain current I_D is taken with respect to. The second column indicates the nominal value of that element as it appears in the Spice input file. The third column indicates the sensitivity quantity $\frac{\partial I_D}{\partial x}$ where x is the corresponding element appearing in the leftmost column. The units of this sensitivity quantity are the units of the output variable specified on the .SENS statement divided by the units of the element x. For instance, in the case of element R_{G1}, the sensitivity quantity $\frac{\partial I_D}{\partial R_{G1}}$ is expressed in A/Ω. The final column that appears on the right, is a normalized sensitivity measure. It simply expresses the sensitivity in more convenient units of A/%. Mathematically, it is written as $\frac{\partial I_D}{(\partial R_{G1}/R_{G1})\%}$ where $(\partial R_{G1}/R_{G1})\%$ is the relative accuracy of R_{G1} expressed in per-cent.

Returning to the results of the two sensitivity analysis, let us consider one interpretation of these results: If we were to build the two amplifiers shown in Fig. 5.29 using resistors that have a tolerance of $\pm 5\%$, then according to the principle of a total derivative, the total change in the drain current due to variations in the biasing resistors (ie. R_{G1}, R_{G2}, R_D and R_S — in the case of the amplifier in Fig. 5.29(b)) can be approximated by the following:

$$\Delta I_D \approx \frac{\partial I_D}{(\partial R_{G1}/R_{G1})\%} \times (\Delta R_{G1}/R_{G1})\% + \frac{\partial I_D}{(\partial R_{G2}/R_{G2})\%} \times (\Delta R_{G2}/R_{G2})\% + \frac{\partial I_D}{(\partial R_S/R_S)\%} \times (\Delta R_S/R_S)\% + \frac{\partial I_D}{(\partial R_D/R_D)\%} \times (\Delta R_D/R_D)\% \quad (5.14)$$

Now, if we assume the worst-case situation, where each resistor undergoes the same amount of change with a sign that contributes to the total sum, as opposed to reduce the sum, then we can write the above equation as,

$$\Delta I_D \approx \left[\left| \frac{\partial I_D}{(\partial R_{G1}/R_{G1})\%} \right| + \left| \frac{\partial I_D}{(\partial R_{G2}/R_{G2})\%} \right| + \left| \frac{\partial I_D}{(\partial R_S/R_S)\%} \right| + \left| \frac{\partial I_D}{(\partial R_D/R_D)\%} \right| \right] \times (\Delta R/R)\% \quad (5.15)$$

5 Field-Effect Transistors (FETs)

where we denote the relative magnitude change of each resistor as $(\Delta R/R)_\%$. Now, under the assumed worst-case condition, $(\Delta R/R)_\% = 5\%$, we can substitute the sensitivities computed above by Spice into Eqn. 5.15 and determine the expected change in the drain current of the MOSFET in each amplifier of Fig. 5.29. On doing so, we find that for the amplifier with the fixed gate voltage biasing scheme (Fig.5.29(a)) the expected change in the drain current will be about 984.4 μA. Whereas, in the amplifier with a negative feedback biasing scheme (Fig. 5.29(b)), the worst-case change in the drain current is expected to be only 260 μA. This is about 4 times less change than the previous case.

This same type of analysis can be repeated with regards to a variation in the supply voltage V_{DD}. If we consider a 1% change in the supply voltage, then we can expect that the drain current of the MOSFET in the fixed-bias amplifier will experience a 131 μA change. This is contrasted against the amplifier with a negative feedback biasing scheme whose MOSFET drain current will only change by 40 μA.

Unfortunately, the sensitivity analysis seen above does not list the sensitivity relative to the MOSFET. To obtain this information, we use a simple brute-force approach. Consider changing the device parameters K and V_t separately and observe their effect on the drain current through an operating point (.OP) command.

For instance, in the case of the fixed-bias MOSFET amplifier stage shown in Fig. 5.29, consider changing the value of $kp=2$ mA/V^2 on the MOSFET model statement seen in Fig. 5.30 to $kp=2.2$ mA/V^2, a positive change of +10%. Re-running the Spice job, we obtain the following operating point information for the MOSFET:

```
**** MOSFETS

NAME        M1
MODEL       nmos
ID          3.37E-03
VGS         3.75E+00
VDS         8.26E+00
VBS         0.00E+00
VTH         2.00E+00
VDSAT       1.75E+00
```

Thus, the drain current is now 3.37 mA. Comparing this to the drain current prior to the change in kp, listed previously at 3.06 mA, we see that a +10% change in the process transconductance kp results in a +10.1% change in the drain current.

Similarly, if we repeat this same experiment with the MOSFET amplifier having a feedback biasing scheme, that is, consider changing kp of 2 mA/V^2 to 2.2 mA/V^2 in the Spice deck seen listed in Fig. 5.31 and re-simulating the circuit, we get the following operating point results:

```
**** MOSFETS

NAME      M1
MODEL     nmos
ID        2.90E-03
VGS       3.62E+00
VDS       2.83E+00
VBS      -6.38E+00
VTH       2.00E+00
VDSAT     1.62E+00
```

Thus, the drain current becomes 2.90 mA. Comparing this to the drain current prior to the change in kp, listed previously at 2.87 mA, we see that a +10% change in the process transconductance kp results in a +1.04% change in the drain current. This is about 10 times less sensitive to a process change in kp than the fixed biasing scheme.

A similar approach can be taken for changes in the threshold voltage. If we assume that due to a process variation, the threshold voltage of the MOSFET changes by −5%, then we would find for the fixed biasing scheme that the drain current would change by +11.8%. In contrast, the drain current in the MOSFET of the amplifier having a negative feedback biasing scheme would only change by +1.39%.

The benefits of an amplifier biasing scheme that incorporates negative feedback should now be self-evident.

5.6 Integrated-Circuit MOS Amplifiers

In this section we shall investigate the behavior of several different types of fully-integratable amplifiers that are constructed with MOSFETs only.

5.6.1 Enhancement-Load Amplifier Including The Body Effect

Fig. 5.32 shows an enhancement-load NMOS amplifier with the substrate connections clearly shown. This arrangement would be typical of an amplifier implemented in an NMOS fabrication process. One important drawback to this amplifier is that its voltage gain is reduced because of the presence of the MOSFET body-effect in transistor M_2. To see this, let us consider that the two MOSFETs in the circuit of Fig. 5.32 have the following device parameters: a process transconductance coefficient ($\mu_n C_{OX}$) of 0.25 mA/V^2, a zero-bias threshold voltage of 1 V, a channel-length modulation factor λ of 0.02 V^{-1}, and a body-effect coefficient γ of 0.9 V$^{1/2}$. Transistor M_1 will have a length − width dimension of 10 μm by 100 μm whereas the dimensions of transistor M_2 will be reciprocated at 100 μm by 10 μm. The Spice input file describing this arrangement is listed in Fig. 5.33. A DC

5 Field-Effect Transistors (FETs)

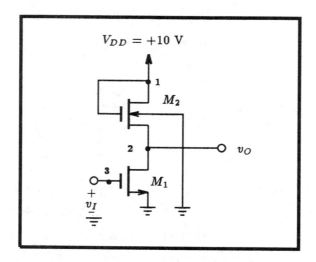

Figure 5.32 An enhancement-load amplifier.

An Enhancement-Load Amplifier Including Body Effect

```
** Circuit Description **
* dc supplies
Vdd 1 0 DC +10V
* input signal
Vi 3 0 DC 0V
* amplifier circuit
M1 2 3 0 0 nmos L=10u W=100u
M2 1 1 2 0 nmos L=100u W=10u
* mosfet model statement (by default, level 1)
.model nmos nmos (kp=0.25m Vto=+1.0V lambda=0.02 gamma=0.9)
** Analysis Requests **
.OP
.DC Vi 0V 10V 100mV
** Output Requests **
.PLOT DC V(2)
.Probe
.end
```

Figure 5.33 The Spice input file for calculating the DC transfer characteristic of the enhancement-load amplifier shown in Fig. 5.32. Each MOSFET is modeled with the effect of transistor body-effect included.

sweep of the input voltage level between ground and V_{DD} is requested. For comparison, we shall repeat the same analysis just described on an identical circuit, with identical device parameters except that the body-effect coefficient will be set to zero. The Spice deck for this particular case is simply concatenated on the end of the Spice file shown in Fig. 5.33 and both are submitted to Spice for analysis.

The results of the analysis are shown in Fig. 5.34. Here the DC transfer characteristics

5.6 Integrated-Circuit MOS Amplifiers

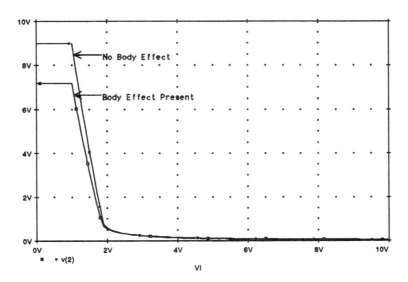

Figure 5.34 The DC transfer characteristic of the enhancement-load amplifier shown in Fig. 5.32 with and without the MOSFET body-effect present.

of the enhancement-load amplifier with and without the transistor body-effect present are shown. As is clearly evident, transistor body-effect alters the transfer characteristics of the enhancement-load amplifier significantly. With the input level below one-volt, corresponding to the threshold of M_1, the output voltage is held constant at either 9 V or 7.2 V, depending on which characteristic curve one is looking at. In the case of the transfer characteristic curve for the enhancement-load circuit with the body-effect present, with input levels increasing above the one-volt level, the output voltage decreases linearly at a rate of about −7.9 volt-per-volt from its initial 7.2 V level until the input exceeds approximately 1.8 V. At an input of 1.8 V the output is 0.75 V. Above this input voltage level, transistor M_1 leaves the saturation-region and enters the triode region, resulting in the amplifier characteristics becoming nonlinear.

In the case of the amplifier with the body-effect eliminated, with inputs above 1 V, the output level (beginning at 9 V) decreases linearly at a rate of −9.2 volt-per-volt. Like the previous case, when the input voltage level exceeds 1.8 V (and the output at 0.75 V), transistor M_1 enters the triode region and the amplifier characteristic curve becomes nonlinear. Comparing these details to that of the enhancement-load amplifier with the body-effect included suggest that the presence of transistor body-effect has decreased the effective gain of this enhancement-load amplifier.

To further demonstrate this, let us compute the voltage gain of this amplifier, with and

5 Field-Effect Transistors (FETs)

without the body-effect present, using the transfer function (.TF) analysis command of Spice with the amplifier biased in its linear region. For illustrative purposes, we shall bias the input to the amplifier at 1.5 V since this input level maintains both amplifiers in their linear region. Modifying each of the two Spice decks used previously, by changing the source statement to read as follows

```
Vi 3 0 DC +1.5V
```

and including the following .TF command

```
.TF V(2) Vi.
```

The results of the two small-signal transfer function analyses are then found in their respective output files. In the case of the enhancement-load amplifier with body-effect present, the results are:

```
****     SMALL-SIGNAL CHARACTERISTICS
         V(2)/Vi =  -7.316E+00
         INPUT RESISTANCE AT Vi =   1.000E+20
         OUTPUT RESISTANCE AT V(2) =   5.508E+03
```

For the case of the enhancement-load amplifier with the body-effect eliminated, the results are:

```
****     SMALL-SIGNAL CHARACTERISTICS
         V(2)/Vi =  -9.027E+00
         INPUT RESISTANCE AT Vi =   1.000E+20
         OUTPUT RESISTANCE AT V(2) =   6.677E+03
```

The above two sets of results, therefore, re-confirm our earlier claim that the MOSFET body-effect acts to decrease the effective-gain of an enhancement-load amplifier.

Before we leave this section it would be instructive to confirm that the small-signal formula for the voltage gain of this amplifier including transistor body-effect is accurate. According to the development provided in Section 5.9 of Sedra and Smith, the voltage gain of the amplifier shown in Fig. 5.32 is given by the following equation:

$$A_V = -\frac{g_{m_1}}{g_{m_2} + g_{m_{b2}} + g_{ds_1} + g_{ds_2}} \tag{5.16}$$

Through an operating point (.OP) analysis command (included in the previous analysis), we found the following values for the small-signal model parameters of each transistor:

5.6 Integrated-Circuit MOS Amplifiers

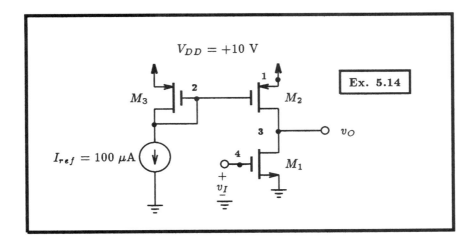

Figure 5.35 A CMOS amplifier with current source biasing.

```
**** MOSFETS

NAME      M1          M2
MODEL     nmos        nmos
ID        3.32E-04    3.32E-04
VGS       1.50E+00    6.87E+00
VDS       3.13E+00    6.87E+00
VBS       0.00E+00   -3.13E+00
VTH       1.00E+00    2.04E+00
VDSAT     5.00E-01    4.83E+00
GM        1.33E-03    1.37E-04
GDS       6.25E-06    5.84E-06
GMB       7.72E-04    3.20E-05
```

Substituting the appropriate values into Eqn. (5.16), we find $A_V = -7.344$ V/V. This is very close to the value predicted directly by Spice above (ie. $A_V = -7.316$ V/V). If the output conductances are neglected in this calculation, then we would find $A_V = -7.869$ V/V. This result, for most practical applications, would be more than adequate.

5.6.2 CMOS Amplifier

As an example of an amplifier that is fully integratable using MOS technology, we display in Fig. 5.35 a CMOS amplifier with an active current-source load. Using Spice we would like to compute and plot the amplifier DC transfer characteristic (ie. v_O vs. v_I). This particular circuit was presented in Example 5.14 of Sedra and Smith where it was assumed that $V_{tn} = |V_{tp}| = 1$ V, $\mu_n C_{OX} = 2\mu_p C_{OX} = 20$ μA/V^2, and $\lambda = 0.01$ V^{-1} for n and p devices. Furthermore, for all devices it was assumed that $W = 100$ μm and $L = 10$ μm.

The Spice input file corresponding to this CMOS amplifier is listed in Fig. 5.36. A zero-valued DC voltage source V_I is initially applied to the input of the amplifier, and will be varied over a range of values beginning at ground potential and increasing to V_{DD} in

5 Field-Effect Transistors (FETs)

Example 5.14: A CMOS Amplifier

```
** Circuit Description **
* dc supplies
Vdd 1 0 DC +10V
Iref 2 0 DC 100uA
* input signal
Vi 4 0 DC 0V
* amplifier circuit
M1 3 4 0 0 nmos L=10u W=100u
M2 3 2 1 1 pmos L=10u W=100u
M3 2 2 1 1 pmos L=10u W=100u
* mosfet model statements (by default, level 1)
.model nmos nmos (kp=20u Vto=+1V lambda=0.01)
.model pmos pmos (kp=10u Vto=-1V lambda=0.01)
** Analysis Requests **
* calculate DC transfer characteristics
.DC Vi 0V +10V 10mV
** Output Requests **
.PLOT DC V(3)
.probe
.end
```

Figure 5.36 The Spice input file for calculating the DC transfer characteristic of the CMOS amplifier circuit shown in Fig. 5.35.

small 10 mV steps. The output voltage (V(3)) will then be plotted as a function of the input voltage V_I.

The DC transfer characteristic of the CMOS amplifier, as calculated by Spice, is shown in Fig. 5.37. Here we see that the output voltage is very nearly 10 V for input signals less than about +1.0 V and near ground potential when the input level exceeds +2.5 V. Between these two values, the output voltage begins to change value in a somewhat gradual fashion, except around the 2 V input level. Here the output level changes quite dramatically, albeit in a linear manner. Thus, the CMOS amplifier experiences large voltage gain in the vicinity of the 2 V input level.

To see the high-gain region of this amplifier more closely, we will repeat the previous DC sweep analysis and evaluate the amplifier transfer characteristic ranging between +1.9 V and +2.1 V. A very small step-size of 100 μV is used to obtain a smooth curve through the high-gain region of the amplifier. The results are shown in Fig. 5.38. The linear region of the amplifier is clearly visible. It is bounded between input voltages of 1.955 V and 2.027 V. Correspondingly, the output voltage varies between 8.589 V and 0.9966 V. This suggests that the gain of this amplifier in this linear region is approximately $\frac{8.589-0.9966}{1.955-2.027} = -118.9$ V/V.

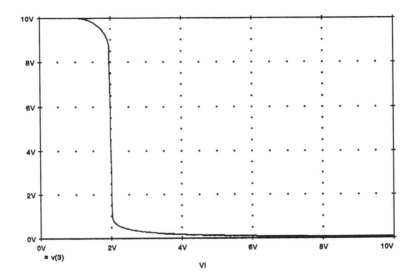

Figure 5.37 The DC transfer characteristic of the CMOS amplifier shown in Fig. 5.35 as calculated by Spice.

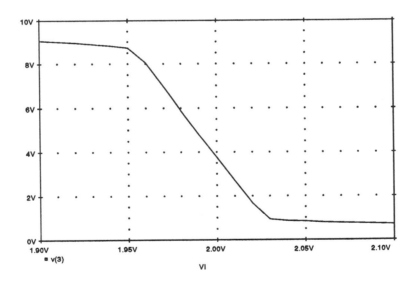

Figure 5.38 An expanded view of the large-signal transfer characteristic of the CMOS amplifier shown in Fig. 5.35 in its high-gain region.

A similar result is also obtained when the small-signal gain is evaluated around a single operating point inside this linear region. Consider this point to be midway between the extremes of the linear region; this would correspond to an input bias level of about 2 V.

Modifying the Spice input deck shown in Fig. 5.36 so that the input is biased at 2 V and replacing the DC sweep command seen there by a .TF analysis command, one would find in the Spice output file the following small-signal DC transfer function information:

```
****    SMALL-SIGNAL CHARACTERISTICS

        V(3)/Vi = -1.050E+02

        INPUT RESISTANCE AT Vi =  1.000E+20

        OUTPUT RESISTANCE AT V(3) =  5.059E+05
```

Thus, the small-signal gain of the CMOS amplifier is −105 V/V at an input DC voltage level of 2 V.

It is interesting to note that the simple formula for the voltage gain in the linear region of the CMOS amplifier shown in Fig. 5.35 derived in Section 5.9 of Sedra and Smith, and repeated here below,

$$A_V = -\frac{\sqrt{(1/2)\mu_n C_{OX}(W/L)_1} \; |V_A|}{\sqrt{I_{ref}}}, \qquad (5.17)$$

results in a gain value that is quite close to that computed by Spice above. Substituting the appropriate device parameter values into Eqn. (5.17), we get $A_V = -100$ V/V.

5.7 MOSFET Switches

MOSFETs are commonly used as switches for both analog and digital signals. In analog circuit applications, a switch is used to control the passage of an analog current signal between two nodes in a circuit, in either direction, without distortion or attenuation. In digital applications, a switch is commonly used as a transmission gate to realize specific logic functions when combined with other logic gates.

An ideal mechanical switch is depicted in Fig. 5.39(a). In the "on" state, ie. switch closed, a direct connection is made between nodes 1 and 2. Thus, a signal applied to node 1 will also appear at node 2. The reverse is also true if nodes 1 and 2 are interchanged. In practise, a signal passing through a switch will experience a signal attenuation due to the electrical resistance of the switch. Thus, we can model this behavior by adding a resistance R_{ON} in series with the ideal switch, as illustrated in Fig. 5.39(b). Clearly then, the larger R_{ON}, the more loss a signal will experience as it passes through the switch.

To judge the on-resistance of a single n-channel MOSFET switch, let us create a Spice input file that represents the situation shown in Fig. 5.40(a). In this circuit, we shall sweep the input voltage v_I from $V_{SS} = -5$ V to $V_{DD} = +5$ V and compute the current that

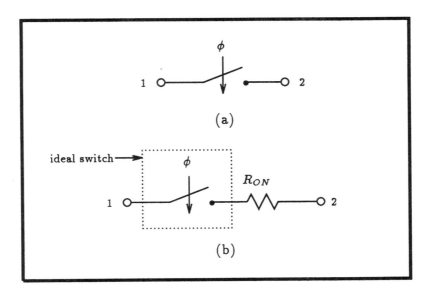

Figure 5.39 (a) An ideal mechanical switch (b) Electrical equivalent of a real switch with on-resistance R_{ON}.

is supplied to the switch by v_I. The other terminal of the MOSFET switch is connected directly to ground, thus the resistance of the switch R_{ON} is given by the ratio of the input voltage to the current supplied by this voltage source. The gate of the MOSFET, being the control terminal of the switch, will be held at $V_{DD} = +5$ V to ensure that the switch is turned-on. The resulting Spice input file is shown in Fig. 5.41. The NMOS transistor will be assumed to have a process transconductance parameter $\mu_n C_{OX}$ equal to 0.25 mA/V^2, a zero-bias threshold voltage of 1 V, a channel-length modulation factor λ of 0.02 V^{-1}, and a body-effect coefficient γ of 0.9 V$^{1/2}$. The dimensions of the MOSFET will be 100 μm by 100 μm.

It should be pointed out that the DC sweep was selected to have a voltage step of 10.01 mV instead of a more even 10 mV. This is to ensure that when we compute the ratio of input voltage to input current, we don't try and divide 0 by 0. Moreover, our choice of which terminal of the MOSFET is the source, and which is the drain, was arbitrary. The results should be same regardless of our choice. (If there is any doubt, the reader should repeat the simulation with the source and drain terminals interchanged).

On completion of Spice, we divide the input voltage by the input current and plot the results as a function of the input voltage. The curve representing this result is shown in Fig. 5.42 and is labeled as the NMOS switch. As is evident, the on-resistance of the switch is lowest at $R_{ON} = 560$ Ω when the input equals $V_{SS} = -5$ V. The switch on-resistance steadily increases as the input voltage increases towards $V_{DD} = +5$ V. In fact, as the input voltage approaches V_{DD}, the switch on-resistance increases substantially, eg. at $v_I = +5$ V,

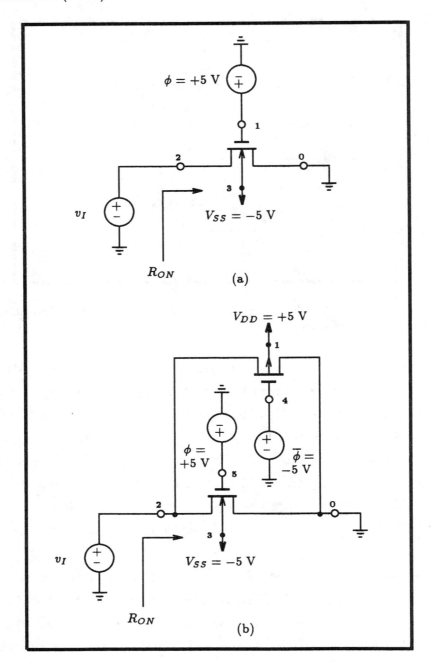

Figure 5.40 MOSFET switch realizations: (a) a single n-channel MOSFET (b) a CMOS transmission gate.

$R_{ON} = 5.6$ kΩ. We can conclude that the on-resistance of this switch is dependent on the input signal level, which introduces undesirable nonlinearity.

If we repeat the above experiment on a p-channel MOSFET, similar results would result. These are also shown in Fig. 5.42. Clearly, the PMOS switch has similar behavior, but in a complementary manner. Observing these results, we can conclude that a single MOSFET

```
Computing Switch On-Resistance Of A NMOS Switch

** Circuit Description **
* Input signal
Vi 2 0 DC 0V
* Substrate Bias
Vss 3 0 DC -5V
* Control Signal
Vphi 1 0 DC +5V
* Switch: NMOS Transistor
M1 2 1 0 3 nmos_enhancement_mosfet L=100u W=100u
* mosfet model statement (by default, level 1)
.model nmos_enhancement_mosfet nmos (kp=0.25m Vto=+1.0V lambda=0.02 gamma=0.9)
** Analysis Requests **
* sweep the input voltage from Vss to Vdd (skip over 0,0 point)
.DC Vi -5V +5V 10.01mV
** Output Requests **
.PLOT DC I(Vi) V(2)
.Probe
.end
```

Figure 5.41 The Spice input file for calculating the input current as a function of input voltage for the n-channel MOSFET switch of Fig. 5.39(a). Post-processing will be used to compute the on-resistance of the switch.

does not make for a very effective switch.

A better approach, and one that is extensively used in analog IC design when a high-quality switch is required, is to connect both the n-channel and the p-channel MOSFETs in parallel as illustrated in Fig. 5.40(b). The gate control of each MOSFET is driven by complementary signals. To see why this transistor arrangement makes for an effective switch, let us compute the on-resistance of this switch. To turn on the switch, control signal ϕ is set to $V_{DD} = +5$ V and $\overline{\phi}$ is set to $V_{SS} = -5$ V. The Spice input file describing the situation is listed in Fig. 5.43. The MOSFETs are assumed to have the same device parameters used in the previous example. For easy comparisons, the Spice results for this CMOS switch are shown superimposed on the same graph of on-resistance of the NMOS and PMOS switches. Clearly, the on-resistance of the CMOS transmission gate is much more constant than was the case for a single MOSFET switch. Furthermore, the on-resistance of the CMOS switch is much lower than that of the previous two switches. The on-resistance is seen to vary between 500 Ω and 800 Ω, with the maximum on-resistance occurring when the input signal is zero.

This example did not show another limitation of the single transistor switch, namely, its limited range of operation which is caused by its non-zero threshold voltage. To illustrate

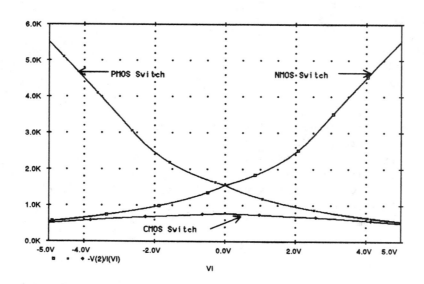

Figure 5.42 The on-resistance R_{ON} as a function of the input signal level of an analog switch realized as: (a) a single n-channel MOSFET, or (b) a single p-channel MOSFET, (c) a parallel combination of an n-channel and p-channel MOSFETs.

this limitation, consider the switch arrangement shown in Fig. 5.44. Here a single n-channel MOSFET is connected between the voltage source v_I and a load consisting of a 100 kΩ resistor and a 10 pF capacitor. The n-channel MOSFET is assumed identical to that just described above. Moreover, the gate of this MOSFET is biased at +5 V, intended to turn the switch fully on. The Spice deck for this case is provided in Fig. 5.44. Now if we simulate the action of this switch using Spice with a 10 V step input beginning at −5 V, then, instead of the voltage at the output following the input in its entirety, we see that the output voltage actually clamps at a level of about 1.85 V as seen in Fig. 5.46. It is easily shown that this clamping limit depends on the threshold voltage of the MOSFET (see Sedra and Smith), but it may not be quite so obvious that this clamping limit also depends on the i-v characteristics of the switch. To convince one self of this, consider increasing the size of the transistor so that its i-v characteristics change. For example, let us increase the width of the MOSFET to 500 μm. On re-running the Spice simulation we find that the output voltage has similar behavior as in the previous case except that the output voltage clamps at a higher voltage of 2.0 V. This is also shown in Fig. 5.46. The reason for this behavior is because the MOSFET must supply a continuous current to the RC load in order to sustain the output voltage.

```
Computing Switch On-Resistance Of A CMOS Switch

** Circuit Description **
* Input signal
Vi 2 0 DC 0V
* Substrate Bias
Vdd 1 0 DC +5V
Vss 3 0 DC -5V
* Control Signal
Vphi    4 0 DC -5V
Vphibar 5 0 DC +5V
* Switch: NMOS + PMOS Transistors
M1 2 5 0 3 nmos_enhancement_mosfet L=100u W=100u
M2 2 4 0 1 pmos_enhancement_mosfet L=100u W=100u
* mosfet model statement (by default, level 1)
.model nmos_enhancement_mosfet nmos (kp=0.25m Vto=+1.0V lambda=0.02 gamma=0.9)
.model pmos_enhancement_mosfet pmos (kp=0.25m Vto=-1.0V lambda=0.02 gamma=0.9)
** Analysis Requests **
* sweep the input voltage from Vss to Vdd (skip over 0,0 point)
.DC Vi -5V +5V 10.01mV
** Output Requests **
.PLOT DC I(Vi) V(2)
.Probe
.end
```

Figure 5.43 The Spice input file for calculating the input current as a function of input voltage for the CMOS transmission gate of Fig. 5.39(b). Post-processing will be used to compute the on-resistance of the switch.

5.8 Describing MESFETs To PSpice

Built-in models for metal-semiconductor FETs (MESFETs) do not exist in Spice (versions 2G6 and earlier). Instead, we shall rely on the built-in MESFET model of PSpice to carry out our simulation of circuits containing MESFETs.

MESFETs are described to PSpice in the exact same way as any other semiconductor device is described to Spice; that is, using an element statement and a model statement. The following outlines the syntax of these two statements, including the details of the built-in MESFET model of PSpice.

5.8.1 MESFET Element Description

MESFETs are described in a PSpice listing using an element statement beginning with a unique name prefixed with the letter B. This is followed by a list of nodes that the drain, gate, and source of the MESFET are connected to. Subsequently, in the next field, a name of a model characterizing the particular MESFET is given – more on this in a moment. The

5 Field-Effect Transistors (FETs)

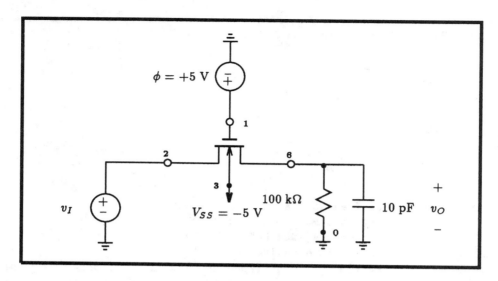

Figure 5.44 Circuit arrangement illustrating the effect of MOSFET threshold voltage on the switch range of operation.

```
The Effect Of The Threshold Voltage On Switch Operation

** Circuit Description **
* Input signal
Vi 2 0 DC 0V PWL (0 -5V 100ns -5V 200ns 5V 10ms 5V)
* Substrate Bias
Vss 3 0 DC -5V
* Control Signal
Vphi 1 0 DC +5V
* Switch: NMOS Transistor
M1 2 1 6 3 nmos_enhancement_mosfet L=100u W=100u
* Load resistor
Rload 6 0 100k
Cload 6 0 10pF
* mosfet model statement (by default, level 1)
.model nmos_enhancement_mosfet nmos (kp=0.25m Vto=+1.0V lambda=0.02 gamma=0.9)
** Analysis Requests **
.TRAN 1ns 500ns 0ns 1ns
** Output Requests **
.PLOT TRAN V(6)
.Probe
.end
```

Figure 5.45 The Spice input file for calculating the step response of the NMOS switch arrangement shown in Fig. 5.44.

final field of this statement is optional and its purpose is to allow one to scale the size of the device by declaring the number of MESFETs connected in parallel. A summary of the element statement syntax for the n-channel MESFET is provided in Fig. 5.47. Included

5.8 Describing MESFETs To PSpice

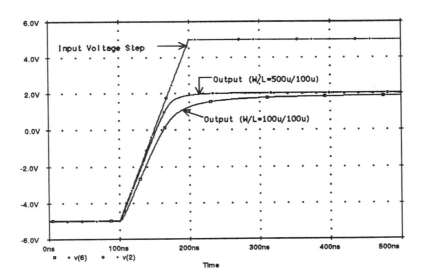

Figure 5.46 The step response of a NMOS switch arrangement shown in Fig. 5.44 for two different sized MOSFETs.

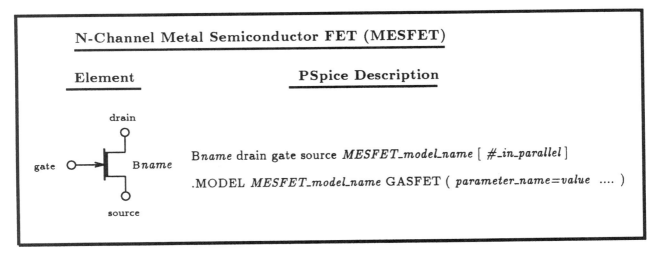

Figure 5.47 PSpice element description for the n-channel depletion-mode GaAs MESFET. Also listed is the general form of the associated MESFET model statement. A partial listing of the parameters applicable to the n-channel MESFET is given in Table 5.5.

in this list is the corresponding syntax for the MESFET model statement (.MODEL) that must be present whenever a MESFET is made reference to. This statement defines the terminal characteristics of the MESFET by specifying the values of particular parameters in the MESFET model. Parameters not specified in the model statement are assigned default values by PSpice. Details of this model will be discussed briefly next.

249

5 Field-Effect Transistors (FETs)

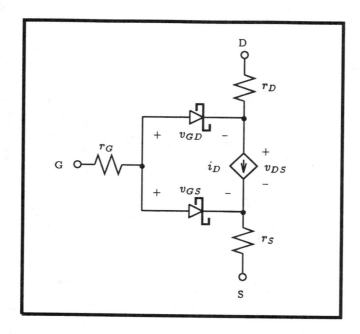

Figure 5.48 The PSpice large-signal MESFET model under static conditions.

5.8.2 MESFET Model Description

As is evident from Fig. 5.47, the model statement for the n-channel MESFET begins with the keyword .MODEL and is followed by the name of the model used by a MESFET element statement, the nature of the MESFET (ie. GASFET), and a list of the parameters characterizing the terminal behavior of the MESFET is enclosed between brackets. The parameters describing the terminal characteristics of the MESFET are quite similar to those used to describe the JFET above. An additional parameter (α) has been added to properly account for the early saturation phenomenon of the MESFET [Hodges and Jackson, 1983].

The general form of the DC Spice model for an n-channel MESFET is illustrated schematically in Fig. 5.48. The bulk resistance of the drain, gate, and source regions of the MESFET are lumped into three linear resistances r_D, r_G, and r_S, respectively. The DC characteristic of the intrinsic MESFET is determined by the nonlinear dependent current source i_D and the two Schottky-barrier diodes. These two Schottky diodes represent the metal semiconductor junctions that define the channel region. The functional description of the drain current i_D can take on three different forms depending on the value assigned to the parameter LEVEL that appears on the MESFET model statement. PSpice defaults to a LEVEL 1 model when none is specified. The LEVEL 1 MESFET model is probably the most widely used of the three and will be the only one discussed here. Details pertaining to the other models can be found in the *PSpice Users' Manual*.

The equation describing the drain current for the LEVEL 1 MESFET model is given

5.8 Describing MESFETs To PSpice

Symbol	PSpice Name	Model Parameter	Units	Default
	Level	Model type		1
β	beta	Transconductance coefficient	A/V^2	0.1
α	alpha	Saturation voltage parameter	V^{-1}	2.0
V_t, V_P	Vto	Pinch-off voltage	V	-2.5
λ	lambda	Channel-length modulation	V^{-1}	0
r_G	Rg	Gate ohmic resistance	Ω	0
r_D	Rd	Drain ohmic resistance	Ω	0
r_S	Rs	Source ohmic resistance	Ω	0

Table 5.5 A partial listing of the PSpice parameters for the LEVEL 1 MESFET model.

below:

$$i_D = \begin{cases} 0 & \text{for } v_{GS} < V_t \\ \beta(v_{GS} - V_t)^2(1 + \lambda v_{DS})\tanh(\alpha v_{DS}) & \text{for } v_{GS} > V_t \end{cases} \quad (5.18)$$

Here the same equation is used to describe both the triode and saturation regions of the device; the distinction between the two is provided by the factor $\tanh(\alpha v_{DS})$. This factor does also accounts for the early saturation phenomenon observed in MESFETs. See the i-v characteristics in Fig. 5.49 for a typical MESFET having device parameters $V_t = -1$ V, $\beta = 10^{-4}$ A/V^2, $W = 100$ μm, $L = 1$ μm, $\lambda = 0.05$ V^{-1}, and α taking on values of 0.5, 1.0, and 2.0. As an extreme limit, the i-v characteristic for $\alpha = 10^6$ is also included. In these equations, β and λ have the same meaning as in the Spice JFET model. That is, β is the device transconductance parameter, and λ is the channel-length modulation coefficient.

A partial listing of the parameters associated with the Spice MESFET model under static conditions is given in Table 5.5. Also listed are the associated default values which a parameter assumes if a value is not specified for it on the .MODEL statement. To specify a parameter value one simply writes, for example: level=1, beta=20u, Vto=-1V, etc.

5.8.3 Small-Signal MESFET Model

The linear small-signal model of the n-channel MESFET is identical to that seen previously for the JFET in Fig. 5.26(b). It consists of a single voltage-controlled current source having transconductance g_m. Here g_m is the MESFET transconductance and is related to the DC bias according to the following:

5 Field-Effect Transistors (FETs)

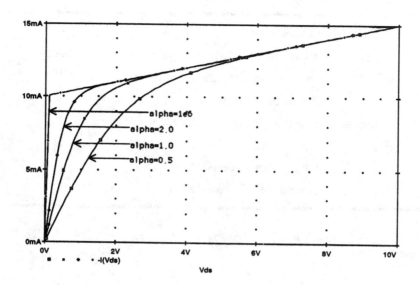

Figure 5.49 Illustrating the dependence of the i-v characteristic of a single MESFET on the value of α.

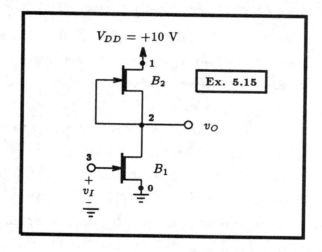

Figure 5.50 A MESFET amplifier with MESFET load.

$$g_m = \left.\frac{\partial i_D}{\partial v_{GS}}\right|_{OP} = 2\beta(V_{GS} - V_t)(1 + \lambda V_{DS})\tanh(\alpha V_{DS}) \quad (5.19)$$

Likewise, the output conductance g_{ds}, which is also equal to $1/r_o$, is computed according to the following:

$$g_{ds} = \left.\frac{\partial i_D}{\partial v_{DS}}\right|_{OP} \quad (5.20)$$

252

5.8.4 A MESFET Biasing Example

As an example of an application of GaAs MESFETs, in Fig. 5.50 we display a simple MESFET amplifier circuit. The lengths of the two devices are assumed to be equal to the minimum value for a particular technology (1 μm in this case). Whereas, the width dimensions of B_1 and B_2 are $W_1 = 100$ μm and $W_2 = 50$ μm, respectively. The characteristics of the GaAs process is specified in terms of the electrical parameters of minimum sized devices (ie. $L = 1$ μm and $W = 1$ μm), and will be assumed to be: $V_t = -1$ V, $\beta = 10^{-4}$ A/V^2 and $\lambda = 0.1$ V^{-1}. To determine the parameters of a particular device, it is a simple matter to scale the appropriate parameters of the unit-sized device. Alternatively, we can let PSpice perform this scaling operation for us. We shall illustrate the latter, since PSpice is less likely to make a mistake than we are.

Using the device parameters for the minimum-sized device, we can establish the following PSpice model statement for the unit-sized n-channel MESFET:

```
.model n_mesfet gasfet (beta=0.1m Vto=-1.0V lambda=0.1)
```

Parameters not specified will, as usual, assume default values.

To account for the different widths of the two transistors, we simply specify on the element statement of each MESFET the ratio of the transistor width to the unit-sized width. In this case, the unit-sized width is 1 μm, so for B_1 with a device width of 100μm one would specify a scale factor of 100. Of course, this has identical meaning as connecting 100 unit-sized transistors in parallel. Likewise, one would specify a scale factor of 50 for the other transistor. To illustrate, consider the element statement for transistor B_1 is

```
B1 2 3 0 n_mesfet 100.
```

With the aid of Spice we would like to determine the DC transfer characteristics of this amplifier. This is easily performed using the PSpice listing given in Fig. 5.51. Here we are sweeping the input voltage level between –10 V and +1 V in 10 mV steps. The input voltage should normally remain negative, otherwise the input metal-semiconductor junction of B_1 (the gate-source Schottky diode) becomes forward biased. For demonstration purposes, we are allowing the input voltage to go slightly positive in order to observe the effect of forward biasing this junction.

The resulting DC transfer characteristics of this MESFET amplifier are on display in Fig. 5.52. Here we see that the amplifier has inverter-like characteristics, with its high gain region in the vicinity of a –0.3 V input level. When the input signal level goes more positive than 0.7 V we see that the output voltage begins to rise, corresponding to the fact the input

5 Field-Effect Transistors (FETs)

```
Example 5.15: A MESFET Amplifier

** Circuit Description **
* dc supplies
Vdd 1 0 DC +10V
* input signal
Vi 3 0 DC -0.3V
* amplifier circuit
B1 2 3 0 n_mesfet 100
B2 1 2 2 n_mesfet 50
* mesfet model statements (by default, level 1)
.model n_mesfet gasfet (beta=0.1m Vto=-1.0V lambda=0.1)
** Analysis Requests **
* calculate DC transfer characteristics
.DC Vi -10V +1V 0.01V
** Output Requests **
.PLOT DC V(2)
.probe
.end
```

Figure 5.51 The Spice input file for calculating the DC transfer characteristic of the MESFET amplifier circuit shown in Fig. 5.50.

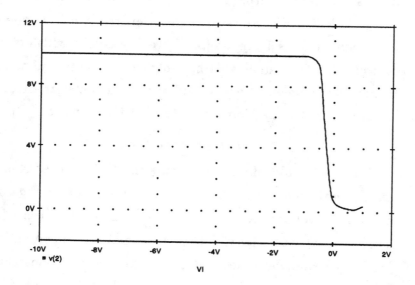

Figure 5.52 The DC transfer characteristics of the MESFET amplifier shown in Fig. 5.50 as calculated by PSpice.

metal-semiconductor junction of the MESFET has become forward biased and that the gate voltage no longer controls the drain-to-source current. Using the transfer function (.TF) command we can determine the voltage gain in the high-gain region of the amplifier. This requires that we bias the input to the amplifier at −0.3 V, as is already shown in the Spice deck of Fig. 5.51. Further, we need to change the analysis request from a DC sweep to the following small-signal transfer function command:

.TF V(2) Vi

We also include an operating point command (.OP) to obtain the small-signal model parameters.

On completion of Spice, we find the following small-signal characteristics associated with the MESFET amplifier:

```
****  SMALL - SIGNAL CHARACTERISTICS

V(2) / Vi = -2.143E+01

INPUT RESISTANCE AT Vi = 4.269E+10

OUTPUT RESISTANCE AT V(2) = 1.010E+03
```

Here we see that the gain in the linear region of the MESFET amplifier is −21.43 V/V, about a factor of 5 times less than the gain available with the simple CMOS amplifier discussed previously in Section 5.6. This result seems to correspond directly with the result that the output resistance for the MESFET amplifier is about five times lower than the output resistance of the CMOS amplifier.

The small-signal model parameters of the two MESFETs as calculated by PSpice are found in the output file as follows:

```
****  GASFETS

NAME        B1          B2
MODEL       n_mesfet    n_mesfet
ID          7.42E-03    7.42E-03
VGS         -3.00E-01   0.00E+00
VDS         5.15E+00    4.85E+00
GM          2.12E-02    1.48E-02
GDS         4.90E-04    5.00E-04
```

It is re-assuring that when these values are substituted into the formula for the small-signal voltage gain given by $A_V = -g_m(r_{o1} \| r_{o2})$ we get very similar results to those computed above using the .TF command (ie. $A_V = -21.41$ V/V).

5.9 Spice Tips

- FETS are described to Spice using an element statement and a model statement. The element statement describes the connections that the FET makes to the rest of the network and the name of the model that specifies its terminal behavior. The model statement assigns particular values to internal parameters of the built-in FET model of Spice. Parameters whose values are not specified on the model statement are assigned default values.

- Spice has built-in models for MOSFETs and JFETs.
- PSpice has built-in models for MOSFETs and JFETs, and in addition MESFETs.
- V_{t0} is positive for enhancement-mode n-channel MOSFETs and depletion-mode p-channel MOSFETs.
- V_{t0} is negative for depletion-mode n-channel MOSFETs and enhancement-mode p-channel MOSFETs.
- V_t is negative for both n-channel and p-channel JFETs.
- V_t is negative for an n-channel depletion MESFET and is positive for the n-channel enhancement MESFET.
- The sensitivity analysis command of Spice (.SENS) does not compute the sensitivities of the circuit variables to the parameters of the MOSFET model. These must be computed directly by perturbing each parameter of the MOSFET model and observing its effect.
- A small-signal analysis of an amplifier should always be computed around a known operating point and one that is normally within the linear region of the amplifier. The best way to determine the appropriate operating point is by first performing a DC sweep of the input voltage over the range defined by the power supplies and locating the linear region of the amplifier. The input to the amplifier is then biased inside this region where Spice can compute the small-signal behavior.

5.10 Bibliography

A. Vladimirescu and S. Liu, *The simulation of MOS integrated circuits using SPICE2*, Memorandum no. M80/7, February 1980, Electronics Research Laboratory, University of California, Berkeley.

R.L. Geiger, P. E. Allen and N. R. Strader, *VLSI Design Techniques For Analog And Digital Circuits*, New York: McGraw-Hill, 1990.

Staff, *PSpice Users' Manual*, MicroSim Corporation, Irvine, California, Jan. 1991.

D.A. Hodges and H. Jackson, *Analysis and Design of Digital Integrated Circuits*, New York: McGraw-Hill, 1983.

5.11 Problems

5.1 Let an n-channel enhancement transistor for which $\mu C_{OX} = 50$ $\mu A/V^2$, $W = L = 25$ μm and $V_t = 2$ V be operated with $v_{GS} = 6$ V. Find i_D for $v_{DS} = 2$ V using Spice. Repeat with $v_{DS} = 6$ V.

5.2 An enhancement PMOS transistor has $\mu C_{OX} = 40$ $\mu A/V^2$, $L = 25$ μm, $W = 50$ μm, $V_t = -1.5$ V, and $\lambda = 0.02$ V^{-1}. The gate is connected to ground and the source to +5 V. Find the drain current using Spice for (a) $v_D = 4$ V, (b) $v_D = +1.5$ V, (c) $v_D = 0$ V, and (d) $v_D = -5$ V.

5.3 A depletion-type n-channel MOSFET with $\mu C_{OX} = 4$ mA/V^2 and $V_t = -3$ V has its source and gate grounded. With the aid of Spice, find the region of operation and the drain current for (a) $v_D = 0.1$ V, (b) $v_D = 1$ V, (c) $v_D = 3$ V, and (d) $v_D = 5$ V. How do these results compare with hand analysis? Repeat with $\lambda = 0.05$ V^{-1}.

5.4 A depletion-type PMOS transistor has $I_{DSS} = 8.8$ mA, $V_t = +2.2$ V and $\lambda = 0.04$ V^{-1}. The source is connected to ground, the gate is connected to -2 V and a 2 mA current is pulled from the drain, using Spice determine the source-drain voltage. Assume that the width of the MOSFET is twice that of its length. Compare your results with a hand analysis.

5.5 Design the circuit of Fig. 5.3 to establish a drain current of 1 mA and a drain voltage of 0 V. The MOSFET has $V_t = 2$ V, $\mu C_{OX} = 20$ $\mu A/V^2$, $L = 10$ μm and $W = 400$ μm. Confirm your design using Spice. How much does the drain current change (in percent) when the effect of channel-length modulation is included in the Spice model of the MOSFET (assume $\lambda = 0.04$ V^{-1})?

5.6 Using Spice as a curve tracer, plot the i_D - v_{DS} characteristics of an enhancement-mode n-channel MOSFET having $\mu C_{OX} = 50$ $\mu A/V^2$, $L = 25$ μm, $W = 50$ μm, $V_t = +1.5$ V, and $\lambda = 0.04$ V^{-1}. Provide curves for $v_{GS} = 1, 2, 3, 4$ and 5 volts. Show the characteristics for v_{DS} up to 10 V.

5.7 Using Spice as a curve tracer, plot the i_D - v_{DS} characteristics of a depletion-mode p-channel MOSFET having $\mu C_{OX} = 50$ $\mu A/V^2$, $L = 25$ μm, $W = 50$ μm, $V_t = +1.2$ V, and $\lambda = 0.04$ V^{-1}. Provide curves for $v_{GS} = -1, -2, -3, -4$ and -5 volts. Show the characteristics for v_{DS} from 0 to -10 V.

5.8 Consider an n-channel JFET with $I_{DSS} = 4$ mA and $V_p = -2$ V. If the source is grounded and a -1 V dc voltage source is applied to the gate, using Spice find the drain current that corresponds to the minimum drain voltage that results in pinch-off operation.

5.9 For a JFET having $V_p = -2$ V and $I_{DSS} = 8$ mA operating at $v_{GS} = -1$ V and a very small v_{DS}, with the aid of Spice, determine the value of r_{DS}.

5.10 Using Spice as a curve tracer, plot the i_D - v_{DS} characteristics of an n-channel JFET with $I_{DSS} = 8$ mA and $V_p = -4$ V and $\lambda = 0.01$ V^{-1}. Provide curves for $v_{GS} = -5, -4, -3, -2$ and 0 volts. Show the characteristics for v_{DS} up to 10 V.

5.11 The NMOS transistor in the circuit of Fig. P5.11 has $V_t = 1$ V, $\mu C_{OX} = 1$ mA/V^2, and $\lambda = 0.02$ V^{-1}. If v_G is a pulse with 0 and 5 V levels and having a 1 ms pulse width, simulate the circuit using Spice and determine the pulse signal that appears at the output.

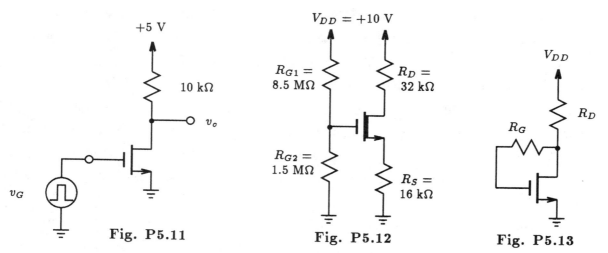

Fig. P5.11   Fig. P5.12   Fig. P5.13

5.12 Simulate the circuit in Fig. P5.12 to determine the drain current and the drain voltage. Assume that the depletion MOSFET has $V_t = -1$ V, $\mu C_{OX} = 1$ mA/V^2, and $\lambda = 0.03$ V^{-1}. If K of the MOSFET increases by a factor of 2, determine the new level of drain current and drain voltage. Compare this to the previous case and note whether the biasing scheme is very effective.

5.13 A MOSFET having $\mu C_{OX} = 2$ mA/V^2 and $V_t = 1$ V operates in a feedback bias arrangement such as that shown in Fig. P5.13, from a 10 V supply with $R_D = 8$ kΩ and $R_G = 10$ MΩ. What value of I_D results? If the FET is replaced by another with (a) $\mu C_{OX} = 1$ mA/V^2 and $V_t = 1$ V, and (b) $\mu C_{OX} = 2$ mA/V^2 and $V_t = 2$ V, what percentage change in I_D results?

5.14 The JFET in the amplifier circuit in Fig. P5.14 has $V_P = -4$ V, $I_{DSS} = 12$ mA and $\lambda = 0.003$ V^{-1}. Using Spice, answer the following:

(a) Determine the dc bias quantities V_G, I_D, V_{GS}, and V_D associated with the JFET.

(b) Determine the overall voltage gain v_o/v_i and input resistance R_{in}.

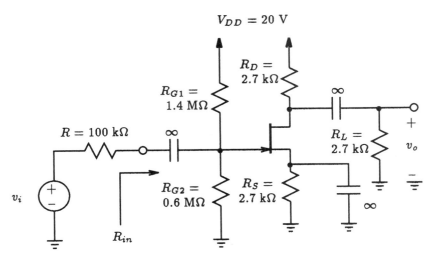

Fig. P5.14

5.15 In the circuit of Fig. P5.15 all devices are matched and assumed to have the following parameters: $\mu C_{OX} = 50\ \mu A/V^2$, $W = L = 25\ \mu m$ and $V_t = 2$ V. Using Spice, plot the transfer characteristics v_O vs. v_I between V_{DD} and ground. What is the small-signal voltage gain in the linear region of this amplifier? What is the corresponding output resistance?

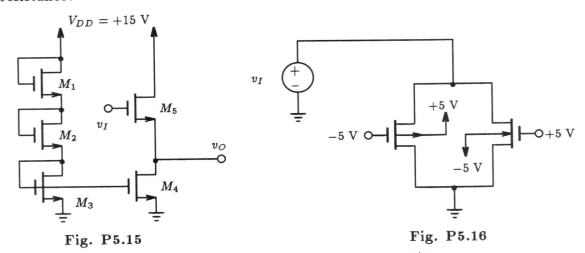

Fig. P5.15   **Fig. P5.16**

5.16 A CMOS switch for which $\mu C_{OX} = 50\ \mu A/V^2$, $\gamma = 0.3\ V^{1/2}$ and $|V_{t0}| = 2$ V is placed in the circuit setup shown in Fig. P5.16 for measuring its on-resistance. Sweep the input voltage signal v_I from -5 V to $+5$ V using Spice and determine the resistance of the switch over this range of signals. Compare these results to those obtain from a switch consisting of a single n-channel MOSFET.

5.17 A CMOS switch is used to connect a sinusoidal source of $0.1\ \sin(2\pi 10^3 t)$ to a load consisting of a 10 kΩ resistor and a 5 pF capacitor. If the control terminals of the switch are driven by complementary signals of ± 5 V at a frequency of 10 kHz, simulate the

5 Field-Effect Transistors (FETs)

transient behavior of the switch using Spice. Assume that the model parameters of the NMOS and PMOS devices are $\mu_n C_{OX} = 20$ μA/V^2, $\mu_p C_{OX} = 10$ μA/V^2, $|V_{t0}| = 1$ V, $\gamma = 0.3$ V$^{1/2}$, $L_n = L_p = 5$ μm, $W_n = 15$ μm, and $W_p = 30$ μm.

5.18 Using Spice as a curve tracer, plot the i_D - v_{DS} characteristics of an n-channel MESFET with $\beta = 10^{-4}$ A/V^2, $V_t = -1$ V, $I_S = 10^{-15}$ A and $\lambda = 0.1$ V^{-1}. Provide curves for $v_{GS} = -5, -4, -3, -2$ and 0 volts. Show the characteristics for v_{DS} up to 10 V.

5.19 The n-channel MESFET in the circuit of Fig. 5.50 has $V_t = -1$ V, $\beta = 0.1$ mA/V^2, and $\lambda = 0.1$ V^{-1}. If v_i is a pulse with -5 and 0 V levels of 1 μs pulse width, simulate the circuit using Spice and plot the pulse signal that appears at the output.

Chapter 6

Differential And Multistage Amplifiers

Up to this point we have mainly been looking at transistor circuits driven by unbalanced inputs (ie. one terminal of the signal source is grounded). A more important transistor arrangement is one that receives differential or balanced input signals. The differential pair is one such example and is used extensively in present day monolithic IC operational amplifiers. In this chapter we shall demonstrate how one generates the appropriate input signals for investigating the behavior of differential amplifiers. Following this, we shall investigate several circuits involving differential and multistage amplifiers. This will also include an investigation of various types of current mirrors and current sources.

6.1 Input Excitation For The Differential Pair

The differential pair, shown in Fig. 6.1, is the most widely used circuit building block in analog integrated circuits. Its operation is based on the fact that only the difference between the signals appearing at the two inputs is amplified. The signals appearing as common-mode at the amplifier inputs are (ideally) not amplified. In the following we would like to use Spice to investigate the effect of varying the input common-mode and differential-mode signal levels on the collector currents. The focus of this problem is not so much on the behavior of the

6 Differential And Multistage Amplifiers

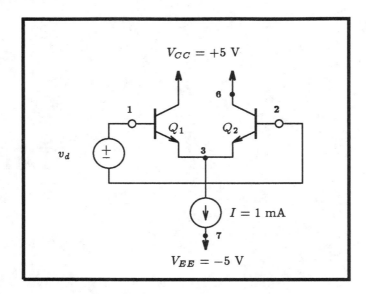

Figure 6.1 A BJT differential pair driven by a differential input signal without a defined input common-mode level. This approach is not recommended.

```
A BJT Differential Pair

** Circuit Description **
* power supply
Vdd 6 0 DC +5V
Vee 7 0 DC -5V
* input differential signal source
Vd 1 2 DC 0V
* differential pair
Q1 6 1 3 npn_transistor
Q2 6 2 3 npn_transistor
* bias source
I 3 7 DC 1mA
* transistor model statements
.model npn_transistor npn ( Is=14fA Bf=100 VAf=100V )
** Analysis Requests **
.OP
** Output Requests **
.end
```

Figure 6.2 The Spice input file for illustrating the problem of applying a single ungrounded voltage between the input terminals of a differential amplifier (see Fig. 6.1).

differential pair, but rather on how we generate the input signals for differential amplifiers within Spice.

For example, to analyze the behavior of a differential pair subject to a differential input signal, first-time users of Spice are often tempted to apply a single ungrounded source to

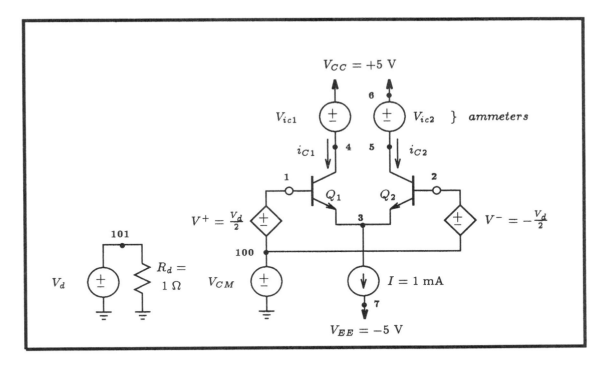

Figure 6.3 A BJT differential pair driven by a set of input voltage sources arranged such that the common-mode and differential voltage components can be independently varied. But, more importantly, both components will always exist, and are clearly defined. Voltage sources V_{ic1} and V_{ic2} are used to monitor the collector current of each transistor.

the amplifier input as illustrated in Fig. 6.1. Unfortunately, due to the lack of a defined common-mode input level, erroneous data is generated by Spice. To see this, we perform a DC analysis of the differential pair shown in Fig. 6.1 assuming the BJTs had the following parameters: $I_S = 14$ fA, $\beta_F = 100$ and $V_{AF} = 100$ V. The Spice input file is listed in Fig. 6.2.

The Spice results of the DC analysis are then found in the output file as follows:

```
A BJT Differential Pair

****    SMALL SIGNAL BIAS SOLUTION      TEMPERATURE =   27.000 DEG C
*******************************************************************

 NODE    VOLTAGE     NODE    VOLTAGE     NODE    VOLTAGE     NODE    VOLTAGE

 (  1)-22.19E+03  (  2)-22.19E+03  (  3)-22.19E+03  (  6)    5.0000
 (  7)   -5.0000
```

As is evident from these results, we see that the DC voltage appearing at the two inputs to the differential pair (nodes 1 and 2), and at the emitters of the two transistors (node 3), are very large negative levels, outside the limits of the supply voltages. Certainly, these levels would not be observed in any real circuit. The reason for this, as just mentioned, is the lack of a common-mode input level, and can be corrected by revising the input excitation to

6 Differential And Multistage Amplifiers

```
A BJT Differential Pair

** Circuit Description **
* power supply
Vdd 6 0 DC +5V
Vee 7 0 DC -5V
* differential-mode signal level
Vd 101 0 DC 0V
Rd 101 0 1
EV+ 1 100 101 0 +0.5
EV- 2 100 101 0 -0.5
* common-mode signal level
Vcm 100 0 DC 0V
* monitor collector currents of Q1 and Q2
Vic1 6 4 0
Vic2 6 5 0
* differential pair
Q1 4 1 3 npn_transistor
Q2 5 2 3 npn_transistor
* bias source
I 3 7 DC 1mA
* transistor model statements
.model npn_transistor npn ( Is=14fA Bf=100 VAf=100V )
** Analysis Requests **
.DC Vcm -5V +6V 100mV
** Output Requests **
.PLOT DC I(Vic1) I(Vic2)
.probe
.end
```

Figure 6.4 The Spice input file for analyzing the effect of common-mode input signals on the collector currents of a BJT differential pair (see Fig. 6.3).

include a common-mode level amongst the differential component.

The circuit shown in Fig. 6.3 accomplishes this in a convenient way. It allows the input common-mode level to be adjusted by varying the value of V_{CM}, independent of the value of the differential-mode component being established by the two VCVSs connected across the input terminals of the differential pair. The level of each VCVS is one-half the voltage value set by the isolated voltage source V_d. This voltage source is loaded arbitrarily in a 1 Ω resistor in order to satisfy the Spice requirement that every node in a circuit has at least two connections.

To demonstrate the versatility of this input-source arrangement, we shall investigate the effect of separately varying the input common-mode and differential signal components on the collector currents of the differential pair of Fig. 6.3. The Spice input file describing this circuit is listed in Fig. 6.4. The parameters of the *npn* BJTs are assumed to be the same

6.1 Input Excitation For The Differential Pair

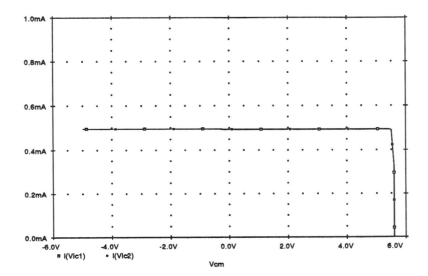

Figure 6.5 The collector currents of the BJT differential pair for a range of input common-mode levels.

as before (ie. $I_S = 14$ fA, $\beta_F = 100$ and $V_{AF} = 100$ V). As our first analysis request, we are asking that Spice perform a DC sweep of the input common-mode voltage level V_{CM} beginning at −5 V and ending at +6 V in increments of 100 mV. The differential input component V_d is set to zero during this analysis, thus also making V^+ and V^- equal zero. On completion of Spice we observe in Fig. 6.5 the behavior of the two collector currents as a function of V_{CM}. Here we see that both Q_1 and Q_2 are conducting equal currents of 0.5 mA for most input common-mode levels; However, when V_{CM} exceeds 5.6 V the collector current of both Q_1 and Q_2 fall off very rapidly to 0 mA. Any further increase in V_{CM} causes the collector currents to go negative. This observed behavior is easily accounted for by the fact that the base-collector junctions of both Q_1 and Q_2 become forward-biased when V_{CM} exceeds 5.6 V and conduct appreciable currents in the opposite direction to the normal flow of collector current.

To investigate the effect of a differential signal appearing at the input terminals of the differential pair, we simply revise the Spice input file given in Fig. 6.4 by replacing the DC sweep command given there by the following one:

```
.DC Vd -500mV +500mV 10mV
```

Here we are requesting that Spice vary the level of V_d between −500 mV and +500 mV in 10 mV increments. The common-mode input level V_{CM} is to remain at 0 V throughout this analysis. Submitting the revised input deck to Spice, results in the display of the collector

6 Differential And Multistage Amplifiers

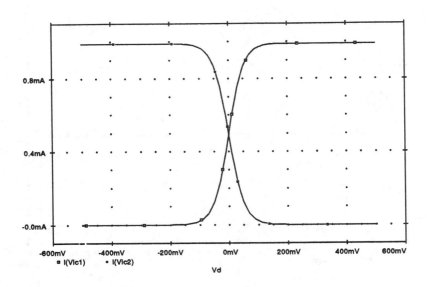

Figure 6.6 The collector currents of the BJT differential pair versus the level of differential input signal.

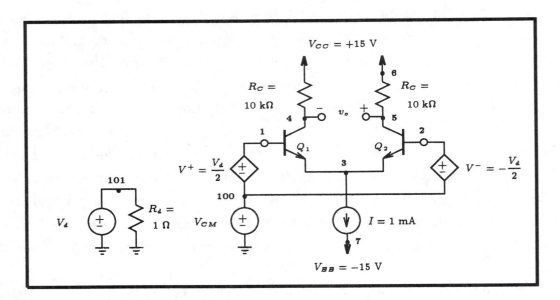

Figure 6.7 The basic BJT differential-pair amplifier configuration driven with the multiple-source arrangement described in Section 6.1.

currents in Fig. 6.6. Here the results are as expected: Depending on the level of input differential signal, the biasing current of 1 mA is *steered* between the two transistors.

6.2 Small-Signal Analysis Of The Differential Amplifier: Symmetric Conditions

In this section we investigate the small-signal behavior of the BJT differential-pair amplifier configuration shown in Fig. 6.7 using Spice. We shall assume throughout this section that the circuit remains symmetric, ie. the resistances in collectors are equal and transistors Q_1 and Q_2 are matched. Our purpose here is two fold: to illustrate how one uses Spice to determine the small-signal behavior of a differential amplifier, such as that shown in Fig. 6.7, using the multiple source arrangement discussed in the previous section, and to determine the accuracy of the expressions that are derived by Sedra and Smith in Sections 6.2 and 6.3 of their text.

It is important to point out that when one compares results computed by Spice to those computed through the application of closed-form expressions derived by hand analysis from a small-signal circuit model of the transistor circuit, the accuracy of the results depend on two factors: (1) how precise the estimate of the values of the parameters of the small-signal model is, and (2) how accurate the small-signal expressions are; given that certain simplifying assumptions are made in their development. Within this section, we are simply addressing the second issue; that being, the accuracy of the expressions derived by Sedra and Smith for the BJT differential amplifier under various circuit conditions. This is accomplished by making use of the small-signal model parameters computed by Spice directly through a DC operating point analysis command rather than estimating them ourselves from the DC circuit conditions.

According to the development provided by Sedra and Smith, the small-signal analysis of the differential-pair shown in Fig. 6.7 is relatively straight forward. In Table 6.1 we summaries the results of their analysis. This table includes expressions for both the differential-mode and common-mode voltage gain (A_d and A_{CM}) and the input resistances (R_{id} and R_{iCM}). As well, it includes expressions for both the input-referred offset voltage, input bias current and input offset current (V_{OS}, I_B and I_{OS}) under asymmetric circuit conditions.

Let us consider using Spice to calculate the differential-mode voltage gain and input resistance of the differential amplifier shown in Fig. 6.7. We shall assume that each transistor has device parameters: $I_S = 14$ fA and $\beta = 100$. For the moment we shall neglect the effect of the transistors Early Voltage (ie. $V_A = \infty$). This is equivalent to stating on the transistor model statement that VAf=0. The Spice input file describing the differential amplifier of Fig. 6.7 is seen listed in Fig. 6.8. The input to the differential amplifier is arranged using the multiple-source set-up described in the last section. It consists of two voltage controlled

6 Differential And Multistage Amplifiers

Differential Amplifier	
Parameter	Formula
A_d	$-g_m(R_C \| r_o)$
R_{id}	$2r_\pi$
A_{CM}	$\frac{R_C}{2R}\frac{\Delta R_C}{R_C}$
R_{iCM}	$(\frac{r_\mu}{2}) \| [(\beta+1)R] \| [(\beta+1)\frac{r_o}{2}]$
$\|V_{OS}\|$	$V_T\sqrt{(\frac{\Delta R_C}{R_C})^2 + (\frac{\Delta I_S}{I_S})^2}$
I_B	$\frac{I/2}{\beta+1}$
$\|I_{OS}\|$	$I_B(\frac{\Delta \beta}{\beta})$

Table 6.1 General expressions for estimating the small-signal behavior of the differential amplifier shown in Fig. 6.7 (derived in Sections 6.2 and 6.3 in Sedra/Smith). Note that R is the output resistance of the current source I.

sources (EV+ and EV−) and two independent input voltage signals (Vd and Vcm). The DC level of both of these voltage sources are set equal to 0 V since we already know that the two transistors of the differential pair will remain in the active region under these bias conditions. That is, with a transistor quiescent current of 0.5 mA, the voltage at the collector and emitter of each transistor will be approximately 10 V and −0.7 V, respectively. The differential input voltage source Vd also includes an AC component of 1 V. The reason for this component will be made clear in a moment.

Our first analysis request is a .TF command written as

.TF V(5,4) Vd.

This analysis command will provide us with an equivalent circuit representation of the differential amplifier evaluated around 0 V DC as seen looking into the port made by the input differential signal source V_d and the output port denoted between nodes 5 and 4 in Fig. 6.7. Unfortunately, with this multiple source arrangement, the resistance seen by the input source V_d is just the 1 Ω resistor connected across V_d. Recall that this resistor was added just to increase the number of connections made at node 101 and serves no circuit function. We therefore require another means of obtaining the resistance seen looking into this differential amplifier.

If we consider applying a 1 V AC signal across R_d, then this signal will also appear across the input terminals of the differential pair. By calculating the current that flows into the terminals of the differential amplifier due to this AC signal, we can obtain the input

6.2 Small-Signal Analysis Of The Differential Amplifier: Symmetric Conditions

```
A BJT Differential Pair

** Circuit Description **
* power supply
Vdd 6 0 DC +15V
Vee 7 0 DC -15V
* differential-mode signal level
Vd 101 0 DC 0V AC 1V
Rd 101 0 1
EV+ 1 100 101 0 +0.5
EV- 2 100 101 0 -0.5
* common-mode signal level
Vcm 100 0 DC 0V
* differential pair
Q1 4 1 3 npn_transistor
Q2 5 2 3 npn_transistor
* load resistors
Rc1 6 4 10k
Rc2 6 5 10k
* bias source
I 3 7 DC 1mA
* transistor model statements
.model npn_transistor npn ( Is=14fA Bf=100 VAf=0 )
** Analysis Requests **
.TF V(5,4) Vd
.AC LIN 1 1Hz 1Hz
.OP
** Output Requests **
.PRINT AC Im(EV+) Im(EV-) Vm(1,2)
.probe
.end
```

Figure 6.8 The Spice input file for calculating the 2-port equivalent of the differential amplifier shown in Fig. 6.7.

resistance of this amplifier as the reciprocal of this current. In the Spice deck listed in Fig. 6.8, we have already indicated this 1 V AC component on the statement describing the differential input voltage source. The AC current that flows through the input terminals of the differential amplifier can then be monitored with the two dependent voltage sources in series with the input terminals (ie. EV+ and EV−). To obtain this information, we have added the following two command statements to the Spice deck:

```
.AC LIN 1 1Hz 1Hz
.PRINT AC Im(EV+) Im(EV-) Vm(1,2).
```

The first command describes to Spice that an AC analysis is to be performed at only one frequency point of 1 Hz. The second statement instructs Spice to send to the output file the magnitude of the AC current that flows into the input terminals of the differential amplifier.

6 Differential And Multistage Amplifiers

As a means of insuring that the correct AC voltage appears at the input terminals of the amplifier, we are also requesting that Spice print out the voltage that appears there.

Finally, as our last analysis request we have included an operating point command (.OP) in the Spice deck. This command will provide us with the parameters of the small-signal model of each transistor.

On completion of Spice, we obtain the following parameters of the 2-port equivalent circuit of the differential amplifier shown in Fig. 6.7:

```
****     SMALL-SIGNAL CHARACTERISTICS

         V(5,4)/Vd =   1.914E+02

         INPUT RESISTANCE AT Vd =   1.000E+00

         OUTPUT RESISTANCE AT V(5,4) =   2.000E+04
```

We see here that this particular differential amplifier has a small-signal voltage gain of 191.4 V/V, and an output resistance of 20 kΩ. The input resistance indicated here is not the input resistance of the amplifier but, rather, the 1 Ω resistance shunting the input signal generator V_d. To determine the input resistance of the amplifier, we use the results of the AC analysis found in the output file,

```
****     AC ANALYSIS

  FREQ        IM(EV+)      IM(EV-)      VM(1,2)

  1.000E+00   9.570E-05    9.570E-05    1.000E+00
```

The input differential resistance to the amplifier is then found to be 10.44 kΩ.

At this point it would be interesting to check if the results obtained by hand analysis agree with those computed by Spice. To determine whether this is the case, consider that the parameters of the small-signal model of each transistor, as computed by Spice, are as follows:

```
**** BIPOLAR JUNCTION TRANSISTORS

NAME      Q1               Q2
MODEL     npn_transistor   npn_transistor
IB        4.95E-06         4.95E-06
IC        4.95E-04         4.95E-04
VBE       6.28E-01         6.28E-01
VBC       -1.01E+01        -1.01E+01
VCE       1.07E+01         1.07E+01
BETADC    1.00E+02         1.00E+02
GM        1.91E-02         1.91E-02
RPI       5.22E+03         5.22E+03
RX        0.00E+00         0.00E+00
RO        1.00E+12         1.00E+12
```

Using the expressions given in Table 6.1, we can expect that this amplifier will have a

6.2 Small-Signal Analysis Of The Differential Amplifier: Symmetric Conditions

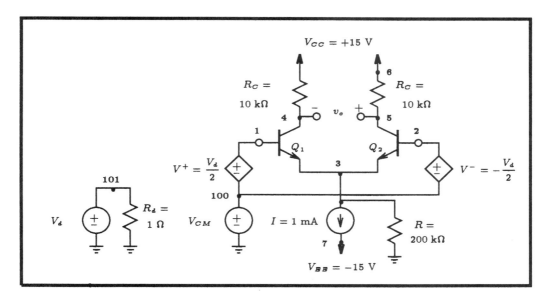

Figure 6.9 Including a 200 kΩ current source resistance in the differential amplifier of Fig. 6.7.

differential gain of 191.0 V/V and an input resistance of 10.44 kΩ, values that are very close to those computed by Spice.

To further analyze the differential amplifier shown in Fig. 6.7, let us compute the common-mode voltage gain and common-mode input resistance using Spice. One proceeds in exactly the same way that the differential-mode analysis was performed above using the .TF command, except that we change the source reference from Vd to Vcm as shown below:

```
.TF V(5,4) Vcm
```

The results of this analysis are:

```
****      SMALL-SIGNAL CHARACTERISTICS

          V(5,4)/Vcm =  0.000E+00

          INPUT RESISTANCE AT Vcm =  4.927E+11

          OUTPUT RESISTANCE AT V(5,4) =  2.000E+04
```

Here we see that the common-mode voltage gain of this amplifier is exactly zero, which agrees with what one would expect given that the resistance in the two collectors are equal. Note that the input resistance is 492.7 GΩ rather than the theoretically expected value of infinity; a numerical artifact of Spice. We also note that the output resistance computed during this analysis is identical to that found during the previous analysis for differential-mode behavior.

In the above analysis, we assumed that the differential pair was biased by an infinite-output-resistance current source I. In practise, an infinite-output-resistance current source

6 Differential And Multistage Amplifiers

can not be achieved. It is therefore imperative to repeat the above analysis including this current-source resistance, such as that shown in Fig. 6.9, and thus verify whether the expressions seen listed in Table 6.1 remain accurate. After all, in the development of most of these formula, the current-source resistance was assumed infinite. Here we shall assume a current-source resistance of 200 kΩ.

We begin by modifying the Spice input file of the differential amplifier circuit in Fig. 6.8 by including a Spice statement for the current-source resistance as:

```
R  3  0  200k.
```

Submitting the revised file to Spice, results in the following pertinent small-signal differential-mode information:

```
****      SMALL-SIGNAL CHARACTERISTICS

    V(5,4)/Vd =  1.908E+02

    INPUT RESISTANCE AT Vd =  1.000E+00

    OUTPUT RESISTANCE AT V(5,4) =  2.000E+04


****      AC ANALYSIS
  FREQ         IM(EV+)      IM(EV-)      VM(1,2)

  1.000E+00    9.540E-05    9.540E-05    1.000E+00
```

From the above information we find that the differential-mode voltage gain is 190.8 V/V and the input differential resistance is 10.48 kΩ. These results are very close to the previous case when the resistance of the current source was assumed infinite. The small differences in these small-signal results are due not directly to the presence of the source resistance R, but rather, because the addition of the source resistance alters the DC bias conditions of the two transistors.

We can proceed and modify the Spice deck to compute the common-mode input equivalent circuit, as was done before, and arrive at the following Spice results:

```
****      SMALL-SIGNAL CHARACTERISTICS

    V(5,4)/Vcm =  0.000E+00

    INPUT RESISTANCE AT Vcm =  2.020E+07

    OUTPUT RESISTANCE AT V(5,4) =  2.000E+04
```

Using the formulas given in Table 6.1, together with the small-signal model parameters for Q_1 and Q_2, we would obtain exactly the same results as those computed by Spice.

In each of the analyses performed above in this section, the Early effect was neglected.

6.2 Small-Signal Analysis Of The Differential Amplifier: Symmetric Conditions

Let us now consider the situation where the Early voltage of each transistor of the differential pair is 100 V. We shall maintain the current-source resistance at 200 kΩ. The Spice input file for this situation is very similar to that shown in Fig. 6.8, with the model statement for each transistor modified according to:

.model npn_transistor npn (Is=14fA Bf=100 VAf=100V).

The results of both the differential-mode and common-mode analysis computed by Spice are:

Differential-Mode Analysis:

```
****      SMALL-SIGNAL CHARACTERISTICS

          V(5,4)/Vd =  1.827E+02

          INPUT RESISTANCE AT Vd =  1.000E+00

          OUTPUT RESISTANCE AT V(5,4) =  1.914E+04

****      AC ANALYSIS

     FREQ        IM(EV+)      IM(EV-)     VM(1,2)

     1.000E+00   8.676E-05    8.676E-05   1.000E+00
```

Common-Mode Analysis:

```
****      SMALL-SIGNAL CHARACTERISTICS

          V(5,4)/Vcm = -7.733E-15

          INPUT RESISTANCE AT Vcm =  7.825E+06

          OUTPUT RESISTANCE AT V(5,4) =  1.914E+04
```

As is evident from these results, the differential-mode voltage gain is 182.7 V/V. The input differential resistance is calculated from the input current of 86.76 μA to be 11.52 kΩ. The common-mode voltage gain is, for all practical purposes, zero and the input common-mode resistance is 7.825 MΩ.

As a means of verifying the formulae given in Table 6.1 under the conditions of finite Early voltage, we also list the parameters of small-signal model computed by Spice below:

```
****  BIPOLAR JUNCTION TRANSISTORS

      NAME      Q1               Q2
      MODEL     npn_transistor   npn_transistor
      IB        4.49E-06         4.49E-06
      IC        4.94E-04         4.94E-04
      VBE       6.26E-01         6.26E-01
      VBC      -1.01E+01        -1.01E+01
      VCE       1.07E+01         1.07E+01
      BETADC    1.10E+02         1.10E+02
      GM        1.91E-02         1.91E-02
```

6 Differential And Multistage Amplifiers

Condition	Amplifier Parameter	Hand Analysis	Spice	% Error
$V_A = \infty$ $R = \infty$	A_d R_{id} R_o A_{CM} R_{iCM}	191.0 V/V 10.44 kΩ 20 kΩ 0 ∞	191.4 V/V 10.45 kΩ 20 kΩ 0 4.927×10^{11} Ω	0.21 % 0.1 % 0 % 0 % –
$V_A = \infty$ $R = 200$ kΩ	A_d R_{id} R_o A_{CM} R_{iCM}	191.0 V/V 10.48 kΩ 20 kΩ 0 20.20 MΩ	190.8 V/V 10.48 kΩ 20 kΩ 0 20.20 MΩ	−0.1 % 0 % 0 % 0 % 0 %
$V_A = 100$ v $R = 200$ kΩ	A_d R_{id} R_o A_{CM} R_{iCM}	182.8 V/V 11.52 kΩ 19.14 kΩ 0 6.75 MΩ	182.7 V/V 11.53 kΩ 19.14 kΩ 0 7.825 MΩ	−0.05 % 0.09 % 0 % 0 % 13.7 %

Table 6.2 Comparing the small-signal parameters of the differential amplifier shown in Figs. 6.7 and 6.9 as calculated by hand analysis and those computed by Spice.

```
RPI     5.76E+03            5.76E+03
RX      0.00E+00            0.00E+00
RO      2.23E+05            2.23E+05
```

Substituting the above parameter values into the expressions for A_d, R_{id}, A_{CM} and R_{iCM} seen listed in Table 6.1, and approximating r_μ by $10 r_o \beta$, we obtain the following estimates of the amplifiers small-signal behavior: $A_d = 182.8$ V/V, $R_{id} = 11.52$ kΩ, $A_{CM} = 0$ and $R_{iCM} = 6.75$ MΩ. Comparing these with those computed directly by Spice we see that we are in good agreement with all of them except R_{iCM}. The formula for R_{iCM} seen listed in Table 6.1 seems to under estimate the actual input common-mode resistance by about 14%. One possible source of error may be due to our approximation of r_μ by $10 r_o \beta$.

To summarize the results of this section we have compiled a list in Table 6.2 that compares the results computed directly by Spice to those computed using the formulae presented in Table 6.1. Furthermore, the rightmost column of this table lists the relative error in percent. As is evident, the results predicted by the formulae of Table 6.1 agree quite well with the results computed by Spice. We can therefore conclude that when given good estimates of the small-signal model parameters, the formulae of Table 6.1 will predict quite accurately the small-signal differential and common-mode voltage gain of a differential amplifier and its corresponding input resistances.

6.3 Small-Signal Analysis Of The Differential Amplifier: Asymmetric Conditions

The previous section assumed that many of the components in the differential amplifier of Figs. 6.7 or 6.9 were matched. In practise this is rarely the case. In this section we shall investigate the effect of asymmetric circuit conditions on amplifier behavior. Specifically, we are interested in observing the effect of variations, or mismatches, in collector resistances, transistor saturation (scale) currents, and transistor β's on circuit behavior.

Input Offset Voltage

To begin with, let us consider using Spice to determine the input offset voltage V_{OS} for the differential amplifier shown in Fig.6.9 assuming that each collector resistor undergoes a change of 5%, one positive, the other negative – resulting in a net change of 10%. The Spice deck for this particular example is shown listed in Fig. 6.10. The analysis that we are requesting here is a DC sweep of the input differential voltage V_d. The range of our sweep is limited to be between -10 mV and +10 mV. The range of this sweep was determined by considering the formula for input offset voltage V_{OS} in Table 6.1. Given that $\Delta R_C / R_C = 10\%$, we can expect an input offset voltage having a magnitude of 2.6 mV. Thus, we decided that our sweep should not exceed this by very much so that details, such as zero crossings, can be easily seen. Also included in the Spice deck is a .TF command to compute the common-mode voltage gain. As we can see from Table 6.1, the magnitude of the common-mode voltage gain A_{CM} will no longer be zero but approximately 2.5 mV/V.

The results of the DC sweep of the input differential voltage are shown in Fig. 6.11 for a 10% variation in the collector resistance. Here we see that the transfer function curve v_o vs. v_d for this particular case (as there are other cases also shown in this figure), no longer passes through the origin. Instead, careful probing using the cursor feature of the PROBE facility of PSpice indicates that the output offset voltage for zero input is -472.7 mV. Conversely, the input voltage that corresponds to zero output voltage is +2.6 mV. This, then, is the negative of the input offset voltage (ie. $V_{OS} = -2.6$ mV). Interestingly enough, this corresponds exactly with the magnitude of the value predicted by the formula given in Table 6.1, as demonstrated above.

The common-mode voltage gain of the differential amplifier shown in Fig. 6.9 when the resistances in the two collectors differ by 10% is found by Spice to be:

```
****    SMALL-SIGNAL CHARACTERISTICS

          V(5,4)/Vcm =  -2.042E-03

        INPUT RESISTANCE AT Vcm =  7.825E+06
```

6 Differential And Multistage Amplifiers

```
                Differential Amplifier: Asymmetric Collector Resistance

        ** Circuit Description **
        * power supply
        Vdd 6 0 DC +15V
        Vee 7 0 DC -15V
        * differential-mode signal level
        Vd 101 0 DC 0V AC 1V
        Rd 101 0 1
        EV+ 1 100 101 0 +0.5
        EV- 2 100 101 0 -0.5
        * common-mode signal level
        Vcm 100 0 DC 0V
        * differential pair
        Q1 4 1 3 npn_transistor
        Q2 5 2 3 npn_transistor
        * unequal collector resistors (10% different)
        Rc1 6 4 9.50k
        Rc2 6 5 10.5k
        * bias source
        I 3 7 DC 1mA
        R 3 0 200k
        * transistor model statements
        .model npn_transistor npn ( Is=14fA Bf=100 VAf=100V )
        ** Analysis Requests **
        .OP
        .DC Vd -10mV +10mV 100uV
        .TF V(5,4) Vcm
        ** Output Requests **
        .PLOT DC V(5,4)
        .probe
        .end
```

Figure 6.10 The Spice input file for calculating the input offset voltage V_{OS} of the differential amplifier shown in Fig. 6.9 when the collector resistors undergo a 10% relative variation. Also included is a .TF command to compute the common-mode voltage gain.

```
            OUTPUT RESISTANCE AT V(5,4) =  1.914E+04
```

Spice reports that the common-mode voltage gain is –2.042 mV/V. The magnitude of this value is reasonably close to the value that was predicted by the formula given in Table 6.1 at 2.5 mV/V.

We can repeat the above analysis and determine the effect that a 5% difference between the saturation currents I_S of the two transistors on the amplifiers input offset voltage. According to the formula given in Table 6.1, we can expect an input offset voltage of 1.3 mV. The Spice input file for this particular case is provided in Fig. 6.12. The results of this

6.3 Small-Signal Analysis Of The Differential Amplifier: Asymmetric Conditions

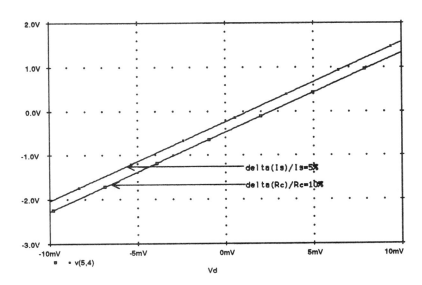

Figure 6.11 Highlighting the input and output offset voltage of the differential amplifier shown in Fig. 6.9 subject to two different mismatches: (i) $\Delta R_C/R_C = 10\%$ and (ii) $\Delta I_S/I_S = 5\%$.

analysis are shown superimposed in the previous graph of output voltage versus input differential voltage shown in Fig. 6.11. Using PROBE, we can determine that the output offset voltage is −236.4 mV and the input offset voltage is −1.3 mV. Clearly our estimate of the input offset voltage agrees with that obtained with Spice.

Input Bias And Offset Currents

Mismatches in transistor β's result in different base currents, which, in turn, give rise to differences in the input bias currents to the amplifier. To demonstrate this, consider altering the β of Q_1 in the differential amplifier shown in Fig. 6.9 by −5%, and the β of Q_2 by +5%. Specifically, for the example used above, the following model statements would be used to describe this situation to Spice:

```
.model npn_transistor1 npn ( Is=14fA Bf=95  VAf=100V )
.model npn_transistor2 npn ( Is=14fA Bf=105 VAf=100V )
```

These two statements can be used to replace the transistor model statements shown in the Spice file listed above in Fig. 6.12. Only the .OP command is necessary to see the effect of β variation on input bias currents. The input bias current to the amplifier can be seen directly from the bias information generated by Spice for the two transistors.

Submitting this input file to Spice, results in the following DC operating point informa-

6 Differential And Multistage Amplifiers

Differential Amplifier: Asymmetric Saturation Current

```
** Circuit Description **
* power supply
Vdd 6 0 DC +15V
Vee 7 0 DC -15V
* differential-mode signal level
Vd 101 0 DC 0V AC 1V
Rd 101 0 1
EV+ 1 100 101 0 +0.5
EV- 2 100 101 0 -0.5
* common-mode signal level
Vcm 100 0 DC 0V
* differential pair
Q1 4 1 3 npn_transistor1
Q2 5 2 3 npn_transistor2
* load resistors
Rc1 6 4 10k
Rc2 6 5 10k
* bias source
I 3 7 DC 1mA
R 3 0 200k
* transistor model statements
.model npn_transistor1 npn ( Is=13.65fA Bf=100 VAf=100V )
.model npn_transistor2 npn ( Is=14.35fA Bf=100 VAf=100V )
** Analysis Requests **
.DC Vd -10mV +10mV 100uV
.OP
.TF V(5,4) Vcm
** Output Requests **
.PLOT DC V(5,4)
.probe
.end
```

Figure 6.12 The Spice input file for calculating the input offset voltage V_{OS} of the differential amplifier shown in Fig. 6.9 when the saturation currents of the two transistors differ by 5%.

tion for the two transistors:

```
**** BIPOLAR JUNCTION TRANSISTORS

NAME         Q1               Q2
MODEL        npn_transistor1  npn_transistor2
IB           4.72E-06         4.27E-06
IC           4.94E-04         4.94E-04
VBE          6.26E-01         6.26E-01
VBC          -1.01E+01        -1.01E+01
VCE          1.07E+01         1.07E+01
BETADC       1.05E+02         1.16E+02
GM           1.91E-02         1.91E-02
RPI          5.48E+03         6.05E+03
RX           0.00E+00         0.00E+00
RO           2.23E+05         2.23E+05
```

As we can see, the base currents of Q_1 and Q_2 are not equal but differ by 450 nA. The input bias current to the differential amplifier is the average of these two currents, being 4.50 μA. Correspondingly, the offset current for the amplifier is 450 nA. It is interesting to note that the formula provided in Table 6.1 generates the same value observed from the Spice simulation.

It is also interesting to note that according to the above list of small-signal model parameters, the BETADC of Q_1 is 105 and for Q_2 it is 116. This is quite different from the β that was assigned on the transistor model statement. This difference arises from the definition Spice uses to compute BETADC. This parameter is computed by Spice as the ratio of the collector current to base current, which includes the effect of transistor output resistance. This is contrary to what β represents in the model statement, ie. short-circuit current gain.

6.4 Current-Mirror Circuits

Current mirror circuits play a very important role in the design of IC current sources and current-steering circuits. A current mirror circuit consists of two or more transistors arranged in such a way that at least two transistors in the circuit have their bases and emitters connected together causing them to have equal v_{BE}'s. There are many different current mirror circuits, each suitable for a different application. In the following we shall investigate several widely used current mirror circuits: A two-transistor current mirror, herein referred to as the simple current mirror, the simple current mirror with base-current compensation, and the Wilson current mirror. Each of these is depicted in Fig. 6.13. In our investigation, we shall judge the behavior of these current mirrors based on their: (a) current gain accuracy, (b) output resistance, (c) minimum output voltage, and (d) input current range. Another important aspect of current mirror circuits is their frequency response behavior. However, we defer discussion of this topic until the next chapter.

Current-Gain Accuracy:

The current-gain accuracy is an indication of how significantly the transistor base currents affect the operation of the current mirror circuit. The current gain is the ratio of the output current to the input current (I_{out}/I_{in}). In hand analysis, the current gain is usually evaluated with the Early effect neglected. Obviously, the more ideal a current mirror is, the closer this ratio is to unity. For the simple current mirror, it has been shown by Sedra and Smith in Section 6.4 of their text that the current transfer ratio is $\frac{1}{1+2/\beta}$. Conversely, the current transfer ratio for the remaining two current-mirror circuits shown in Fig. 6.13 can

6 Differential And Multistage Amplifiers

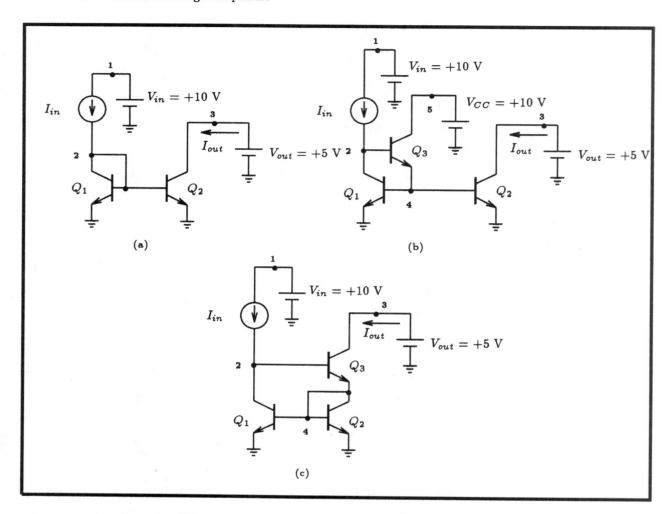

Figure 6.13 Various current mirror circuits: (a) A simple two-transistor current-mirror circuit. (b) A current mirror with base-current compensation. (c) The Wilson current mirror circuit. Various voltage sources are used to monitor different branch currents.

be approximated by $\frac{1}{1+2/\beta^2}$. For $\beta = 100$, the simple current mirror has a current transfer ratio of 0.9803 A/A, whereas the other two mirror circuits have a ratio much closer to unity, of value 0.9998 A/A. Clearly, the simple current mirror is not very accurate.

With the aid of Spice, let us observe the current transfer ratio of the current-mirror circuits shown in Fig. 6.13 for various output voltage levels. We shall consider that the transistor is ideal except that it has a finite β equal to 100. Modeling the transistor in this way will enable us to compare the result generated by Spice with that calculated by hand. Furthermore, we shall assume that each transistor of the mirror circuit is integrated on a common p-type substrate.† This is usually necessary in order to obtain matched devices

---

† Whenever an IC transistor is employed and its model is obtained from a library, the substrate connection must be defined.

```
Wilson Current Mirror

** Circuit Description **
Vin 1 0 DC +10V
Iin 1 2 1mA
Q1 4 4 0 0 npn
Q2 2 4 0 0 npn
Q3 3 2 4 0 npn
Vout 3 0 DC 5V
* simple transistor model
.MODEL npn npn (Is=14fA Bf=100 VAf=0V)
** Analysis Requests **
.OP
* sweep the output voltage Vout between 0V and +5V
.DC Vout 0V 5V 10mV
** Output Requests **
.PLOT DC I(Vout) I(Vin)
.Probe
.end
```

Figure 6.14 The Spice input file for calculating the current transfer ratio I_{out}/I_{in} of the Wilson current mirror circuit shown in Fig. 6.13(c).

without trimming. To ensure proper isolation, the pn junction formed by the substrate is reverse biased by connecting the substrate to the lowest potential in the circuit. In this particular case it would be the ground node (node 0).

Let us consider creating Spice decks for each of the three current mirror circuits shown in Fig. 6.13. Since they are all quite similar, we shall only show the Spice deck for the Wilson current mirror. The Spice input file describing the Wilson current mirror shown in Fig. 6.13(c) is provided in Fig. 6.14. The current mirror is supplied with a 1 mA input current source connected in series with a 10 V DC voltage source. This voltage source will enable us to monitor the input current with a Spice print or plot command. The value of this voltage source is not important, but its level is set typically at V_{CC}. The output terminal of the current mirror is connected to a 5 V DC voltage source. This voltage source acts to simulate various load conditions, as well as allowing us to monitor the output current of the mirror. A DC sweep of the output voltage source is to be performed over a voltage range varying between 0 and 5 V in increments of 10 mV. A DC operating point analysis is also included in this Spice deck.

The results of the Spice analysis are shown in Fig. 6.15. Here we have plotted the output current normalized to the input current level of 1 mA. This operation was performed directly with the Probe facility of PSpice; unfortunately, Spice can not perform this operation directly. Furthermore, with the cursor feature in Probe, we were able to determine from the

6 Differential And Multistage Amplifiers

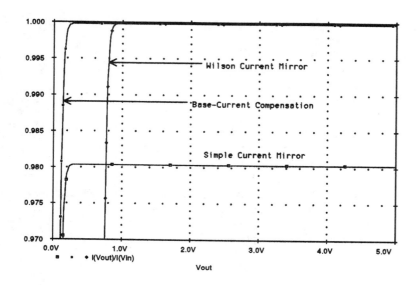

Figure 6.15 Comparing the current transfer characteristics of the various current mirror circuits shown in Fig. 6.13 as a function of the output voltage. Only the effect of transistor base currents are considered in this analysis (ie. $V_A = \infty$).

graph in Fig. 6.15 that the base-current compensated mirror circuit and the Wilson current mirror circuit have nearly ideal current gain at 0.9998 A/A, whereas, the current transfer ratio for the simple current mirror was found to be 0.9804 A/A. All three sets of results agree with those found with the above hand calculations. We also notice from the current transfer characteristics shown in Fig. 6.15 that when the output voltage goes too low, the current transfer ratio drops significantly towards zero. This suggests that the current mirror has a limited range of operation. We shall have more to say about this in a moment when we consider a more sophisticated model for the transistor.

Output Resistance:

Based on the above analysis, one might be tempted to conclude that both the base-current compensated current mirror and the Wilson current mirror could be used interchangeably, given that their current gain accuracies are identical. Unfortunately, this is not the complete picture because the analysis above ignored the Early effect. As we shall see, the current gain ratio of a current mirror can be adversely affected by the Early voltage of a transistor, thus altering our conclusion about which circuit is the better current mirror.

To see this, let us modify the model statement for the *npn* transistor given in the Spice deck for each of the current mirror circuits by the following transistor model statement:

```
.MODEL npn NPN (IS=5E-17 BF=147 VAF=80 IKF=4.3E-3 ISE=8E-18 NE=1.233
```

6.4 Current-Mirror Circuits

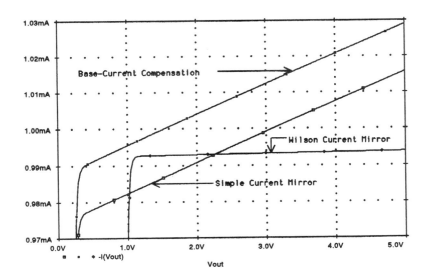

Figure 6.16 The I_{out} versus V_{out} characteristics of the various current mirror circuits shown in Fig. 6.13. The effect of transistor Early voltage has been included in this analysis.

```
+           BR=1.9 VAR=11 IKR=6E-4 ISC=5E-16 NC=1.08 RE=12 RB=1200 RBM=200 RC=25
+           CJE=58E-15 VJE=0.83 MJE=0.35 CJC=133E-15 VJC=0.6 MJC=0.44 XCJC=1
+           CJS=830E-15 VJS=0.6 MJS=0.4 ISS=1E-16 FC=0.85 TF=60P XTF=48 ITF=3E-2
+           TR=10N EG=1.16 XTI=3 XTB=1.6)
```

This Spice model was chosen here because it is representative of a typical small-sized integrated *npn* transistor found on a bipolar semi-custom analog transistor array manufactured by the Gennum Corporation [Gennum Data Book, 1991]. We shall maintain the input current to each mirror at 1 mA. Typically, these transistors have a forward β_{ac} of approximately 90 at a bias current of 1 mA and an Early voltage in the neighborhood of 80 V. One should bear in mind that current mirror circuits are generally called on to mirror a wide range of currents and not just a single current level. We shall address this issue below in the subsection entitled: Input Current Range.

Submitting the revised decks for the three current mirrors to Spice, we see on completion of Spice in Fig. 6.16 a plot of the output current as a function of the output voltage. As is evident, the behavior of the base-current compensated mirror circuit and the Wilson current mirror circuit now differ significantly. The Wilson current mirror has an $i - v$ behavior that is very much independent of the output voltage, ie. near zero slope. On the other hand, both the base-current compensated mirror and the simple current mirror circuit have behavior that depends strongly on the output voltage. To quantify the output resistance of each of these current mirrors, we can simply estimate this from the reciprocal of the slopes

6 Differential And Multistage Amplifiers

of the $i - v$ characteristics of each mirror in their linear regions. The Probe facility of PSpice will be used to obtain these slopes directly from the graph shown in Fig. 6.16. In the case of the simple current mirror and the base-current compensated mirror circuit, the output resistance is approximately 120 kΩ. The Wilson current mirror has a much higher output resistance of about 3.38 MΩ.

According to small-signal analysis, the output resistance of the simple current mirror, and the base-compensated current mirror, is simply the incremental resistance r_o of the output transistor (ie. Q_2 in the circuits of Fig. 6.13(a) and (b)). However, according to the Early voltage and the transistor bias level, we estimate the incremental output resistance of this transistor to be V_A / I = 80 V / 1 mA = 80 kΩ. (The results of a .OP command also confirm something very close to this value). However, the Spice results indicate mirror output resistance of 120 kΩ. Thus, some other effect must be playing a major role in increasing the output resistance of these current mirrors. A detailed investigation reveals that the integrated npn transistor used in our simulations has a 12 Ω resistor in series with the emitter (See the model statement for this transistor, given above). Although, the value of this resistor seems small, its effect on the output resistance of the current mirror is significant.

According to small-signal analysis, the resistance seen looking into the collector terminal of a transistor, denoted as R_o, with emitter resistance R_E is given by

$$R_o \approx \left[1 + g_m(R_E \| r_\pi)\right] r_o \qquad (6.1)$$

In the situation described above for the two simple current mirror circuits, R_E of 12 Ω dominates the parallel combination of $R_E \| r_\pi$ because r_π is in the kΩ range (ie. $r_\pi = \beta/g_m$ where $g_m = I_C/V_T$ =1 mA /25 mV = 40 mA/V). Thus, according to Eqn. (6.1), the output resistance of this mirror is calculated as follows:

$$R_o = \left[1 + (40 \times 10^{-3})(12)\right](80 \times 10^3) = 118.4 \text{ k}\Omega.$$

Clearly, our estimate of the output resistance is now in-line with that observed through computer simulation of 120 kΩ.

For the case of the Wilson current mirror, its output resistance is given by

$$R_o \approx \frac{\beta r_o}{2}. \qquad (6.2)$$

Attaching a numerical value of 90 to β and 80 kΩ to r_o, suggests that the output resistance is approximately 3.6 MΩ. This, then, agrees quite closely with the value of the output resistance obtain through Spice of 3.38 MΩ. The effect of the emitter resistance of each

6.4 Current-Mirror Circuits

transistor is much greatly reduced due to the feedback action of the three transistor loop which forms the Wilson current mirror circuit.

Based on the above observations of current mirror accuracy and output resistance, one would probably prefer the Wilson current mirror over the other two current mirror circuits shown in Fig. 6.13

Minimum Output Voltage:

A drawback to the Wilson current mirror is that it requires a relatively high output voltage to operate in the linear region (1.14 V as opposed to 0.36 V for the other two circuits; see Fig. 6.16). When used as an active load in an amplifier configuration, the required voltage reduces the range of the output voltage swing and is therefore not desirable.

Input Current Range:

Current mirrors are expected to operate over a wide range of input current levels. In the above analysis, the various attributes of the different current mirror circuits were all evaluated at a single input current level of 1 mA. We shall now explore the performance of the Wilson mirror at various input current levels. Specifically we shall evaluate its $i - v$ characteristic at input current levels of 1 μA, 10 μA, 100 μA and 1 mA.

The Spice input file is a concatenation of four separate Spice files of the type seen listed in Fig. 6.14 with only the input current level altered according to the list described above. Once again, this allows us to plot all the results together on a single graph using the Probe facility of PSpice.[†]

The results of these analyses are shown plotted in Fig. 6.17. Rather than plot the wide range of output currents on a single graph, where much detail would be lost with the scale used there, we instead plot the ratio of the output current to the input current, I_{out}/I_{in}, as a function of the output voltage. In this way, the same scale can be used for all input current levels and comparisons can easily be made. The output resistance and the minimum output voltage can be derived from the data contained in this graph.

Over an input current range of 1 μA to 1 mA, the minimum output voltage can be seen to vary between 0.97 V and 1.2 V. In addition, over this same current range, we see that the slopes of I_{out}/I_{in} versus V_{out} curves are quite similar ranging between 210 to 292 μA/A per volt. Multiplying each one of these slopes by the corresponding input current level, we can convert these slopes to output conductances, and then to output resistances. On doing so,

---

[†] Recent versions of Spice, such as PSpice, allow the user more efficient ways of implementing functions of this sort. See for example the .STEP command of PSpice.

6 Differential And Multistage Amplifiers

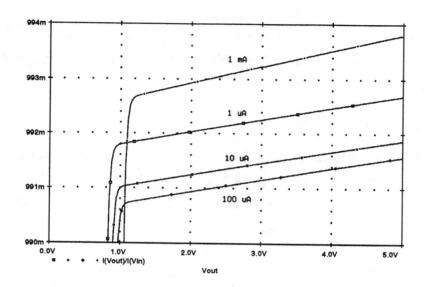

Figure 6.17 Demonstrating the variations in the current transfer ratio of the Wilson current mirror over an input current range of 1 μA to 1 mA.

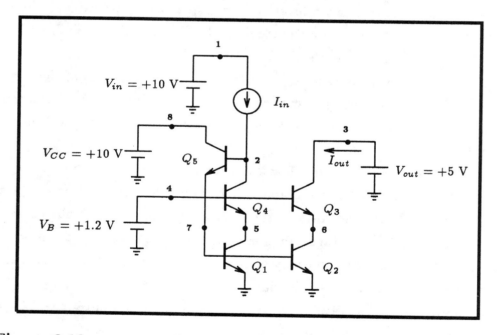

Figure 6.18 A high-swing cascode current mirror circuit with base-current compensation. Various voltage sources are used to separately monitor different branch currents.

we find that for input current levels of 1 μA, 10 μA, 100 μA and 1 mA, the output resistance is 4.5 GΩ, 474 MΩ, 47 MΩ, and 3.4 MΩ, respectively. According to the output resistance formula given in Eqn. (6.2), with $r_o = V_A/I$, these values are in close agreement.

6.5 A High-Performance Current Mirror

One means of decreasing the minimum output voltage of a current mirror circuit whose output is derived from the transistor action of two transistors stacked one above the other is to reduce the voltage that appears at the base of the top transistor. In this way, the voltage that can appear at the output of the current mirror can be reduced without the top transistor saturating. The circuit shown in Fig. 6.18 accomplishes this task while maintaining the high output resistance through the cascode output. Here the voltage at the base of Q_3 and Q_4 is set by the external voltage source V_B. Correspondingly, the voltage at their respective emitters is one diode drop lower at about $(V_B - 0.7)$ V. Clearly, then, if V_B is set between 1.0 V and 1.4 V, then the voltage at the emitter of Q_3 will fall somewhere between 0.3 V and 0.7 V. Thus, the voltage at the output terminal of the current mirror can be reduced to about 0.6 – 1.0 V before Q_3 saturates (assuming Q_3 saturates at 300 mV). The other nodes in the circuit are set at levels that ensure that the other transistors are operating in their active regions. Transistor Q_5 is used to compensate for the base current of Q_1 and Q_2.

We shall now compare the behavior of this high-swing cascode current mirror circuit shown in Fig. 6.18 with the Wilson current mirror circuit shown in Fig. 6.13(c). The input to the current mirror will be set to 1 mA. The Spice input file for the high-swing cascode circuit is shown in Fig. 6.19. The Spice input file for the Wilson current mirror was already shown in Fig. 6.14. These two files will be concatenated together and submitted to Spice. A DC sweep of the output voltage beginning at 0 V and increased +5 V in increments of 10 mV is requested. The output current will then be plotted as a function of the output voltage.

The results of the Spice analysis are shown plotted in Fig. 6.20. As can be seen, the output current from the high-swing cascode current mirror is much closer to the input current of 1 mA than in the case of the Wilson current mirror, ie. the new circuit behaves in a more ideal fashion. Although it is not readily apparent, the high-swing cascode current-mirror has an output resistance about twice that of the Wilson current mirror of about 6.5 MΩ. This output resistance value was determined with the aid of the cursor facility of Probe. With regards to the minimum output voltage, we see that the high-swing cascode mirror circuit does indeed succeed at it's primary objective of reducing the minimum output voltage below that of the Wilson current mirror, specifically it can operate with a voltage as low as 0.7 V.

6.6 Current-Source Biasing In Integrated Circuits

A typical approach for implementing a current source in IC technology is with current

6 Differential And Multistage Amplifiers

```
High-Swing Cascode Current Mirror

** Circuit Description **
Vcc 8 0 DC +10V
Vb  4 0 DC +1.2V
Vin 1 0 DC +10V
Iin 1 2 1mA
Q1 5 7 0 0 npn
Q2 6 7 0 0 npn
Q3 3 4 6 0 npn
Q4 2 4 5 0 npn
Q5 1 2 7 0 npn
Vout 3 0 DC 5V
* transistor model statement for integrated NPN transistor by Gennum Corp.
.MODEL npn NPN (IS=5E-17 BF=147 VAF=80 IKF=4.3E-3 ISE=8E-18 NE=1.233
+               BR=1.9 VAR=11 IKR=6E-4 ISC=5E-16 NC=1.08 RE=12 RB=1200 RBM=200 RC=25
+               CJE=58E-15 VJE=0.83 MJE=0.35 CJC=133E-15 VJC=0.6 MJC=0.44 XCJC=1
+               CJS=830E-15 VJS=0.6 MJS=0.4 ISS=1E-16 FC=0.85 TF=60P XTF=48 ITF=3E-2
+               TR=10N EG=1.16 XTI=3 XTB=1.6)
** Analysis Requests **
.OP
* sweep the output voltage Vout between 0V and +5V
.DC Vout 0V 5V 10mV
** Output Requests **
.PLOT DC I(Vout) I(Vin)
.Probe
.end
```

Figure 6.19 The Spice input file for calculating the output current from the high-swing cascode current-mirror circuit shown in Fig. 6.18 as a function of the output terminal voltage. The input is biased at a 1 mA current level.

mirror circuits. Figure 6.21 illustrates a simple current mirror arranged as a current source. The input terminal of the current mirror is fed with a 942 kΩ resistor connected to the positive power supply. This results in an input bias current of about 10 μA. The output terminal is connected to a variable DC voltage source V_{out} to simulate the action of a load. With the aid of Spice, we would like to determine the range of output voltages that can appear across the output port of the current mirror before the circuit ceases to operate as a current source. In addition, we would also like to determine the Norton equivalent circuit representation of the output port of this current mirror. We shall assume that the transistors are matched and are typical of the small *npn* integrated transistor variety manufactured by the Gennum Corporation. As mentioned before, these transistors have a forward β_{ac} of approximately 90 at a bias current of 1 mA and an Early voltage in the neighborhood of 80 V.

The Spice input file describing this circuit is given in Fig. 6.22 where we have requested

6.6 Current-Source Biasing In Integrated Circuits

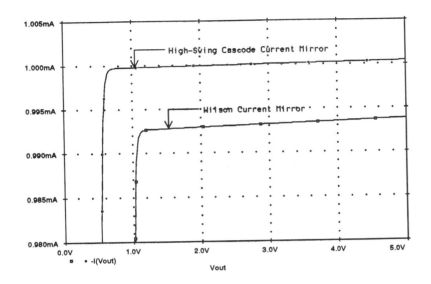

Figure 6.20 Comparing the $I_{out} - V_{out}$ behavior of the high-swing cascode current mirror with that of the Wilson current mirror. Each transistor is modeled after a small npn integrated transistor manufactured by Gennum Corp.

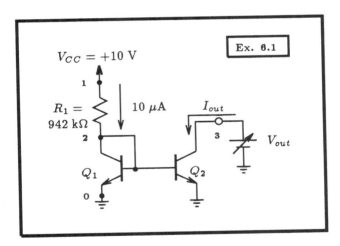

Figure 6.21 A simple current-mirror circuit setup as a current source.

a DC sweep of the output voltage V_{out} beginning at 0 V and ending at 5 V. The output collector current of Q_2 will then be plotted as a function of V_{out}. The results of the Spice analysis are shown in Fig. 6.23. For output voltages larger than 220 mV, we see that the output current remains relatively constant around 10 μA, increasing slightly at a rate of 0.13 μA per volt. When the output voltage drops below 220 mV the output current drops quickly to zero amps. Thus, the circuit operates as an effective current source provided the voltage appearing across the output remains above 220 mV.

6 Differential And Multistage Amplifiers

```
A Simple Bipolar Current Source

** Circuit Description **
Vcc 1 0 DC +10V
R1 1 2 942k
Q1 2 2 0 0 npn
Q2 3 2 0 0 npn
Vout 3 0 DC 5V
* transistor model statement for integrated NPN transistor by Gennum Corp.
.MODEL npn NPN (IS=5E-17 BF=147 VAF=80 IKF=4.3E-3 ISE=8E-18 NE=1.233
+               BR=1.9 VAR=11 IKR=6E-4 ISC=5E-16 NC=1.08 RE=12 RB=1200 RBM=200 RC=25
+               CJE=58E-15 VJE=0.83 MJE=0.35 CJC=133E-15 VJC=0.6 MJC=0.44 XCJC=1
+               CJS=830E-15 VJS=0.6 MJS=0.4 ISS=1E-16 FC=0.85 TF=60P XTF=48 ITF=3E-2
+               TR=10N EG=1.16 XTI=3 XTB=1.6)
** Analysis Requests **
.OP
* sweep the output voltage Vout between 0V and +5V
.DC Vout 0V 5V 10mV
** Output Requests **
.PLOT DC I(Vout)
.Probe
.end
```

Figure 6.22 The Spice input file for computing the I_{out}-V_{out} characteristics of the simple current-mirror circuit shown in Fig. 6.21.

We can go further and characterize the output port of this circuit (assuming $V_{out} > 220$ mV) using the Norton equivalent circuit as shown in Fig. 6.24. The level of the current source of 9.64 μA was found by finding the y-axis intersection of the extrapolated line joining the points on the $i-v$ curve in Fig. 6.21 for $V_{out} > 220$ mV. The output resistance $R_o = 7.8$ MΩ is simply the inverse of the slope of this line found using the Probe facility of PSpice. It is interesting to note that a similar resistance value is found through a .TF command with the output voltage biased somewhere inside the linear region of the current source.

It is also reassuring that hand analysis also generates a value of the output resistance for the current source that is in the neighborhood of that predicted by Spice, ie. $V_A/I = 80$ V / 10μA $= 8$ MΩ. The effect of the emitter resistance of 12 Ω is not very significant at a current level of 10 μA because the transistor transconductance is small, ie. 0.4 mA/V. Thus, from Eqn. 6.1, we see that R_o does not change by much, eg. $R_o \approx [1 + (0.4 \times 10^{-3})(12)](8 \times 10^6)$ $= 8.04$ MΩ.

6.7 A CMOS Differential Amplifier With Active Load

Differential pairs, current mirrors and current sources are usually combined in MOS

6.7 A CMOS Differential Amplifier With Active Load

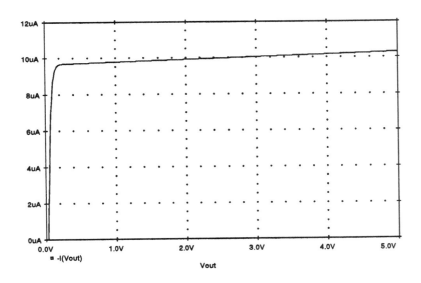

Figure 6.23 I_{out} vs. V_{out} for the current source implementation shown in Fig. 6.21.

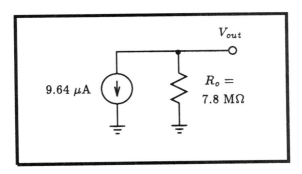

Figure 6.24 Norton equivalent representation of the current source in Fig. 6.21 for $V_{out} > 220$ mV.

technology to form differential amplifiers. The current source is used to bias the differential pair and the current mirror acts as a large resistive load, thus providing large voltage gain. An example of a CMOS differential amplifier is shown in Fig. 6.25. It is easily shown by hand analysis that the small-signal differential-mode voltage gain A_d of this stage is given approximately by $A_d = -g_{m_1}(r_{o_2} \| r_{o_4})$. The corresponding common-mode voltage gain A_{CM} is, to a first order approximation, zero. An expression for A_{CM} can be derived through a small-signal analysis, however, the result is usually too complex to be insightful.

Using Spice let us compute the differential-mode and common-mode voltage gains of the differential amplifier shown in Fig. 6.25, and thus, its common-mode rejection ratio (CMRR).

6 Differential And Multistage Amplifiers

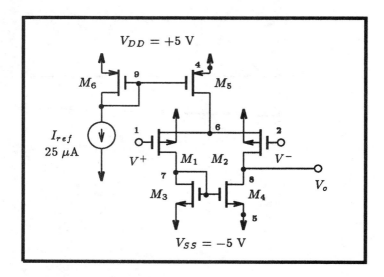

Figure 6.25 A CMOS differential pair with active load and current source biasing.

Transistor	M_1	M_2	M_3	M_4	M_5	M_6
W/L	120/8	120/8	50/10	50/10	150/10	150/10

Table 6.3 Transistor dimensions of the CMOS amplifier in Fig. 6.25.

We shall assume that the NMOS and PMOS transistors are fabricated with a CMOS process which can be characterized by the following Spice model parameters: $\mu_n C_{OX} = 20 \ \mu A/V^2$, $\mu_p C_{OX} = 10 \ \mu A/V^2$, $|V_t| = 1$ V, and $\lambda = 0.04 \ V^{-1}$. For the time being we shall neglect the body effect of the transistor (ie. $\gamma = 0$). The length and width dimension of each transistor are listed in Table 6.3.

The Spice description of this circuit is seen in Fig. 6.26. The input is excited using the multiple-source arrangement depicted in Fig. 6.3. The first analysis that is requested is a DC sweep of the input differential voltage. The input differential voltage is swept between the two voltage supply limits (ie. -5 V and +5 V) with a voltage increment of 50 mV. This Spice analysis will then be followed by a DC sweep of the input common-mode voltage. These two DC sweeps are necessary to locate the high-gain linear region of the amplifier.

The large-signal differential-input transfer characteristic of the CMOS amplifier as calculated by Spice is shown in Fig. 6.27. Here we see that the high-gain region is in the vicinity of 0 V. However, pertinent details of this region are not clearly evident. We shall therefore re-run the Spice job using a more refined step size for V_d. Specifically, we shall replace the DC sweep command given earlier in Fig. 6.26 by the following one:

.DC Vd -100mV +100mV 1mV

6.7 A CMOS Differential Amplifier With Active Load

A CMOS Differential Amplifier With Current Source Biasing

```
** Circuit Description **
* power supplies
Vdd 4 0 DC +5V
Vss 5 0 DC -5V
* differential-mode signal level
Vd 101 0 DC 0V
Rd 101 0 1
EV+ 1 100 101 0 +0.5
EV- 2 100 101 0 -0.5
* common-mode signal level
Vcm 100 0 DC 0V
* front-end stage
M1 7 1 6 4 pmos_transistor L=8u W=120u
M2 8 2 6 4 pmos_transistor L=8u W=120u
M3 7 7 5 5 nmos_transistor L=10u W=50u
M4 8 7 5 5 nmos_transistor L=10u W=50u
* current source biasing stage
M5 6 9 4 4 pmos_transistor L=10u W=150u
M6 9 9 4 4 pmos_transistor L=10u W=150u
Iref 9 5 25uA
* transistor model statements
.model pmos_transistor pmos (kp=10u Vto=-1V lambda=0.04 gamma=0)
.model nmos_transistor nmos (kp=20u Vto=+1V lambda=0.04 gamma=0)
** Analysis Requests **
.OP
.DC Vd -5 5 50mV
** Output Requests **
.PLOT DC V(8)
.probe
.end
```

Figure 6.26 The Spice input file for calculating the large- and small-signal transfer characteristics of the CMOS amplifier shown in Fig. 6.25.

On completion of Spice, we observe the expanded view of the CMOS amplifier high-gain region in Fig. 6.28. As is evident, this particular amplifier has an output DC offset of -3.5 V, or equivalently an input offset voltage of -50 mV. The linear region of this amplifier is between $V_d = -10$ mV and $+65$ mV. To obtain the small-signal differential gain in this region we can either estimate it from the slope of the line forming the amplifier linear region, or calculate it directly using a .TF command. We shall choose the latter and enter the following .TF command in the Spice deck listed in Fig. 6.26:

.TF V(8) Vd

6 Differential And Multistage Amplifiers

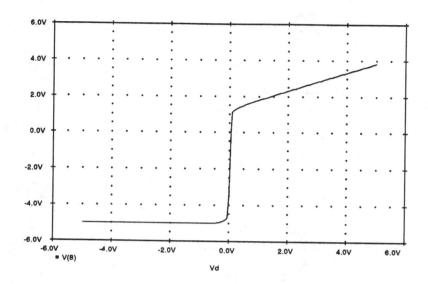

Figure 6.27 The large-signal differential-input transfer characteristics of the CMOS amplifier shown in Fig. 6.25.

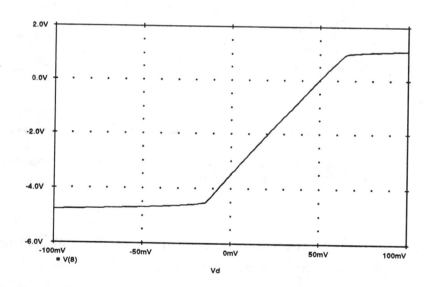

Figure 6.28 An expanded view of the high-gain differential region of the CMOS amplifier shown in Fig. 6.25.

Most applications of differential amplifiers involve using them in a negative feedback configuration where the negative feedback forces the input offset voltage towards zero. It seems reasonable, therefore, to evaluate the differential gain and other characteristics with the transfer characteristics shifted so that the input referred offset voltage is zero. This can

6.7 A CMOS Differential Amplifier With Active Load

be achieved by applying a differential offset voltage of +50 mV to the input of the amplifier, ie. modify the Vd source statement to:

```
Vd 101 0 DC 50mV.
```

An operating point command (.OP) is included to obtain information about the small-signal transistor model parameters.

The results of this analysis are then found in the Spice output file as follows:

```
****    SMALL-SIGNAL CHARACTERISTICS

        V(8)/Vd =    6.784E+01

        INPUT RESISTANCE AT Vd =    1.000E+00

        OUTPUT RESISTANCE AT V(8) =    9.536E+05
```

The small-signal differential gain A_d is therefore 67.84 V/V. To compare this quantity with that estimated by hand, we recall that $A_d = -g_{m_1}(r_{o_2} \| r_{o_4})$ and

$$g_{m_1} = \sqrt{2\mu_n C_{OX}(W/L)_1 I_{D_1}}.$$

Thus, we can estimate g_{m_1} to be 86.6 mA/V by assuming $I_{D_1} = 12.5\ \mu$A. Similarly, the output resistance of M_2 and M_4 is given by $r_o = 1/\lambda I_D = 1/(0.04 * 12.5 \times 10^{-6})$ which gives $r_o = 2$ MΩ. Substituting these values into the expression for A_d results in $A_d = 86.6$ V/V. When compared to the gain computed by Spice, our estimate here has a relative error of about 28%. The reason for this error is largely due to the inaccuracy in estimating the drain bias current of each transistor, which in turn is the result of neglecting the Early effect.

A better estimate of the differential voltage gain can be obtained by using the bias point and thus the small-signal model parameters generated by Spice. These are listed below:

```
**** MOSFETS
```

NAME	M1	M2	NAME	M3	M4
MODEL	pmos_transistor	pmos_transistor	MODEL	nmos_transistor	nmos_transistor
ID	-1.26E-05	-1.43E-05	ID	1.26E-05	1.43E-05
VGS	-1.38E+00	-1.43E+00	VGS	1.49E+00	1.49E+00
VDS	-4.91E+00	-1.39E+00	VDS	1.49E+00	5.01E+00
VBS	3.60E+00	3.60E+00	VBS	0.00E+00	0.00E+00
VTH	-1.00E+00	-1.00E+00	VTH	1.00E+00	1.00E+00
VDSAT	-3.75E-01	-4.25E-01	VDSAT	4.88E-01	4.88E-01
GM	6.73E-05	6.73E-05	GM	5.17E-05	5.86E-05
GDS	4.22E-07	5.42E-07	GDS	4.76E-07	4.76E-07
GMB	0.00E+00	0.00E+00	GMB	0.00E+00	0.00E+00

NAME	M5	M6
MODEL	pmos_transistor	pmos_transistor
ID	-2.69E-05	-2.50E-05
VGS	-1.56E+00	-1.56E+00
VDS	-3.60E+00	-1.56E+00
VBS	0.00E+00	0.00E+00
VTH	-1.00E+00	-1.00E+00

6 Differential And Multistage Amplifiers

```
VDSAT      -5.60E-01           -5.60E-01
GM          9.61E-05            8.93E-05
GDS         9.41E-07            9.41E-07
GMB         0.00E+00            0.00E+00
```

Using these, we compute the differential voltage gain A_d to be 66.4 V/V. This is obviously much closer to the value computed by Spice at 67.84 V/V.

Once again we see that the small-signal gain expression derived by Sedra and Smith results in accurate gain prediction provided that good estimates of the bias points and hence of the small-signal model parameters of the transistors are used.

In a similar fashion, the large-signal common-mode transfer characteristic of the amplifier is computed by replacing the DC sweep command in the Spice deck listed in Fig. 6.25 by one that sweeps the input common-mode voltage (V_{CM}) between −5 V and +5 V in 50 mV increments. The syntax of such a Spice statement would appear as follows:

```
.DC Vcm -5V +5V 50mV.
```

The revised Spice input file is then re-run and the effect of the input common-mode signal level on the output is then graphically displayed as in Fig. 6.29. Here we see that an input common-mode level ranging between −1.2 V and +3 V has little effect on the output signal. For instance, a 1 V change in the input common-mode level causes a +180 mV change in the output voltage level and this is fairly consistent over the −1.2 V to +3 V range. In other words, the common-mode gain A_{CM} is approximately 180 mV/V. This also can be confirmed by using a .TF command. However, the estimate we obtained directly from the transfer characteristic shown in Fig. 6.29 is sufficiently accurate for our purposes here.

We should note here that once the common-mode range is known we must check to see whether the V_{CM} used to determine the large-signal differential characteristic is valid. Specifically, the large-signal differential characteristic computed earlier was obtained with $V_{CM} = 0$ V. Fortunately, this lies within the common-mode range of the amplifier, and the differential characteristics obtained are therefore valid.

Combining the above estimate of the amplifier common-mode gain with the small-signal differential gain calculated earlier, we can compute the CMRR of the amplifier to be $67.84/180 \times 10^{-3} = 376.9$ or 51.5 dB.

In the above analysis we ignored the presence of transistor body effect (ie. $\gamma = 0$). In the following we shall repeat the above analysis with $\gamma = 0.9$ V$^{1/2}$. This requires that we alter the two MOS model statement provided in the Spice deck seen listed in Fig. 6.26 according to:

```
.model pmos_transistor pmos (kp=10u Vto=-1V lambda=0.04 gamma=0.9 )
```

6.8 GaAs Differential Amplifiers

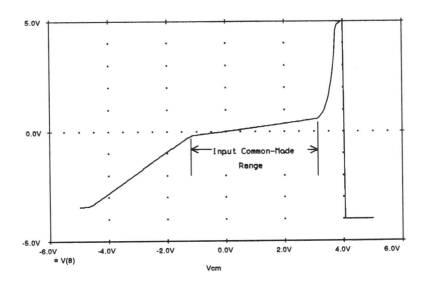

Figure 6.29 The large-signal common-mode DC transfer characteristics of the CMOS differential amplifier shown in Fig. 6.25. An input differential offset voltage of +50 mV is applied to the amplifier input to ensure that the amplifier is biased in its linear region.

```
.model nmos_transistor nmos (kp=20u Vto=+1V lambda=0.04 gamma=0.9 )
```

Repeating each Spice analysis suggested above, we then find the following differential-mode and common-mode voltage gains:

```
****     SMALL-SIGNAL CHARACTERISTICS

         V(8)/Vd =   7.118E+01

         INPUT RESISTANCE AT Vd =  1.000E+00

         OUTPUT RESISTANCE AT V(8) =  9.989E+05


****     SMALL-SIGNAL CHARACTERISTICS

         V(8)/Vcm =  1.416E-01

         INPUT RESISTANCE AT Vcm = 1.000E+20

         OUTPUT RESISTANCE AT V(8) =  9.989E+05
```

Comparing these results with those computed without the body effect taken into account, we see that both A_d and A_{CM} have changed little. The new CMRR then becomes 54.0 dB; an increase of 3.5 dB.

6 Differential And Multistage Amplifiers

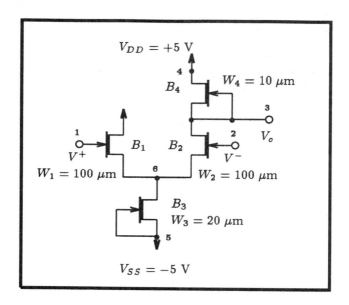

Figure 6.30 A simple MESFET differential amplifier.

6.8 GaAs Differential Amplifiers

In Fig. 6.30 we show a simple GaAs MESFET differential amplifier. Each MESFET is assumed to have minimum length (1 μm in this case), but their widths vary and are specified alongside each transistor. Using PSpice (recall that Spice does not have a built-in model for MESFETs) we would like to compute the large-signal differential and common-mode DC transfer characteristics of this amplifier. From this, we would like to determine its common-mode rejection ratio (CMRR). We shall assume that the MESFETs are characterized by the following parameters: $\beta = 0.1$ mA/V^2 (for a 1-μm wide device), $V_t = -1.0$ V, and $\lambda = 0.05$ V^{-1}.

The PSpice input file describing the circuit of Fig. 6.30 is listed in Fig. 6.31. The inputs to this amplifier are assumed to be driven by the same multiple voltage source combination shown connected to the input terminals of the circuit in Fig. 6.3. To obtain the large-signal DC differential transfer characteristic we shall perform a DC sweep of the input differential voltage between the voltage limits of the two supplies (ie. $V_{DD} = +5$ V and $V_{SS} = -5$ V) using a 50 mV step. The input common-mode level V_{CM} shall be set equal to zero, as it is usually assumed that $V_{CM} = 0$ V is an input that will be within the linear range of the amplifier.

On completion of the PSpice run the large-signal differential characteristic shown in Fig. 6.32 is found. Here we see a rather strange large-signal differential transfer characteristic. Instead of the output voltage swinging between the limits of the two power supplies, the output only swings between V_{DD} and ground — about one-half the normal voltage swing.

6.8 GaAs Differential Amplifiers

```
A MESFET Differential Amplifier

** Circuit Description **
* dc supplies
Vdd 4 0 DC +5V
Vss 5 0 DC -5V
* differential-mode signal level
Vd 101 0 DC 0V
Rd 101 0 1
EV+ 1 100 101 0 +0.5
EV- 2 100 101 0 -0.5
* common-mode signal level
Vcm 100 0 DC 0V
* amplifier circuit
B1 4 1 6 n_mesfet 100
B2 3 2 6 n_mesfet 100
B3 6 5 5 n_mesfet 20
B4 4 3 3 n_mesfet 10
* mesfet model statements (by default, level 1)
.model n_mesfet gasfet (beta=0.1m Vto=-1.0V lambda=0.05)
** Analysis Requests **
.OP
* calculate the large-signal DC transfer characteristics
.DC Vd -5V +5V 50mV
** Output Requests **
.PLOT DC V(3)
.probe
.end
```

Figure 6.31 A PSpice input file for calculating the large-signal DC transfer characteristic of the MESFET differential amplifier shown in Fig. 6.30.

We also notice that below the high-gain region of the amplifier, for input levels less than −1 V, the output level does not saturate, but instead begins to rise in a very linear manner. This suggests that this amplifier has a limited range of operation as a differential amplifier and therefore the input differential voltages must be restricted to be greater than −1 V.

To obtain a better view of the linear region of this amplifier, we shall re-sweep the input differential voltage between −200 mV and +200 mV using a step size of 10 mV. This requires that we alter the DC sweep command seen listed in the Spice deck of Fig. 6.31 to the following one:

```
.DC Vd -200mV +200mV 10mV.
```

Re-running the Spice analysis, we obtain the large-signal characteristic of the amplifier shown in Fig. 6.33. We see that the slope of the linear region is rather low, approximately 40 V/V, extending between 0 and +50 mV along the Vd axis. Such low voltage gain can be attributed

6 Differential And Multistage Amplifiers

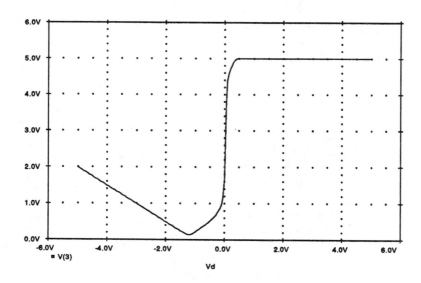

Figure 6.32 The large-signal differential DC transfer characteristics of the MESFET differential amplifier shown in Fig. 6.30 with $V_{CM} = 0V$.

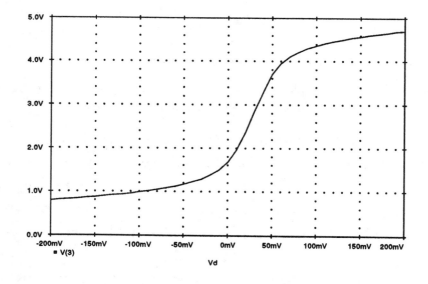

Figure 6.33 An expanded view of the linear region of the differential DC transfer characteristics of the MESFET differential amplifier shown in Fig. 6.30 with $V_{CM} = 0V$.

to the rather low channel-length modulation factor of the MESFET (ie. $\lambda = 0.05$ V^{-1}). As an estimate of the small-signal voltage gain in the high gain region of this amplifier, consider evaluating the small-signal transfer characteristic of this amplifier using the .TF command

6.8 GaAs Differential Amplifiers

of PSpice around an input differential voltage of 25 mV. This point was chosen because it lies approximately midway between the two extremes of the linear region. To accomplish this, the following .TF command is placed in the PSpice deck shown in Fig. 6.31,

```
.TF V(3) Vd.
```

and the statement for the input differential voltage is modified to include the input +25 mV offset voltage according to

```
Vd 101 0 DC +25mV.
```

The results of this analysis are then found in the Spice output file as follows:

```
****    SMALL-SIGNAL CHARACTERISTICS

        V(3)/Vd =  4.723E+01

        INPUT RESISTANCE AT Vd =  1.000E+00

        OUTPUT RESISTANCE AT V(3) =  1.238E+04
```

Thus, we see that this particular amplifier has a somewhat low differential voltage gain of 47.2 V/V. This low gain can be attributed to the rather low output resistance of the amplifier (ie. $R_o = 12.38$ kΩ).

According to a hand analysis the voltage gain and output resistance of the differential amplifier shown in Fig. 6.30 are given by $A_d = g_{m_2} R_o$ and $R_o = (r_{o_2} \| r_{o_4})$. To check these expressions against the values computed by PSpice, we list below the parameters of the small-signal model of each MESFET as computed by Spice:

```
**** GASFETS
```

NAME	B1	B2	B3	B4
MODEL	n_mesfet	n_mesfet	n_mesfet	n_mesfet
ID	1.45E-03	1.12E-03	2.57E-03	1.12E-03
VGS	-6.55E-01	-6.80E-01	0.00E+00	0.00E+00
VDS	4.33E+00	1.92E+00	5.67E+00	2.41E+00
GM	8.39E-03	7.01E-03	5.13E-03	2.24E-03
GDS	5.95E-05	5.52E-05	1.00E-04	5.06E-05
CGS	0.00E+00	0.00E+00	0.00E+00	0.00E+00
CGD	0.00E+00	0.00E+00	0.00E+00	0.00E+00
CDS	0.00E+00	0.00E+00	0.00E+00	0.00E+00

Substituting the appropriate parameter values into the expression for R_o and A_d above, we get $R_o = 9.45$ kΩ and $A_d = 66.25$ V/V. When comparing these with those generated by PSpice (ie. $R_o = 12.38$ kΩ and $A_d = 47.2$ V/V), we see that our small-signal hand calculations are not as accurate as we've seen previously for Bipolar and CMOS technologies. The reason for this is that the large-signal model for the MESFET described by Sedra and

6 Differential And Multistage Amplifiers

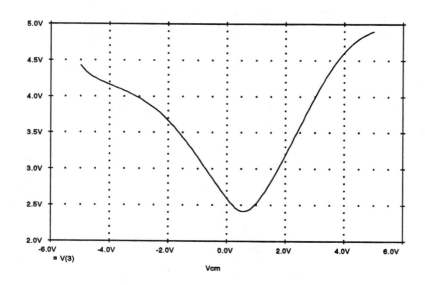

Figure 6.34 The large-signal common-mode DC transfer characteristics of the MESFET differential amplifier shown in Fig. 6.30. The input differential offset voltage is set equal to 25 mV.

Smith in section 6.9 is different than that used by PSpice. As a result, the small-signal models are slightly different. This was not the case for the other technologies. Here it should be noted that the development of accurate GaAs MESFET models is still a subject of current research.

To determine the large-signal DC common-mode transfer characteristics of the MESFET amplifier in Fig. 6.30 we require that PSpice compute the output voltage as a function of the input common-mode voltage. This is easily achieved by modifying the PSpice input file given in Fig. 6.31 by replacing the DC sweep command given there by the following one:

```
.DC Vcm -5V +5V 100mV.
```

The input differential voltage V_d will remain offset by 25 mV in order to keep the amplifier in its linear region.

Re-submitting this revised input file to PSpice results in the large-signal common-mode transfer characteristic shown in Fig. 6.34. Here we see that the common-mode characteristic consists of two almost linear regions with opposite signed slopes. For input common-mode voltages less than 0.5 V, the slope of the large-signal characteristic is negative with a magnitude estimated directly from the graph to be approximately 60 mV/V. Whereas, for input common-mode voltages larger than 0.5 V, the slope is estimated to be about +75 mV/V. This rather unusual looking characteristic is a result of the low output resistance of MESFETs.

6.8 GaAs Differential Amplifiers

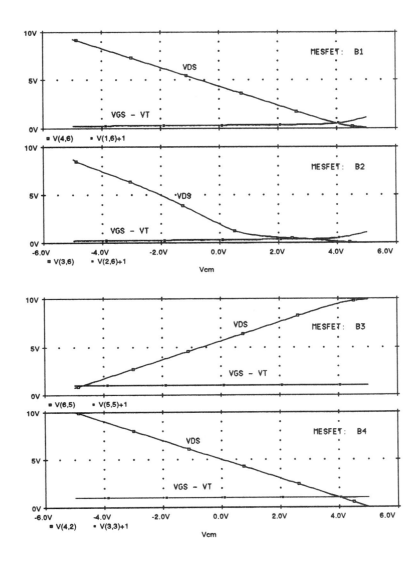

Figure 6.35 Investigating the mode of operation of each MESFET in the amplifier of Fig. 6.30. Here we are comparing V_{DS} with $V_{GS} - V_t$ of each MESFET and determining the range of input common-mode voltage that maintains each transistor in its linear region.

The input common-mode range of this amplifier is not obvious from the graph of the output voltage as a function of the input common-mode voltage shown in Fig. 6.34. But it can be determined by plotting the drain-source voltage V_{DS} of each MESFET as a function of the input common-mode voltage and compare it to the corresponding gate-source voltage minus the threshold voltage $V_{GS} - V_t$ of each MESFET. If $V_{DS} \geq V_{GS} - V_t$, then the MESFET is in its linear region, otherwise it is not. The same PSpice input file can be used here without any modifications. Recall that V_t is equal to –1 V. The results, as further calculated and displayed by Probe, are shown in Fig. 6.35. From these results we see that

303

6 Differential And Multistage Amplifiers

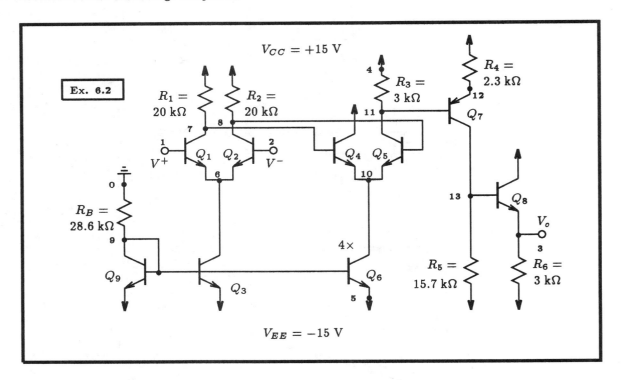

Figure 6.36 A simple operational amplifier consisting of 3 stages.

the upper linear region of this amplifier is determined by MESFET B_2 entering the triode region for V_{CM} exceeding +3 V. The lower common-mode input range limit is determined solely by B_3: When the V_{CM} decreases below −4.7 V, this transistor enters triode. Therefore, the common-mode input range (CMR) for this amplifier is between −4.7 and +3 V. We note that a common-mode input voltage of 0 V is within the CMR of this amplifier and thus our previous calculation of differential-mode gain is valid.

To estimate the CMRR of this amplifier, we have to consider that when the input common-mode voltage is less than 0.5 V, the common-mode gain A_{CM} is about −60 mV/V and when the input common-mode voltage is greater than this amount, $A_{CM} = +75$ mV/V. Thus, using the worst-case situation, that is, when $|A_{CM}|$ is largest, the CMRR is computed according to $20 \log \frac{|A_d|}{|A_{CM}|}$ to be 56 dB.

6.9 A BJT Multistage Amplifier Circuit

Fig. 6.36 illustrates the circuit of a simple operational amplifier. It consists of a cascade of several gain stages, two of which are made from differential pairs, and an output buffer. The positive and negative input terminals to the amplifier are labeled as V^+ and V^-, respectively. The output terminal is denoted as V_o. This particular operational amplifier circuit is presented in Examples 6.2 and 6.3 in Sedra and Smith and the results of their

6.9 A BJT Multistage Amplifier Circuit

analysis will be summarized at the end of this discussion. For the purpose of our analysis we shall assume that both the *npn* and *pnp* transistors have the following device parameters: $I_S = 18$ fA, $\beta_F = 100$, $V_{AF} = 100$ V. In addition, Q_6 is considered to have 4× the area of Q_9 and Q_3. Using Spice we would like to analyze this operational amplifier to determine its DC operating point which includes such information as output offset voltage, input bias currents and quiescent power dissipation. In addition, we would also like to compute the small-signal differential voltage gain (in the high gain region of the amplifier) and the input common-mode range.

The Spice input file describing the make-up of the operational amplifier shown in Fig. 6.36, together with the multiple input voltage source arrangement discussed in section 6.1, is given in Fig. 6.37. Notice how Q_6 is specified to have 4× the area of Q_3 and Q_9. Two analysis request commands are listed here: a DC operating point command and a DC sweep command. Also seen listed in the analysis request section of the Spice deck is a .NODESET command. We shall explain its role in a moment.

The DC operating point command will, in addition to calculating the DC operating point of the circuit and the small-signal model parameters of each transistor, determine the power dissipated by the amplifier and the input bias currents. The DC sweep command will be used to compute the large signal transfer characteristic of the amplifier by varying the DC level of the input differential source V_d between –15 V and +15 V in 100 mV increments. The common-mode input V_{CM} will be held at zero volts throughout this DC sweep. One may be tempted to add a small-signal transfer function analysis request here; however, this should be deferred until one sees the large signal differential transfer characteristic of the amplifier and can determine what DC input conditions are required so that Spice linearizes the amplifier around a known operating point.

To assist in the DC bias calculation, we include in the Spice input file a list of initial guesses of some node voltages in the circuit on the statement beginning with the keyword *.NODESET*. If one provides good estimates of the circuit's DC bias point, then the time required by Spice to compute the DC bias point is reduced, but more importantly, including a .NODESET statement can make the difference between obtaining a DC solution and Spice terminating the analysis with no solution because of a non-convergence problem. In older versions of Spice (version 2G6 and less), this step is usually required but newer versions, such as PSpice, have improved algorithms that have more robust convergence capabilities and rarely require guesses of the node voltages.

Submitting the Spice input file listed in Fig. 6.37 to Spice, results partially in the following DC analysis output:

6 Differential And Multistage Amplifiers

Example 6.2: A Simple Operational Amplifier

```
** Circuit Description **
* power supplies
Vcc 4 0 DC +15V
Vee 5 0 DC -15V
* differential-mode signal level
Vd 101 0 DC 0V
Rd 101 0 1
EV+ 1 100 101 0 +0.5
EV- 2 100 101 0 -0.5
* common-mode signal level
Vcm 100 0 DC 0V
* 1st stage
R1 4 7 20k
R2 4 8 20k
Q1 7 1 6 npn_transistor
Q2 8 2 6 npn_transistor
Q3 6 9 5 npn_transistor
* 2nd stage
R3 4 11 3k
Q4 4 7 10 npn_transistor
Q5 11 8 10 npn_transistor
Q6 10 9 5 npn_transistor 4
* 3rd or output stage
R4 4 12 2.3k
Q7 13 11 12 pnp_transistor
R5 13 5 15.7k
Q8 4 13 3 npn_transistor
R6 3 5 3k
* biasing stage
Rb 0 9 28.6k
Q9 9 9 5 npn_transistor
* transistor model statements
.model npn_transistor npn ( Is=18fA Bf=100 VAf=100V )
.model pnp_transistor pnp ( Is=18fA Bf=100 VAf=100V )
** Analysis Requests **
* compute DC operating point using the following initial guesses
.OP
.NODESET V(3)=0V V(6)=-0.7V V(7)=+10V V(8)=+10V V(9)=-14.3V V(10)=+9.3V
+       V(11)=+12V V(12)=+12.7V V(13)=+0.7V
* compute large-signal differential-input transfer characteristics of amplifier
.DC Vd -15V +15V 100mV
** Output Requests **
.PLOT DC V(3)
.probe
.end
```

Figure 6.37 The Spice input file for computing the DC operating point and the large-signal differential transfer characteristic of the amplifier shown in Fig. 6.36.

6.9 A BJT Multistage Amplifier Circuit

Transistor	Hand	Spice	% Error
Q_1	0.25	0.267	6.4%
Q_2	0.25	0.267	6.4%
Q_3	0.5	0.540	7.4%
Q_4	1.0	1.18	15.3%
Q_5	1.0	1.14	12.3%
Q_6	2.0	2.34	14.5%
Q_7	1.0	1.18	15.3%
Q_8	5.0	5.64	11.3%
Q_9	0.5	0.47	−5.5%

Table 6.4 DC collector currents of the operational amplifier shown in Fig. 6.36 expressed in mA as computed by hand analysis and Spice.

```
****    SMALL SIGNAL BIAS SOLUTION       TEMPERATURE =   27.000 DEG C
******************************************************************************

 NODE   VOLTAGE     NODE   VOLTAGE     NODE   VOLTAGE     NODE   VOLTAGE

(   1)    0.0000  (   2)    0.0000  (   3)    2.0677  (   4)   15.0000
(   5)  -15.0000  (   6)    -.6035  (   7)    9.4295  (   8)    9.4295
(   9)  -14.3790  (  10)    8.7868  (  11)   11.6170  (  12)   12.2590
(  13)    2.7493  ( 100)    0.0000  ( 101)    0.0000


        VOLTAGE SOURCE CURRENTS
        NAME         CURRENT

        Vcc         -9.691E-03
        Vee          1.020E-02
        Vd           0.000E+00
        Vcm         -4.887E-06

        TOTAL POWER DISSIPATION   2.98E-01  WATTS

****    VOLTAGE-CONTROLLED VOLTAGE SOURCES

NAME         EV+         EV-
V-SOURCE     0.000E+00   0.000E+00
I-SOURCE    -2.443E-06  -2.443E-06
```

From these results, we see that this amplifier has an output DC offset of +2.0677 V and input bias currents of 2.443 μA. The static power dissipated by the amplifier is 0.298 W.

The collector currents of each device have also been computed by Spice; these can be seen listed in Table 6.4. Also shown in this table are the currents computed by hand analysis (performed in Example 6.2 of Sedra and Smith assuming that $\beta \gg 1$ and ignoring the effect of transistor Early voltage). A third column has also been added showing the relative error (in percent) between these two currents. As we can see, our hand estimates are reasonably

6 Differential And Multistage Amplifiers

Transistor	Hand	Spice			
	$\beta \gg 1$ $V_A = \infty$	$\beta = 10^6$ $V_A = \infty$	$\beta = 100$ $V_A = \infty$	$\beta = 100$ $V_A = 100$ V	$\beta = 100$ $V_A = 35$ V
Q_1	0.25	0.251	0.235	0.267	0.328
Q_2	0.25	0.251	0.235	0.267	0.328
Q_3	0.5	0.503	0.474	0.540	0.661
Q_4	1.0	1.01	0.939	1.18	1.61
Q_5	1.0	1.01	0.939	1.14	1.45
Q_6	2.0	2.01	1.90	2.34	3.08
Q_7	1.0	1.03	0.926	1.18	1.57
Q_8	5.0	5.18	4.35	5.64	7.58
Q_9	0.5	0.503	0.474	0.474	0.474

Table 6.5 Observing the variation in the DC collector currents (in mA) of the operational amplifier for different β's and V_A's as computed by Spice. These can be compared with those computed by simple hand analysis.

close to the Spice results; the largest error in our estimates never exceeds 15.3%. It is reassuring that reasonably good estimates of the DC collector currents of a complicated transistor circuit can be obtained by assuming ideal transistor behavior (ie. $\beta \gg 1$ and $V_A = \infty$).

As a further check on this, we compiled a table of collector current values that were computed by Spice for various combinations of β and V_A values (I_S remains at 18 fA). As one can see from this table, as β and V_A approach infinity (ie. the transistors become more ideal), the results approach those computed by the simplified hand analysis.

The large-signal differential transfer characteristic of this amplifier is displayed in Fig. 6.38. This figure illustrates a view of the operational amplifier differential-input transfer characteristics between -15 V and $+15$ V. We recognize that the high-gain region of the amplifier is in the vicinity of 0 V; however, the resolution of the input voltage axis does not enable us to be certain of the boundaries of this high-gain region. Therefore, we shall re-run the Spice input file with the DC sweep command modified to include an expanded view of this high-gain region between -5 mV and $+5$ mV. This will require that we replace the DC sweep command in the previous Spice input file with the following one:

```
.DC Vd -5mV +5mV 10uV
```

The results of this analysis are on display in Fig. 6.39. For inputs less than -2 mV, the

6.9 A BJT Multistage Amplifier Circuit

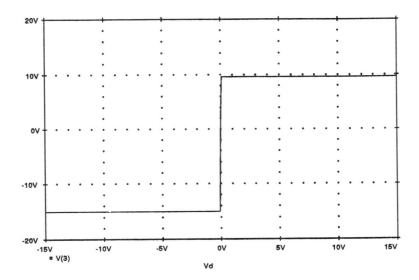

Figure 6.38 The large-signal differential transfer characteristic of the operational amplifier shown in Fig. 6.36. The input common-mode voltage V_{CM} is set to zero.

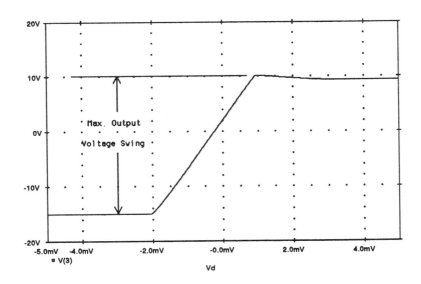

Figure 6.39 An expanded view of the high-gain differential region of the operational amplifier shown in Fig. 6.36.

output remains saturated at −15 V. In the region between −2 mV and +1 mV, the output levels changes from −15 V to +10 V in a linear manner. Thus, the output voltage swing for this amplifier is bounded between −15 V and +10 V, a somewhat unsymmetrical voltage

309

6 Differential And Multistage Amplifiers

swing. The gain experienced by input signals in this linear region is approximately (10 − (−15))/3 mV = 8.33 kV/V. For inputs greater than +2 mV, the output levels' off at +8.5 V. We also see from these results that the input offset voltage V_{OS} for this amplifier is +230.0 μV. This is of course the systematic offset of the amplifier and does not include the components due to various imbalances in the circuit (see Section 6.3).

Now that we know the boundaries of the high-gain region of the amplifier, we can use the transfer function command of Spice to compute the small-signal equivalent circuit parameters of the amplifier in its linear region. But first we must decide on which point inside the linear region of the amplifier we should linearize about. Consider that, for most applications, a high-gain amplifier such as that shown in Fig. 6.36 is usually used in conjunction with negative feedback, and, as a result, the output potential of the amplifier is held close to ground potential (when no input signal is applied). Thus, the small-signal parameters of the amplifier should be obtained around the bias point that has the amplifier output voltage close to 0 V. This is easily obtained by applying the negative of the amplifier input-referred offset voltage (ie. $-V_{OS}$) across the input terminals of the amplifier. For the multiple source arrangement suggested in Fig. 6.3, this is easily achieved by setting V_d equal to $-V_{OS}$.

Returning to the example at hand, we shall modify the Spice statement for V_d according to

```
Vd 101 0 DC -230.0uV AC 1V.
```

A 1 V AC voltage has also been appended to this input voltage source statement. This will enable us to compute the AC current that circulates around the input terminals of the op-amp when a 1 V AC voltage signal is placed across its input terminals. From this we can then calculate the input differential resistance. For more details on this approach, the reader should refer back to Section 6.2. Two analysis requests and a .PRINT command will be added to the input file in order to compute the small-signal parameters of the amplifier. These appear as follows:

```
.TF V(3) Vd
.AC LIN 1 1Hz 1Hz
.PRINT AC Im(EV+) Im(EV-) Vm(1,2).
```

The revised Spice input deck would then be re-submitted to Spice. Some of the small-signal circuit parameters of this amplifier are then found in the output file as follows:

```
****    SMALL-SIGNAL CHARACTERISTICS

        V(3)/Vd =  8.834E+03

        INPUT RESISTANCE AT Vd =  1.000E+00
```

6.9 A BJT Multistage Amplifier Circuit

```
OUTPUT RESISTANCE AT V(3) =   1.337E+02
```

Here we see that the actual voltage gain in the amplifier linear region is 8.834 kV/V, very close to the 8.83 kV/V we estimated from the amplifier large-signal differential transfer characteristic displayed in Fig. 6.39. The input resistance listed here is not for the amplifier but rather the resistance seen by the voltage source V_d. In this particular case, this is the 1 Ω resistor connected in series with V_d. In contrast, the output resistance of 133.7 Ω listed above is the actual output resistance for this amplifier. As an estimate of the input differential resistance of the amplifier, we use the results of the AC analysis, given below:

```
****      AC ANALYSIS

    FREQ        IM(EV+)     IM(EV-)     VM(1,2)

    1.000E+00   4.729E-05   4.718E-05   1.000E+00
```

We see that current IM(EV+) is not quite the same as the current IM(EV-), but very close. The reason for this difference lies in the amplifiers systematic offset that arises because of transistor Early voltage. To determine the input differential resistance of the amplifier under such asymmetric conditions, we shall work with the average of the two base currents. In this way, we can eliminate the presence of the offset current in the resistance calculation. Thus, we compute the average input base current to be 47.235 μA and therefore obtain the input differential resistance at 21.17 kΩ.

The final analysis that we would like to perform on the operational amplifier shown in Fig. 6.36 is to determine its input common-mode range. Consider performing a DC sweep of the input common-mode voltage V_{CM} between the rails of the two power supplies. This requires that we replace the DC sweep command stated in the Spice listing in Fig. 6.37 by the following one:

```
.DC Vcm -15V +15V 0.1V
```

We also maintain an input differential offset voltage of $V_d = -230$ μV. This is necessary to ensure that the amplifier is biased inside its linear region.

Re-submitting the revised input file to Spice, and observing the results in the output file, we obtain the amplifier large-signal common-mode transfer characteristic given in Fig. 6.40. As we can see from these results, the transfer characteristic is linear behavior over the range of V_{CM} between −14.3 V and +9.6 V. Outside these limits the characteristic becomes nonlinear. Thus, the input common-mode range for this amplifier is between −14.3 V and +9.6 V. We should also note that our large-signal differential transfer characteristic computed earlier is valid since it was obtained with an input common-mode voltage that falls within the input common-mode range of the amplifier.

6 Differential And Multistage Amplifiers

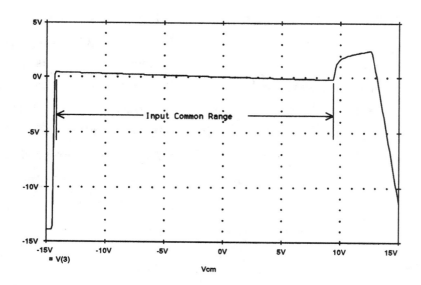

Figure 6.40 The large-signal common-mode DC transfer characteristic of the BJT amplifier shown in Fig. 6.36. An input differential offset voltage of −230 μV is applied to the amplifier input to prevent premature saturation.

It is interesting to correlate the limits of the amplifier common-mode range with the mode of operation of the transistors in the front-end stage. Specifically, Sedra and Smith mention in their text that the upper limit to the common-mode range is determined by Q_1 or Q_2 saturating, and that the lower limit is determined by Q_3 saturating. We can determine when these transistors saturate by observing the voltage across the base-collector junction of each transistor and recalling that a transistor enters its saturation region when the base-collector junction becomes forward biased. For instance, in Fig. 6.41 we display the voltage across the base-collector junctions of Q_1 and Q_3 and observe that Q_1 saturates when V_{CM} exceeds +9.6 V. In contrast, transistor Q_3 saturates when V_{CM} goes below −14.3 V.

Finally, as a summary of what we have learned about this op-amp through the application of Spice, Table 6.6 presents a collection of the results. Moreover, these results are compared with the simplified hand analysis performed by Sedra and Smith in examples 6.2 and 6.3 of their text.

6.10 Spice Tips

- Spice can be conveniently used to compute both the large and small-signal characteristics of amplifier circuits.
- Inputs to differential amplifiers should consist of both a differential and common-mode

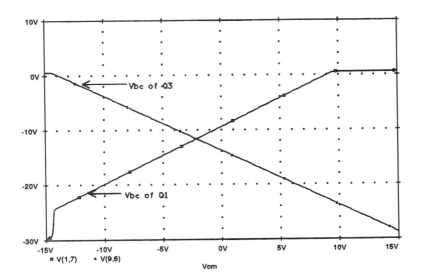

Figure 6.41 The effect of a common-mode input voltage V_{CM} on the linearity of the input stage of the operational amplifier shown in Fig. 6.36. Here we illustrate the base-collector voltage of Q_1 and Q_3 as a function of V_{CM}. The first-stage of the amplifier leaves the active region when the base-collector junction of either Q_1 or Q_3 becomes forward bias.

Parameter	Units	Hand Calculation	Spice
Output-Referred Offset Voltage	V	0	2.07
Input-Referred Offset Voltage	μV	0	230.0
Input Bias Currents	μA	2.5	2.443
Quiescent Power Dissipation	mW	262.5	298
Differential Voltage Gain	kV/V	8.513	8.834
Input Differential Resistance	kΩ	20.1	21.17
Output Resistance	Ω	152	133.7
Input Common-Mode Range	V	-13.6 to $+10.0$	-14.3 to $+9.6$
Output Voltage Swing	V	$-$	-15.0 to $+10.0$

Table 6.6 Comparison of the results of the analysis of the operational amplifier shown in Fig. 6.36 by hand (in Sedra and Smith) and using Spice.

level. An interesting arrangement of several voltage sources was given in this chapter (see Fig. 6.3, for instance), illustrating how the differential and common-mode levels can be independently adjusted.

- The high-gain linear region of an amplifier is located by first sweeping the input differential voltage v_d between the limits of the power supplies with the common-mode voltage

6 Differential And Multistage Amplifiers

V_{CM} equal to 0 V. This analysis is repeated with a reduced sweep range centered more closely around the high-gain region of the amplifier until a smooth transition through the high-gain region is achieved. Following this, one must check to see whether the input common-mode voltage of 0 V keeps the amplifier in its linear region.

- Care must be exercised when computing the small-signal characteristics of an amplifier using Spice. One should first decide what DC input conditions are required so that the amplifier is linearized around an appropriate DC operating point. Generally, selecting an input differential offset voltage that forces the output voltage to 0 V will bias the amplifier at an operating point that is quite close to the operating point that results when some external negative feedback connection is made around the amplifier.
- Any time an IC transistor is used in a circuit and its model is obtained from a library, the substrate connection must be defined.
- Spice version 2G6 does not have a built-in model for the MESFET but PSpice does.
- At times, Spice will not be able to compute the DC node voltages in a nonlinear circuit due to a DC convergence problem. Sometimes this convergence problem can be alleviated by providing Spice with a set of initial estimates of the DC node voltages of the circuit. These are entered into the Spice deck using the .NODESET command of Spice.
- The small-signal input resistance to a differential amplifier is computed using Spice by applying a known AC voltage across the input terminals of the differential amplifier and computing the AC currents that flows into the amplifier terminals. In many practical amplifier situations, these currents will not be equal. So, instead, the average of these two currents is used in the input resistance calculation.

6.11 Bibliography

Staff, *1990-1991 IC Data Book*, Gennum Corporation, Burlington, Ontario, Canada.

6.12 Problems

6.1 A BJT differential amplifier is biased from a 2 mA constant-current source and includes a 100 Ω resistor in each emitter. The collectors are connected to +10 V via 5 kΩ resistors. A differential input signal of 0.1 V is applied between the two bases. Assume that the transistors are matched and have $\beta = 100$ and $I_S = 14$ fA.

(a) With the aid of Spice, determine the signal current in the emitters (i_e) and the base-emitter voltage v_{be} for each BJT.

(b) What is the total emitter current in each BJT?

(c) What is the signal voltage at each collector?

(d) What is the voltage gain realized when the output is taken between the two collectors?

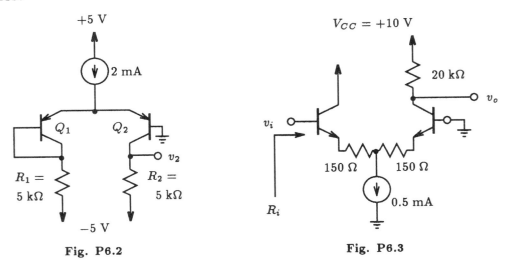

Fig. P6.2

Fig. P6.3

6.2 For the circuit in Fig. P6.2 in which the transistors have high β, with the aid of Spice, determine the value of v_2. If the resistor R_1 is reduced to 2.5 kΩ, what does v_2 become?

6.3 Find the voltage gain and the input resistance of the amplifier in Fig. P6.3 using Spice assuming that $\beta = 100$.

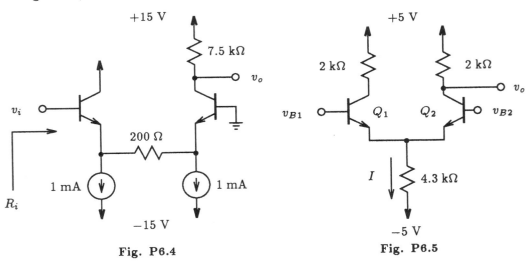

Fig. P6.4

Fig. P6.5

6.4 Find the voltage gain and the input resistance of the amplifier shown in Fig. P6.4 using Spice assuming $\beta = 100$.

6.5 The differential amplifier circuit of Fig. P6.5 utilizes a resistor connected to the negative power supply to establish the bias current I.

(a) For $v_{B1} = v_d/2$ and $v_{B2} = -v_d/2$, where v_d is a small signal with zero average, find the magnitude of the differential gain, $|v_o/v_d|$ using Spice.

(b) For $v_{B1} = v_{B2} = v_{CM}$, find the magnitude of the common-mode gain, $|v_o/v_{CM}|$ using Spice.

(c) Calculate the CMRR.

(d) If $v_{B1} = 0.1\sin(2\pi t) + 0.005\sin(2\pi t)$ volts, $v_{B2} = 0.1\sin(2\pi t) - 0.005\sin(2\pi t)$ volts, plot the output voltage v_O using the .PLOT command of Spice.

6.6 A BJT differential amplifier is biased from a 300 μA constant-current source and the collectors are connected to +7.5 V via 50 kΩ resistors. If the scale currents I_S of the two transistors have a nominal value of 10 fA but differ by 10 %, what is the resulting input offset voltage?

6.7 A BJT differential amplifier is biased from a 1 mA constant-current source and the collectors are connected to +15 V via a 10 kΩ resistors. If the β's of the two transistors are 100 and 200, what is the resulting input offset voltage?

6.8 A BJT differential amplifier is biased from a 1 mA constant-current source and the collectors are connected to +15 V via a 10 kΩ resistors. If V_A of the two transistors are 100 and 200, what is the resulting input offset voltage?

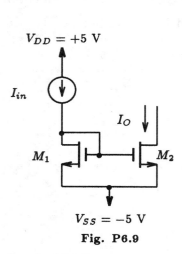

Fig. P6.9

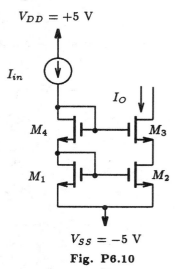

Fig. P6.10

6.9 For the simple MOS current mirror shown in Fig. P6.9, the devices have $V_t = 1$ V, $\mu_n C_{OX} = 200$ μA/V^2 and $V_A = 20$ V. Further, $V_{DD} = +5$ V and $V_{SS} = -5$ V. Using Spice, determine the Norton equivalent of this circuit when the input current is 100 μA. Also, what is the minimum voltage that can appear at the output of this mirror circuit while maintaining linear operation?

6.10 For the cascode current mirror shown in Fig. P6.10, with $V_t = 1$ V, $\mu_n C_{OX} = 200$ μA/V^2 and $V_A = 20$ V. Further, $I_{in} = 100$ μA, $V_{DD} = -5$ V, $V_{SS} = -5$ V, and $V_O = +5$ V. Using Spice, determine the Norton equivalent of this circuit when the input current is 100 μA. Also, what is the minimum voltage that can appear at the output of this mirror circuit while maintaining linear operation? Note that, although, the output resistance of this current source is much larger than the simple current mirror of Problem 6.9, its linear region is somewhat reduced.

6.11 In Fig. P6.11(a) we present a NMOS version of the Wilson current mirror. In (b), an additional transistor has been added to the Wilson circuit to make it more symmetrical. Assuming that the MOS devices have Spice model parameters $V_t = 1$ V, $\mu_n C_{OX} = 200$ $\mu A/V^2$ and $V_A = 10$ V, compare the input - output current behavior of these two mirror circuits.

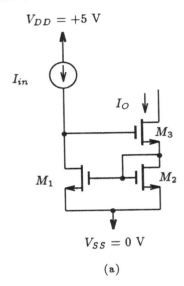 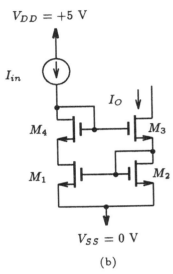

Fig. P6.11

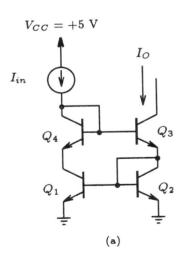 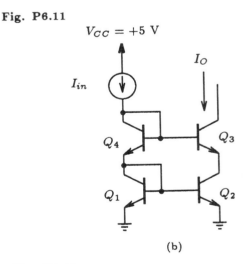

Fig. P6.12

6.12 Compare the input – output current behavior (ie. accuracy, output resistance, minimum output voltage and input current range) of the modified Wilson current mirror circuit shown in Fig. P6.12(a) with the behavior of the cascode current mirror shown in Fig. P6.12(b). Model each bipolar transistor after the Gennum Corporation integrated *npn* transistor described in section 6.4. Which is the better current mirror? How do these two current mirrors compare with the high-swing cascode current mirror circuit shown in Fig. 6.18.

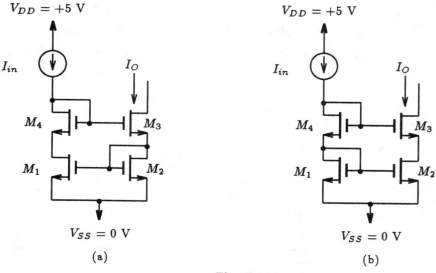

Fig. P6.13

6.13 Compare the input – output current behavior (ie. accuracy, output resistance, minimum output voltage and input current range) of the modified Wilson current mirror circuit shown in Fig. P6.13(a) with the behavior of the cascode current mirror shown in Fig. P6.13(b). Assume that the MOS devices have Spice parameters: $V_t = 1$ V, $\mu_n C_{OX} = 200\ \mu A/V^2$ and $V_A = 70$ V. Which is the better current mirror? Why are your conclusion here different than that found above in Problem 6.12 for the bipolar case?

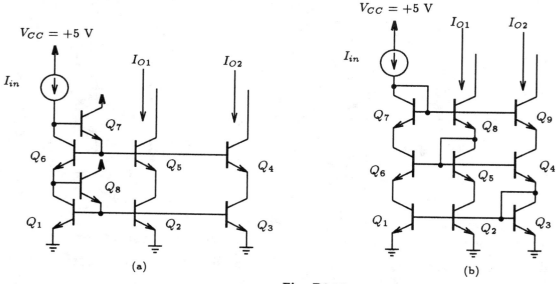

Fig. P6.15

6.14 For the high-swing cascode current mirror shown in Fig. 6.18, determine the maximum and minimum voltage limits of V_B that maintain the current mirror in its linear region. Assume that the transistors are modeled after the integrated *npn* transistors first suggested in Section 6.4.

6.15 In Fig. P6.15 we present two different ways of realizing a two-output current mirror circuit. The circuit in part (a) is a simple extension of the two transistor cascode current mirror circuit. The circuit in part (b) is a generalization of the Wilson current mirror circuit. Assuming that the transistors are modeled after the integrated *npn* transistor first suggested in Section 6.4, determine using Spice which is the better current mirror. Base your judgment on current transfer function accuracy, output resistance, minimum output voltage and input current range.

6.16 Compare your results found in Problem 6.15, with a multiple-output current mirror circuit created by extension of the high-swing cascode current mirror circuit of Fig. 6.18. Which makes the better multiple-output current mirror?

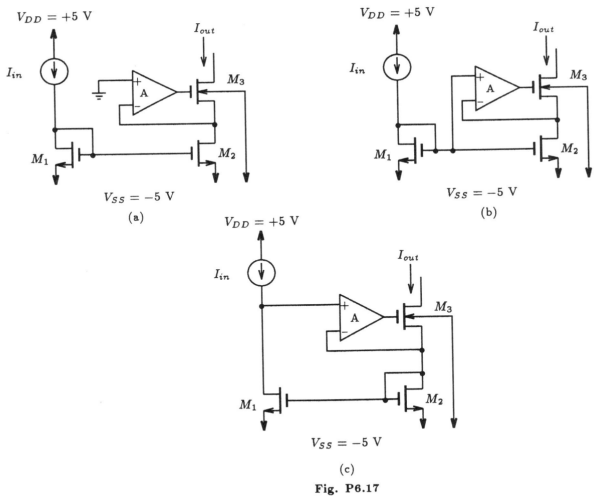

Fig. P6.17

6.17 To improve the quality of a current mirror, an op-amp of gain A is sometimes used. In Fig.

6 Differential And Multistage Amplifiers

P6.17 we present several possible designs of a MOS current mirror circuit that incorporate an op-amp. For each design, determine using Spice how the op-amp gain affects the quality of the current mirror, ie. (i) current gain accuracy, (ii) output resistance, (iii) minimum output voltage and (iv) input current range. Of the three circuits shown in Fig. P6.17, which behaves closer to the ideal current mirror? Assume the following parameter values for each MOSFET: $W = 100 \ \mu m$, $L = 10 \ \mu m$, $\mu_n C_{OX} = 20 \ \mu A/V^2$, $|V_{t0}| = 1$ V and $\lambda = 0.04$ V^{-1}, and $\gamma = 0.9$ V$^{1/2}$.

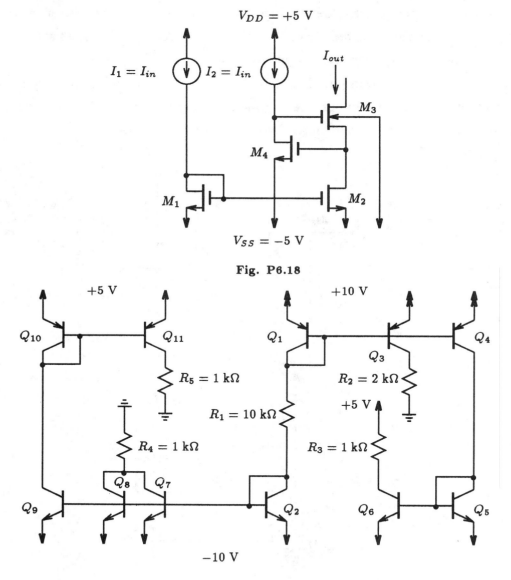

Fig. P6.18

Fig. P6.19

6.18 In Fig. P6.18 we present a high-performance current mirror that is referred to in the literature as a regulated cascode current mirror. It is assumed that the two input current signals track one another. For an input current level of 1 μA, 10 μA, and 100 μA, determine (a) the current gain accuracy, (b) the output resistance, and (c) the minimum output voltage. Assume the following parameter values for each MOSFET: $W = 100$ μm, $L = 10$ μm, $\mu_n C_{OX} = 20$ μA/V^2, $|V_{t0}| = 1$ V, $\lambda = 0.04$ V^{-1}, and $\gamma = 0.9$ V$^{1/2}$.

6.19 Compute the voltages at all nodes and the currents through all branches in the circuit of Fig. P6.19 assuming β is very large using Spice. Compare this to the case when $\beta = 100$.

6.20 Compute the voltages at all nodes and the currents through all branches in the circuit of Fig. P6.19 assuming V_A is very large using Spice. (Setting $V_A = 0$ in the model statement for the BJT is equivalent to setting $V_A = \infty$). Compare this to the case when $|V_A| = 100$ V.

6.21 Consider the effect of power-supply variation on the dc bias of the op-amp circuit of Fig. 6.36: If +15 V is lowered to +14 V, what is the effect on the amplifier parameters listed in Table 6.6? If, separately, −15 V is raised to −14 V, what is effect on the amplifier parameters in Table 6.6?

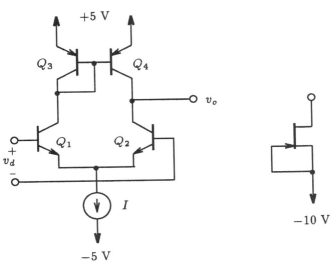

Fig. P6.22   Fig. P6.26

6.22 The differential amplifier in Fig. P6.22 is operated with $I = 100$ μA, with devices for which $V_A = 200$ V and $\beta = 100$. What is the differential input resistance, output resistance, equivalent transconductance, and open-circuit voltage gain of this amplifier?

6.23 In the multistage amplifier of Fig 6.36, 100 Ω resistors are introduced into the emitter lead of each transistor in the differential pair of the first-stage (Q_1 and Q_2) and 25 Ω for each of the second-stage transistors (Q_4 and Q_5). With the aid of Spice, find the effect that these additions have on the input resistance, the voltage gain of the first stage, and the overall voltage gain.

6.24 If, in the multistage amplifier of Fig. 6.36, the resistor R_5 is replaced by a constant-current source of the same value that flows through R_5 (ie. maintain the same bias

6 Differential And Multistage Amplifiers

situation), calculate the overall voltage gain of this amplifier using Spice.

6.25 A differential amplifier, utilizing JFETs for which $I_{DSS} = 2$ mA, $|V_P| = 2$ V, and $V_A = 100$ V, is biased at a constant current of 2 mA. For drain resistors of 10 kΩ, what is the gain of the amplifier for differential output? If the drain resistors have $\pm 1\%$ tolerance, what is the worst-case common-mode gain and CMRR?

6.26 The JFET circuit shown in Fig. P6.26 can be used to implement the current-source bias in a differential amplifier. With the aid of Spice, determine the output resistance if $I_{DSS} = 2$ mA, $V_P = -2$ V, and $V_A = 100$ V.

6.27 A JFET differential amplifier is loaded with the basic BJT current mirror. The JFETs have $V_P = -2$ V, $I_{DSS} = 4$ mA, and $V_A = 100$ V. The BJTs have $|V_A| = 100$ V and β is large. The bias current $I = 2$ mA. Find R_i, G_m, R_o, and the open-circuit voltage gain.

6.28 An NMOS differential pair is to be used in an amplifier whose drain resistors are 100 k$\Omega \pm 1\%$. For the pair, $\mu_n C_{OX} = 200$ μA/V^2 and $V_t = 1$ V. If the output is taken differentially, contrast the voltage gain and input offset voltage of this amplifier for a bias current of 100 μA and 200 μA.

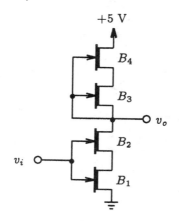

Fig. P6.29

6.29 For the amplifier circuit in Fig. P6.29 let $W_1 = 100$ μm $W_2 = 500$ μm, $W_3 = 50$ μm, and $W_4 = 250$ μm. With the aid of Spice, compute the voltage gain of this circuit. Assume that the MESFETs are modeled with parameters: $V_t = -1$ V, $\beta = 0.1$ mA/V^2, and $\lambda = 0.1$ V^{-1}. Assume that the length of each device is 1 μm.

Chapter 7

Frequency Response

In this chapter we shall investigate the frequency response behavior of various amplifier circuits using Spice. Among other things, this will enable us to verify the accuracy of the gain and bandwidth expressions derived by Sedra and Smith in Chapter 7 of their text.

7.1 Investigating Transfer Function Behavior Using PSpice

A useful feature of PSpice, not available in Spice, is the ability to specify the frequency-dependent gain of a current or voltage dependent-source as a Laplace Transform function. Although one could envision many new applications of this feature; here we are mainly concerned with it as a means for investigating transfer-function behavior as a function of frequency.

Specifying the gain of a dependent source as a Laplace transform function is restricted in PSpice to only the VCVS and VCCS (designated with the letters E and G, respectively). The general form of the element statement used to specify the gain of either one of these controlled sources in terms of the Laplace variable s is illustrated in Fig. 7.1. The first part of this statement, before the keyword Laplace, is identical to that given previously for either the VCVS or VCCS (ie. element type with a unique name and the nodes the dependent source is connected to). After the keyword Laplace, the variable that the controlled source depends on is specified between braces (eg. { V(1) }). This is then followed by an equal sign (=) and an expression written in terms of the Laplace variable and enclosed between braces.

Having the ability to specify gain as an arbitrary expression in terms of the Laplace variable provides a simple means for investigating transfer-function frequency behavior. Consider

7 Frequency Response

$$\begin{Bmatrix} \text{E}name \\ \text{G}name \end{Bmatrix} \quad \text{n+} \quad \text{n-} \quad \text{Laplace} \quad \{\ Controlling\_Variable\ \} \quad = \quad \{\ Laplace\_Transform\ \}$$

Figure 7.1 General form of the PSpice element statement for a voltage or current dependent-source with its gain written as a function of the complex frequency variable.

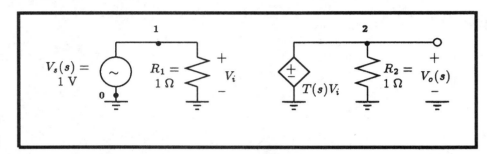

Figure 7.2 A simple circuit arrangement for investigating transfer function behavior.

the simple circuit in Fig. 7.2. Here a VCVS having a voltage gain expressed in the form of a transfer function $T(s)$ is excited by an input AC voltage source of 1 V of arbitrary frequency. As a result, the voltage appearing across the output port is equal directly to the transfer function, ie. $V_o(s) = T(s)$. Thus, using the AC analysis command of PSpice, one can investigate the behavior of the transfer function $T(s)$ for physical frequencies defined by $s = j\omega$.

Resistors R_1 and R_2 are included in the circuit of Fig. 7.2 in order to satisfy the PSpice requirement of having two or more element connections made at each and every node. They do not affect the circuit's overall transfer function. A similar approach can be adopted for a VCCS. We leave this to the reader.

To illustrate the idea, let us consider having PSpice compute the magnitude and phase behavior of the following transfer function for physical frequencies:

$$T(s) = \frac{10s}{(1 + s/10^2)(1 + s/10^5)}. \tag{7.1}$$

This same transfer function is presented in Example 7.1 of Sedra and Smith where they graphically display its Bode plot. A PSpice description of the circuit shown in Fig. 7.2 with $T(s)$ given in Eqn. (7.1) is listed in Fig. 7.3. An AC analysis command is provided requesting that PSpice evaluate the transfer function using a logarithmic frequency sweep of 10 points per decade beginning at 0.01 Hz and ending at 100 MHz. As the output request

7.1 Investigating Transfer Function Behavior Using PSpice

```
Example 7.1: Bode Plot

** Circuit Description **
* input signal
Vin 1 0 AC 1V
* VCVS with transfer function:
*                         10s
*          T(s) = ---------------------------
*                  ( 1+s/10^2 ) ( 1 + s/10^5 )
*
E1 2 0 Laplace {V(1)} = {(10*s) / ( (1+s/1e2) * (1+s/1e5) )}
R1 1 0 1Ohm
R2 2 0 1Ohm
** Analysis Requests **
.AC DEC 10 1e-2 1e8
** Output Requests **
.PLOT AC VdB(2) Vp(2)
.probe
.end
```

Figure 7.3 The PSpice input file for computing the magnitude and phase of the transfer function given in Eqn. (7.1).

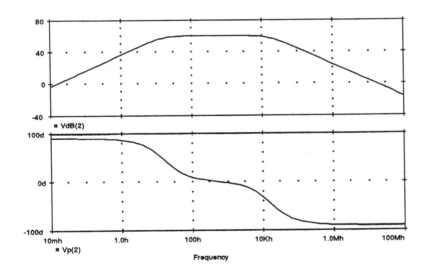

Figure 7.4 Magnitude and phase response of the transfer function given in Eqn. (7.1) as calculated by PSpice.

command, we are asking that PSpice plot both the magnitude and phase of the output voltage V(2) which is equal to the transfer function specified as the gain of the VCVS.

The results of this analysis are shown graphically in Fig. 7.4. These results can be

7   Frequency Response

compared to the Bode plot provided by Sedra and Smith in Example 7.1 of their text.

7.2   Modeling Dynamic Effects In Semiconductor Devices

Spice performs frequency domain analysis on diode and transistor circuits in much the same way that one performs this analysis by hand. Each diode or transistor in the circuit is replaced by it's small-signal frequency-dependent equivalent circuit, and all DC voltage sources are shorted to ground and all DC current sources are open circuited. Standard frequency-domain circuit analysis techniques are then applied to the equivalent small-signal circuit to determine the signal levels at various nodes or branches of the circuit.

The small-signal frequency domain model of either a diode or a transistor is similar to the small-signal model of the semiconductor device under static conditions with the exception that capacitors are included to account for the frequency effects. In Fig. 7.5 we illustrate the small-signal dynamic models of the semiconductor junction diode, the BJT, and the MOSFET as generally used by Spice. The BJT small-signal model applies equally well to both *npn* and *pnp* devices. Similarly, the FET model applies to both n- and p-channel devices. Except for the additional capacitors included in these models, they are identical to those introduced in previous chapters. Table 7.1 provides a cross-reference of the names of each capacitor in each of these models with those used by Spice.

To gain insight into circuit frequency-domain behavior, it is sometimes useful to know the value of the parameters of the small-signal model used by Spice in AC analysis. This information is sent to the Spice output file when an .OP (operating point) command is included in the Spice input file.

Consider, for example, that we are biasing a 2N2222A transistor with a 10 μA base current and a collector-emitter voltage of 5 V, as illustrated in Fig. 7.6. The Spice input file for this circuit arrangement with only an .OP analysis request would appear as follows:

```
Small-Signal Model For Q2N2222A (Ib=10uA, Vce=5V)

** Circuit Description **
* bias conditions
Vce 1 0 DC +5V
Ib 0 2 10uA
* transistor under test
Q1 1 2 0 Q2N2222A
* transistor model statement for the 2N2222A
.model Q2N2222A NPN (Is=14.34f Xti=3 Eg=1.11 Vaf=74.03 Bf=255.9 Ne=1.307
+                   Ise=14.34f Ikf=.2847 Xtb=1.5 Br=6.092 Nc=2 Isc=0 Ikr=0 Rc=1
+                   Cjc=7.306p Mjc=.3416 Vjc=.75 Fc=.5 Cje=22.01p Mje=.377 Vje=.75
+                   Tr=46.91n Tf=411.1p Itf=.6 Vtf=1.7 Xtf=3 Rb=10)
** Analysis Requests **
```

7.2 Modeling Dynamic Effects In Semiconductor Devices

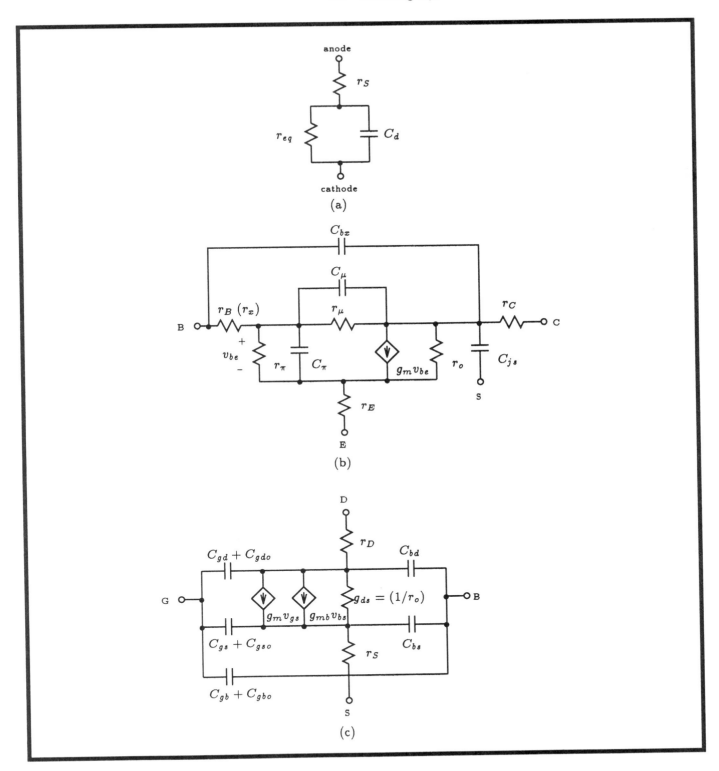

Figure 7.5 Dynamic Spice models of the small-signal behavior of: (a) junction diode (b) BJT and (c) MOSFET.

7 Frequency Response

Device	Symbol	Spice Name	Capacitance Description
Diode	C_d	CAP	Junction capacitance
BJT	C_π	CBE	Base-emitter junction capacitance
	C_μ	CBC	Base-collector junction capacitance
	C_{bx}	CBX	Extrinsic base–instrinsic collector capacitance
	C_{js}	CJS	Substrate junction capacitance
MOSFET	C_{bd}	CBD	Bulk-drain junction capacitance
	C_{bs}	CBS	Bulk-source junction capacitance
	C_{gso}	CGSOV	Gate-source overlap capacitance
	C_{gdo}	CGDOV	Gate-drain overlap capacitance
	C_{gbo}	CGBOV	Gate-bulk overlap capacitance
	C_{gs}	CGS	Gate-source instrinsic capacitance
	C_{gd}	CGD	Gate-drain instrinsic capacitance
	C_{gb}	CGB	Gate-bulk instrinsic capacitance

Table 7.1 Capacitances associated with the small-signal models of the diode, BJT and MOSFET shown in Fig. 7.5.

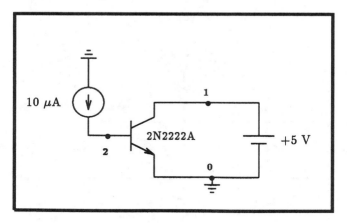

Figure 7.6 One particular bias arrangement for the commercial transistor 2N2222A.

```
.OP
** Output Requests **
* none required
.end
```

On completion of Spice, one finds in the Spice output file the following list of parameters pertaining to the small-signal BJT model shown in Fig. 7.5(b):

7.2 Modeling Dynamic Effects In Semiconductor Devices

```
****     OPERATING POINT INFORMATION        TEMPERATURE =   27.000 DEG C

***************************************************************************

**** BIPOLAR JUNCTION TRANSISTORS

    NAME        Q1
    MODEL       Q2N2222A
    IB          1.00E-05
    IC          1.63E-03
    VBE         6.57E-01
    VBC         -4.34E+00
    VCE         5.00E+00
    BETADC      1.63E+02
    GM          6.26E-02
    RPI         2.85E+03
    RX          1.00E+01
    RO          4.82E+04
    CBE         6.24E-11
    CBC         3.80E-12
    CBX         0.00E+00
    CJS         0.00E+00
    BETAAC      1.78E+02
    FT          1.50E+08
```

Here we see a list describing the transistor DC operating point and small-signal model parameters. Most parameters of the small-signal model should be self-evident from the above list. Capacitors C_{BE} and C_{BC} are equivalent to capacitances C_π and C_μ, respectively, of the hybrid-pi model described in Sedra and Smith. Several capacitances of this particular transistor, specifically C_{BX} and C_{JS}, are zero because they represent capacitive coupling to the substrate of the device. Since the 2N2222A is a discrete transistor, constructed from a sandwich of three separate layers of semiconductor material, it has no substrate layer. As a result, Spice simply assumes that these substrate capacitances do not exist and are set equal to zero.

In addition to printing the hybrid-pi model parameters, Spice also computes and prints the values β_{dc}, β_{ac}, and f_T, which are related to the transistor's operating point according to the following equations:

$$\beta_{dc} = \frac{I_C}{I_B}\bigg|_{OP} \tag{7.2}$$

$$\beta_{ac} = \frac{i_c}{i_b}\bigg|_{OP} = g_m r_\pi \tag{7.3}$$

and

$$f_T = \frac{1}{2\pi}\frac{g_m}{C_{BE} + C_{BC}}. \tag{7.4}$$

For this particular case, this transistor has a DC current gain of 163 and an AC current gain of 178. It is important to note that in general these values are different, but more

7 Frequency Response

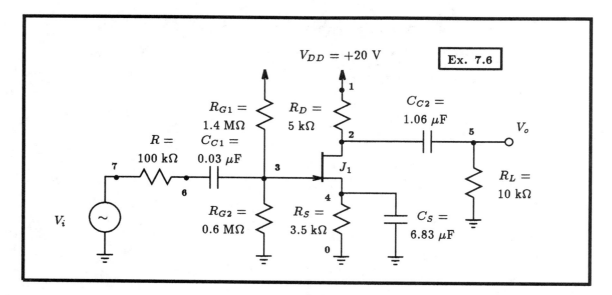

Figure 7.7 A capacitive coupled common-source JFET amplifier.

importantly, that they can be quite different than the value of β_F specified on the model statement used to describe the device in the Spice input deck (ie. 255.9). This arises because the dependence of the transistor current gain on collector current is captured by the Spice model for the 2N2222A. The β_F value specified on a BJT model statement represents the peak value of β_{dc}. The unity-gain bandwidth f_T for this transistor is estimated to be about 150 MHz.

Similar ideas extend to JFETs, MOSFETs, MESFETs and junction diodes.

7.3 The Low-Frequency Response Of The Common-Source Amplifier

In this section we shall investigate the low-frequency response of the classical common-source JFET amplifier shown in Fig. 7.7. The component values for this particular circuit were selected in Example 7.6 of Sedra and Smith such that the low-frequency response of this amplifier is dominated by a pole at 100 Hz and that the nearest pole or zero will be at least a decade away at a lower frequency. Using Spice we shall verify whether this is indeed the case.

The Spice input file describing this circuit is given Fig. 7.8. The device parameters of the n-channel JFET are assumed to be as follows: $V_{t0} = -2$ V, $\beta = 2$ mA/V^2 (This is the parameter K in Sedra and Smith; equal to I_{DSS}/V_P^2) and $\lambda = 0$ V^{-1}. This corresponds to the device parameters used by Sedra and Smith in Example 7.6, that is: $V_P = -2$ V, $I_{DSS} = 8$ mA and $r_o = \infty$. The amplifier input is excited by a 1 V AC input signal

7.3 The Low-Frequency Response Of The Common-Source Amplifier

---

Example 7.6: Frequency Response Of A JFET Amplifier

```
** Circuit Description **
* power supply
Vdd 1 0 DC +20V
* input signal and source resistance
Vi 7 0 AC 1V
R 7 6 100k
* JFET amplifier circuit
J1 2 3 4 n_jfet
Rg1 1 3 1.4Meg
Rg2 3 0 0.6Meg
Rd 1 2 5k
Rs 4 0 3.5k
* by-pass and coupling capacitors
Cs 4 0 6.83uF
Cc1 6 3 0.03uF
Cc2 2 5 1.06uF
* load
Rl 5 0 10k
* n-channel jfet model statement
.model n_jfet NJF (beta=2m Vto=-2V lambda=0)
** Analysis Requests **
.OP
.AC DEC 10 1Hz 1kHz
** Output Requests **
.PLOT AC VdB(5)
.probe
.end
```

Figure 7.8 The Spice input file for computing the low-frequency response behavior of the capacitively coupled common-source JFET amplifier shown in Fig. 7.7.

and an AC frequency sweep command is included in order to compute the small-signal frequency response of the amplifier. A logarithmic sweep of 10 points per decade varies the input frequency of excitation beginning at 1 Hz and ending at 1 kHz. A corresponding plot command as the output request command will then produce a plot of the magnitude of the amplifier output voltage as a function of frequency. Also included in this input file is an .OP command. The results of this analysis are used to verify that the amplifier is operating correctly in its linear region (ie. J_1 is in pinch-off). This is left for the reader to confirm.

The frequency response behavior of the JFET amplifier, as calculated by Spice, is shown in Fig. 7.9. The midband voltage gain is found to be +20.64 dB, and the 3 dB frequency is located very near to 100 Hz. Interestingly enough, these values correspond almost exactly to the desired design values suggested by Sedra and Smith in Example 7.6 of their text.

The magnitude response of the JFET amplifier shown in Fig. 7.9 does not have a simple

7 Frequency Response

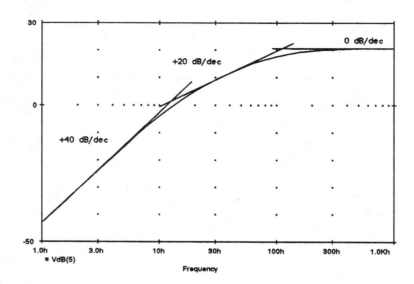

Figure 7.9 The low-frequency magnitude response of the capacitively coupled common-source JFET amplifier shown in Fig. 7.7. Three straight lines having slopes of +40 dB/dec, +20 dB/dec, and 0 dB/dec, are superimposed on the magnitude response of the amplifier to illustrate the approximate locations of the different break frequencies.

one-pole response – instead, the magnitude response increases at a rate of +40 dB/dec for low frequencies, much like a two-pole response. We see that the magnitude response increases at this rate until about 10 Hz when the rate of rise in magnitude decreases to +20 dB/dec. This continues until 100 Hz when the magnitude levels off at a constant +20.64 dB for increasing frequencies. This is clarified by the addition of the three straight-lines seen superimposed on the graph containing the magnitude response of the amplifier in Fig. 7.9.

It appears as though the dominant pole is located at about 100 Hz and the nondominant pole is located at 10 Hz. The effect of the zero (putting aside the two zeros at DC) is nullified by the presence of another pole located quite close to it. We thus confirm that the dominant pole and the next closest pole or zero are approximately one decade apart.

7.4 The High Frequency Analysis Of A Common-Source Amplifier

Miller's theorem is commonly used to simplify the task of analyzing electronic circuits. It is especially useful when estimating the upper 3 dB bandwidth of a common-source or common-emitter amplifier. Consider the small-signal equivalent circuit of a common-source amplifier, shown in Fig. 7.10(a). This circuit would be representative of the small-signal

7.4 The High Frequency Analysis Of A Common-Source Amplifier

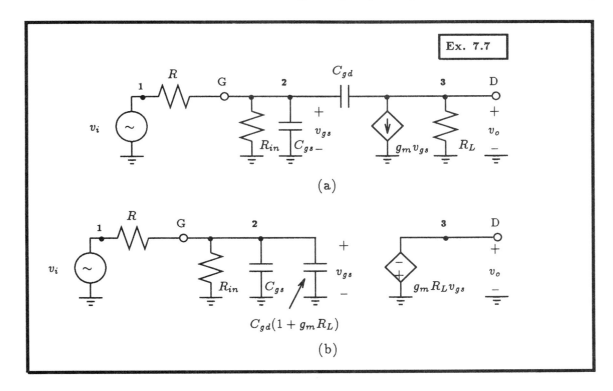

Figure 7.10 (a) The small-signal high-frequency equivalent circuit of a common-source FET amplifier (b) Simplified circuit using the Miller approximation to eliminate the bridging capacitor C_{gd}.

equivalent circuit derived for the common-source amplifier shown previously in Fig. 7.7. Calculating the upper 3 dB bandwidth of this amplifier is complicated by the presence of bridging capacitor C_{gd}. Through the application of Miller's theorem, and assuming that the voltage gain from the gate to drain of J_1 is known (ie. $v_o \approx -g_m R_L v_{gs}$), we can approximate the low-frequency behavior of the circuit shown in Fig. 7.10(a) with that shown in part (b). Clearly, the transfer function v_o/v_i of the circuit shown in Fig. 7.10(b) is dominated by a single pole located at frequency ω_{P_1} given by,

$$\omega_{P_1} = \frac{1}{(R||R_{in})[C_{gs} + C_{gd}(1+g_m R_L)]}. \tag{7.5}$$

With the aid of Spice, and some numerical values, we would like to investigate the validity of this Miller approximation by comparing the frequency response behavior of each circuit shown in Fig. 7.10, and determine whether the estimate of 3 dB bandwidth is in agreement with that computed directly from the original circuit. Assuming the same component values as used in the previous example of the last section (with some simplifications), ie. $R = 100$ kΩ, $R_{in} = 420$ kΩ, $R_L = 3.33$ kΩ, $C_{gs} = 1$ pF, $C_{gd} = 1$ pF and $g_m = 4$ mA/V, the input file used to describe these two circuits to Spice is listed in Fig. 7.11. Here two Spice circuit

7 Frequency Response

Example 7.7: Frequency Response Of Small-Signal Model Of A FET Amplifier

```
** Circuit Description **
* input signal and source resistance
Vi 1 0 AC 1V
R 1 2 100k
* small-signal equivalent circuit
Rin 2 0 420k
Cgs 2 0 1pF
Cgd 2 3 1pF
Gds 3 0 2 0 4m
* load
Rl 3 0 3.33k
** Analysis Requests **
.AC DEC 10 1e3 1e11
** Output Requests **
.PLOT AC VdB(3) Vp(3)
.probe
.end
Application of the Miller Theorem

** Circuit Description **
* input signal and source resistance
Vi 1 0 AC 1V
R 1 2 100k
* small-signal equivalent circuit
Rin 2 0 420k
Cgs 2 0 1pF
Cgd_in 2 0 14.3pF
Eo 0 3 2 0 13.2  ; Vo=-gm*RL*Vgs
* load
Rload 3 0 1
** Analysis Requests **
.AC DEC 10 1e3 1e11
** Output Requests **
.PLOT AC VdB(3) Vp(3)
.probe
.end
```

Figure 7.11 The Spice input file consisting of two concatenated Spice decks used for comparing the high-frequency response behavior of the circuit shown in Fig. 7.10(a) with the circuit shown in part (b) of the same figure. The latter circuit is a simplified version obtained using Miller's theorem.

description are concatenated together in order to compare their respective outputs using the Probe facility of PSpice. A frequency response calculation beginning at 1 kHz and ending at 100 GHz is to be performed using a logarithmic sweep of 10 points per decade. Such a wide frequency sweep is necessary to capture all of the interesting frequency domain behavior of the two circuits shown in Fig. 7.10.

7.5 High-Frequency Response Comparison Of The Common-Emitter and Cascode Amplifiers

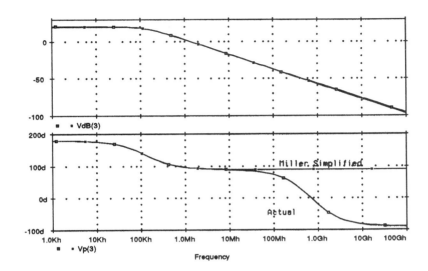

Figure 7.12 Comparing the high-frequency response behavior of the circuit shown in Fig. 7.10(a) with the circuit shown in part (b) of the same figure simplified using Miller's theorem.

On completion of Spice, we display the magnitude and phase of the output voltage V_o as a function of frequency for the two circuits shown in Fig. 7.12. This, of course, corresponds directly to the transfer functions of the two circuits because a 1 V AC input is applied to each circuit. As is evident from the magnitude response of the two circuits, both have very similar behavior for frequencies below 100 MHz. Above this frequency, the phase characteristics of the actual circuit undergoes a change of 180 degrees, but it's magnitude behavior remains relatively unchanged. Detailed analysis reveals the presence of a right-half-plane zero located near a second nondominant pole around 500 MHz, thus nullifying the effect of the second (nondominant) pole on the magnitude characteristics while adding to the overall phase. The 3 dB frequency for both networks is located at exactly the same location, at approximately 130 kHz. Interestingly enough, this frequency agrees quite closely with the value obtained through Eqn. (7.5) at 128.6 kHz. Thus, the Miller approximation appears to be quite accurate in estimating the upper 3 dB frequency of this amplifier.

7.5 High-Frequency Response Comparison Of The Common-Emitter and Cascode Amplifiers

Figure 7.13 illustrates two different styles of high-frequency amplifier configurations; these being, the common-emitter amplifier and the cascode amplifier. In the following we would like to use Spice to analyze the high-frequency operation of these two amplifier config-

7 Frequency Response

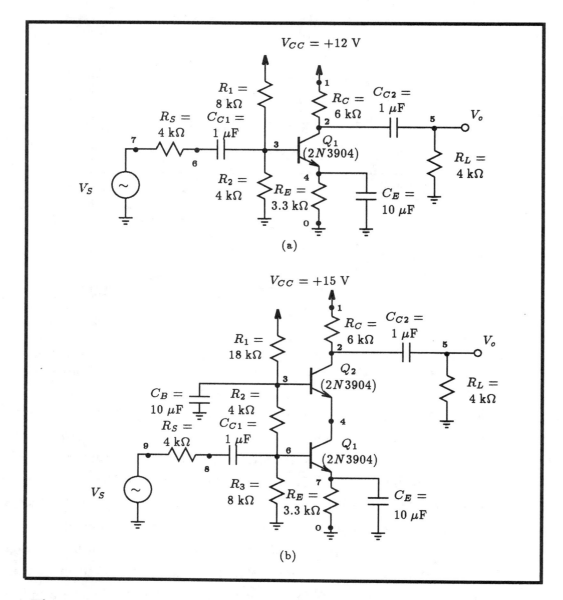

Figure 7.13 Two high-frequency amplifier configurations: (a) common-emitter (b) cascode.

urations, confirm the formulas used to estimate amplifier midband gain and 3 dB bandwidth, and in the process, determine which amplifier configuration is more suited for high-frequency operation. For a realistic comparison, we shall use identical transistors in both amplifiers and model them after the commercially available 2N3904 *npn* transistor. Furthermore, we note that fair comparisons can be made between the different amplifiers because the load and source resistances are identical in each case.

The particulars of the two amplifier designs shown in Fig. 7.13, specifically the midband gain and 3 dB frequencies, were presented by Sedra and Smith in Sections 7.6 and 7.7 of

7.5 High-Frequency Response Comparison Of The Common-Emitter and Cascode Amplifiers

Common Emitter Amplifier		
Midband Gain	3 dB Frequencies	
	lower 3 dB Frequency	upper 3 dB Frequency
$A_M = -\dfrac{R_{in}}{R_{in} + R_S} g_m R'_L$ where $R_{in} = R_1 \| R_2 \| (r_\pi + r_x)$ $R'_L = R_C \| R_L \| r_o$	$\omega_L \approx \dfrac{1}{C_{C_1} R_{C_1}} + \dfrac{1}{C_E R'_E} + \dfrac{1}{C_{C_2} R_{C_2}}$ where $R_{C_1} = R_S + [R_1 \| R_2 \| (r_\pi + r_x)]$ $R'_E = R_E \| \left(\dfrac{(R_1 \| R_2 \| R_S) + r_\pi + r_x}{\beta_{AC} + 1} \right)$ $R_{C_2} = R_L + (R_C \| r_o)$	$\omega_H \approx \dfrac{1}{(R_S \| R_{in})[C_\pi + C_\mu(1 + g_m R'_L)]}$ where $R_{in} = R_1 \| R_2 \| (r_\pi + r_x)$ $R'_L = R_C \| R_L \| r_o$

Cascode Amplifier		
Midband Gain	3 dB Frequencies	
	lower 3 dB Frequency	upper 3 dB Frequency
$A_M = -\dfrac{R}{R + R_S} g_m R'_L$ $\times \dfrac{r_\pi}{r_\pi + r_x + R_2 \| R_3 \| R_S}$ where $R = R_2 \| R_3$ $R'_L = R_C \| R_L$	$\omega_L \approx \dfrac{1}{C_{C_1} R_{C_1}} + \dfrac{1}{C_E R'_E} + \dfrac{1}{C_{C_2} R_{C_2}}$ where $R_{C_1} = R_S + [R_2 \| R_3 \| (r_\pi + r_x)]$ $R'_E = R_E \| \left(\dfrac{(R_2 \| R_3 \| R_S) + r_\pi + r_x}{\beta_{AC} + 1} \right)$ $R_{C_2} = R_L + R_C$	$\omega_H \approx \dfrac{1}{R'_S (C_{\pi_1} + 2 C_{\mu_1})}$ where $R'_S = r_{\pi_1} \| [r_{x_1} + R_2 \| R_3 \| R_S]$

Table 7.2 General expressions for estimating the midband gain (A_M) and 3 dB frequencies (ω_L and ω_H) of the common-emitter and cascode amplifiers as derived by Sedra and Smith in Sections 7.6 and 7.7 of their text.

their text. In Table 7.2 we summarize the hand-analysis formulas for later use where we shall verify their accuracy by comparing the results predicted by these expressions with those results computed by Spice.

Figures 7.14 and 7.15 show the Spice listings for the two amplifier configurations presented in Fig. 7.13, with the appropriate model parameters for the 2N3904 *npn* transistor. Both an operating point (.OP) command and an AC analysis (.AC) command are included in these two Spice decks. The operating point command is included in order to obtain the small-signal model parameters for the 2N3904 transistor, and the AC command is necessary to obtain the frequency response behavior of the amplifier. In order to compare the frequency response results for the two amplifiers, we shall concatenate the above two Spice files into one file before submitting them to Spice. In this way the Probe facility of PSpice can be

7 Frequency Response

```
A Common-Emitter Amplifier

** Circuit Description **
* power supplies
Vcc 1 0 DC +12V
* input signal source
Vs 7 0 AC 1V
Rs 7 6 4k
* CE stage
Cc1 6 3 1uF
R1 1 3 8k
R2 3 0 4k
Q1 2 3 4 Q2N3904
Re 4 0 3.3k
Rc 1 2 6k
Ce 4 0 10uF
Cc2 2 5 1uF
* output load
Rl 5 0 4k
*
* transistor model statement for 2N3904
.model Q2N3904   NPN (Is=6.734f Xti=3 Eg=1.11 Vaf=74.03 Bf=416.4 Ne=1.259
+                    Ise=6.734f Ikf=66.78m Xtb=1.5 Br=.7371 Nc=2 Isc=0 Ikr=0 Rc=1
+                    Cjc=3.638p Mjc=.3085 Vjc=.75 Fc=.5 Cje=4.493p Mje=.2593 Vje=.75
+                    Tr=239.5n Tf=301.2p Itf=.4 Vtf=4 Xtf=2 Rb=10)
** Analysis Requests **
.OP
.AC DEC 10 1Hz 100MegHz
** Output Requests **
.PLOT AC VdB(5)
.probe
.end
```

Figure 7.14 The Spice input file for computing the frequency response of the common-emitter amplifier circuit shown in Fig. 7.13(a).

used to graphically display the frequency behavior of the two amplifiers on the same graph.

On completion of Spice, the results of the AC analysis are displayed in Fig. 7.16. Here we see the magnitude response of the two amplifiers. As is evident, both amplifiers have very similar low frequency response, but quite different high frequency behavior. Each amplifier has a lower 3 dB frequency of about 436 Hz and a midband gain of approximately 28 dB. In the case of the common-emitter amplifier, the upper 3 dB frequency is limited to 575 kHz. This is in contrast to the cascode amplifier which has an upper 3 dB frequency of more than 5.8 MHz. Clearly then, the cascode amplifier has a 3 dB bandwidth more than 10 times that of the common-emitter amplifier and would therefore be the preferred choice as a wideband amplifier.

7.5 High-Frequency Response Comparison Of The Common-Emitter and Cascode Amplifiers

```
A Cascode Amplifier

** Circuit Description **
* power supplies
Vcc 1 0 DC +15V
* input signal source
Vs 9 0 AC 1V
Rs 9 8 4k
* CE stage (input stage)
Cc1 6 8 1uF
R1 1 3 18k
R2 3 6 4k
R3 6 0 8k
Q1 4 6 7 Q2N3904
Re 7 0 3.3k
Ce 7 0 10uF
* CB stage (upper stage)
Q2 2 3 4 Q2N3904
Rc 1 2 6k
Cb 3 0 10uF
Cc2 2 5 1uF
* output load
Rl 5 0 4k
*
* transistor model statement for 2N3904
.model Q2N3904  NPN (Is=6.734f Xti=3 Eg=1.11 Vaf=74.03 Bf=416.4 Ne=1.259
+               Ise=6.734f Ikf=66.78m Xtb=1.5 Br=.7371 Nc=2 Isc=0 Ikr=0 Rc=1
+               Cjc=3.638p Mjc=.3085 Vjc=.75 Fc=.5 Cje=4.493p Mje=.2593 Vje=.75
+               Tr=239.5n Tf=301.2p Itf=.4 Vtf=4 Xtf=2 Rb=10)
** Analysis Requests **
.OP
.AC DEC 10 1Hz 100MegHz
** Output Requests **
.PLOT AC VdB(5)
.probe
.end
```

Figure 7.15 The Spice input file for computing the frequency response of the cascode amplifier circuit shown in Fig. 7.13(b).

To illustrate the accuracy in our hand analysis, we present in Table 7.3 a comparison of the midband gain and 3 dB frequencies of each amplifier as calculated using the equations presented in Table 7.2 and as computed by exact analysis using Spice. In the hand calculations, we make use of the small-signal model parameters generated by Spice for each transistor of each amplifier. The actual operating point information found in the Spice output file is as follows:

7 Frequency Response

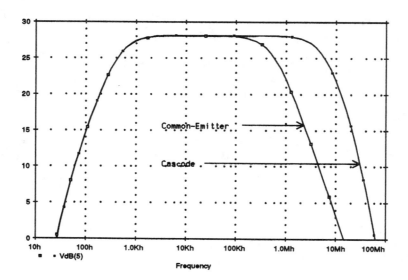

Figure 7.16 A comparison of the frequency response of the two amplifier configurations presented in Fig. 7.13.

Amplifier Type	Amplifier Parameter	Hand Analysis	Spice	% Error
Common Emitter	A_M	−25.5 V/V	−25.4 V/V	−0.2 %
	f_L	487.4 Hz	436 Hz	−11.8 %
	f_H	587.6 kHz	575 kHz	−2.2 %
Cascode	A_M	−25.8 V/V	−25.6 V/V	−0.9 %
	f_L	478.8 Hz	436 Hz	−9.8 %
	f_H	5.97 MHz	5.84 MHz	−2.2 %

Table 7.3 The midband gain and 3 dB frequencies of the common-emitter and cascode amplifiers as computed by hand and as obtained by exact analysis using Spice. The rightmost column presents the relative error (in percent) between the value predicted by hand and Spice.

```
Common-Emitter Amplifier        Cascode Amplifier

NAME      Q1                    NAME      Q1         Q2
MODEL     Q2N3904               MODEL     Q2N3904    Q2N3904
IB        7.34E-06              IB        7.29E-06   7.08E-06
IC        9.97E-04              IC        9.80E-04   9.73E-04
VBE       6.65E-01              VBE       6.65E-01   6.64E-01
VBC      -2.04E+00              VBC      -1.33E+00  -3.25E+00
VCE       2.70E+00              VCE       1.99E+00   3.91E+00
BETADC    1.36E+02              BETADC    1.35E+02   1.37E+02
GM        3.80E-02              GM        3.74E-02   3.71E-02
```

7.6 High-Frequency Response Of The CC-CE Amplifier

```
RPI      4.10E+03        RPI      4.13E+03    4.25E+03
RX       1.00E+01        RX       1.00E+01    1.00E+01
R0       7.63E+04        R0       7.69E+04    7.94E+04
CBE      1.79E-11        CBE      1.77E-11    1.76E-11
CBC      2.43E-12        CBC      2.66E-12    2.17E-12
CBX      0.00E+00        CBX      0.00E+00    0.00E+00
CJS      0.00E+00        CJS      0.00E+00    0.00E+00
BETAAC   1.56E+02        BETAAC   1.54E+02    1.58E+02
FT       2.97E+08        FT       2.92E+08    2.98E+08
```

To illustrate the accuracy of the expressions found in Table 7.2, we created a third column in Table 7.3 which expresses the relative error in the hand calculation as compared to the results computed by Spice. As is clearly evident, hand analysis provides very reasonable estimates of the amplifiers midband gain and 3 dB frequencies, with at most, a 11.8 per-cent error.

Although the above hand analysis resulted in very reasonable numbers when compare with those obtained directly by Spice, it should be pointed out here that these results should not suggest to the reader that Spice is not really necessary to compute the frequency response behavior of circuits. In fact, Spice is useful in several ways: (1) Spice can provide more accurate results than that which is possible by hand analysis; (2) Spice provides a very easy way of investigating component trade-offs in a design by allowing the designer to simply change component values and observe the effect on the frequency response; (3) Unlike hand analysis, Spice is not limited by the complexity of the circuit. However, one has to be sensible in the application of Spice, otherwise one might end up swamped in a morass of results that are difficult to interpret or derive any design insight from.

7.6 High-Frequency Response Of The CC-CE Amplifier

Another important high frequency amplifier configuration is the common-collector common-emitter (CC-CE) cascade. An example of this type of amplifier is shown in Fig. 7.17. Here, a common-collector stage isolates the source resistance R_S from the Miller capacitance of the common-emitter stage, thus extending the high frequency operation. In the following we would like to use Spice to analyze the high-frequency operation of this amplifier configuration and investigate the validity of the formulae used to estimate the amplifier's midband gain and 3 dB frequencies in Sedra and Smith. For easy reference, we summarize these expressions in Table 7.4.

The Spice input file describing the amplifier shown in Fig. 7.17 is seen listed in Fig. 7.18. Each transistor is modeled after the commercially available 2N3904 transistor. Both a DC operating point and AC analysis command are included in this Spice deck.

The frequency response of the CC-CE amplifier as calculated by Spice is shown in Fig.

7 Frequency Response

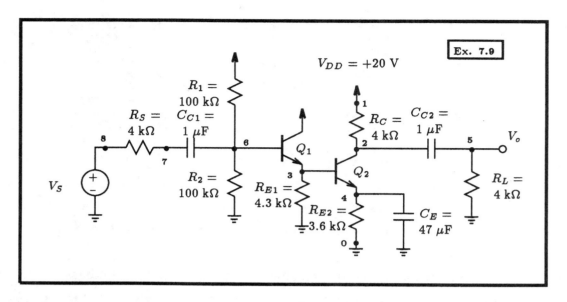

Figure 7.17 The common-collector common-emitter cascade amplifier configuration.

7.19. Using the Probe facility of PSpice, we can read off this graph and find that the midband gain is 35.8 dB or 61.9 V/V. In addition, we also find that the lower 3 dB frequency of this amplifier is 122 Hz and its upper 3 dB frequency is 9.2 MHz. Also evident in this magnitude plot is signs of high frequency resonance (ie. magnitude peaking). As we shall see when we compare our hand analysis results with those obtained using Spice, this resonant peaking extends somewhat the upper 3 dB frequency of this amplifier which is not accounted for by the simple theory presented in Sedra and Smith. The gain-bandwidth product of this amplifier is approximately 570 MHz, which is the largest of any of the amplifiers that we have seen so far (eg. The CE and Cascode stage presented in the last section had gain-bandwidth product of 14.6 MHz and 149 MHz, respectively.)

To investigate the accuracy of the expressions listed in Table 7.4 for the CC-CE amplifier stage, we compare the midband gain and 3 dB frequencies as calculated by hand with those obtained using Spice in Table 7.5. For the hand calculations, we made use of the small-signal model parameters for Q_1 and Q_2 as generated by Spice and repeated here below:

Common-Collector Common-Emitter Cascade Amplifier

NAME	Q1	Q2
MODEL	Q2N3904	Q2N3904
IB	6.66E-06	6.83E-06
IC	9.32E-04	9.22E-04
VBE	6.62E-01	6.63E-01
VBC	-5.33E+00	-2.31E+00
VCE	6.00E+00	2.97E+00
BETADC	1.40E+02	1.35E+02
GM	3.55E-02	3.52E-02
RPI	4.52E+03	4.41E+03
RX	1.00E+01	1.00E+01

7.7 Frequency Response Of The Differential Amplifier

Common-Collector Common-Emitter Cascade Amplifier		
Midband Gain	3 dB Frequencies	
	lower 3 dB Frequency	upper 3 dB Frequency
$A_M = -\dfrac{R_{in}}{R_{in}+R_S} g_{m_2} R'_L$ $\times \dfrac{R_{E1}\|r_{\pi_2}}{(R_{E1}\|r_{\pi_2}) + r_{e_1}}$ where $R_{in} = R_1\|R_2\|R'_{B1}$ $R'_{B1} = r_{\pi_1} + (\beta_1+1)(R_{E1}\|r_{\pi_2})$ $R'_L = R_C\|R_L\|r_o$	$\omega_L \approx \dfrac{1}{C_{C_1} R_{C_1}} + \dfrac{1}{C_E R'_E} + \dfrac{1}{C_{C_2} R_{C_2}}$ where $R_{C_1} = R_S + (R_1\|R_2\|R'_{B1})$ $R'_{B1} = r_{\pi_1} + (\beta_1+1)(R_{E1}\|r_{\pi_2})$ $R'_E = R_{E2}\|\left(\dfrac{R'_{E1}}{\beta_2+1}\right)$ $R'_{E1} = \left(\dfrac{(R_1\|R_2\|R_S)+r_{\pi_1}}{\beta_1+1}\right)\|R_{E1}$ $\quad\quad + r_{\pi_2}$ $R_{C_2} = R_L + (R_C\|r_{o_2})$	$\omega_H \approx \dfrac{1}{C_{\mu_1} R_{\mu_1} + C_{\pi_1} R_{\pi_1} + C_T R_T + C_{\mu_2} R'_L}$ where $R_{\mu_1} = R_S\|R_{in}$ $R_{in} = R_1\|R_2\|R'_{B1}$ $R'_{B1} = r_{\pi_1} + (\beta_1+1)(R_{E1}\|r_{\pi_2})$ $R'_{E1} = R_{E1}\|r_{\pi_2}$ $R_{\pi_1} = r_{\pi_1}\|\left(\dfrac{R'_S + R'_{E1}}{1 + g_{m_1} R'_{E1}}\right)$ $R'_S = R_S\|R_1\|R_2$ $R'_L = R_C\|R_L$ $R_T = R'_{E1}\|\left(\dfrac{R'_S + r_{\pi_1}}{\beta_1+1}\right)$ $C_T = C_{\pi_2} + C_{\mu_2}(1 + g_{m_2} R'_L)$

Table 7.4 General expressions for estimating the midband gain (A_m) and 3 dB frequencies (ω_L and ω_H) of the common-collector common-emitter cascode amplifier as derived by Sedra and Smith in Sections 7.3 and 7.9 of their text.

```
         RO      8.52E+04     8.28E+04
         CBE     1.72E-11     1.70E-11
         CBC     1.91E-12     2.36E-12
         CBX     0.00E+00     0.00E+00
         CJS     0.00E+00     0.00E+00
         BETAAC  1.61E+02     1.55E+02
         FT      2.97E+08     2.88E+08
```

As can been seen from Table 7.5, the value of the midband gain calculated by hand is in very good agreement with that obtained from Spice (−0.3% error). Both the lower and upper 3 dB frequencies (f_L and f_H) computed by the hand analysis differ from the results computed by Spice with an absolute error of about 20 per-cent. In the case of the upper 3 dB frequency, we attribute most of this error to the high-frequency magnitude peaking that seems to extend the high-frequency operation of the CC-CE amplifier stage. The formula derived for the upper 3 dB frequency and used in the hand analysis above assumes that all circuit poles are real; magnitude peaking is an indication that some of the poles are complex.

7 Frequency Response

```
A Common-Collector Common-Emitter Cascade Stage

** Circuit Description **
* power supplies
Vcc 1 0 DC +10V
* input signal source
Vs 8 0 AC 1V
Rs 8 7 4k
* CE stage (input stage)
Cc1 6 7 1uF
R1 1 6 100k
R2 6 0 100k
Q1 1 6 3 Q2N3904
Re1 3 0 4.3k
* CC stage (output stage)
Q2 2 3 4 Q2N3904
Re2 4 0 3.6k
Rc 1 2 4k
Ce 4 0 47uF
Cc2 2 5 1uF
* output load
Rl 5 0 4k
*
* transistor model statement for 2N3904
.model Q2N3904  NPN (Is=6.734f Xti=3 Eg=1.11 Vaf=74.03 Bf=416.4 Ne=1.259
+               Ise=6.734f Ikf=66.78m Xtb=1.5 Br=.7371 Nc=2 Isc=0 Ikr=0 Rc=1
+               Cjc=3.638p Mjc=.3085 Vjc=.75 Fc=.5 Cje=4.493p Mje=.2593 Vje=.75
+               Tr=239.5n Tf=301.2p Itf=.4 Vtf=4 Xtf=2 Rb=10)
** Analysis Requests **
.OP
.AC DEC 10 1Hz 100MegHz
** Output Requests **
.PLOT AC VdB(5)
.probe
.end
```

Figure 7.18 The Spice input file for computing the frequency response of the common-collector common-emitter cascade amplifier circuit shown in Fig. 7.17.

7.7 Frequency Response Of The Differential Amplifier

The differential pair is the work-horse of modern day analog integrated circuit design. In the following we shall investigate the high-frequency operation of the differential pair in several different circuit arrangements. We shall assume that the transistors of each amplifier are modeled after the commercial 2N3904 *npn* transistor. This will give us some sense of realistic amplifier behavior.

In Fig. 7.20 we present three different configurations for the differential pair. Each is driven in a single-ended fashion. In part (a) of this figure, we display a typical amplifier

7.7 Frequency Response Of The Differential Amplifier

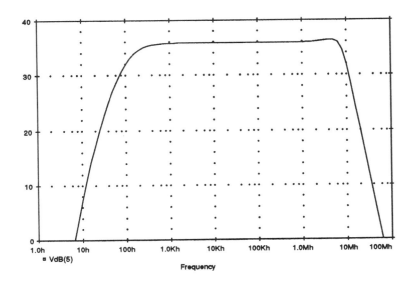

Figure 7.19 The magnitude response of the common-collector common-emitter cascade shown in Fig. 7.17 as computed by Spice.

Amplifier Type	Amplifier Parameter	Hand Analysis	Spice	% Error
Common-Collector Common-Emitter Cascade	A_M	−62.0 V/V	−61.9 V/V	−0.3 %
	f_L	143.1 Hz	122 Hz	−17.3 %
	f_H	7.17 MHz	9.2 MHz	+22.1 %

Table 7.5 The midband gain and 3 dB frequencies of the common-collector common-emitter cascade amplifier as computed by hand and exact analysis using Spice. The rightmost column presents the relative error (in percent) between the value predicted by hand and Spice.

incorporating a differential pair with the output taken differentially. In part (b), resistors of 100 Ω value are added in series with the emitters of each transistor. These resistors act to reduce the gain of the amplifier while increasing the 3 dB bandwidth. We shall illustrate more of this in the next section. The final amplifier shown in Fig. 7.20(c) is a modified version of the differential amplifier shown in part (a): The collector resistance of Q_1 is eliminated and the output is then taken at the collector of Q_2. These modifications act to increase the bandwidth of the amplifier. This configuration is sometimes referred to as the common-collector common-base (CC–CB) cascade. In the following we shall compare the frequency response of these amplifier as calculated by Spice, and investigate the accuracy of the expressions derived by Sedra and Smith in Section 7.10 of their text for these three

7 Frequency Response

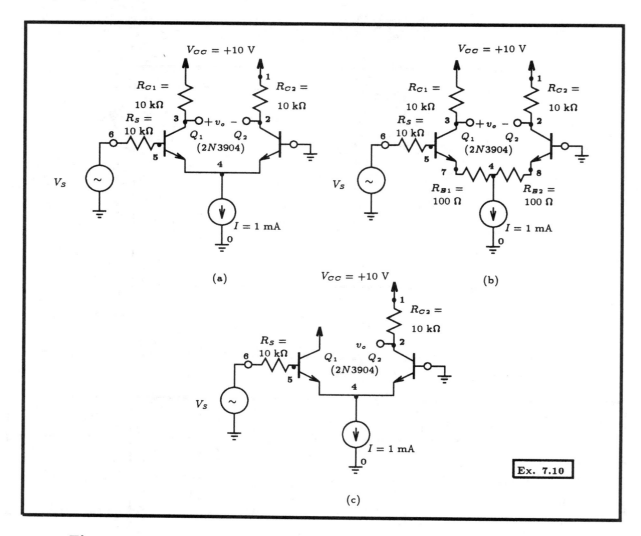

Figure 7.20 Various differential amplifier configurations with single-ended excitation (a) differential amplifier (b) differential amplifier including emitter resistance (c) differential amplifier in common-collector common-base configuration.

different amplifier configurations.

Spice descriptions of the three differential-pair amplifier configurations shown in Fig. 7.20 are given in Figures 7.21, 7.22 and 7.23. Both a DC operating point and an AC analysis are requested. The AC analysis command will be used to calculate the frequency response of each amplifier between 1 Hz and 100 MHz. Whereas, the operating point analysis will provide us with information about the small-signal model parameters of each transistor. Each amplifier is driven by a 1 V AC input signal so that the gain of each amplifier can be read directly from its output voltage.

The magnitude response of the three amplifier circuits shown in Fig. 7.20 as calculated by Spice are on display in Fig. 7.24. The differential amplifier in Fig. 7.20(a) has a low-

```
Example 7.10a: A Differential Amplifier

** Circuit Description **
* power supplies
Vcc 1 0 DC +10V
* input signal source
Vs 6 0 AC 1V
Rs 6 5 10k
* differential pair
Rc1 1 2 10k
Rc2 1 3 10k
Q1 3 5 4 Q2N3904
Q2 2 0 4 Q2N3904
* current biasing
Ibias 4 0 1mA
*
* transistor model statement for 2N3904
.model Q2N3904   NPN (Is=6.734f Xti=3 Eg=1.11 Vaf=74.03 Bf=416.4 Ne=1.259
+                Ise=6.734f Ikf=66.78m Xtb=1.5 Br=.7371 Nc=2 Isc=0 Ikr=0 Rc=1
+                Cjc=3.638p Mjc=.3085 Vjc=.75 Fc=.5 Cje=4.493p Mje=.2593 Vje=.75
+                Tr=239.5n Tf=301.2p Itf=.4 Vtf=4 Xtf=2 Rb=10)
** Analysis Requests **
.OP
.AC DEC 10 1Hz 100MegHz
** Output Requests **
.PLOT AC VdB(3,2) Vp(3,2)
.probe
.end
```

Figure 7.21 The Spice input file for computing the magnitude response of the differential amplifier shown in Fig. 7.20(a) with single-ended excitation.

frequency gain of 39.7 dB and a 3 dB bandwidth of 170 kHz. With the addition of emitter resistors to the differential pair, as shown in Fig. 7.20(b), the low-frequency voltage gain drops by 5.5 dB, or by a factor of 1.9, and the 3 dB bandwidth increases by a similar factor to 304 kHz. The magnitude response for the common-collector common-base cascade amplifier shown in Fig. 7.20(c) has a low-frequency gain of 34.2 dB and a 3 dB bandwidth of 2.56 MHz. Clearly then, the CC–CB amplifier stage has a higher gain-bandwidth product than the other two amplifier stages; although, it's output has to be taken single-ended.

Expressions that relate the low frequency gain and 3 dB frequency of each differential amplifier configuration shown in Fig. 7.20 are tabulated in Table 7.6. These expressions were derived by Sedra and Smith assuming that the two transistors of the differential pair have identical small-signal model parameters. As we shall see, this is not quite true when the differential pair is driven by single-ended excitation.

To see this, let us first consider the case of the differential amplifier with single-ended

7 Frequency Response

Example 7.10b: Differential Amplifier With Emitter Resistors

```
** Circuit Description **
* power supplies
Vcc 1 0 DC +10V
* input signal source
Vs 6 0 AC 1V
Rs 6 5 10k
* differential pair
Rc1 1 2 10k
Rc2 1 3 10k
Q1 3 5 7 Q2N3904
Q2 2 0 8 Q2N3904
* emitter resistors
Re1 7 4 100
Re2 8 4 100
* current biasing
Ibias 4 0 1mA
*
* transistor model statement for 2N3904
.model Q2N3904   NPN (Is=6.734f Xti=3 Eg=1.11 Vaf=74.03 Bf=416.4 Ne=1.259
+                Ise=6.734f Ikf=66.78m Xtb=1.5 Br=.7371 Nc=2 Isc=0 Ikr=0 Rc=1
+                Cjc=3.638p Mjc=.3085 Vjc=.75 Fc=.5 Cje=4.493p Mje=.2593 Vje=.75
+                Tr=239.5n Tf=301.2p Itf=.4 Vtf=4 Xtf=2 Rb=10)
** Analysis Requests **
.OP
.AC DEC 10 1Hz 100MegHz
** Output Requests **
.PLOT AC VdB(3,2) Vp(3,2)
.probe
.end
```

Figure 7.22 The Spice input file for computing the magnitude response of the differential amplifier with emitter resistors, shown in Fig. 7.20(b).

excitation shown in Fig. 7.20(a). The small-signal model parameters associated with transistors Q_1 and Q_2 as computed by Spice are as follows:

Differential Pair With Single-Ended Excitation:

NAME	Q1	Q2
MODEL	Q2N3904	Q2N3904
IB	2.41E-06	5.33E-06
IC	2.92E-04	7.00E-04
VBE	6.31E-01	6.55E-01
VBC	-7.10E+00	-3.00E+00
VCE	7.73E+00	3.66E+00
BETADC	1.21E+02	1.31E+02
GM	1.13E-02	2.68E-02
RPI	1.26E+04	5.66E+03
RX	1.00E+01	1.00E+01
RO	2.78E+05	1.10E+05
CBE	9.72E-12	1.45E-11
CBC	1.76E-12	2.21E-12

7.7 Frequency Response Of The Differential Amplifier

```
Example 7.10c: A Modified Differential Amplifier

** Circuit Description **
* power supplies
Vcc 1 0 DC +10V
* input signal source
Vs 6 0 AC 1V
Rs 6 5 10k
* differential pair
Rc1 1 2 10k
Q1 1 5 4 Q2N3904
Q2 2 0 4 Q2N3904
* current biasing
Ibias 4 0 1mA
*
* transistor model statement for 2N3904
.model Q2N3904  NPN (Is=6.734f Xti=3 Eg=1.11 Vaf=74.03 Bf=416.4 Ne=1.259
+               Ise=6.734f Ikf=66.78m Xtb=1.5 Br=.7371 Nc=2 Isc=0 Ikr=0 Rc=1
+               Cjc=3.638p Mjc=.3085 Vjc=.75 Fc=.5 Cje=4.493p Mje=.2593 Vje=.75
+               Tr=239.5n Tf=301.2p Itf=.4 Vtf=4 Xtf=2 Rb=10)
** Analysis Requests **
.OP
.AC DEC 10 1Hz 100MegHz
** Output Requests **
.PLOT AC VdB(2) Vp(2)
.probe
.end
```

Figure 7.23 The Spice input file for computing the magnitude response of the modified differential amplifier shown in Fig. 7.20(c).

CBX	0.00E+00	0.00E+00
CJS	0.00E+00	0.00E+00
BETAAC	1.42E+02	1.52E+02
FT	1.56E+08	2.55E+08

Clearly, the small-signal model parameters for Q_1 and Q_2 are quite different. This stems from the fact that the two transistors are biased at different current levels. This suggests that the expressions in Table 7.6 for this amplifier are not applicable to the situation we have here. To salvage the situation, let us consider the accuracy in assuming that the two transistors have equal small-signal model parameters and take on the small-signal parameter values of: (a) Q_1 (b) Q_2, and (c) the average of Q_1 and Q_2. On doing so, we collect the results shown in the upper portion of Table 7.7. To judge the accuracy of each case, we also compare the values to those found using Spice and obtained the relative error expressed in per-cent. This error analysis for the three separate cases are shown listed in the right-most column of Table 7.7. As is evident, the relative error for some of these case is quite high.

7 Frequency Response

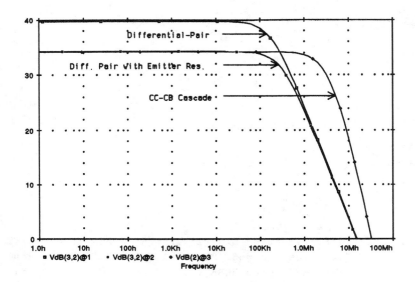

Figure 7.24 A comparison of the high-frequency magnitude response of the three differential amplifier configurations presented in Fig. 7.20.

We can repeat the above analysis on the remaining amplifier configurations shown in Fig. 7.20 and extend Table 7.7 with results for the differential amplifier with emitter degeneration and the common-collector common-base stage. For reference, the small-signal model parameters for the transistor of the other two amplifier configurations as computed by Spice are listed below:

Differential Pair With Emitter Degeneration			Common-Collector Common-Base Configuration:		
NAME	Q1	Q2	NAME	Q1	Q2
MODEL	Q2N3904	Q2N3904	MODEL	Q2N3904	Q2N3904
IB	3.16E-06	4.59E-06	IB	2.40E-06	5.27E-06
IC	3.96E-04	5.96E-04	IC	3.01E-04	6.92E-04
VBE	6.39E-01	6.51E-01	VBE	6.31E-01	6.55E-01
VBC	-6.07E+00	-4.04E+00	VBC	-1.00E+01	-3.08E+00
VCE	6.71E+00	4.69E+00	VCE	1.07E+01	3.74E+00
BETADC	1.25E+02	1.30E+02	BETADC	1.26E+02	1.31E+02
GM	1.52E-02	2.29E-02	GM	1.16E-02	2.65E-02
RPI	9.62E+03	6.58E+03	RPI	1.27E+04	5.72E+03
RX	1.00E+01	1.00E+01	RX	1.00E+01	1.00E+01
RO	2.02E+05	1.31E+05	RO	2.80E+05	1.11E+05
CBE	1.09E-11	1.33E-11	CBE	9.82E-12	1.44E-11
CBC	1.84E-12	2.05E-12	CBC	1.60E-12	2.20E-12
CBX	0.00E+00	0.00E+00	CBX	0.00E+00	0.00E+00
CJS	0.00E+00	0.00E+00	CJS	0.00E+00	0.00E+00
BETAAC	1.46E+02	1.51E+02	BETAAC	1.47E+02	1.51E+02
FT	1.89E+08	2.37E+08	FT	1.61E+08	2.54E+08

On review of the new results that have been added to Table 7.7, we see that the relative error experienced by our estimates of gain and 3 dB bandwidth for the differential amplifier

7.7 Frequency Response Of The Differential Amplifier

Differential Amplifier With Single Excitation		
Condition	Low-Frequency Gain	3 dB Frequency
Diff. Amp.	$A_M = -g_m R_C \dfrac{2r_\pi}{2r_\pi + R_S + 2r_x}$	$\omega_H \approx \dfrac{1}{\left[2r_\pi \| (R_S + 2r_x)\right] \| \left[C_\pi/2 + (g_m R_C)C_\mu/2\right]}$
Diff. Amp. With Emitter Degeneration	$A_M = -\dfrac{(\beta+1)(r_e + R_E)}{R_S/2 + r_x + (\beta+1)(r_e+R_E)}$ $\times \dfrac{(\alpha R_C)}{r_e + R_E}$	$\omega_H \approx \dfrac{1}{C_\pi R_\pi + C_\mu R_\mu}$ where $R_\pi = r_\pi \| \left(\dfrac{R'_S + R_E}{1 + g_m R_E}\right)$ $R'_S = R_S/2 + r_x$ $R_\mu = R_C + \dfrac{1 + R_E/r_e + g_m R_C}{1/r_\pi + (1/R'_S)(1 + R_E/r_e)}$
Common-Collector Common-Base Configuration	$A_M = -\dfrac{2r_\pi}{2r_\pi + R_S} \dfrac{\alpha R_C}{2r_e}$	$\omega_H \approx \sqrt{\tfrac{1}{2}\left[-\omega_{p1}^2 - \omega_{p2}^2 + D\right]}$ where $\omega_{p1} = \dfrac{1}{(R_S \| 2r_\pi)(C_\mu + C_\pi/2)}$ $\omega_{p2} = \dfrac{1}{C_\mu R_C}$ $D = \sqrt{\omega_{p1}^4 + 6\omega_{p1}^2 \omega_{p2}^2 + \omega_{p2}^4}$

Table 7.6 General expressions for estimating the low-frequency gain (A_M) and 3 dB frequency (ω_H) of the differential amplifier with single-ended excitation, differential amplifier with emitter degeneration, and the common-collector common-base configuration. These expressions were derived by Sedra and Smith in Sections 7.9 and 7.10 of their text.

with emitter degeneration are quite good in all three cases. The largest error has a magnitude of only 16 per-cent. This is in contrast to the differential amplifier without emitter degeneration, where as we have seen previously, almost all relative errors are large, having magnitudes that vary between 17 and 46 per-cent. In the CC-CB amplifier configuration our estimates have relative errors that can be as large as −37 per-cent.

In all of the above three configurations, the formulae used to estimate the gain and 3 dB bandwidth do not seem to be very accurate. This stems from the fact that the various differential amplifier configurations are driven asymmetrically and cause the two transistors in the amplifier to be biased at slightly different operating points. Of course, in each of the above cases, we could re-work the small-signal analysis assuming two different sets of

Amplifier Type	Amplifier Parameter	Hand Analysis			Spice	% Error		
		Q_1	Q_2	Average		Q_1	Q_2	Average
Diff. Amp	A_M (V/V)	−80.9	−142.2	−123.3	−97.1	+17 %	−46 %	−27 %
	f_H (kHz)	212.8	98.7	126.1	170	−25 %	+42 %	+26 %
Diff. Amp with Emitter Degeneration	A_M (V/V)	−49.8	−56.4	−52.9	−51.4	+3.2 %	−9.7 %	−2.9 %
	f_H (kHz)	315.9	255.4	272.7	304	−3.9 %	+16 %	+10 %
CC-CB	A_M (V/V)	−41.5	−70.4	−52.4	−51.1	+18 %	−37 %	−2.6 %
	f_H (MHz)	3.09	2.74	2.77	2.56	−21 %	−7.2 %	−8.1 %

Table 7.7 The low-frequency gain and 3 dB frequency of the differential pair, differential pair with emitter degeneration, and the common-collector common-base configuration, as computed by hand and by exact analysis using Spice. The rightmost column presents the relative error (in percent) between the value predicted by hand and that computed by Spice.

small-signal parameters for the two transistors; however, the resulting formulas will be more complicated and more difficult to apply. Therefore if greater precision is thought necessary than that which can be obtained from the small-signal formula given in Table 7.7, then it is more useful to go directly to Spice.

7.8 The Effect Of Emitter Degeneration On Differential Amplifier Characteristics

In this final section of this chapter, we would like to demonstrate the effect of emitter degeneration on differential amplifier frequency response characteristics. According to the gain and bandwidth expressions provided in Table 7.6, we can see that as R_E increases, the low frequency gain decreases but the 3 dB frequency increases. Using Spice, we would like to observe this effect for 4 different values of emitter resistances. Specifically, emitter resistances of 0, 100, 200, and 300 Ω. The Spice input file for each of these cases is identical to that seen previously in Fig. 7.22. In the case of $R_E = 0$, instead of modifying the topology of the Spice deck by removing the two emitter resistors R_{E1} and R_{E2}, we simply set these resistors to a very low value of 0.01 Ω. Concatenating these four Spice files together and submitting to Spice, we obtain the magnitude response shown in Fig. 7.25.

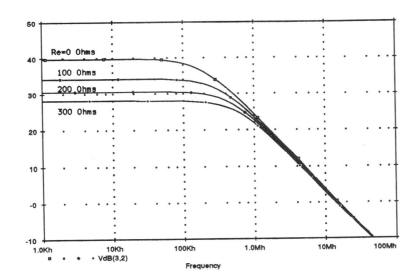

Figure 7.25 Observing the effect of emitter degeneration on the magnitude response of the differential amplifier shown in Fig. 7.20(b).

As can be seen from the magnitude plot in Fig. 7.25, as R_E increases, the low frequency gain of the amplifier decreases, but the corresponding 3 dB frequency increases. For example, for $R_E = 200$ Ω, the low frequency gain is 30.6 dB or 33.9 V/V and the 3 dB frequency is 432.9 kHz. For this particular case, the amplifier has a gain-bandwidth product of 14.7 MHz. When R_E is increased to 300 Ω, the low frequency gain decreases to 28.1 dB or 25.3 V/V, but the 3 dB bandwidth increases to 562.3 kHz. The gain-bandwidth product for this case becomes 14.2 MHz. Similar conclusions can be drawn for the other two cases of emitter resistance. We notice that in all cases, the magnitude response crosses the 0 dB axis at about 14 MHz which corresponds quite closely to the gain-bandwidth product of the amplifier.

7.9 Spice Tips

- Spice is extremely useful for calculating the frequency response of amplifier circuits. Unlike hand-analysis, Spice need not make any assumptions about the location of the poles or zeros, thus providing exact (or as accurate as the model provides) frequency response behavior. Also, Spice is not limited by the circuit complexity.

- PSpice allows the gain of a dependent source to be expressed as an arbitrary function of the complex frequency variable s.

- Only the gain of a VCVS and a VCCS in PSpice (designated with the letters E and G, respectively) can be represented as a Laplace transform function. However, both these controlled-sources can be utilized to describe the behavior of a CCVS or a CCCS using the special syntax of PSpice for describing a VCVS and VCCS – See Fig. 7.1.

- If a circuit is excited by a 1 V AC input signal, then the magnitude and phase of the voltage appearing across the output terminals computed by Spice are directly equal to the magnitude and phase of the transfer function of that circuit evaluated at the frequency of the input signal.

- An AC analysis of Spice is, by definition, a small-signal analysis. The level of the input AC signal will not alter this basic assumption. This is, of course, not true for a transient analysis where the level of the input signal affects the linearity of the circuit.

- Spice not only provides more accurate results than hand analysis, it also allows the user a very easy way of investigating component trade-offs in a design. Changing various component values in a known manner and observing the effects on the frequency response enables the designer to select the components that best suit the design requirements.

- All capacitances used by Spice in the small-signal equivalent circuit of a transistor are available for the user to see. They are placed in the Spice output file under the heading "OPERATING POINT INFORMATION" and are included among the small-signal parameters of each transistor.

7.10 Problems

7.1 If PSpice is available, use the Laplace transform function of either the voltage-controlled voltage or current source to plot the magnitude and phase response of the following transfer functions:

(i) $T(s) = \dfrac{10s}{(1+s/10^1)(1+s/10^4)}$

(ii) $T(s) = \dfrac{10^4(1+s/10^5)}{(1+s/10^3)(1+s/10^4)}$

(iii) $T(s) = \dfrac{5s(1+s/10^5)}{(1+s/10^3)}$

(iv) $T(s) = \dfrac{10^4}{(1+s/10^2)(1+s/10^4)(1+s/10^6)}$

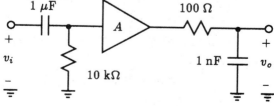

Fig. P7.2

7.2 If in the circuit of Fig. P7.2 is an ideal voltage amplifier of gain 100, use Spice to determine A_M, ω_L and ω_H of this amplifier.

7.3 The low-frequency response of an amplifier is characterized by three poles of frequencies 10 Hz, 3 Hz, and 1 Hz, and three zeros at $\omega = 0$. With the aid of the Laplace transform function capability of specifying the gain of a dependent source in PSpice, calculate the lower 3-dB frequency f_L and compare this result with the value obtained using: (a) the dominant pole approximation method, and (b) the root-sum-of-squares approximation method.

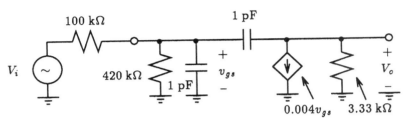

Fig. P7.4

7.4 Fig. P7.4 shows the high-frequency equivalent circuit of a common-source FET amplifier driven by a signal generator having a resistance of 100 kΩ. Using Spice, compute the upper 3-dB frequency f_H using the method of open-circuit time constants. Compare this result with that obtain directly from the input-output voltage transfer characteristics.

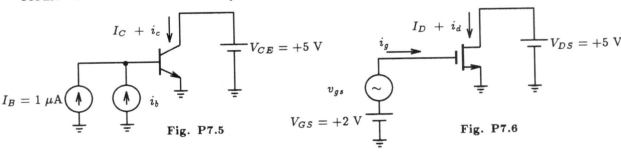

Fig. P7.5   Fig. P7.6

7.5 Using the circuit setup shown in Fig. P7.5, calculate the small-signal current gain i_c/i_b of the commercial 2N3904 npn transistor using Spice. The Spice model parameters for the 2N3904 can be found in the Spice listing of Fig. 7.14. What is the corresponding f_T of this transistor?

7.6 Using the circuit setup shown in Fig. P7.6, have Spice calculate the small-signal current gain i_d/i_g as a function of frequency for an n-channel MOSFET characterized by $V_t = 1$ V, $\mu_n C_{OX} = 20$ μA/V^2, $L = 10$ μm, $W = 30$ μm, $\lambda = 0.04$ V^{-1}, $C_{GSO} = 1$ pF, and $C_{GDO} = 1$ pF. What is the corresponding unity current gain frequency of this transistor?

7.7 A small-signal equivalent circuit of a FET common-source amplifier has $R_{in} = 2$ MΩ, $g_m = 4$ mA/V, $r_o = 100$ kΩ, $R_D = 10$ kΩ, $C_{gs} = 2$ pF, and $C_{gd} = 0.5$ pF. The amplifier is fed from a voltage source with an internal resistance of 500 kΩ and is connected to a 10 kΩ load. With the aid of Spice, find the following:

(a) The midband voltage gain A_M.

(b) The frequency locations of the two poles and zeros by plotting the input-output voltage characteristics as a function of frequency.

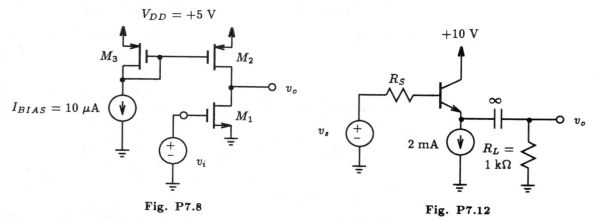

Fig. P7.8 **Fig. P7.12**

7.8 With the aid of Spice, determine the frequency of the pole and zero of the CMOS amplifier shown in Fig. P7.8 in the linear region of the amplifier. For M_1, assume the Spice parameters are: $V_t = 1$ V, $\mu_n C_{OX} = 20$ μA/V^2, $L = 10$ μm, $W = 640$ μm, $\lambda = 0.02$ V^{-1}, $C_{GSO} = 1$ pF, and $C_{GDO} = 1$ pF. For M_2 and M_3, the Spice parameters are: $V_t = -0.8$ V, $\mu_p C_{OX} = 10$ μA/V^2, $L = 10$ μm, $W = 1000$ μm, $\lambda = 0.02$ V^{-1}, $C_{GSO} = 1$ pF, and $C_{GDO} = 1$ pF.

7.9 Consider the common-emitter amplifier of Fig. 7.13(a), with the aid of Spice, plot the amplifiers input impedance Z_{in} (magnitude and phase) as seen by the source excluding R_s.

7.10 For the common-emitter amplifier shown in Fig. 7.13(a), if C_E is increased to 30 μF, what happens to the bandwidth of this amplifier? Quantify your answer using Spice.

7.11 For a 10 kHz, 100 mV sine-wave applied to the input of the CC-CE cascade amplifier shown in Fig. 7.13(b), using Spice, plot the voltage waveform appearing at the output for at least one complete period of the input signal. Compare these results with those obtained for an input signal increased to 5 V of the same frequency.

7.12 For the emitter-follower shown in Fig. P7.12, with the aid of Spice, calculate the midband gain and the upper 3-dB frequency of the amplifier under the following source resistance conditions:

(a) $R_s = 1$ kΩ,

(b) $R_s = 10$ kΩ, and

(c) $R_s = 100$ kΩ.

Assume the transistor is modeled after the commercial 2N3904 *npn* transistor. In addition, for simulation purposes, replace the infinite-valued decoupling capacitor by a very large value (ie. 10^6 pF).

7.13 Plot the waveform appearing at the output of the cascode amplifier shown in Fig. 7.13(b) for a 100 mV input triangular waveform of 1 kHz frequency. Repeat this Spice simulation with the frequency of the input signal increased to 100 kHz. Comment on your results.

7.14 Compare the input-output frequency response behavior of the differential amplifier shown in Fig. 7.20(a) with the transistor modeled after the 2N3904 and 2N2222A. Comment on the results. The Spice parameters for the 2N3904 and 2N2222A can be found in Sections 7.5 and 4.5, respectively.

7.15 If the current source used to bias the differential amplifier in Fig. 7.20(a) is assumed to have an output resistance of 200 kΩ, with the aid of Spice, compute the CMRR of this amplifier as a function of frequency. Assume that the amplifier is driven differentially. How does these results compare to the CMRR of the CC-CB differential amplifier shown in Fig. 7.20(c) if it is also driven differentially?

7.16 Compute the frequency response behavior of a simple bipolar current mirror circuit. Model each bipolar transistor after the Gennum Corporation integrated *npn* transistor described in Section 6.4.

7.17 Compute the frequency response behavior of the modified Wilson current mirror and the cascode current mirror circuit of Problem 6.12 of the previous chapter. Model each bipolar transistor after the Gennum Corporation integrated *npn* transistor described in Section 6.4. Which circuit has the higher frequency response?

Chapter 8   Feedback

In this chapter we shall investigate the accuracy of representing a complicated electronic amplifier circuit by a negative feedback system consisting of two parts: a feedforward network A and a feedback network β. The accuracy of such an approach is to be determined by comparing the performance of the amplifier predicted using feedback theory with that computed directly by Spice. Comparisons will be made based on closed-loop gain, and input and output resistance. As we shall see, the results compare extremely well, providing important justification for viewing a complicated electronic circuit as a single-loop negative feedback structure. In addition to this, we shall demonstrate how Spice can be used to determine the loop gain or transmission $A\beta$ of an arbitrary feedback circuit. Several examples will be presented to illustrate this, from which, the stability behavior of the network can be deduced. Taking this one step further, we shall demonstrate how Spice can be used to aid in the design of the frequency compensation required to stabilize an unstable feedback network. Results in both the time and frequency domains will be given to illustrate the effectiveness of the frequency compensation used.

8.1   The General Feedback Structure

The basic structure of a system including some form of negative feedback is shown in block diagram form in Fig. 8.1. Here the block depicted by A represents the feedforward gain stage of the closed-loop system, and the block denoted by β represents the feedback network. The summing node is used to subtract the signal (x_f) that is fed back from the output from the input signal x_s to create what is known as the error signal x_i. This error signal is then applied to the input of the feedforward gain stage, thus creating the output signal x_o. The closed-loop input – output signal gain A_f can be expressed in terms of A and β according to:

8.2 The Four Basic Feedback Amplifier Topologies

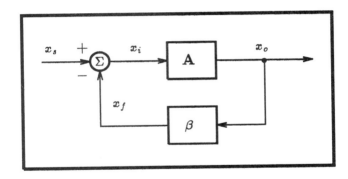

Figure 8.1 The general form of a negative feedback structure.

$$A_f = \frac{A}{1 + A\beta}. \tag{8.1}$$

8.2 The Four Basic Feedback Amplifier Topologies

Amplifiers incorporating some form of feedback can be divided into four general classes. The type of feedback used dictates the category in which the amplifier belongs. In the following we shall present an example of an amplifier belonging to each class and illustrate how one uses Spice to extract both the feedforward gain A and feedback factor β from the closed-loop amplifier. Subsequently, we shall compare the expected closed-loop gain of the amplifier as calculated from feedback theory with that obtained directly from circuit simulation. Likewise, we shall also compare the input and output resistances of the closed-loop amplifier obtained in the same manner.

For easy reference, Table 8.1 provides a summary of the formulae that relate the gain, input and output resistances of the closed-loop circuit for the four different feedback topologies in terms of the open-loop circuit parameters. These were derived in Chapter 8 of Sedra and Smith.

8.2.1 Voltage-Sampling Series-Mixing Topology

As the first example of this chapter we present in Fig. 8.2 the small-signal equivalent circuit of an op-amp connected in the noninverting configuration. This particular amplifier was presented in Example 8.1 of Sedra and Smith as an example of a voltage-sampling series-mixing feedback amplifier topology.

Before one can begin to analyze the amplifier shown in Fig. 8.2 for its feedback structure, the amplifier must be placed in its proper form for which the theory presented by Sedra and Smith applies. This requires that the two input common-mode resistances $2R_{icm}$ of the

8 Feedback

Topology	Closed-Loop Gain A_f	Input Resistance R_{if}	Output Resistance R_{of}
Voltage-Sampling Series-Mixing	$A/(1+A\beta)$	$R_i(1+A\beta)$	$R_o/(1+A\beta)$
Current-Sampling Series-Mixing	$A/(1+A\beta)$	$R_i(1+A\beta)$	$R_o(1+A\beta)$
Voltage-Sampling Shunt-Mixing	$A/(1+A\beta)$	$R_i/(1+A\beta)$	$R_o/(1+A\beta)$
Current-Sampling Shunt-Mixing	$A/(1+A\beta)$	$R_i/(1+A\beta)$	$R_o(1+A\beta)$

Table 8.1 Summary of the closed-loop parameters for the four feedback topologies in terms of the open-loop parameters.

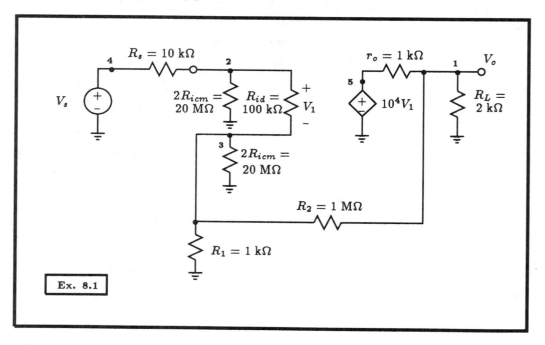

Figure 8.2 The small-signal equivalent circuit of an op-amp connected in the non-inverting configuration. The topology of this particular amplifier is an example of the voltage-sampling series-mixing feedback type.

op-amp be combined with either the source resistance R_s or R_1 as illustrated in Fig. 8.3. This step is necessary to eliminate the ground connections made by $2R_{icm}$, thus causing the input resistance of the feedforward gain stage to float. The feedback network is highlighted by the broken box shown in Fig. 8.3. The remaining circuitry makes up the feedforward gain stage.

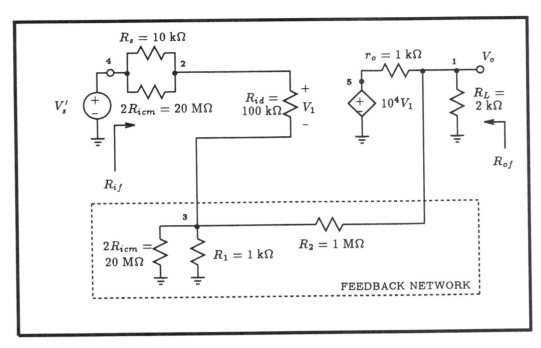

Figure 8.3 Proper form of a voltage-sampling series-mixing feedback amplifier: the input resistance of the feedforward gain stage has no direct connection to ground, the input port is voltage driven, and the output is an opened port.

Now, following the rules outlined in Sedra and Smith for voltage-sampling series-mixing amplifier topologies, we have separated the feedforward amplifier A from the feedback network β as shown in Fig. 8.4. These two networks can then be analyzed using Spice to determine the respective signal gains (A and β), and the input and output resistances R_i and R_o, respectively. For frequency independent circuits, such as that shown in Fig. 8.4, the transfer function command (.TF) in Spice will provide all three of these parameters. The Spice input file for the feedforward gain stage is listed below in Fig. 8.5 with the .TF command included.

On completion of Spice the following transfer function information is found in the Spice output file:

```
****    SMALL-SIGNAL CHARACTERISTICS
        V(1)/VI' =  6.002D+03
        INPUT RESISTANCE AT VI' =  1.110D+05
        OUTPUT RESISTANCE AT V(1) =  6.662D+02
```

One can proceed in the exact same manner for the feedback network and use the transfer function analysis command of Spice. However, for the feedback network shown in Fig. 8.4(b), one could easily carry out the required analysis by hand. We have chosen the latter and the

8 Feedback

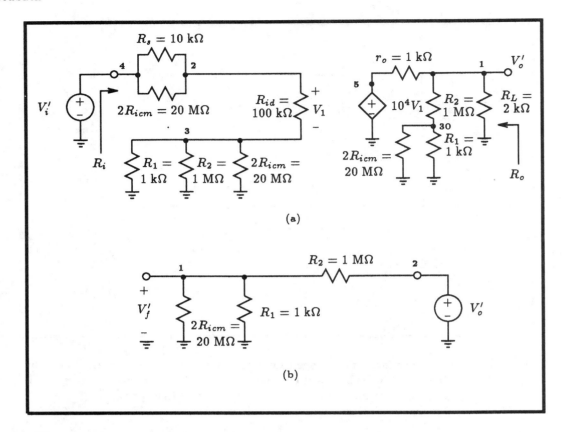

Figure 8.4 (a) Feedforward gain stage of the noninverting amplifier circuit displayed in Fig. 8.3 (b) corresponding feedback network.

results of our hand analysis are included among the feedforward parameter results listed in Table 8.2.

Using the theory developed for voltage-sampling series-mixing amplifiers in Sedra and Smith, and previously summarized in Table 8.1, we can calculate several parameters of the modified closed-loop noninverting amplifier shown in Fig. 8.3 from the open-loop parameters listed in Table 8.2. This would include the input-output voltage gain V_o/V_s', and the input and output resistances R_{if} and R_{of}, respectively. The results of these calculations are listed in Table 8.3. Also included in this table are the corresponding parameters as calculated directly by Spice for the circuit shown in Fig. 8.3. These parameters were also obtained using the .TF command of Spice. The Spice input file for this particular case is quite similar to the one given for the feedforward network listed in Fig. 8.5 and is therefore not given here. As is evident from the results listed in Table 8.3, excellent correlation exists between the results generated using feedback theory and Spice. This example, and the others that follow, demonstrate the validity of the feedback method as an alternative approach to the analysis of feedback amplifiers. It provides important insight into circuit operation that is not always obtainable through direct application of Spice.

8.2 The Four Basic Feedback Amplifier Topologies

Example 8.1: Feedforward Gain Stage Of Noninverting Amplifier

```
** Circuit Description **
* input signal
Vi' 4 0 DC 0V
Rs 4 2 10k
Ricm2 2 4 20Meg
* amplifier
Rid 2 3 100k
Eamp 5 0 2 3 10k
Ro 5 1 1k
* load
Rl 1 0 2k
* input resistance of feedback network
R1i 3 0 1k
R2i 3 0 1Meg
Ricm2i 3 0 20Meg
* output resistance of feedback network
R2o 1 30 1Meg
R1o 30 0 1k
Ricm2o 30 0 20Meg
** Analysis Requests **
* calculate signal gain A=V(1)/Vi', Ri, and Ro.
.TF V(1) Vi'
** Output Requests **
* none required
.end
```

Figure 8.5 The Spice input file for calculating the signal gain A, and the input and output resistances (R_i and R_o), of the feedforward gain stage of the noninverting amplifier displayed in Fig. 8.4(a).

	Parameter	Spice
Feedforward Network	A	6002 V/V
	R_i	111.0 kΩ
	R_o	666.2 Ω
Feedback Network	β	0.999×10^{-3} V/V
	R_{11}	999 Ω
	R_{22}	1.001 MΩ

Table 8.2 Parameters of the feedforward and feedback stages shown in Fig. 8.4 as calculated by Spice.

Checking Basic Feedback Theory Assumptions:

Fundamental to the development of the formulae that apply to voltage-sampling series-

	Parameter	Direct Analysis By Spice	Feedback Theory
Closed-Loop Amplifier	A_f	858.0 V/V	857.2 V/V
	R_{if}	776.5 kΩ	777.2 kΩ
	R_{of}	95.23 Ω	95.14 Ω

Table 8.3 Network parameters of the modified noninverting amplifier circuit shown in Fig. 8.3 as calculated by Spice and through the application of feedback theory.

mixing networks seen listed in Table 8.1 are two assumptions. These are:

(1) Most of the forward signal transmission comes from the feedforward gain stage and not through the feedback network. Using h-parameter terminology, this is expressed as:

$$|h_{21}|_A \gg |h_{21}|_\beta \qquad (8.2)$$

(2) Most of the signal that is fed back and mixed at the input to the amplifier comes from the feedback network and not through the feedforward gain stage. With h-parameter terminology this is expressed as:

$$|h_{12}|_\beta \gg |h_{12}|_A \qquad (8.3)$$

With the aid of Spice let us compute the two-port hybrid-parameters of the feedforward and feedback networks shown in Fig. 8.4, and verify that the above assumptions are indeed satisfied. (We should expect that this is the case because the results generated by the feedback theory compare extremely well with those found directly from Spice.)

Using each circuit setup shown in Fig. 8.6, together with Spice, we can compute the h-parameters of both the feedforward gain stage and the feedback network shown in Fig. 8.4. As an example of one such case, the Spice deck for computing the h-parameters h_{11} and h_{21} for the feedforward gain stage is shown in Fig. 8.7. In this Spice deck the input port of the amplifier is driven by a one-amp AC signal I1 and the output port is short-circuited by a zero-valued voltage source V2. An AC analysis is asked to be performed at a single frequency of 1 Hz. (Any frequency will do, since the network contains no reactive elements). A .PRINT command is used to print out the voltage that appears at the input of the amplifier V(4) and the current that flows into the output port I(V2). Since the input current level is unity, both the voltage V(4) and the output current I(V2) will equal h-parameters h_{11} and h_{21}, respectively. The results of this analysis are found in the output Spice file as follows:

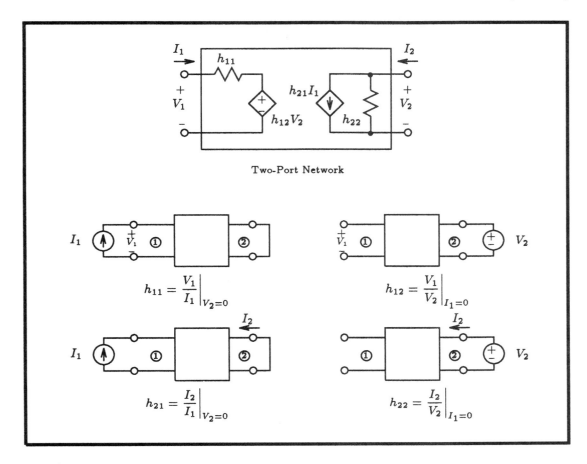

Figure 8.6 Hybrid-parameter representation of an arbitrary two-port network, and circuit setup for computing individual h-parameters.

```
****    AC ANALYSIS                    TEMPERATURE =   27.000 DEG C

   FREQ        VM(4)        VP(4)        IM(V2)       IP(V2)

  1.000E+00   1.110E+05    0.000E+00    1.000E+06    0.000E+00
```

Thus, $h_{11} = 1.11 \times 10^5$ V/A and $h_{21} = -1 \times 10^6$ A/A for the feedforward gain stage. (The sign of h_{21} is negative because the current monitored by V2 is positive and in opposite direction to the current denoted by I_2 in Fig. 8.6. Recall that a current is considered positive when it flows from the positive to negative terminal of a voltage source.)

The same approach is also used to compute the remaining two parameters, h_{12} and h_{22}. One simply modifies the previous Spice deck by altering the two source statements given there. First, the level of current source I1 is set to zero and voltage source V2 is set to have a one-volt level according to the following:

```
I1 0 4 DC 0A AC 0A
V2 1 0 DC 0V AC 1V.
```

8 Feedback

```
Evaluating The h-Parameters Of Feedforward Gain Stage

** Circuit Description **

* input port conditions
I1 0 4 DC 0A AC 1A
V2 1 0 DC 0V AC 0V

* feedforward gain stage
Rs 4 2 10k
Ricm2 2 4 20Meg
* amplifier
Rid 2 3 100k
Eamp 5 0 2 3 10k
Ro 5 1 1k
* load
Rl 1 0 2k
* feedback network input resistance
R1i 3 0 1k
R2i 3 0 1Meg
Ricm2i 3 0 20Meg
* feedback network output resistance
R2o 1 30 1Meg
R1o 30 0 1k
Ricm2o 30 0 20Meg
** Analysis Requests **
.AC LIN 1 1Hz 1Hz
** Output Requests **
.PRINT AC Vm(4) Vp(4) Im(V2) Ip(V2)
.end
```

Figure 8.7 The Spice input file for calculating h-parameters h_{11} and h_{21} of the feedforward gain stage of the noninverting amplifier displayed in Fig. 8.4(a). The remaining two h-parameters h_{12} and h_{22} are found by setting the input level of I1 to zero and the output voltage source V2 to one-volt. Everything else remains the same

The AC analysis request, together with the .PRINT command, remain the same. The results then computed by Spice are as follows:

```
****    AC ANALYSIS                         TEMPERATURE =   27.000 DEG C

  FREQ      VM(4)       VP(4)       IM(V2)      IP(V2)

1.000E+00  1.000E-30   0.000E+00   1.501E-03   1.800E+02
```

In this case, the voltage V(4) corresponds to parameter h_{12} since the voltage driven into the second port is unity. As seen, it has a value of zero. This implies that no signal transmission occurs from the output of the feedforward gain stage back to its input. This is not too surprising given that there does not exist any direct connection between the input and

8.2 The Four Basic Feedback Amplifier Topologies

h-parameter	A	β
h_{11}	1.11×10^5 V/A	$9.99 \times 10^{+2}$ V/A
h_{12}	0 V/V	9.99×10^{-4} V/V
h_{21}	-1×10^6 A/A	-9.99×10^{-4} A/A
h_{22}	1.501×10^{-3} A/V	9.99×10^{-7} A/V

Table 8.4 The h-parameters of the feedforward and feedback networks shown in Fig. 8.4.

output ports of the feedforward gain stage (see Fig. 8.4). Finally, parameter h_{22} is simply equal to the current supplied by V2 and is seen from above to be 1.501×10^{-3} A/V.

We can repeat this same type of analysis for the feedback network and arrive at the following set of h-parameters: $h_{11} = 9.99 \times 10^{+2}$ V/A, $h_{12} = 9.99 \times 10^{-2}$ V/V, $h_{21} = -9.99 \times 10^{-2}$ A/A and $h_{22} = 9.99 \times 10^{-2}$ A/V.

The h-parameters for both the feedforward and feedback networks are compiled in Table 8.4 for easy comparison. Clearly, the above two assumptions, ie. $|h_{21}|_A \gg |h_{21}|_\beta$ and $|h_{12}|_\beta \gg |h_{12}|_A$, are easily satisfied. Finally, it should be noted that resistances R_{11} and R_{22} that represent the loading of the feedback network on the feedforward network are given by $R_{11} = h_{11}|_\beta$ and $R_{22} = 1/h_{22}|_\beta$.

8.2.2 Current-Sampling Series-Mixing Topology

The next example, shown in Fig. 8.8(a), illustrates a cascade of three transistors in a current-sampling series-mixing feedback configuration known as a feedback triple. This particular amplifier was presented as Example 8.2 in Sedra and Smith where the biasing circuit has been purposely left off the schematic. However, we are given enough biasing information from which to calculate the small-signal equivalent circuit of each transistor. We summarize the small-signal equivalent circuit parameters of each transistor below:

Q_1 @ $I_C = 0.6$ mA: $r_{\pi 1} = 4.167$ kΩ $g_{m1} = 24$ mA/V

Q_2 @ $I_C = 1$ mA: $r_{\pi 2} = 2.5$ kΩ $g_{m2} = 40$ mA/V

Q_3 @ $I_C = 4$ mA: $r_{\pi 3} = 625$ Ω $g_{m3} = 160$ mA/V

Using the above information we can replace each transistor in Fig. 8.8(a) by its small-signal equivalent as shown in Fig. 8.8(b). Recognizing that the feedback network includes resistors R_{E1}, R_{E2} and R_F, we can apply the loading rules presented by Sedra and Smith and separate the feedforward circuit from the feedback circuit. These two circuits are given in Fig. 8.9. It is important to note that the current denoted as the output is not the actual current that

8 Feedback

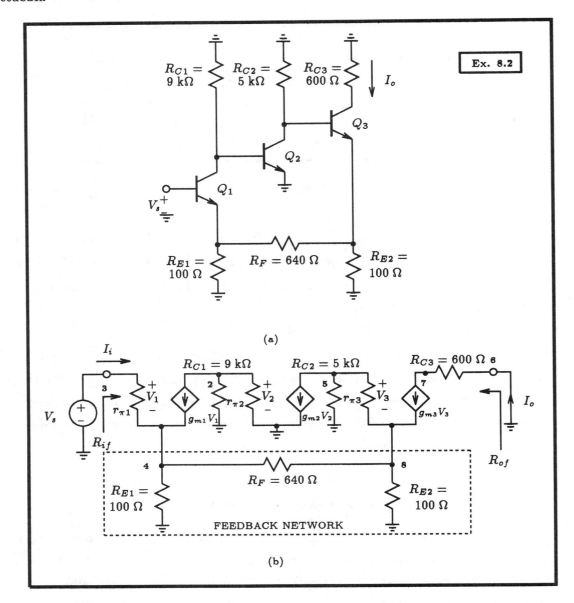

Figure 8.8 (a) A broadband amplifier composed of a feedback triple. The biasing circuitry is not shown (b) small-signal equivalent.

is sampled and fed back to mix with the input signal, but rather a very close approximation to it. As we shall see, this approximation is reasonable for predicting the closed-loop signal gain A_f of the amplifier, but incorrectly predicts the output resistance R_{of}.

The Spice input file describing the feedforward circuit shown in Fig. 8.9 is listed below in Fig. 8.10. Notice here that we are using a zero-valued voltage source to form the short-circuit output. The current flowing through this voltage source can then be monitored. In a manner identical to that used in the voltage-sampling series-mixing case, we can obtain the signal gain I'_o/V'_i, R_i and R_o, directly using the built-in transfer function command (.TF)

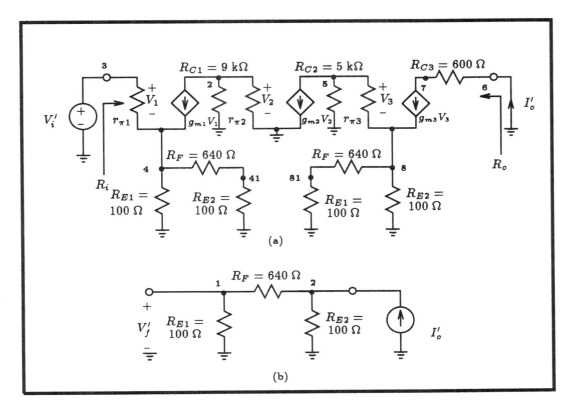

Figure 8.9 Isolated network portions of the broadband amplifier circuit: (a) feedforward network; (b) feedback network.

of Spice. These results, as calculated by Spice, are summarized in Table 8.5, together with the parameters for the feedback network obtained through hand analysis.

Using the results compiled in Table 8.5 and the theory developed for series-series feedback networks in Sedra and Smith (also seen summarized in Table 8.1), we can compute the signal gain A_f and, the input and output resistances (R_{if} and R_{of}) of the small-signal equivalent circuit of the broadband amplifier displayed in Fig. 8.8. These results are listed in Table 8.6 and compared with the results calculated by direct analysis of the circuit shown in Fig. 8.8. The Spice input listing for the direct calculations is very similar to that given in Fig. 8.10 for the feedforward network and is therefore not shown here. The results predicted by feedback theory seem to agree quite closely with those computed directly with Spice.

If we repeat the above analysis with for the output resistance of each transistor included, specifically assuming that each has an Early voltage of 100 V (ie. $r_{o1} = 167$ kΩ, $r_{o2} = 100$ kΩ, and $r_{o3} = 25$ kΩ), then we find using Spice that the open-loop gain $A = 19.68$ A/V, the feedback factor $\beta = 11.9$ V/A, and the input and output resistances are $R_i = 12.96$ kΩ and $R_o = 65.91$ kΩ, respectively. According to feedback theory, the closed-loop gain will then be $A_f = 83.68$ mA/V, the input resistance $R_{if} = 3.048$ MΩ and the output resistance $R_{of} = $

```
Example 8.2: Feedforward Circuit Of Broadband Amplifier

** Subcircuits **
* small-signal equivalent transistor circuits
.subckt Q1 1 2 3
rpi 2 3 4.1667k
Gt 1 3 2 3 24m
.ends
.subckt Q2 1 2 3
rpi 2 3 2.5k
Gt 1 3 2 3 40m
.ends
.subckt Q3 1 2 3
rpi 2 3 625
Gt 1 3 2 3 160m
.ends

** Main Circuit **
* input signal
Vi' 3 0 DC 0V
* broadband amplifier
* 1st stage
Xt1 2 3 4 Q1
Rc1 2 0 9k
* 2nd stage
Xt2 5 2 0 Q2
Rc2 5 0 5k
* 3rd stage
Xt3 7 5 8 Q3
Rc3 6 7 600
* input resistance of feedback network
Re1i 4 0 100
Rfi 4 41 640
Re2i 41 0 100
* output resistance of feedback network
Re1o 81 0 100
Rfo 81 8 640
Re2o 8 0 100
* short-circuit output port
Vout 0 6 0
** Analysis Requests **
* calculate signal gain A=I(Vout)/Vi', Ri and Ro.
.TF I(Vout) Vi'
** Output Requests **
* none required
.end
```

Figure 8.10 The Spice input file for calculating the signal gain A, and the input and output resistances (R_i and R_o), of the feedforward network of the broadband amplifier displayed in Fig. 8.9(a).

8.2 The Four Basic Feedback Amplifier Topologies

	Parameter	Spice
Feedforward Network	A	20.62 A/V
	R_i	13.06 kΩ
	R_o	1×10^{20} Ω
Feedback Network	β	11.9 V/A
	R_{11}	88.1 Ω
	R_{22}	88.1 Ω

Table 8.5 Parameters of the feedforward and feedback circuit shown in Fig. 8.9 as calculated by Spice.

	Parameter	Direct Analysis By Spice	Feedback Theory
Closed-Loop Amplifier	A_f	82.83 mA/V	83.69 mA/V
	R_{if}	3.253 MΩ	3.218 MΩ
	R_{of}	1×10^{20} Ω	∞

Table 8.6 Network parameters of the broadband amplifier circuit shown in Fig. 8.8 as calculated by Spice and through the application of feedback theory.

15.50 MΩ. When compared to the results computed directly by Spice, ie. $A_f = 82.77$ mA/V, $R_{if} = 2.603$ MΩ and $R_{of} = 2.183$ MΩ, we see that the signal gain A_f is quite close to that computed by Spice. This is in contrast with the input resistance R_{if} and output resistance R_{of} predicted by feedback theory. In the case of the input resistance, the value predicted by feedback theory is about 17% larger than the value predicted by Spice. The reason for this discrepancy can be traced back to the fact that the input port of the feedforward network is not truly in series with the input port of the feedback network as was assumed in the development of the feedback theory for current-sampling series-mixing networks. Specifically, note that the current that flows into the input terminal of the feedforward network, I_i seen in Fig. 8.8, is not the same current that is eventually fed into the input port of the feedback network. The latter is instead $(1 + g_{m_1} r_{\pi_1})I_i$ or $(1 + \beta)I_i$. Although these two currents are dramatically different, in the ratio of 1:80 for the parameter values used above for g_{m_1} and r_{π_1}, the error that this causes in the estimate of the input resistance of the closed-loop amplifier is relatively small at 17%.

In the case of the output resistance of the closed-loop amplifier, the value predicted by feedback theory is about 7 times larger than that computed by Spice; albeit, the actual output resistance has increased with the feedback applied, as expected. The reason for this large error is due to the fact that the current-sampling action is not taking place at the

8 Feedback

designated output port, that being in series with the collector of Q_3, instead it is actually taking place on the emitter side of Q_3.

If we re-perform the above feedback analysis on the original amplifier with the output port designated in series with the emitter of Q_3 as shown in Fig. 8.11(a) then we should find that the output resistance computed directly by Spice will better agree with that estimated using feedback theory. Consider the small-signal equivalent circuit of the feedforward portion of the amplifier, shown in Fig. 8.11(b). Let us compute the signal gain, and the input and output resistance using Spice. The corresponding Spice deck is provided in Fig. 8.12. A transfer function (.TF) analysis is requested to be performed and the results of this analysis are provided below:

```
****    SMALL-SIGNAL CHARACTERISTICS
        I(Vout)/Vi =   1.988E+01
        INPUT RESISTANCE AT Vi =   1.296E+04
        OUTPUT RESISTANCE AT I(Vout) =   1.426E+02
```

Here we see that $A = 19.88$ A/V, $R_i = 12.96$ kΩ and $R_o = 142.6$ Ω.

The feedback network portion of the original amplifier is identical to that seen previously. It is shown in Fig. 8.11(c) for completeness. The feedback factor β was previously found to be 11.9 V/A. (Also, $R_{11} = 88.1$ Ω and $R_{22} = 88.1$ Ω.)

According to feedback theory the closed-loop signal gain A_f is then 83.68 mA/V, the input resistance R_{if} equals 3.08 MΩ and the output resistance R_{of} is 33.88 kΩ. The first two parameters, A_f and R_{if} are essentially the same as before when the output port was designated in series with the collector of Q_3. The output resistance, however, is quite different from that seen previously but this should be expected given that the output port is now found on the emitter side of Q_2. When the output resistance of the amplifier seen in Fig. 8.11(a) is computed directly with Spice, we obtain 33.90 kΩ. This result seems to now agree quite closely with the value predicted by feedback theory. Thus, re-assuring us that the feedback theory is consistent with what is observed in practice. The results of this analysis are summarized in Table 8.7.

To further confirm the validity of the above results, the z-parameters of both the feedforward gain stage and feedback network are listed in Table 8.8. Here we see that the forward transmission through the feedback network is much smaller than that through the feedforward circuit, ie. $|z_{21}|_A \gg |z_{21}|_\beta$. Conversely, the transmission from the output to the input of the feedforward gain stage is much less than the corresponding transmission through the feedback network, ie. $|z_{12}|_\beta \gg |z_{12}|_A$. Finally, note that $R_{11} = z_{11}|_\beta$ and $R_{22} = z_{22}|_\beta$.

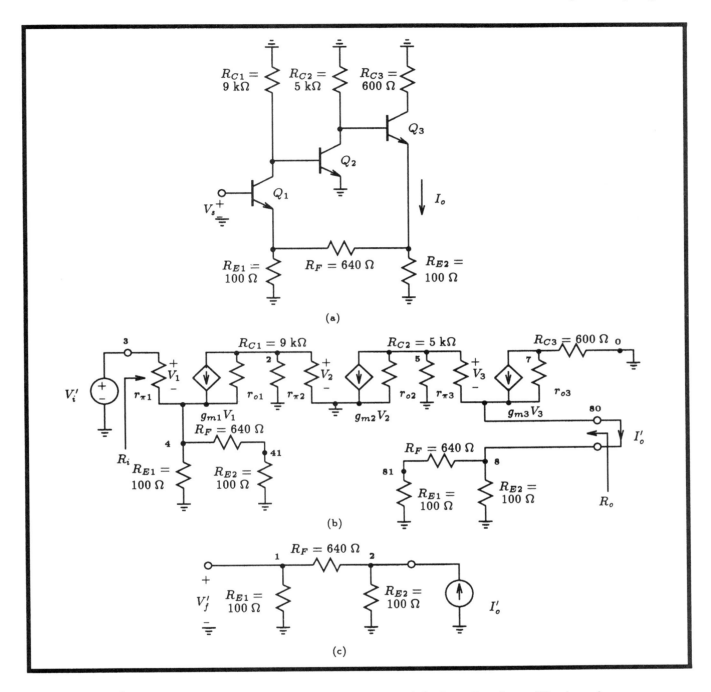

Figure 8.11 (a) Designating the output port of the broadband amplifier in series with the emitter of Q_3 to coincide with the port at which the current-sampling is taking place. (b) Small-signal equivalent circuit of the feedforward portion of the amplifier. (c) Feedback network.

8 Feedback

	Parameter	Spice	
Feedforward Network	A	19.88 A/V	
	R_i	12.96 kΩ	
	R_o	142.6 Ω	
Feedback Network	β	11.9 V/A	
	R_{11}	88.1 Ω	
	R_{22}	88.1 Ω	

	Parameter	Direct Analysis By Spice	Feedback Theory
Closed-Loop Amplifier	A_f	83.63 mA/V	83.68 mA/V
	R_{if}	2.603 MΩ	3.08 MΩ
	R_{of}	33.90 kΩ	33.88 kΩ

Table 8.7 Network parameters of the broadband amplifier circuit shown in Fig. 8.11(a) with the emitter current of Q_3 as the output signal.

z-parameter	A	β
z_{11}	1.296×10^4 V/A	$8.81 \times 10^{+1}$ V/A
z_{12}	1.0×10^{-30} V/A	$1.19 \times 10^{+1}$ V/A
z_{21}	$1.681 \times 10^{+10}$ V/A	$1.19 \times 10^{+1}$ V/A
z_{22}	$6.591 \times 10^{+4}$ V/A	$8.81 \times 10^{+1}$ V/A

Table 8.8 The z-parameters of the feedforward and feedback networks shown in Fig. 8.11.

8.2.3 Voltage-Sampling Shunt-Mixing Topology

In Fig. 8.13 we display a single-stage common-emitter amplifier with feedback resistor R_f. This particular circuit was presented by Sedra and Smith in Example 8.3 of their text as an example of an amplifier having a voltage-sampling shunt-mixing topology. Here we shall decompose the feedback amplifier into its feedforward and feedback portions and analyze the individual circuits to estimate the behavior of the closed-loop amplifier. However, unlike the example in Sedra and Smith, we shall assume that Q_1 is the commercial *npn* discrete transistor 2N2222A. This will complicate our analysis because the small-signal model of this transistor is much more complicated than that assumed by Sedra and Smith. Fortunately, using the following approach, the details of the small-signal equivalent circuit of the transistor

8.2 The Four Basic Feedback Amplifier Topologies

```
Open-Loop Portion Of Broadband Amplifier (Revised Output Port)

** Subcircuits **
* small-signal equivalent transistor circuits
.subckt Q1 1 2 3
rpi 2 3 4.1667k
Gt 1 3 2 3 24m
ro 1 3 167k
.ends
.subckt Q2 1 2 3
rpi 2 3 2.5k
Gt 1 3 2 3 40m
ro 1 3 100k
.ends
.subckt Q3 1 2 3
rpi 2 3 625
Gt 1 3 2 3 160m
ro 1 3 25k
.ends

** Main Circuit **
* input signal
Vi' 3 0 DC 0V
* broadband amplifier
* 1st stage
Xt1 2 3 4 Q1
Rc1 0 2 9k
* 2nd stage
Xt2 5 2 0 Q2
Rc2 0 5 5k
* 3rd stage
Xt3 7 5 80 Q3
Rc3 7 0 600
* input resistance of feedback network
Re1i 4 0 100
Rfi 4 41 640
Re2i 41 0 100
* output resistance of feedback network
Re1o 81 0 100
Rfo 81 8 640
Re2o 8 0 100
* monitor output current
Vout 80 8 0
** Analysis Requests **
.TF I(Vout) Vi'
** Output Requests **
* none required
.end
```

Figure 8.12 The Spice input file for calculating the signal gain A, and the input and output resistances (R_i and R_o), of the feedforward circuit with the emitter current of Q_3 designated as the output signal as depicted in Fig. 8.11.

8 Feedback

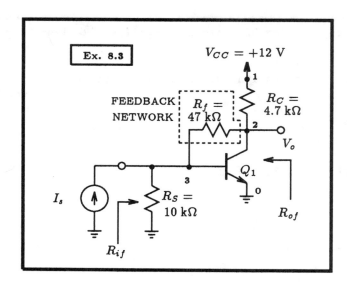

Figure 8.13 A single-stage amplifier circuit. An example of a voltage-sampling shunt-mixing feedback amplifier topology.

will be kept hidden inside the computer – thus avoiding tedious manipulations of the small-signal equivalent circuit and, more importantly, avoiding the necessity of making simplifying approximations and the corresponding loss of accuracy.

Consider applying the separation rules for shunt-shunt amplifier networks presented in Section 8.6 of Sedra and Smith to an assumed small-signal equivalent circuit of the amplifier shown in Fig. 8.13. To avoid complicating the schematics of the feedforward and feedback networks, we shall represent the small-signal equivalent circuit of the transistor by a single transistor symbol. In addition, the DC supplies will also be set to zero, as is always the case when working with small-signal circuit equivalents of transistor circuits. The result is the feedforward gain stage and the feedback network illustrated in Fig. 8.14.

To determine the small-signal model parameters of the transistor we shall first run a DC operating point (.OP) analysis on the original amplifier circuit shown in Fig. 8.13. The corresponding Spice input deck is shown in Fig. 8.13. Here we shall calculate both the DC operating point of Q_1, and for future use, the gain $A = V_o/I_s$, and the input and output resistances R_{if} and R_{of}. The bias conditions and transfer function characteristics as calculated by Spice are provided below:

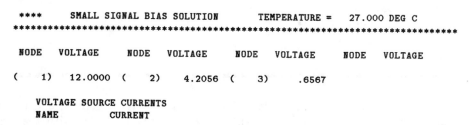

```
                Vcc         -1.658E-03

        TOTAL POWER DISSIPATION   1.99E-02  WATTS


    ****      OPERATING POINT INFORMATION    TEMPERATURE =   27.000 DEG C
    **********************************************************************

    **** BIPOLAR JUNCTION TRANSISTORS

        NAME        Q1
        MODEL       Q2N2222A
        IB          9.84E-06
        IC          1.58E-03
        VBE         6.57E-01
        VBC         -3.55E+00
        VCE         4.21E+00
        BETADC      1.61E+02
        GM          6.09E-02
        RPI         2.90E+03
        RX          1.00E+01
        RO          4.90E+04


    ****        SMALL-SIGNAL CHARACTERISTICS

        V(2)/Is = -4.305E+04

        INPUT RESISTANCE AT Is =  1.807E+02

        OUTPUT RESISTANCE AT V(2) =  3.303E+02
```

Now that the small-signal model of transistor Q_1 is known, we could use it in place of the transistor in the feedforward network of Fig. 8.14, in much the same way that was demonstrated in the previous example. Instead, a more elegant means of performing the required analysis is to externally bias the transistor using a current source to set the base current and a voltage source to set the collector-emitter voltage. Several AC analyses are then performed on the circuit to determine the signal gain, and the input and output resistances. To ensure that the transistor is properly biased and that the external bias sources do not interfere with the circuit small-signal operation, a large-valued decoupling capacitor C_1 is added to the circuit to ensure that all the dc current supplied by the current source goes into the base of the transistor. Similarly, a large-valued inductor L_2 is placed in series with the voltage source to block any AC current that would flow through the zero-source-resistance voltage source. This approach is depicted in Fig. 8.16. This method enables the details of the transistor small-signal model to be kept inside the computer where the greatest numerical precision is maintained.

To convince the reader that this method does indeed maintain the correct transistor bias point, and therefore result in the same small-signal model as before, the small-signal model of the transistor in the circuit arrangement depicted in Fig. 8.16 was computed using Spice. The results found are as follows:

8 Feedback

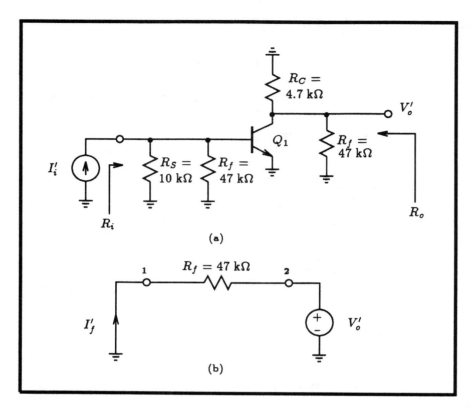

Figure 8.14 Isolated network portion of the amplifier circuit shown in Fig. 8.13: (a) feedforward gain stage without the biasing network shown; (b) feedback network.

```
**** BIPOLAR JUNCTION TRANSISTORS
   NAME        Q1
   MODEL       Q2N2222A
   IB          9.84E-06
   IC          1.58E-03
   VBE         6.57E-01
   VBC        -3.55E+00
   VCE         4.21E+00
   BETADC      1.61E+02
   GM          6.09E-02
   RPI         2.90E+03
   RX          1.00E+01
   RO          4.90E+04
```

Comparing these results with those computed from the original amplifier, we see that they are in perfect agreement to the 3 digits provided.

We are now ready to use Spice to calculate the gain and resistance parameters of the feedforward network of the amplifier. We apply a one-amp AC current signal of 1 Hz frequency and with the aid of Spice compute the output voltage V(2) and input voltage V(3) (see labeled nodes in Fig. 8.16). Because the input current level is unity, the signal gain is simply given by V(2), and likewise, the input resistance is given directly by V(3). The Spice

Example 8.3: Voltage-Sampling Shunt-Mixing Voltage Amplifier Circuit (Closed-Loop)

```
** Circuit Description **
* power supply
Vcc 1 0 DC +12V
* input signal source (set to zero for operating point information)
Is 0 3 DC 0A
Rs 3 0 10k
* amplifier circuit
Rc 1 2 4.7k
Q1 2 3 0 Q2N2222A
* feedback circuit
Rf 2 3 47k
* transistor model statement for 2N2222A
.model Q2N2222A NPN (Is=14.34f Xti=3 Eg=1.11 Vaf=74.03 Bf=255.9 Ne=1.307
+                    Ise=14.34f Ikf=.2847 Xtb=1.5 Br=6.092 Nc=2 Isc=0 Ikr=0 Rc=1
+                    Cjc=7.306p Mjc=.3416 Vjc=.75 Fc=.5 Cje=22.01p Mje=.377 Vje=.75
+                    Tr=46.91n Tf=411.1p Itf=.6 Vtf=1.7 Xtf=3 Rb=10)
** Analysis Requests **
.OP
.TF V(2) Is
** Output Requests **
* none required
.end
```

Figure 8.15 The Spice input file for calculating the DC operating point information of transistor Q_1 in the closed-loop amplifier shown in Fig. 8.13. Also included, for future reference, is a transfer function command (.TF) for calculating the gain A and the input and output resistances R_{if} and R_{of}.

input deck for this situation is listed in Fig. 8.17.

The results of this simulation are:

```
    FREQ        VM(2)       VP(2)       VM(3)       VP(3)

  1.000E+00   5.125E+05   1.800E+02   2.150E+03  -7.889E-04
```

Thus the signal gain is $A = -512.5$ kV/A and the input resistance R_i is 2.15 kΩ.

In a similar manner, the output resistance is computed by applying a one-amp AC current signal to the output port of the feedforward circuit with $I'_i = 0$ A and calculating $R_o = V(2)$. The Spice results of this calculations are:

```
    FREQ        VM(2)       VP(2)

  1.000E+00   3.930E+03  -7.114E-04
```

Thus the output resistance R_o is 3.93 kΩ. The parameters of the feedback network should

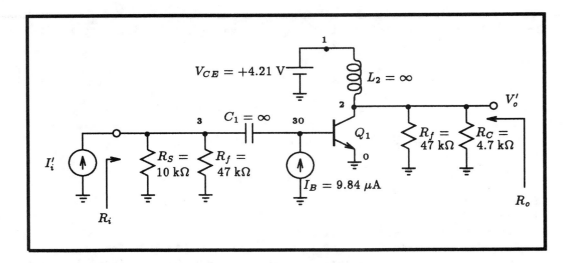

Figure 8.16 Feedforward circuit with transistor Q_1 biased with an external current source I_B and voltage source V_{CE}. Infinite-valued decoupling capacitor C_1 causes all the current supplied by I_B to go into the base of Q_1, but behaves as a short circuit for input AC signals. The infinite-valued inductor L_2 acts as a short-circuit at DC allowing the voltage source V_{CE} to set the voltage at the collector of Q_1, but acts as an open circuit for any AC signal.

be obvious from Fig. 8.14. For easy viewing, we summarize the parameters of both the feedforward and feedback circuits in Table 8.9.

Using the feedback theory developed by Sedra and Smith for shunt-shunt feedback networks, we can calculate A_f, R_{if} and R_{of} for the closed-loop amplifier shown in Fig. 8.13 using the formulae listed in Table 8.1. The results of these calculations are listed in Table 8.10. These results are also compared with those computed directly using Spice for the closed-loop amplifier. As evident, the results compare remarkably well.

To further confirm the validity of the above results, in Table 8.11 are listed the y-parameters of both the feedforward circuit and the feedback network. Here we see that the forward transmission through the feedback network is much smaller than that obtained by the the feedforward network, ie. $|y_{21}|_A \gg |y_{21}|_\beta$. Conversely, the transmission from the output to the input of the feedforward network is much less than the corresponding transmission through the feedback network, ie. $|y_{12}|_\beta \gg |y_{12}|_A$. Finally, note that $R_{11} = 1/y_{11}|_\beta$ and $R_{22} = 1/y_{22}|_\beta$.

8.2.4 Current-Sampling Shunt-Mixing Topology

As the final example of the four feedback amplifier topologies, we display in Fig. 8.18(a) a feedback amplifier circuit that is of the current-sampling shunt-mixing type. We shall carry out our analysis of this example in much the same way as that performed with the previous

8.2 The Four Basic Feedback Amplifier Topologies

```
Example 8.3: Feedforward Portion Of Shunt-Shunt Amplifier

** Circuit Description **
* input signal source
Ii' 0 3 AC 1A
Rs 3 0 10k
* amplifier circuit with isolated biasing network
Rc 2 0 4.7k
Q1 2 30 0 Q2N2222A
C1 3 30 1GF
L2 1 2 1GH
IB 0 30 DC 9.84uA
VCE 1 0 DC +4.21V
* input and output resistance of feedback circuit
Rf1 3 0 47k
Rf2 2 0 47k
* transistor model statement for 2N2222A
.model Q2N2222A  NPN (Is=14.34f Xti=3 Eg=1.11 Vaf=74.03 Bf=255.9 Ne=1.307
+                    Ise=14.34f Ikf=.2847 Xtb=1.5 Br=6.092 Nc=2 Isc=0 Ikr=0 Rc=1
+                    Cjc=7.306p Mjc=.3416 Vjc=.75 Fc=.5 Cje=22.01p Mje=.377 Vje=.75
+                    Tr=46.91n Tf=411.1p Itf=.6 Vtf=1.7 Xtf=3 Rb=10)
** Analysis Requests **
.OP
.AC LIN 1 1Hz 1Hz
** Output Requests **
.Print AC Vm(2) Vp(2) Vm(3) Vp(3)
.end
```

Figure 8.17 The Spice input file for calculating the signal gain $A = V_o'/I_i'$ and input resistance $R_i = V(3)/I_i'$ of the feedforward circuit displayed in Fig. 8.16.

	Parameter	Spice
Feedforward Network	A	-512.5 kV/A
	R_i	2.15 kΩ
	R_o	3.93 kΩ
Feedback Network	β	$-21.28\,\mu$A/V
	R_{11}	47 kΩ
	R_{22}	47 kΩ

Table 8.9 Parameters of the feedforward and feedback circuits shown in Fig. 8.14 as calculated by Spice.

example. However, we shall assume that Q_1 and Q_2 are commercial *npn* transistors of the 2N3904 type. A similar example was presented in Sedra and Smith as Example 8.4 of their text; however, a slight change of the biasing circuitry was necessary to accommodate the

8 Feedback

	Parameter	Direct Analysis By Spice	Feedback Theory
Closed-Loop Amplifier	A_f	-43.05 kV/A	-43.05 kV/A
	R_{if}	180.7 Ω	180.6 Ω
	R_{of}	330.3 Ω	330.1 Ω

Table 8.10 Parameters of the feedback amplifier circuit shown in Fig. 8.13 as calculated by Spice and through the application of feedback theory.

y-parameter	A	β
y_{11}	4.651×10^{-4} A/V	2.128×10^{-5} A/V
y_{12}	2.514×10^{-11} A/V	2.128×10^{-5} A/V
y_{21}	6.065×10^{-2} A/V	2.128×10^{-5} A/V
y_{22}	2.544×10^{-4} A/V	2.128×10^{-5} A/V

Table 8.11 The y-parameters of the feedforward and feedback networks shown in Fig. 8.14.

2N3904 type transistors. Although, the current I_{out} is ultimately the current of primary interest since it is this current that we are trying to stabilize with the application of negative feedback, we shall focus our attention on the collector current of Q_2, denoted by I_o. This is because I_o closely approximates the emitter current of Q_2 which is sampled by the feedback network on which the feedback theory of Sedra and Smith is based. Once I_o is obtained, I_{out} is simply obtained from the following expression:

$$I_{out} = \frac{R_{C2}}{R_L + R_{C2}} I_o. \qquad (8.4)$$

The circuit shown in Fig. 8.18(b) is the same circuit as in part (a) but the output port is re-arranged so that the collector current of Q_2 becomes the output current.

The feedforward circuit, including the loading effects of the feedback network, is shown in Fig. 8.19(a). Here we should notice that because the feedback network is AC coupled, rearranging it does not affect the biasing conditions of either Q_1 or Q_2. Furthermore, because the input signal passes through a decoupling capacitor, we can not obtain the three network parameters usually of interest here using the DC transfer function (.TF) analysis command of Spice. Instead we must use the approach presented in the previous example. That is, to obtain the signal gain I'_o/I'_i and the input resistance R_i, we apply a one-amp AC signal to the input of the feedforward amplifier and calculate the output current I'_o, and voltage appearing across the input port (V(10)). The Spice input file for this particular circuit

8.2 The Four Basic Feedback Amplifier Topologies

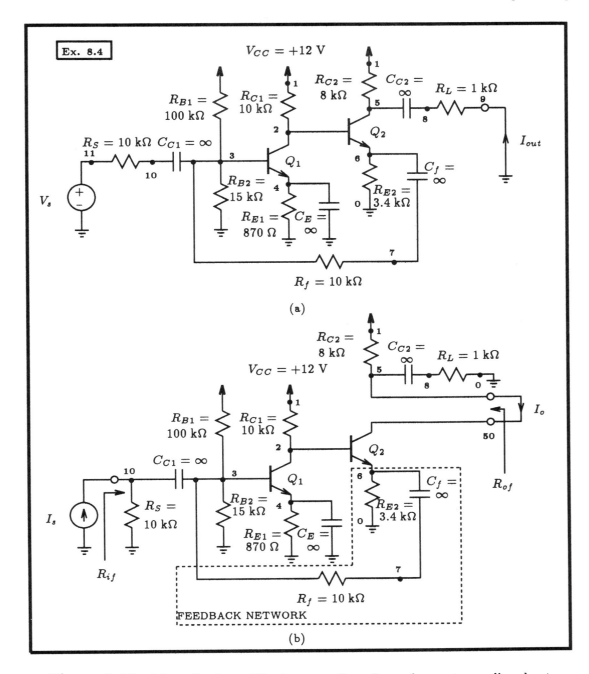

Figure 8.18 (a) Feedback amplifier incorporating a form of current-sampling shunt-mixing feedback. (b) Rearranging the output section of the amplifier so that the collector current of Q_2 becomes the output current.

situation is shown listed in Fig. 8.20. The results of this AC analysis are then found in the output file as follows:

```
    FREQ        IM(Vout)     IP(Vout)      VM(10)      VP(10)

   1.000E+00    2.449E+02    1.800E+02    2.077E+03   -6.311E-04
```

383

8 Feedback

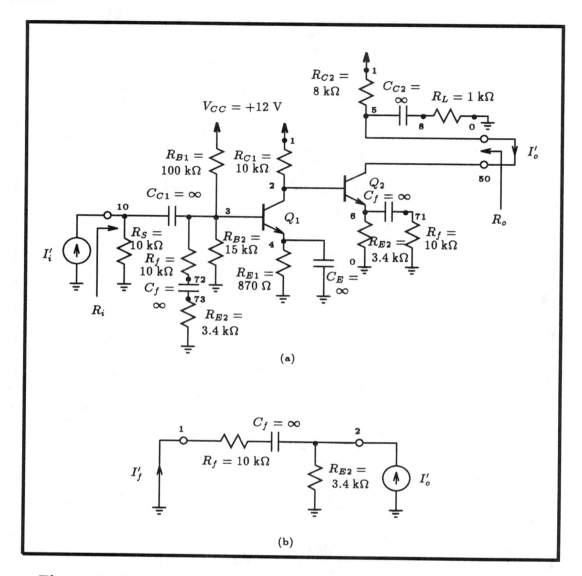

Figure 8.19 Isolated network portions of the amplifier circuit shown in Fig. 8.18: (a) feedforward network with external biasing shown; (b) feedback network.

Here we see that $A = -244.9$ A/A and $R_i = 2.077$ kΩ.

In the case of the output resistance, this is obtained by applying a 1 V AC voltage source in series with the short-circuited output and setting the input current level to zero. The output resistance R_o is then simply the reciprocal of the current supplied by this voltage source. The previous Spice deck can be used to perform this by simply modifying the source statements for Ii' and Vout according to:

```
Ii'   0  10  DC 0A  AC 0A
Vout  5  50  DC 0V  AC 1V.
```

8.2 The Four Basic Feedback Amplifier Topologies

Example 8.4: Feedforward Network Of Current-Sampling Shunt-Mixing Amplifier

```
** Circuit Description **
* input signal source
Ii' 0 10 DC 0A AC 1A
Rs 10 0 10k
* output current (collector current of Q2)
Vout 5 50 DC 0V AC 0V
* power supply
Vcc 1 0 DC +12V
* amplifier circuit
* 1st stage
Q1 2 3 4 Q2N3904
Rc1 1 2 10k
Re1 4 0 870
Ce  4 0 1GF
Rb1 1 3 100k
Rb2 3 0 15k
* 2nd stage
Q2 50 2 6 Q2N3904
Rc2 1 5 8k
* decoupling capacitors
Cc1 10 3 1GF
Cc2 5 8 1GF
* load
Rl 8 0 1k
* open feedback circuit
* input side
Rfi 3 72 10k
Cfi 73 72 1GF
Re2i 73 0 3.4k
* output side
Rfo 71 0 10k
Cfo 6 71 1GF
Re2o 6 0 3.4k
* transistor model statement for 2N3904
.model Q2N3904  NPN (Is=6.734f Xti=3 Eg=1.11 Vaf=74.03 Bf=416.4 Ne=1.259
+                    Ise=6.734f Ikf=66.78m Xtb=1.5 Br=.7371 Nc=2 Isc=0 Ikr=0 Rc=1
+                    Cjc=3.638p Mjc=.3085 Vjc=.75 Fc=.5 Cje=4.493p Mje=.2593 Vje=.75
+                    Tr=239.5n Tf=301.2p Itf=.4 Vtf=4 Xtf=2 Rb=10)
** Analysis Requests **
.OP
.AC LIN 1 1Hz 1Hz
** Output Requests **
.PRINT AC Im(Vout) Ip(Vout) Vm(10) Vp(10)
.end
```

Figure 8.20 The Spice input file for calculating the input – output current gain and input resistance for the feedforward network shown in Fig. 8.19(a).

	Parameter	Spice
Feedforward Network	A	-244.9 A/A
	R_i	2.077 kΩ
	R_o	2.930 MΩ
Feedback Network	β	-0.2537 A/A
	R_{11}	13.40 kΩ
	R_{22}	2.537 kΩ

Table 8.12 Parameters of the feedforward and feedback network shown in Fig. 8.19 as calculated by Spice.

	Parameter	Direct Analysis Of Spice	Feedback Theory
Closed-Loop Amplifier	A_f	-3.853 A/A	-3.879 A/A
	R_{if}	32.65 Ω	32.90 Ω
	R_{of}	18.35 MΩ	184.97 MΩ

Table 8.13 Parameters of the feedback amplifier circuit shown in Fig. 8.18 as calculated by Spice and through the application of feedback theory.

The analysis request and print statement remain unchanged.

On completion of Spice, one finds the following results in the output file:

```
FREQ         IM(Vout)    IP(Vout)     VM(10)       VP(10)

1.000E+00    3.413E-07   -1.800E+02   1.464E-11    -9.469E+01
```

The output resistance of the feedforward network is thus found to be $R_o = 2.930$ MΩ.

The parameters associated with the feedback network shown in Fig. 8.18 are obvious and are summarized in Table 8.12. Also listed are the parameters associated with the feedforward circuit (calculated above).

Finally, in Table 8.13, we compare several parameters of the closed-loop amplifier as calculated directly by Spice with those computed using feedback theory. As is evident, the signal gain and the input resistance are very close. However, the output resistance R_{of} calculated by Spice is about an order of magnitude less than that predicted by feedback theory. The reason for this can be traced back to a very early assumption where the current of interest, the collector current of Q_2, was designated as the output current, where in fact, the emitter current of Q_2 is actually the current that is being sampled and fedback to mix with the input signal. Now, it is generally true that the collector and emitter currents of a transistor are very close owing to the transistor α being nearly unity, and what can be said

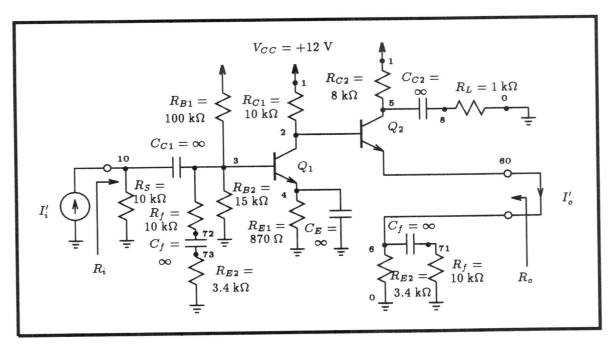

Figure 8.21 Feedforward gain stage of the feedback amplifier of Fig. 8.18 with the output port re-designated to be in series with the emitter of transistor Q_2 instead of in series with the collector of Q_2 as was the case previously.

about one should apply to the other, such as the current gain A_f. It is not true, however, that the theory of feedback networks derived in Sedra and Smith for computing the output resistance R_{of} applies to a port where the output signal is not the actual signal that is being sampled. This is certainly borne out by the results that we see in Table 8.13.

If we re-designate the output port as that which contains the emitter current of Q_2 then we should find that R_{if}, R_{of} and A_f will closely match those predicted by Spice. Consider the revised feedforward gain stage shown in Fig. 8.21. The input resistance, R_i is the same as that seen previously at 2.077 kΩ, the current gain A_i is almost the same as before at -246.6 A/A and the output resistance R_o is 2.639 kΩ. These results were all computed by Spice with an input file that is quite similar to that shown previously in Fig. 8.20 and is therefore not shown here.

Combining the open-loop parameters of the feedforward gain stage with the previously computed feedback network parameters, we obtain $A_f = -3.879$ A/A, $R_{if} = 32.68$ Ω and $R_{of} = 167.7$ kΩ. Comparing these with the results computed directly by Spice, ie. $A_f = -3.879$ A/A, $R_{if} = 32.65$ Ω and $R_{of} = 167.9$ kΩ, we see that all three values agree quite closely with those predicted by feedback theory. Furthermore, we see that the current gain from the input to the emitter current of Q_2 (-3.879 A/A) is exactly the same as the current gain from the input to the collector current of Q_2 (-3.879 A/A), as expected since

8 Feedback

g-parameter	A	β
g_{11}	4.817×10^{-4} A/V	7.463×10^{-5} A/V
g_{12}	9.934×10^{-7} A/A	2.537×10^{-1} A/A
g_{21}	$3.138 \times 10^{+2}$ V/V	2.537×10^{-1} V/V
g_{22}	$2.640 \times 10^{+3}$ V/A	$2.537 \times 10^{+3}$ V/A

Table 8.14 The g-parameters of the feedforward and feedback networks shown in Fig. 8.21 and Fig. 8.19(b), respectively.

transistor α is nearly unity.

To check the validity of the above results, the g-parameters of both the feedforward and feedback networks shown in Fig. 8.21 and Fig. 8.19(b), respectively, are shown listed in Table 8.14. Here we see that the forward transmission through the feedback network is about a 1000 times smaller than the forward gain of the feedforward gain stage, ie. $|g_{21}|_A \gg |g_{21}|_\beta$. Likewise, the reverse transmission through the feedback network is over five orders of magnitude larger than the transmission from the output to the input of the feedforward network, ie. $|g_{12}|_\beta \gg |g_{12}|_A$. Therefore, we are justified in separating the closed-loop amplifier shown in Fig. 8.18 into two separate feedforward and feedback networks. Finally, note that $R_{11} = 1/g_{11}|_\beta$ and $R_{22} = g_{22}|_\beta$.

8.3 Determining Loop Gain With Spice

An important parameter of feedback amplifiers is the loop gain $A\beta$. In this section we shall demonstrate a more direct approach to determining the loop gain using Spice than the network separation method employed in the previous section.

Consider determining the loop gain of the feedback amplifier circuit shown in Fig. 8.22. This is the same circuit presented in our previous example and the results found there can then be compared with the results obtained here. To determine the loop gain, the feedback loop of this amplifier must be broken in order to inject a signal into the loop. However, the circuit conditions of the loop must remain the same as when the loop was closed. This includes maintaining the same bias conditions on each transistor and terminating the feedback loop with the impedance seen by the loop under closed-loop conditions.

With regards to the circuit shown in Fig. 8.22, a convenient location for breaking the feedback loop is in the AC coupled feedback network. By doing so, the DC bias conditions of each transistor are not disturbed. However, in many circuit situations, the feedback network is directly coupled and one therefore requires a method that is amenable to such situations.

8.3 Determining Loop Gain With Spice

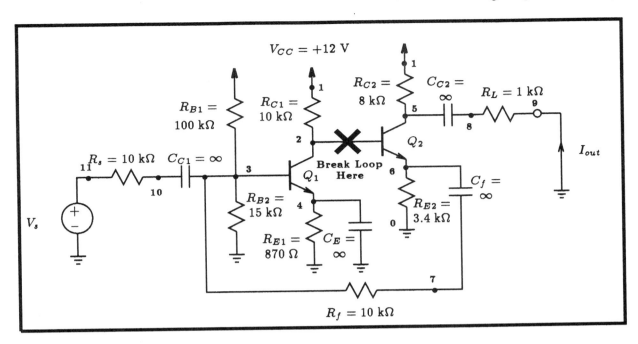

Figure 8.22 Breaking the feedback loop of an amplifier at a point in the network which affects its DC bias.

Consider breaking the feedback loop at a location which disturbs the DC biasing of the circuit. Such a location is highlighted by a big **X** in the circuit diagram of Fig. 8.22. By inserting a very large inductor in series with the terminals of the break, then the DC bias conditions of each transistor would remain invariant because the inductor behaves as a short circuit at DC. However, for AC signals, the large inductor presents a large impedance in series with the loop. For all intents and purposes, selecting a large inductor of say (for Spice purposes) 1 giga-henries opens the loop for almost all AC signals.

To determine the impedance required to terminate the loop, we simply calculate the impedance seen looking into the terminal where the loop was broken. For this particular example, we apply a one-volt AC voltage signal V_t to the base of Q_2 through a DC blocking capacitor as shown in Fig. 8.23(a), and calculate the current supplied by this source. The ratio V_t/I_t then gives the input impedance. Of course these results will be frequency dependent, so we limit our discussion to midband frequencies where the input impedance is purely resistive. In this particular example, a frequency of 1 Hz is considered to be midband. The DC blocking capacitor is necessary to prevent the input voltage source from disturbing the DC bias conditions on Q_2. Its value should be made large to minimize the voltage drop across it (for most purposes, a value of 1 giga-farads should suffice).

The Spice input file for determining the input impedance of the feedback loop is listed in Fig. 8.24. On completion of Spice, the following results are found in the Spice output file,

8 Feedback

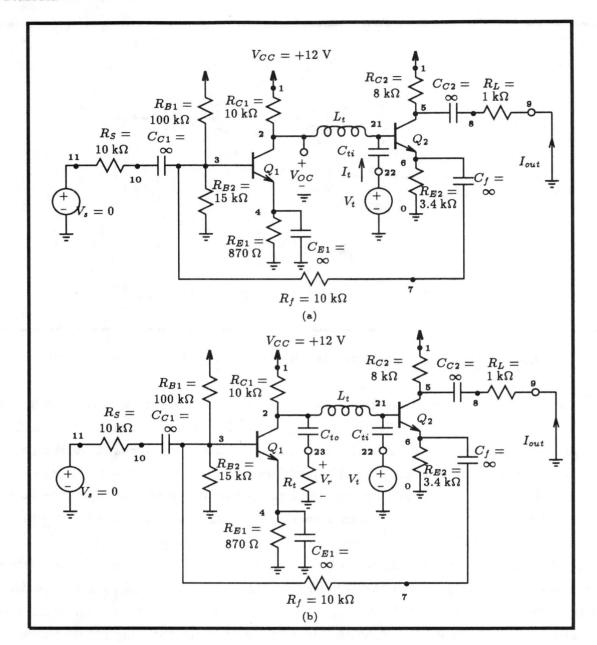

Figure 8.23 (a) Opening a feedback loop with a very large inductor L_t. The impedance seen by the feedback loop, and in which we must terminate the loop is found by calculating the current supplied by the AC coupled test voltage source V_t. (b) Feedback loop terminated in AC coupled load resistance R_t.

FREQ	VM(22)	VP(22)	IM(Vt)	IP(Vt)
1.000E+00	1.000E+00	0.000E+00	2.482E-06	1.800E+02

from which we can determine the input resistance to be 402.9 kΩ.

The output of the feedback loop can now be terminated in a resistance of 402.9 kΩ. A

8.3 Determining Loop Gain With Spice

```
Computing The Loop Terminating Resistance Rt

** Circuit Description **
* input source set to zero for output impedance calculation
Vs 11 0 DC 0 AC 0
Rs 11 10 10k
* short-circuit output port
Vout 0 9 0
* power supply
Vcc 1 0 DC +12V
* amplifier circuit
* 1st stage
Rc1 1 2 10k
Q1 2 3 4 Q2N3904
Re1 4 0 870
Ce1 4 0 1GF
Rb1 1 3 100k
Rb2 3 0 15k
* 2nd stage
Rc2 1 5 8k
Q2 5 21 6 Q2N3904
* decoupling capacitors
Cc1 10 3 1GF
Cc2 5 8 1GF
* load
Rl 8 9 1k
* feedback circuit
Re2 6 0 3.4k
Rf 3 7 10k
Cf 6 7 1GF
* inject signal into feedback loop without disturbing DC bias
Lt 2 21 1GH
Cti 21 22 1GF
Vt 22 0 AC 1V
* transistor model statement for 2N3904
.model Q2N3904  NPN (Is=6.734f Xti=3 Eg=1.11 Vaf=74.03 Bf=416.4 Ne=1.259
+                    Ise=6.734f Ikf=66.78m Xtb=1.5 Br=.7371 Nc=2 Isc=0 Ikr=0 Rc=1
+                    Cjc=3.638p Mjc=.3085 Vjc=.75 Fc=.5 Cje=4.493p Mje=.2593 Vje=.75
+                    Tr=239.5n Tf=301.2p Itf=.4 Vtf=4 Xtf=2 Rb=10)
** Analysis Requests **
.AC LIN 1 1Hz 1Hz
** Output Requests **
* print resistance seen by Vt: Rt=V(22)/I(Vt)
.PRINT AC Vm(22) Vp(22) Im(Vt) Ip(Vt)
.end
```

Figure 8.24 The Spice input file for calculating the input impedance $R_t = V(22)/I(Vt)$ of the feedback loop for the amplifier circuit shown in Fig. 8.23(a).

8 Feedback

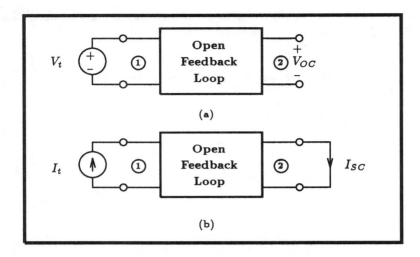

Figure 8.25 (a) Determining the open-circuit voltage transfer function $T_{OC} = V_{OC}/V_t$. (b) Determining the short-circuit current transfer function $T_{SC} = I_{SC}/I_t$.

DC blocking capacitor is necessary, once again, to prevent this resistance from disturbing the amplifier's DC biasing. This is illustrated in the circuit diagram shown in Fig. 8.23(b). Modifying the Spice input file given in Fig. 8.24 by attaching this resistance and capacitor combination to the output of the feedback loop allows us to calculate the loop gain. One simply adds the following two element statements:

```
Cto 2 23 1GF
Rt 23 0 402.9k
```

and revises the output print statement as follows

```
.PRINT AC Vm(22) Vp(22) Vm(23) Vp(23).
```

The results of this Spice calculation are then found in the output file as follows:

```
FREQ        VM(22)      VP(22)      VM(23)      VP(23)

1.000E+00   1.000E+00   0.000E+00   5.954E+01   1.800E+02
```

Hence, one finds the loop gain $A\beta = -V_r/V_t = 59.54$. Comparing this result with that obtained using the separation method performed in the last section, one finds $A\beta = (-244.9) * (-0.2537) = 62.13$. Thus both approaches result in reasonably close values for the loop gain.

An Alternative Method:

An alternative method for determining loop gain, and one that better lends itself to Spice, is due to Rosenstark [Rosenstark, 1986]. Without proof, the method of Rosenstark simply determines the loop gain by two independent calculations. The first involves calculating the voltage transfer function from the input of the broken loop to its output with the loop unterminated (ie. $R_t = \infty$). This we denote as the open-circuit transfer function, T_{OC}. The other calculation involves finding the current transfer function from the input of the loop to its short-circuited output. This we call the short-circuit transfer function, T_{SC}. We depict these two situations for an arbitrary broken feedback loop in Fig. 8.25. These two results are then combined to compute the loop transmission according to

$$A\beta = -T_{OC} \| T_{SC} = -\frac{1}{\dfrac{1}{T_{OC}} + \dfrac{1}{T_{SC}}}. \qquad (8.5)$$

It is obvious from this formula that the smaller of the two transfer functions will dominate the loop gain.

The advantage of this approach over the previous method of terminating the feedback loop is that both T_{OC} and T_{SC} can be calculated over a wide range of frequencies without worrying about whether the loop is properly terminated with the correct impedance.

To demonstrate this approach we shall return to the opened feedback loop of the last example shown in Fig. 8.23(a). Here the loop is shown unterminated and driven by a voltage source V_t. The Spice file for this particular situation was shown in Fig. 8.24 and can be used to calculate the open-circuit voltage V_{OC}. Adding the following print statement to this file,

.PRINT AC Vm(22) Vp(22) Vm(2) Vp(2),

and re-submitting this file to Spice will result in the following open-circuit voltage in the Spice output file:

```
    FREQ        VM(22)      VP(22)      VM(2)       VP(2)

    1.000E+00   1.000E+00   0.000E+00   6.086E+01   1.800E+02
```

Thus we obtain $T_{OC} = -60.86$.

To determine the short-circuit transfer function, we must remove the input voltage source V_t and replace it with an AC current source. In addition, we must also AC short-circuit the output port by adding a zero-valued voltage source in series with a large capacitor. The zero-valued voltage source provides the means of monitoring the short-circuit current. We depict this circuit arrangement in Fig. 8.26 and the corresponding Spice input file in Fig. 8.27. Submitting this file to Spice, results in the following short-circuit output current:

```
    FREQ        IM(Vsc)     IP(Vsc)

    1.000E+00   2.753E+03   1.800E+02
```

8 Feedback

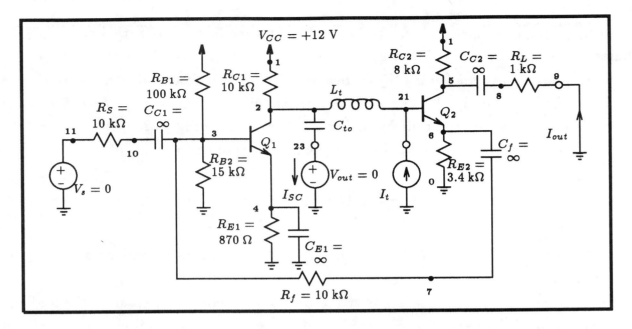

Figure 8.26 Determining the short-circuit current transfer function $T_{SC} = I_{SC}/I_t$.

Therefore we get $T_{SC} = -2753$.

Now from Eqn. (8.5), we calculate the loop gain to be 59.54 which is exactly the same as the loop gain calculated by the method which terminates the loop.

8.4 Stability Analysis Using Spice

From the loop gain, or more appropriately the loop transmission, one can determine the stability of a feedback amplifier. Consider the filter circuit shown in Fig. 8.28(a) where the feedback loop is frequency dependent and the amplifier has unity gain ($K = 1$). One possible point for breaking the loop is at the amplifier's input. This is probably the most convenient point in the loop because the loop is not loaded by the amplifier input and therefore the loop does not need to be terminated when it is opened. The loop transmission is then found by calculating the return voltage V_r, as indicated in Fig. 8.28(b), with a one-volt AC signal applied to the amplifier input. The sign of the input signal is made negative so that the loop transmission $A\beta$ directly equals V_r (instead of $-V_r$). A Spice description of the circuit shown in Fig. 8.28(b) is given in Fig. 8.29. Here we are requesting an AC analysis of the circuit beginning at 1 Hz and extending up to 1 MHz. This range of frequency should be large enough to capture the important behavior of the loop transmission.

On completion of Spice, a plot of both the magnitude and phase of the loop transmission will be found in the output file. These results, as drawn by the Probe facility of PSpice,

Computing The Short-Circuit Current Transfer Function

```
** Circuit Description **
* power supply
Vcc 1 0 DC +12V
* input source set to zero for output impedance calculation
Vs 11 0 DC 0 AC 0
Rs 11 10 10k
* amplifier circuit
* 1st stage
Rc1 1 2 10k
Q1 2 3 4 Q2N3904
Re1 4 0 870
Ce1 4 0 1GF
Rb1 1 3 100k
Rb2 3 0 15k
* 2nd stage
Rc2 1 5 8k
Q2 5 21 6 Q2N3904
* decoupling capacitors
Cc1 10 3 1GF
Cc2 5 8 1GF
* load
Rl 8 9 1k
* output current
Vout 0 9 DC 0
* feedback circuit
Re2 6 0 3.4k
Rf 3 7 10k
Cf 6 7 1GF
* inject signal into feedback loop
Lt 2 21 1GH
It 0 21 AC 1A
Cto 2 23 1GF
Vsc 23 0 0
* transistor model statement for 2N3904
.model Q2N3904   NPN (Is=6.734f Xti=3 Eg=1.11 Vaf=74.03 Bf=416.4 Ne=1.259
+                Ise=6.734f Ikf=66.78m Xtb=1.5 Br=.7371 Nc=2 Isc=0 Ikr=0 Rc=1
+                Cjc=3.638p Mjc=.3085 Vjc=.75 Fc=.5 Cje=4.493p Mje=.2593 Vje=.75
+                Tr=239.5n Tf=301.2p Itf=.4 Vtf=4 Xtf=2 Rb=10)
** Analysis Requests **
.AC LIN 1 1Hz 1Hz
** Output Requests **
* print short-circuit current gain
.PRINT AC Im(Vsc) Ip(Vsc)
.end
```

Figure 8.27 The Spice input file for calculating the short-circuit transfer function $T_{SC} = I(V_{SC})/I_t$ in the opened feedback loop shown in Fig. 8.26.

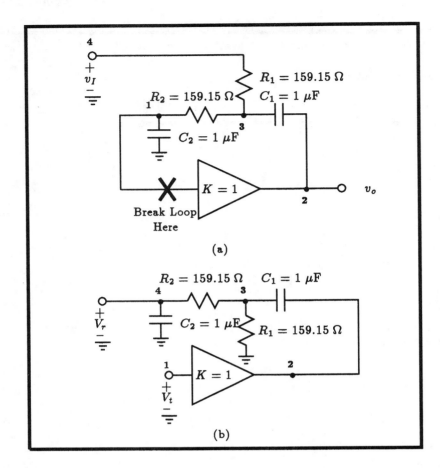

Figure 8.28 (a) A 2nd-order filter circuit with a frequency-dependent feedback loop. (b) Opened feedback loop for calculating loop transmission.

are on display in Fig. 8.30(a). Reviewing these results, we see that the gain margin is $\simeq 10$ dB, suggesting that the filter circuit will be stable when the loop is closed. Equivalently, we can use these results to draw a Nyquist plot and show that the critical point (−1,0) is not encircled. This we have done in Fig. 8.30(b), thus confirming our intuition.

As an example of an unstable filter circuit, consider increasing the gain of the amplifier in the filter circuit of Fig. 8.28(a) to 5 (ie. $K = 5$). Modifying the Spice input file listed in Fig. 8.29 to reflect this change and re-running this file, will result in the new Bode plot shown in Fig. 8.31(a). Here we observe that at the 180°-phase frequency, the loop gain will be greater than unity, indicating unstable performance. Alternatively, a Nyquist plot can also be drawn using this same frequency information. The result is shown in Fig. 8.31(b). Here we see that the critical point (−1,0) is encircled. Also confirming that the filter circuit will be unstable when the loop is closed.

Before moving on, it would be highly instructive to investigate the closed-loop transient behavior of the two filter circuits discussed above, ie. one stable and the other unstable. For

8.4 Stability Analysis Using Spice

Example 8.5: Calculating Loop Transmission Of 2nd Order Active Filter (K=1)

```
** Circuit Description **
* input signal source
Vt 1 0 AC 1V 180degrees
* filter circuit
Egain 2 0 1 0 1
Inull 1 0 0 ; redundant connection
* feedback network
C1 2 3 1uF
R1 3 0 159.15
R2 3 4 159.15
C2 4 0 1uF
** Analysis Requests
.AC DEC 10 1Hz 1MegHz
** Output Requests **
.PLOT AC VdB(4) Vp(4)
.PROBE
.end
```

Figure 8.29 The Spice input file for calculating the frequency-dependent loop transmission of the filter circuit shown in Fig. 8.28.

the original filter circuit seen in Fig. 8.28(a) with $K = 1$, a Spice input file is provided in Fig. 8.32. A one-volt peak amplitude sinewave signal of 1 kHz frequency is applied to the input terminal of the filter using the following Spice transient voltage source statement:

$$\text{Vi 4 0 sin(0V 1V 1kHz)}.$$

A transient analysis is requested to be performed over 5 periods of the input signal (ie. 5 ms) with 100 points collected per period of the input signal. The Spice command appears in the input file as follows:

$$\text{.TRAN 10us 5ms 0ms 10us}.$$

Finally, a .PLOT statement is used to observe both the input and output voltage signal:

$$\text{.PLOT TRAN V(2) V(4)}.$$

Another Spice input file is created for the unstable filter circuit, ie. $K = 5$. This Spice file is almost identical to that seen in Fig. 8.32 with the simple modification of changing the amplifier gain to 5. This Spice file is concatenated on the end of the Spice file seen in Fig. 8.32 and then submitted as one file to Spice for processing. On completion of Spice, the transient response of each filter circuit are shown in Fig. 8.33. The top curve illustrates the sinewave signal applied to the input of each filter circuit. The middle waveform represents the signal that appears at the output of the stable filter circuit. As is evident, except for an amplitude and phase shift, this signal is identical to the input signal shown above it. In

397

8 Feedback

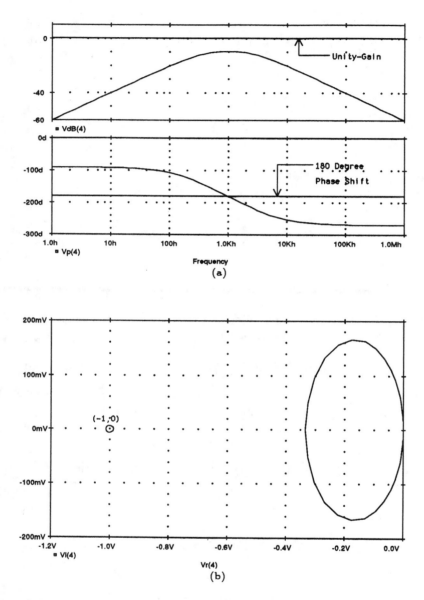

Figure 8.30 Various forms of representing the loop transmission: (a) Bode plot (b) Nyquist plot.

contrast, the signal appearing at the output of the unstable filter in the bottom-most graph bears no resemblance to the input sinewave signal at all. Moreover, the level of this output signal has approached 10^{10} V in about 3 ms. Clearly, this is unstable behavior. If Spice were left running much longer than 3 ms, numbers stored within the computer would get so large that they would overflow the internal registers of the computer. This then causes Spice to terminate the present Spice job.

8.4 Stability Analysis Using Spice

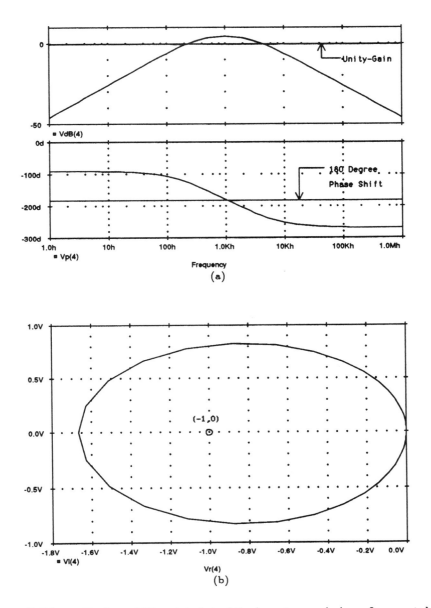

Figure 8.31 A Bode and Nyquist plot of the loop transmission of an unstable filter ($K = 5$). Here the Nyquist plot encircles the critical point of (−1,0).

As a final point, novice users of Spice sometimes think that if Spice can produce a frequency response plot of a closed-loop circuit, then that circuit must be stable. This is certainly not true. To demonstrate this, in Fig. 8.34 the magnitude and phase response of both the stable and the unstable closed-loop filter circuits are shown. These results were computed using a revised version of the Spice deck shown in Fig. 8.32. The input signal source was replaced by an AC signal source having the following form:

399

8 Feedback

Closed-Loop Transient Response Of Stable Filter Circuit (K=1)

** Circuit Description **

* input signal source
Vi 4 0 sin(0V 1V 1kHz)

* filter circuit
Egain 2 0 1 0 1
* feedback stage
C1 2 3 1uF
R1 3 4 159.15
R2 3 1 159.15
C2 1 0 1uF

** Analysis Requests
.TRAN 10us 5ms 0ms 10us
** Output Requests **
.PLOT TRAN V(2) V(4)
.probe
.end

Figure 8.32 The Spice input file for calculating the transient response of the filter circuit shown in Fig. 8.28(a) to a one-volt 1 kHz sinewave input.

Vi 4 0 AC 1V,

and the transient analysis command seen there was replaced with the following AC analysis command:

.AC DEC 10 1Hz 1MegHz.

The following .PLOT command was also added:

.PLOT AC VdB(2).

Returning to Fig. 8.34, we see that both circuits, regardless of stability, have a corresponding magnitude and phase response. The magnitude response behavior of each filter is quite similar, except for a vertical shift as a result of the different amplifier gains. The greatest difference in the behavior of the two filters appears in their phase. The stable filter has a phase response that goes from 0 degrees to −180 degrees as the input signal frequency increases from DC. In contrast, the phase response of the unstable filter goes from 0 degrees to +180 degrees with increasing input signal frequency. The reason for this behavior should be obvious; in the first case, the poles appear on the lefthand side of the $j\omega$ axis in the

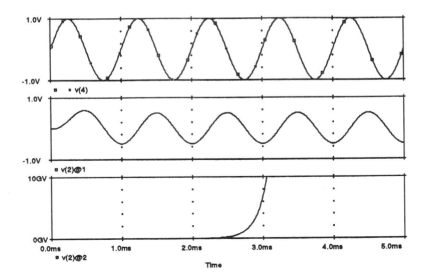

Figure 8.33 The transient response to a one-volt 1 kHz sinewave signal applied to the input of the filter circuit shown in Fig. 8.28(a) with $K = 1$ and $K = 5$. The top curve represents the input sinewave signal; the middle curve represents the output signal from the stable filter circuit ($K = 1$); the bottom curve represents the output signal from the unstable filter circuit ($K = 5$).

s-plane, and in the latter, they appear on the righthand side. From the location of the poles, the stability of the network can be easily deduced. Unfortunately, in more complicated circuits having many more poles and zeros, the phase behavior of the amplifier would not so clearly identify poles in the righthand side of the s-plane. In general, without any special insight into circuit behavior, the stability of a circuit cannot be determined from the frequency response of the closed-loop circuit. It is therefore recommended that the stability of a closed-loop circuit be investigated using a Spice transient analysis (preferably, using a step input signal that is rich in a wide-band of signal frequencies, as opposed to a single sinewave input signal as used in this example).

8.5 Investigating The Range Of Amplifier Stability

In this section we shall investigate the range of stability of an amplifier with complex frequency response behavior using Bode plots generated by Spice. The same amplifier characteristics that were used in Section 8.10 of Sedra and Smith will be used here. This will enable our readers to correlate the results given here with those calculated by hand. Furthermore, a Spice transient analysis will be used to verify circuit stability.

As an example, consider the differential-input amplifier shown in Fig. 8.35(a) with an

8 Feedback

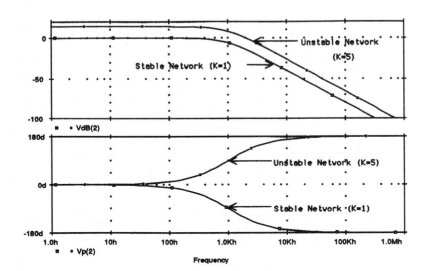

Figure 8.34 Comparing the magnitude and phase response of a stable and an unstable filter circuits. The top graph displays the magnitude responses and the bottom graph displays the corresponding phase responses.

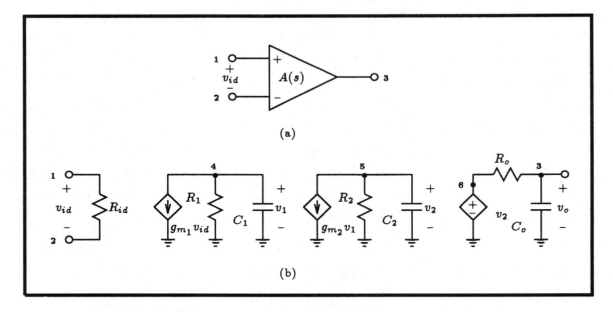

Figure 8.35 (a) A differential-input amplifier having an open-loop response $A(s)$ with three separate poles. (b) Small-signal equivalent circuit.

input–output transfer function $A(s)$. Let us assume that $A(s)$ is characterized by three real poles distributed along the negative axis of the s-plane at frequency locations of: 0.1 MHz, 1 MHz and 10 MHz. Further, let the DC gain of the amplifier be 10^5 V/V or 100 dB. The open-loop transfer function $A(s)$ can then be described by the following:

$$A(s) = \frac{10^5}{(s/0.2\pi \times 10^6 + 1)(s/2\pi \times 10^6 + 1)(s/20\pi \times 10^6 + 1)} \quad (8.6)$$

Figure 8.35(b) shows a small-signal equivalent circuit of a typical three stage amplifier configuration. It consists of two transconductance gain stages and an output buffer stage. With the following component values, the transfer function of this small-signal equivalent circuit is described by that given in Eqn. 8.6:

1st Stage:
$$R_{id} = 1 \text{ M}\Omega, \; g_{m_1} = 4.93 \text{ mA/V}, \; R_1 = 15.915 \text{ k}\Omega, \; C_1 = 100 \text{ pF}$$

2nd Stage:
$$g_{m_2} = 40 \text{ mA/V}, \; R_2 = 31.83 \text{ k}\Omega, \; C_2 = 5 \text{ pF}$$

3rd Stage:
$$R_o = 100 \; \Omega, \; C_o = 159.15 \text{ pF}$$

To observe the frequency response behavior of this amplifier, we use the Spice input file shown listed in Fig. 8.36. The amplifier input is excited with a one-volt AC signal applied to the positive input terminal. The negative input terminal is simply set to 0 V. An AC analysis is then requested between 1 Hz and 100 MHz. Both the magnitude and phase of the signal appearing at the amplifier output are to be plotted. Since the input signal is set at one-volt, the magnitude and phase of this output signal also represent the magnitude and phase of the input–output transfer function $A(s)$.

The results of the Spice analysis have been collected and plotted in Fig. 8.37. Here we see that the gain of the amplifier is 100 dB at low frequencies and has a 3 dB frequency located quite close to 100 kHz. Near the unity-gain frequency of 46 MHz, the slope of the magnitude response is seen to be about −60 dB-per-decade. The phase response of this amplifier is provided in the plot shown below the magnitude response. The majority of the phase shift occurs between 10 kHz and 100 MHz; one decade below the location of the first pole and one decade above the highest frequency pole. The frequency at which the phase shift equals 180° is 3.31 MHz; this was found using the cursor facility of Probe.

For a unity feedback factor independent of frequency, ie. $\beta = 1$, the loop gain $A\beta$ would have a gain margin of $GM = -58.4$ dB, or alternatively, a phase margin of $PM = -76.5°$. Clearly, connecting this amplifier in a unity-gain configuration would result in unstable operation. In fact, it is not until the loop gain drops below 41.6 dB (eg. $A_o - GM = 100$ dB - 58.4 dB), would the closed-loop amplifier configuration result in stable operation. This simply corresponds to the point at which the gain margin reduces to zero for a feedback factor $20\log(1/\beta) = 58.4$ dB, or $\beta = 1.202 \times 10^{-3}$. Thus, stable circuit operation using the

8 Feedback

```
Open-Loop Amplifier Characteristics

** Circuit Description **

* input signals

Vin+ 1 0 AC 1V
Vin- 2 0 AC 0V

* open-loop amplifier configuration
* connections: 1 2 3
*              | | |
*          in+ | |
*            in- |
*              out

* first stage
Rid 1 2 1MegOhm
Gm1 4 0 1 2 4.93m
R1 4 0 15.915k
C1 4 0 100pF
* second stage
Gm2 5 0 4 0 40m
R2 5 0 31.83k
C2 5 0 5pF
* output buffer stage
E3 6 0 5 0 1
Ro 6 3 100
Co 3 0 159.15pF

** Analysis Requests
.AC DEC 10 1Hz 100MegHz
** Output Requests **
.PLOT AC VdB(3) Vp(3)
.PROBE
.end
```

Figure 8.36 The Spice input file for calculating the open-loop frequency response behavior of the three-pole amplifier circuit shown in Fig. 8.35.

amplifier described above is maintained in a closed-loop configuration for frequency independent feedback factors β being less than 1.202×10^{-3}. Correspondingly, since the closed-loop gain under large loop gain conditions, is approximately $1/\beta$, the amplifier described above is limited to applications requiring a closed-loop gain greater than 831.82 V/V or 58.4 dB.

To verify the above result, let us investigate the step response of the noninverting amplifier configuration shown in Fig. 8.38 having a closed-loop gain of +2 V/V and +1001 V/V. The differential amplifier will be assumed to be modeled after the small-signal equivalent

8.5 Investigating The Range Of Amplifier Stability

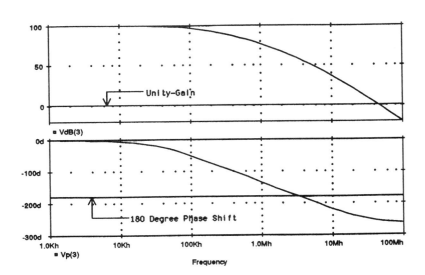

Figure 8.37 Magnitude and phase response of the differential-input amplifier shown in Fig. 8.35 with $A(s) = \dfrac{10^5}{(s/0.2\pi \times 10^6 + 1)(s/2\pi \times 10^6 + 1)(s/20\pi \times 10^6 + 1)}$.

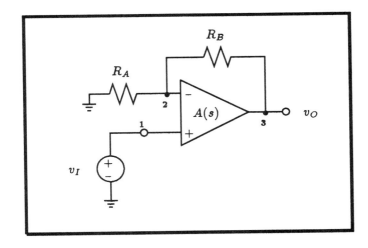

Figure 8.38 A noninverting amplifier configuration with closed-loop gain $A_f = (1 + R_B/R_A)$. The differential-input amplifier has transfer function $A(s)$ defined by Eqn. 8.6 and $\beta = R_A/(R_A + R_B)$.

circuit shown in Fig. 8.35(b). In the first case, we shall select R_A to be 100 kΩ, and in the second case, R_A will be assigned a value of 100 Ω. For both cases, R_B will be set equal to 100 kΩ. The Spice deck for the amplifier having a gain of +2 V/V is shown listed in Fig. 8.39. The Spice deck for the other amplifier is identical except for the change in the resistor value of R_A. A 1 mV step input is applied to the input of the closed-loop amplifier. Hence,

8 Feedback

```
Step Response Of A Noninverting Amplifier With A Gain Of +2 V/V

** Subcircuits **

.subckt diffopamp 1 2 3
* open-loop amplifier configuration
* connections: 1 2 3
*              | | |
*           in+ | |
*             in- |
*                out
* first stage
Rid 1 2 1MegOhm
Gm1 4 0 1 2 4.93m
R1 4 0 15.915k
C1 4 0 100pF
* second stage
Gm2 5 0 4 0 40m
R2 5 0 31.83k
C2 5 0 5pF
* output buffer stage
E3 6 0 5 0 1
Ro 6 3 100
Co 3 0 159.15pF
.ends diffopamp

** Main Circuit **

* one-volt step input signal
Vstep 1 0 PWL (0 0V 1ns 0V 2ns 1mV 1s 1mV)
* noninverting amplifier
Xdiffamp 1 2 3 diffopamp
RA 2 0 100k
RB 2 3 100k

** Analysis Requests
.TRAN 100ns 12us 0s 100ns
** Output Requests **
.PLOT TRAN V(3) V(1)
.PROBE
.end
```

Figure 8.39 The Spice input file for calculating the step response of a noninverting amplifier having a closed-loop gain of $+2$ V/V. The differential-input amplifier is modeled by the small-signal equivalent circuit shown in 8.35(b).

we should expect a 2 mV output signal if the amplifier is stable.

The step input signal is approximated by a series of piece-wise linear segments described to Spice using the PWL source statement. It appears in the Spice input file as follows:

8.5 Investigating The Range Of Amplifier Stability

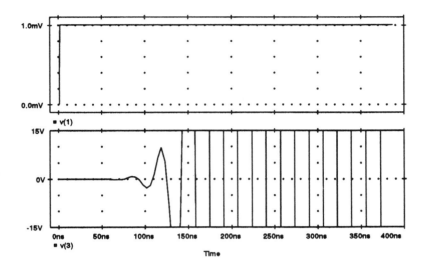

Figure 8.40 The 1 mV step response of a noninverting amplifier having a closed-loop gain of +2 V/V. The top graph displays the 1 mV step input. The bottom graph displays the corresponding output signal from the amplifier. Clearly, this output signal bears no resemblance to the input signal and indicates unstable behavior.

```
Vstep 1 0 PWL ( 0 0V 1ns 0V 2ns 1mV 1s 1mV)
```

Here the input voltage signal is held low for 1 ns and then made to rise to 1 mV with a rise-time of 1 ns, and then held at 1 mV for one complete second. A transient analysis is requested to be performed over a 12 μs interval with a point collected every 100 ns. This should provide sufficient time resolution of the output signal to see most of its important transient behavior.

On completion of the Spice analysis, the step response of the noninverting amplifier with a gain of +2 V/V is shown plotted in Fig. 8.40. The top curve displays the 1 mV step input and the graph below it illustrates the corresponding output signal. Clearly, the output signal bears no resemblance to the input signal. Moreover, the output signal is seen to oscillate between +15 V and −15 V. This, therefore, confirms that the closed-loop amplifier configuration is unstable as was predicted by the Bode analysis performed above.

If we repeat the above step analysis on the noninverting amplifier having a gain of +1001 V/V, then, according to the Bode analysis, our amplifier should be stable. This implies that with a 1 mV step input signal, the output signal should eventually, on the completion of its transient response, settle to an output voltage level of 1 V. The results of the Spice analysis shown in Fig. 8.41 confirm that this is the case. Thus, this closed-loop amplifier configuration

8 Feedback

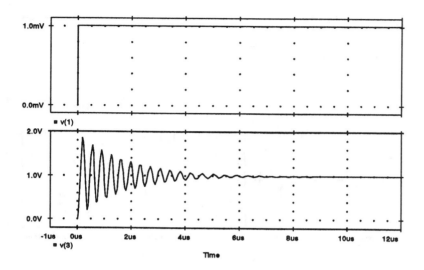

Figure 8.41 The 1 mV step response of a noninverting amplifier having a closed-loop gain of +1001 V/V. The top graph displays the 1 mV step input. The bottom graph displays the corresponding amplifier output signal. Here the output signal settles to a 1 V output level as expected for stable amplifier operation.

is indeed stable. The excessive ringing seen in the transient response is indicative of a closed-loop amplifier having a low phase margin. For this particular amplifier (ie. $A_f = +1001$) the phase margin is only about $PM = 4.3°$. Phase margins greater than 45° are usually desired in feedback amplifier design to obtain reasonable settling times and avoid excessive ringing.

8.6 The Effect Of Phase Margin On Transient Response

In this section we shall investigate the step-response of the noninverting amplifier shown in Fig. 8.38 for a wide range of phase margins. The open-loop frequency response behavior of the differential-input amplifier will be identical to that given previously in Eqn. 8.6. The phase margin of this noninverting amplifier can be altered by simply changing its feedback factor β. For instance, for a phase margin varying between 10° to 100° in steps of 10°, we list the corresponding values of the two feedback resistors R_A and R_B in Table 8.15. The values of these two resistors were obtained indirectly from the frequency response behavior of the open-loop amplifier shown in Fig. 8.37. First, the value of $20\log(1/\beta)$ that corresponds to the desired phase margin was found (listed in the middle column of Table 8.15), then, using the expression $\beta = R_A/(R_A + R_B)$, values for R_A and R_B were found.

8.6 The Effect Of Phase Margin On Transient Response

Phase Margins	$20\log(1/\beta)$	Feedback Resistors	
		R_A	R_B
10°	61.8 dB	81.5 Ω	100 kΩ
20°	66.3 dB	48.5 Ω	100 kΩ
30°	70.5 dB	29.8 Ω	100 kΩ
40°	74.2 dB	19.4 Ω	100 kΩ
50°	77.4 dB	13.4 Ω	100 kΩ
60°	80.6 dB	9.4 Ω	100 kΩ
70°	83.4 dB	6.8 Ω	100 kΩ
80°	86.1 dB	4.9 Ω	100 kΩ
90°	88.7 dB	3.7 Ω	100 kΩ
100°	91.1 dB	2.8 Ω	100 kΩ

Table 8.15 Feedback resistors R_A and R_B of the noninverting amplifier shown in Fig. 8.38 for different amplifier phase margins.

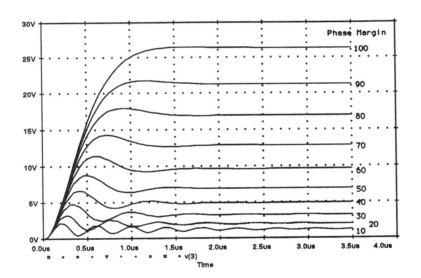

Figure 8.42 Transient response of closed-loop amplifier having a wide range of phase margins. The larger the phase margin, the less the ringing in the output signal. A voltage step signal of 1 mV is applied to the input of each amplifier.

Using the different values for R_A and R_B provided in Table 8.15, together with the Spice deck shown previously in Fig. 8.39, we can change the value of RA and RB in the Spice deck accordingly, and compute the step response of the closed-loop amplifier for different phase margins. In order to compare the different step responses, we shall concatenate all ten files corresponding to the different phase margins into one file and submit it to Spice for analysis.

8 Feedback

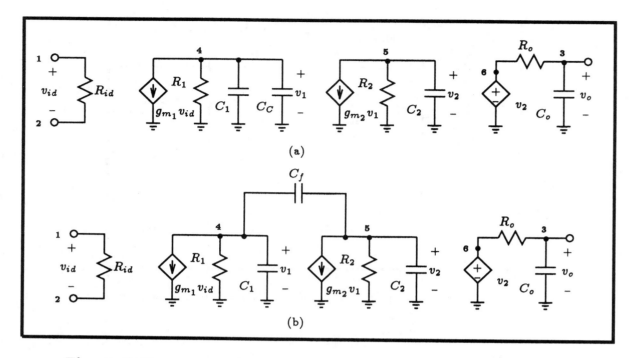

Figure 8.43 Two different versions of a dominate-pole compensation: (a) adding a shunt capacitor C_C across the output of the first stage; (b) introducing a pole-splitting compensation capacitor C_f between the first and second stage.

The step response of the closed-loop amplifier for different phase margins are shown in Fig. 8.42. Clearly, the larger the phase margin, the lower the amount of ringing that occurs in the output signal.

8.7 Frequency Compensation

In this section we shall demonstrate using Spice how an amplifier whose closed-loop behavior is unstable can be made stable by modifying the open-loop transfer function of the amplifier. This technique is referred to as frequency compensation. There are many ways to perform frequency compensation. Here we shall address only the dominant-pole method, but in two different forms: the shunt-capacitor and the pole-splitting techniques.

Consider the differential-input amplifier used in the two previous sections (Fig. 8.35). It was found there that this amplifier was stable only when connected in closed-loop configurations using resistive feedback with gains greater than 58.4 dB, or conversely, with a feedback factor β less than 1.202×10^{-3}. Let us consider modifying the open-loop frequency response behavior of this amplifier so that it has a positive gain or phase margin. The amplifier could then be used in closed-loop configurations with frequency-independent feedback factors as large as unity. Thus, the resulting differential-input amplifier is considered to be *unity-gain stable.*

The first frequency compensation method that we shall consider is the shunt-capacitor method. Consider placing a 1 μF capacitor C_C across the output terminal of the first internal stage of the differential-input amplifier as seen in Fig. 8.43(a). Adding this shunt capacitor will alter the pole originally formed by R_1 and C_1 (located at 0.1 MHz) and move it down in frequency to 10 Hz (ie. $f'_D = 1 \big/ 2\pi (C_1 + C_C) R_1$). The remaining poles formed by the other resistors and capacitors will remain unchanged. Alternatively, the pole-splitting technique requires that a bridging capacitor C_f be connected between the output of the first and second stage of the amplifier as shown in Fig. 8.43(b). A bridging capacitor of 78.5 pF was found in Example 8.6 of Sedra and Smith to move the first pole to a frequency location of 100 Hz (ie. $f'_{P1} \simeq 1 \big/ 2\pi g_{m_2} R_2 C_f R_1$) and the second pole to about 57 MHz ($f'_{P2} \simeq g_{m_2} \big/ 2\pi (C_1 + C_2)$).

To see the effect of the addition of these two compensation capacitors, we simulated the behavior of these two circuits using Spice. The Spice decks used to perform this are identical to that seen previously in Fig. 8.36 with the addition of the appropriate compensation capacitor. In the case of the circuit shown in Fig. 8.35(a), we add the element statement:

```
* compensation capacitor
Cc 4 0 1uF
```

and, the case of the circuit shown in Fig. 8.35(b), we added the statement:

```
* compensation capacitor
Cf 4 5 78.5pF
```

The Spice decks of these amplifiers were combined into one file, together with the Spice deck for the uncompensated amplifier, and then submitted to Spice for processing. The magnitude and phase results are shown plotted in Fig. 8.44.

In Fig. 8.44 we see that the two compensated amplifiers have very different frequency response behavior than the original uncompensated amplifier. The frequency response for each of the two compensated amplifiers is dominated by a low frequency pole. Furthermore, the second pole of each of these amplifiers occurs very near the unity-gain frequency. In the case of the shunt-capacitor compensated amplifier, it has its first break frequency at 10 Hz and a unity-gain frequency of about 800 kHz. The phase margin associated with this amplifier is $PM = 47.7°$, the gain margin is $GM = +20.8$ dB. In the case of the pole-splitting compensated amplifier, its first break frequency occurs at 100 Hz and it has a unity-gain frequency of about 8 MHz. The phase margin for this amplifier is $PM = 38.1°$ and its gain margin is $GM = +11.7$ dB. Since the gain and phase margins of each amplifier are positive, they are both unity-gain stable.

8 Feedback

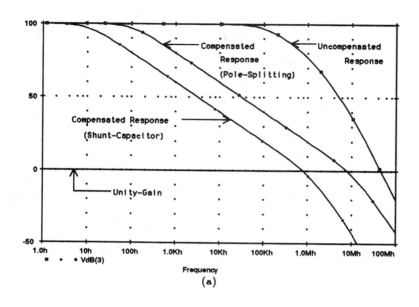

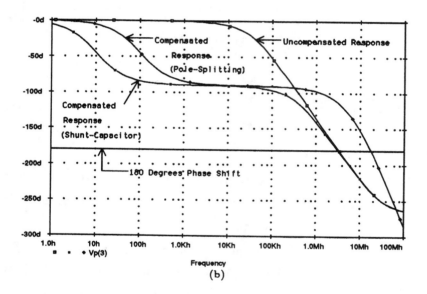

Figure 8.44 Comparing the open-loop frequency response behavior of an amplifier compensated using a shunt-capacitor, and another compensated by the pole-splitting technique. The open-loop frequency response behavior of the uncompensated amplifier is also shown. (a) Magnitude response. (b) Phase response.

To further convince one-self that the amplifiers are indeed unity-gain stable, consider computing the step response of each compensated amplifier connected in the unity-gain configuration shown in Fig. 8.45. The Spice input file for the shunt-capacitor compensated amplifier in this circuit setup is provided in Fig. 8.46. The Spice deck for the pole-splitting compensated amplifier connected in a unity-gain configuration is almost identical to that

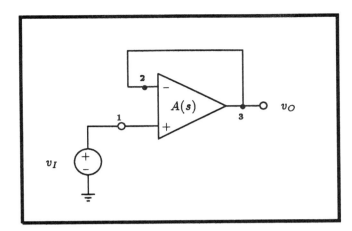

Figure 8.45 Unity-gain configuration. A zero-valued voltage source is used to create the short-circuit connection between the negative input terminal of the amplifier and its output.

given in Fig. 8.46. The element statement for the compensation capacitor C_f should appear in the Spice deck as indicated above and the transient analysis request statement should read as follows:

.TRAN 10ns 500ns 0s 10ns.

The results of this transient analysis are shown in Figs. 8.47 and 8.48. Both the input and the output signals are shown plotted in each graph. Clearly, both responses indicate stable behavior. We also notice that the settling time of the of the pole-splitting compensated amplifier is much shorter than that of the shunt-capacitor compensated amplifier. (Note the time scale on the x-axis of each graph). This result is not too surprising given that the 3 dB bandwidth of the pole-splitting compensated amplifier is about ten times larger than the other.

8.8 Spice Tips

- Separating a feedback amplifier into a feedforward circuit and a feedback circuit can cause transistor bias conditions to be disturbed. The correct bias conditions can be re-established for each transistor using two additional DC sources: a current source to set the proper base current and a voltage source to set the collector voltage. Two additional large-valued reactive elements are used to eliminate the effect of these dc sources on the small-signal equivalent circuit.
- When a feedback loop is opened for investigative purposes (loop gain, stability, etc.), the DC conditions of the loop must remain unchanged. This can be accomplished using

8 Feedback

```
Step Response Of A Unity-Gain Amplifier Compensated With A Shunt-Capacitor

** Subcircuits **

.subckt diffopamp 1 2 3
* open-loop amplifier configuration
* connections: 1 2 3
*              | | |
*           in+ | |
*             in- |
*                out

* first stage
Rid 1 2 1MegOhm
Gm1 4 0 1 2 4.93m
R1 4 0 15.915k
C1 4 0 100pF
* second stage
Gm2 5 0 4 0 40m
R2 5 0 31.83k
C2 5 0 5pF
* output buffer stage
E3 6 0 5 0 1
Ro 6 3 100
Co 3 0 159.15pF
* compensation capacitor
Cc 4 0 1uF
.ends diffopamp

** Main Circuit **

* one-volt step input signal
Vstep 1 0 PWL (0 0V 1ns 0V 2ns 1V 1s 1V)
* noninverting amplifier
Xdiffamp 1 2 3 diffopamp
Vshort 2 3 0

** Analysis Requests
.TRAN 100ns 5us 0s 100ns
** Output Requests **
.PLOT TRAN V(3) V(1)
.end
```

Figure 8.46 The Spice input file for calculating the step response of the shunt-capacitor compensated amplifier shown in Fig. 8.43(a) connected in a unity-gain configuration.

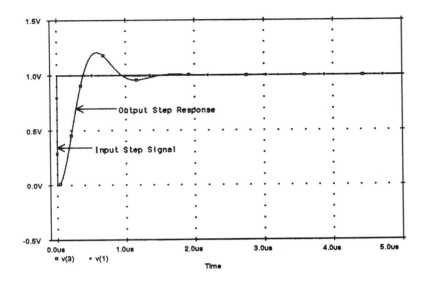

Figure 8.47 A one-volt step response of a shunt-capacitor compensated amplifier connected in a unity-gain configuration. Both the input and output signals are shown. Notice that the time scale ranges between 0 and 5 μs.

a large-valued inductor in series with the feedback loop and a large-valued capacitor in series with the signal source that is used to inject the signal into the loop.

- Loop transmission $A\beta(s)$ in a single-loop feedback amplifier can be determine from the open-circuit and short-circuit transfer functions ($T_{OC}(s)$ and $T_{SC}(s)$, respectively) according to the equation:

$$A\beta = -T_{OC} \| T_{SC} = -\frac{1}{\frac{1}{T_{OC}} + \frac{1}{T_{SC}}}.$$

- The stability of a closed-loop linear circuit can be determined from the magnitude and phase behavior of the loop transmission. Spice can compute and display both these quantities in the form of a Bode plot. Using the Probe facility of PSpice, one can produce either a Bode or a Nyquist plot of this information.

- The stability of a closed-loop circuit can not be determined from the frequency response behavior of the closed-loop circuit. Instead, the closed-loop stability can be observed from the output behavior of the closed-loop circuit when subjected to a fast-rising or falling step input using the transient analysis capability of Spice.

8 Feedback

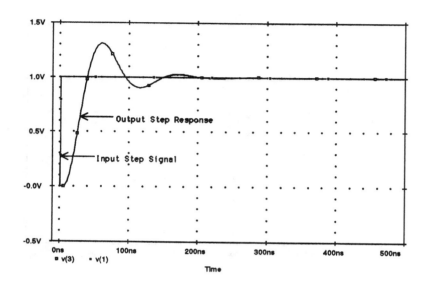

Figure 8.48 A one-volt step response of a pole-splitting capacitor compensated amplifier connected in a unity-gain configuration. Both the input and output signals are shown. Notice that the time scale ranges between 0 and 500 ns.

8.9 Bibliography

S. Rosenstark, *Feedback Amplifier Principles*, New York: Macmillan Publishing Company, 1986.

8.10 Problems

8.1 A series-series feedback amplifier employs a transconductance amplifier having $G_m = 100$ mA/V, input resistance of 10 kΩ, and output resistance of 100 kΩ. The feedback network has $\beta = 0.1$ V/mA, an input resistance (with the port 1 open circuited) of 100 Ω, and an output resistance (with port 2 open circuited) of 10 kΩ. The amplifier operates with a signal source having a resistance of 10 kΩ and with a load of 10 kΩ resistance. With the aid of Spice, compute A_f, R'_{if} and R'_{of} using the feedback method. Check your answers with those obtained directly from simulation.

8.2 For the circuit in Fig. P8.2, $|V_t| = 1$ V, $\mu_n C_{OX} = 1.0$ mA/V^2, $h_{fe} = 100$, and the Early voltage magnitude for all devices (including those that implement the current sources) is 100 V. The signal source V_s has a zero dc component. Using the feedback method, together with Spice, determine the values of A, β, A_f, R'_{if}, and R'_{of}. Compare the closed-loop values obtained with those derived directly from the network using Spice.

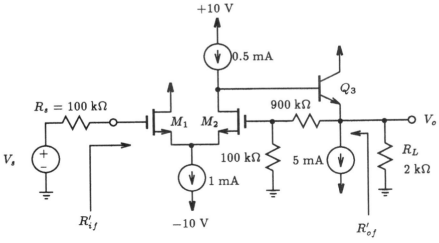

Fig. P8.2

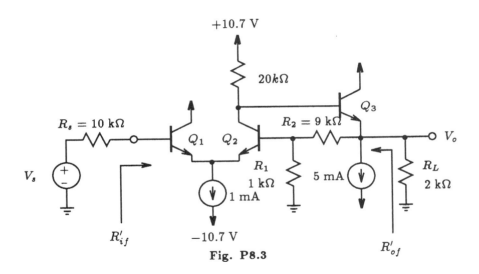

Fig. P8.3

8.3 The circuit shown in Fig. P8.3 consists of a differential stage followed by an emitter follower, with series-shunt feedback supplied by the resistors R_1 and R_2. Assuming that the dc component of V_s is zero, determine the dc operating point of this circuit. Subsequently, using the feedback method in conjunction with Spice, determine the values of A, β, $A_f = V_o/V_s$, R'_{if} and R'_{of}. Assume that the transistors have $\beta = 100$. Check your closed-loop results with those derived directly from the network using Spice.

8.4 The circuit of Fig. P8.3 is modified by replacing R_2 with a short circuit. Using Spice, determine the values of A, β, A_f, R'_{if} and R'_{of}.

8.5 The circuit in Fig. P8.3 is modified by replacing the 20 kΩ resistor with a 0.5 mA current source having an equivalent output resistance of 1 MΩ. With the aid of Spice, find the value of A, β, A_f, R'_{if} and R'_{of}.

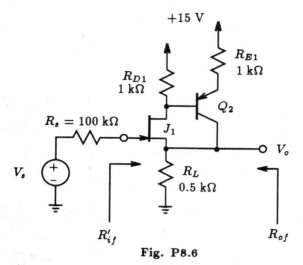

Fig. P8.6

8.6 The transistors of the circuit in Fig. P8.6 have the following parameters: For J_1, $I_{DSS} = 4$ mA, $V_P = -2$ V; for Q_2, $I_s = 10$ fA, $h_{fe} = 100$. Using Spice, determine the values of A, β, A_f, R'_{if} and R_{of}. Hint: Consider transistors J_1 and Q_2 as parts of the feedforward portion of the closed-loop amplifier. As well, consider the output in current form.

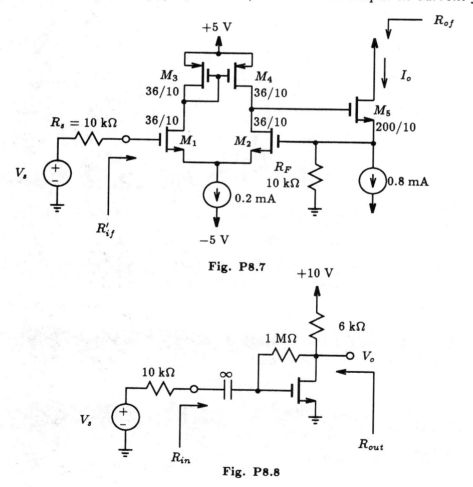

Fig. P8.7

Fig. P8.8

8.7 Fig. P8.7 shows a circuit for a voltage-to-current converter employing series-series feedback via resistor R_F. The MOSFETs have the dimensions shown and $\mu_n C_{OX} = 20\ \mu\text{A}/\text{V}^2$, $|V_t| = 1$ V, and $|V_A| = 100$ V. Use feedback analysis, together with Spice, determine A, β, $A_f = I_o/V_s$, R'_{if} and R_{of}. Compute the value of I_o/V_s using Spice and compare this to A_f.

8.8 For $V_t = 2$ V and $\mu_n C_{OX} = 0.5$ mA/V^2, find the voltage gain V_o/V_s, the input resistance R_{in}, and the output resistance R_{out} of the circuit in Fig. P8.8 using feedback analysis in conjunction with Spice. Verify your results by direct analysis.

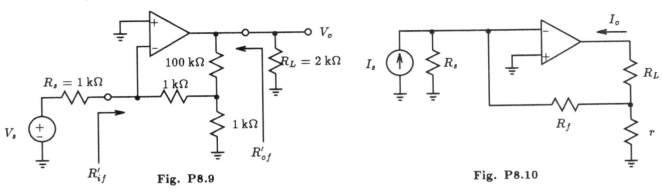

Fig. P8.9   Fig. P8.10

8.9 For the circuit of Fig. P8.9, use the feedback method, in conjunction with Spice, to find the voltage gain V_o/V_s, the input resistance R'_{if}, and the output resistance R'_{of}. The op-amp has open-loop gain $\mu = 10^4$ V/V, $R_{id} = 100$ kΩ, $R_{icm} = \infty$ and $r_o = 1$ kΩ.

8.10 Figure P8.10 shows how shunt-series feedback can be employed to design a current amplifier utilizing an op-amp. Using the feedback analysis method in conjunction with Spice, find the closed loop gain I_o/I_s, the input resistance (excluding R_s), and the output resistance (excluding R_L) for the case: open-loop voltage gain of op-amp $= 10^4$ V/V, $R_{id} = 100$ kΩ, op-amp output resistance $= 1$ kΩ, $R_s = R_L = 10$ kΩ, $r = 100$ Ω, and $R_f = 1$ kΩ.

8.11 For the amplifier circuit in Fig. P8.11, assuming that V_s has a zero dc component find the dc voltages at all nodes and the dc emitter currents of Q_1 and Q_2. Let the $\beta = 100$. With the aid of Spice, use feedback analysis to find V_o/V_s and R_{in}.

8.12 Determine the loop gain of the amplifier in Fig. P8.2 by breaking the loop at the gate of M_2, and finding the return voltage across the 100 kΩ resistor (while setting V_s to zero). The devices have $|V_t| = 1$ V, $\mu_n C_{OX} = 1.0$ mA/V^2, and $h_{fe} = 100$. The Early voltage magnitude for all devices (including those that implement the current sources) is 100 V. Determine the output resistance R'_{of}.

8 Feedback

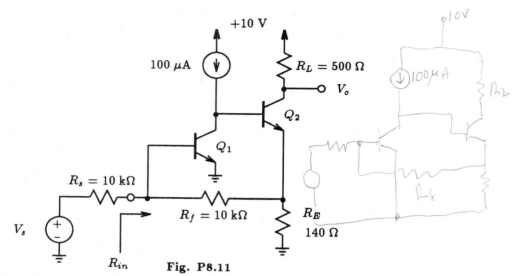

Fig. P8.11

8.13 Consider repeating Problem 8.12 above, but this time break the loop at the base of Q_3 and terminate the loop by the resistance seen looking into the base of Q_3. How does the loop gain compare to that found in Problem 8.12.

8.14 For the voltage amplifier seen in Fig. P8.3, determine the loop gain using the method of open- and short-circuit transfer functions.

8.15 Determine the loop gain of the amplifier circuit shown in Fig. P8.7 using: (a) the method of breaking the loop at the gate of M_2 and terminating the loop with the proper impedance, and (b) the method of open- and short-circuit transfer functions. How do the results compare?

8.16 For the circuit in Fig. P8.11 calculate the loop gain and then find the input resistance R_{in}. Assume that the BJTs have $h_{fe} = 100$.

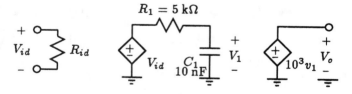

Fig. P8.17

8.17 A single-pole dc amplifier has an equivalent circuit representation shown in Fig. P8.17. If this amplifier is operated in a loop whose frequency-independent feedback factor is 0.1, with the aid of Spice, determine the low-frequency gain, the 3-dB frequency, and the unity-gain frequency of the closed-loop amplifier. Compare these results to those of the open-loop amplifier. By what factor does the single pole shift?

8.18 For the feedback amplifier described above in Problem 8.17, plot both the magnitude and phase of the loop gain. Is the amplifier stable? Confirm your conclusions by performing a transient analysis of the closed-loop amplifier subjected to a one-volt step input signal.

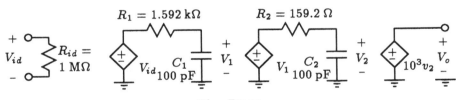

Fig. P8.19

8.19 An amplifier has a dc gain of 10^5 and poles at 10^5 Hz, 3.16×10^5 Hz, and 10^6 Hz. An equivalent circuit representation of this amplifier is shown in Fig. P8.19. Using Spice, plot the magnitude and phase of the voltage gain of this amplifier, then answer the following questions. What is the value of β, and the corresponding closed-loop gain, for which a phase margin of $45°$ is obtained?

8.20 Consider synthesizing an amplifier whose open-loop gain $A(s)$ is given by

$$A(s) = \frac{1000}{(1 + s/10^4)(1 + s/10^5)^2}.$$

Plot its magnitude and phase response using PSpice. If this amplifier is to be used in a feedback configuration where the feedback factor β is independent of frequency, find the frequency at which the phase shift is $180°$, and find the critical value of β at which oscillation will commence. Confirm your conclusion using a Spice transient analysis.

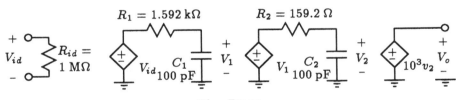

Fig. P8.21

8.21 A two-pole amplifier for which $A_o = 10^3$ and having poles at 1 MHz and 10 MHz has a small-signal equivalent circuit model shown in Fig. P8.21. Assuming that this amplifier is to be connected as a differentiator, what is the smallest differentiator time-constant for which the operation is stable? Base your decision on the 6 dB/octave rate-of-closure rule. What are the corresponding gain and phase margins?

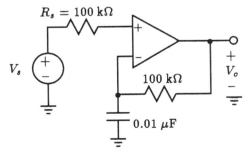

Fig. P8.22

8 Feedback

8.22 The op-amp in the circuit of Fig. P8.22 has an open-loop gain of 10^5 and a single-pole rolloff with $\omega_{3dB} = 10$ rad/sec.

(a) Plot the Bode plot for the loop gain using Spice.

(b) Find the frequency at which $|A\beta| = 1$, and find the corresponding phase margin.

8.23 An op-amp designed to have a low-frequency gain of 10^5 and a high-frequency response dominated by a single pole at 100 rad/sec, acquires, through a manufacturing error, a pair of additional poles at 10,000 rad/sec. Using Spice, plot the magnitude and phase response of the open-loop gain. At what frequency does the total phase shift reach 180°? At this frequency, for what value of β, assumed to be frequency independent, does the loop gain reach a value of unity? What is the corresponding value of closed-loop gain at low frequencies?

8.24 For the situation described in Problem 8.23, with the aid of Spice, obtain Nyquist plots of the loop gain for $\beta = 0.1$ and 10^{-3} between the frequency range of 1 Hz to 100 MHz.

Chapter 9

Output Stages And Power Amplifiers

This chapter is concerned with the study of output stages and power amplifiers. Output stages are used to supply large amounts of power to a load with very little signal distortion and with maximum efficiency. Power amplifiers are simply amplifiers with a high-power output stage. In this chapter we shall demonstrate how one utilizes Spice to compute the various performance measures of an output stage. This will include such measures as power efficiency and harmonic distortion. In addition, we shall also investigate the role of short-circuit protection circuits that are commonly incorporated into the circuit of an output stage.

9.1 Emitter-Follower Output Stage

In Fig. 9.1 we display an emitter-follower output stage biased with a constant current source of $I_{bias} \simeq (15 - 0.7)/5k = 2.86$ mA realized by transistors Q_2, Q_3 and resistor R. The output terminal of this stage is connected to a load resistance of 1 kΩ. With the aid of Spice we would like to determine the transfer characteristic of the emitter-follower assuming that each transistor is modeled after the widely available commercial 2N2222A *npn* transistor. The Spice input listing of the emitter-follower shown in Fig. 9.1 is given below in Fig. 9.2. A DC sweep of the input voltage source is requested to be performed between -20 V and $+20$ V. Although this particular voltage sweep extends beyond the power supply limits of the output stage, it is intended to highlight the saturation limits of the emitter-follower.

9 Output Stages And Power Amplifiers

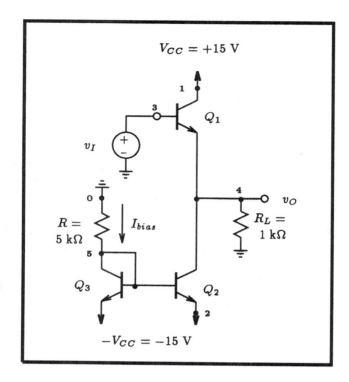

Figure 9.1 An emitter-follower output stage with current-source bias.

Submitting the input file listed in Fig. 9.2 to Spice results in the emitter-follower transfer characteristic shown in Fig. 9.3. As is evident from these results the output voltage saturates at −3.25 V for inputs below −2.7 V, and saturates at +14.9 V for input levels beyond +15.6 V. This output stage also exhibits an input offset voltage of +0.67 V. The slope of the transfer characteristic within the linear region of operation is very nearly unity.

It is interesting to note that the lower saturation limit of the output voltage is somewhat greater than that predicted by the simple equation $I_{bias} R_L$. Specifically, according to this simple expression, the lower saturation limit should be −2.86 mA × 1k Ω = −2.86 V, instead of the simulation result of −3.25 V. The reason for this deviation from the simple theory is due to the fact that the output current of the current source Q_2 varies with output voltage. This is confirmed by the plot of the collector current of Q_2 versus the output voltage v_o, shown in Fig. 9.4. (This is found by sweeping the voltage that appears across the collector-emitter terminals of Q_2 and monitoring its collector current). Notice that at an output voltage of −14.3 V, the collector current of Q_2 equals 2.86 mA and increases with increased output voltages.

To better understand the behavior of the voltage-follower output stage, let us consider applying several sine-wave signals of 1 kHz frequency having various amplitudes to the input of the circuit. In this way we can visualize the expected output signals from the

9.1 Emitter-Follower Output Stage

```
A Class A Output Stage

** Circuit Description **
* power supplies
Vcc+ 1 0 DC +15V
Vcc- 2 0 DC -15V
* input signal source
Vi 3 0 DC 0V
* output transistor
Q1 1 3 4 Q2N2222A
* load resistance
Rl 4 0 1k
* bias circuit
R 0 5 5k
Q2 4 5 2 Q2N2222A
Q3 5 5 2 Q2N2222A
* transistor model statement for the 2N2222A
.model Q2N2222A  NPN (Is=14.34f Xti=3 Eg=1.11 Vaf=74.03 Bf=255.9 Ne=1.307
+                     Ise=14.34f Ikf=.2847 Xtb=1.5 Br=6.092 Nc=2 Isc=0 Ikr=0 Rc=1
+                     Cjc=7.306p Mjc=.3416 Vjc=.75 Fc=.5 Cje=22.01p Mje=.377 Vje=.75
+                     Tr=46.91n Tf=411.1p Itf=.6 Vtf=1.7 Xtf=3 Rb=10)
** Analysis Requests **
.DC Vi -20V +20V 100mV
** Output Requests **
.PLOT DC V(4)
.Probe
.end
```

Figure 9.2 The Spice input file for computing the transfer characteristics of the emitter-follower output stage shown in Fig. 9.1.

emitter-follower circuit as one would see in the laboratory using a function generator and an oscilloscope. More specifically, let us consider applying three different sine-waves having amplitudes of 1 V, 2.7 V and 10 V to the input of the amplifier. According to the transfer characteristic of the emitter-follower circuit shown in Fig. 9.3, the 1 V sine-wave and the 2.7 V sine-wave signal should pass undistorted (although, the 2.7 V signal should just barely pass undistorted). On the other hand, the 10 V signal should become clipped on the negative portion of the output waveform. To see this using Spice, we use for each input level a modified version of the Spice deck shown in Fig. 9.2 in which the input voltage source statement is replaced by one that describes a sine-wave input. For example, in the case of the 1 V input signal, the Spice statement would appear as follows:

```
Vi 3 0 sin ( 0V 1V 1kHz ).
```

Similar statements can be written for the other two input levels. In addition to this, the following two Spice statements are included in each Spice deck to compute the transient

9 Output Stages And Power Amplifiers

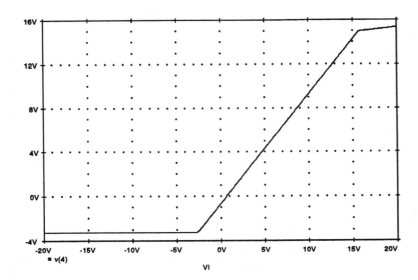

Figure 9.3 The transfer characteristics of the emitter-follower circuit shown in Fig. 9.1. As is evident, the output voltage can swing between −3.25 V and 14.9 V without experiencing significant distortion.

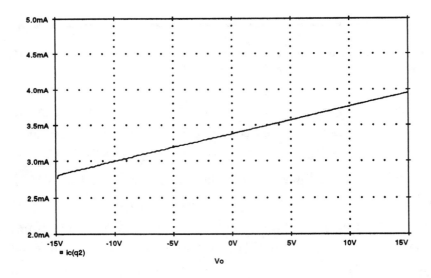

Figure 9.4 The dependence of the collector current of Q_2 on the output voltage v_O.

response of the circuit over three periods of the input signal:

```
.TRAN 10us 3ms 0ms 10us
.PLOT TRAN V(4).
```

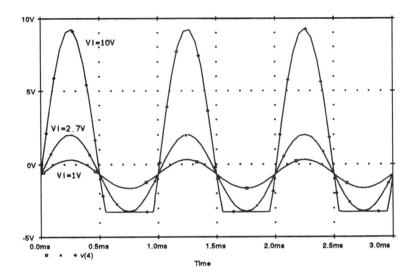

Figure 9.5 The output voltage waveform of the voltage-follower circuit of Fig. 9.1 for three different input levels: 1 V, 2.7 V and 10 V. Excessive voltage clipping occurs on the negative portion of the output voltage waveform corresponding to an input amplitude of 10 V.

One hundred points per period of the output waveform will be collected and plotted. The three Spice decks corresponding to the three input levels are concatenated into one file for easy comparison of the final results.

On completion of the Spice analysis, the output of the emitter-follower for the three different input signals are shown plotted in Fig. 9.5. These results confirm that for sine-wave inputs less than 2.7 V, the output waveform does not reveal any obvious distortion. On the other hand, for an input sine-wave of 10 V peak, the negative portion of the output sine-wave is clipped at a level of −3.25 V.

Other waveforms associated with the emitter-follower as calculated by Spice are shown in Fig. 9.6. The input to this circuit is a 2.7 V sine-wave signal of 1 kHz frequency (the largest input signal that can be applied to the emitter-follower circuit before distortion sets in). The Spice deck for this analysis is identical to that just described; only the plot command is changed according to:

.PLOT TRAN V(4) V(1,4) I(Vcc+).

The top graph shows the voltage waveform that appears across the load resistance. The output signal is sinusoidal with a peak value of 2.62 V riding on a DC level of −0.614

9 Output Stages And Power Amplifiers

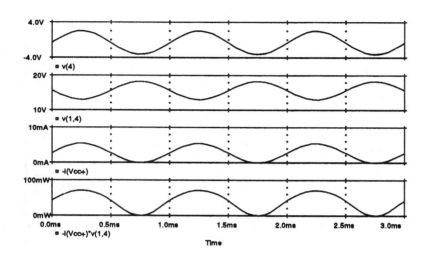

Figure 9.6 Several waveforms associated with the emitter-follower circuit of Fig. 9.1 when excited by a 2.7 V, 1 kHz sinusoidal signal. The top graph displays the voltage across the load resistance, the next graph displays the collector-emitter voltage of Q_1, the third graph from the top displays the collector current of Q_1, and the bottom graph displays the instantaneous power dissipated by Q_1.

V. The second and third graph from the top display both the collector-emitter voltage and the collector current of transistor Q_1, respectively. Both of these waveforms appear sinusoidal with amplitudes of 2.62 V and 2.703 mA, respectively. The average value of the voltage waveform appearing between the collector and emitter terminals is 15.61 V. Similarly, the average value of the collector current is 2.73 mA. The bottom-most graph displays the instantaneous power dissipated by Q_1. This waveform was obtained by multiplying together the collector current and collector-emitter voltage of Q_1 using Probe. The peak value associated with this waveform is 70.58 mW and its average value is about 40 mW.

9.2 Class B Output Stage

In Fig. 9.7 we display a class B output stage loaded by an 8 Ω resistor and driven by voltage source V_i. It consists of a pair of complementary transistors assumed to have similar characteristics. This output stage is noted for its relatively high power conversion efficiency (at most, 78.5%), and its large crossover distortion. In the following we shall investigate these two attributes of the class B output stage using Spice and assuming that the transistors are modeled after National Semiconductor's complementary NA51 and NA52 transistors.

9.2 Class B Output Stage

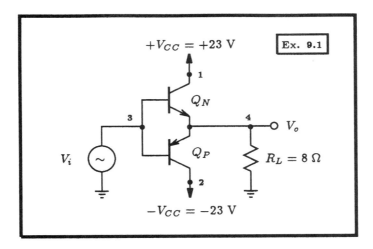

Figure 9.7 A Class B output stage.

Power Conversion Efficiency

With a 17.9 V 1 kHz sinusoidal signal applied to the input of the output stage shown in Fig. 9.7, we would like to determine the average power supplied by the two 23 V power supplies (P_S), the average power provided to the load resistance (P_L) and the power conversion efficiency (ie. $\eta = P_L/P_S$). In addition, we would also like to determine the power that each transistor must dissipate. This problem, in a somewhat different form, was presented in Example 9.1 of Sedra and Smith.

The Spice input file describing the class B output stage of Fig. 9.7 is listed in Fig. 9.8, together with the Spice models of the two complementary transistors. A 1 kHz input signal of 17.9 V amplitude is applied to the input and a transient analysis is to be performed beginning at time 0 ms and ending 3 ms later. This, in effect, will simulate the behavior of the output stage for 3 complete periods of the input signal. As output, we have requested plots of various node voltages and branch currents. Note that we are accessing the current through the load resistance R_L directly using $i(Rl)$ rather than in the usual way of inserting a zero-valued voltage source in series with R_L to sense the current flowing through it. This approach is possible with PSpice and is more convenient to use than the ammeter approach.

The results of this analysis can then be combined to provide the instantaneous power supplied by the dc sources, as well as the power dissipated by the load and by the transistors. Unfortunately, this computation can not be performed directly by Spice (although, one could always add additional circuitry to perform these computations, it is not the most direct approach), so instead, we utilize the Probe facility of PSpice to compute the desired quantities.

In Fig. 9.9 we display several waveforms associated with the output stage shown in Fig.

429

9 Output Stages And Power Amplifiers

Example 9.1: Class B Output Stage

```
** Circuit Description **
* power supplies
Vcc+ 1 0 DC +23V
Vcc- 2 0 DC -23V
* input signal source
Vi 3 0 sin ( 0V 17.9V 1kHz )
* output buffer
Qn 1 3 4 NA51
Qp 2 3 4 NA52
* load resistance
Rl 4 0 8
* transistor model statement for National Semiconductor's
* complementary transistors NA51 and NA52
.model NA51 NPN (Is=10f Xti=3 Eg=1.11 Vaf=100 Bf=100 Ise=0 Ne=1.5 Ikf=0
+                Nk=.5 Xtb=1.5 Br=1 Isc=0 Nc=2 Ikr=0 Rc=0 Cjc=76.97p Mjc=.2072
+                Vjc=.75 Fc=.5 Cje=5p Mje=.3333 Vje=.75 Tr=10n Tf=1n Itf=1 Xtf=0
+                Vtf=10)
.model NA52 PNP (Is=10f Xti=3 Eg=1.11 Vaf=100 Bf=100 Ise=0 Ne=1.5 Ikf=0
+                Nk=.5 Xtb=1.5 Br=1 Isc=0 Nc=2 Ikr=0 Rc=0 Cjc=112.6p Mjc=.1875
+                Vjc=.75 Fc=.5 Cje=5p Mje=.3333 Vje=.75 Tr=10n Tf=1n Itf=1 Xtf=0
+                Vtf=10)
** Analysis Requests **
.Tran 10us 3ms 0ms 10us
** Output Requests **
.Plot Tran V(1) i(Vcc+)
.Plot Tran V(2) i(Vcc-)
.Plot Tran V(4) i(Rl)
.Plot Tran V(1,4) i(Vcc+)
.Plot Tran V(2,4) i(Vcc-)
.probe
.end
```

Figure 9.8 The Spice input file for computing the transient response of the class B output stage shown in Fig. 9.7. A 17.9 V 1 kHz sinusoid is applied to the input.

9.7. The upper-most graph displays the voltage appearing across the load having a peak amplitude of 17.05 V. The middle curve displays the load current having a peak amplitude of 2.13 A. Notice that both the voltage and current waveforms exhibit a small crossover distortion. We shall investigate this further in the next subsection. The bottom graph of Fig. 9.9 displays the instantaneous and average power dissipated by the load as computed by the built-in computational facilities of Probe. The instantaneous power dissipated by the load resistor is computed according to V(4) × i(Rl). A running average of this is then used to determine the average power dissipated by the load resistance (P_L). It is interesting to observe that the average load power has some sort of transient behavior, eventually settling

9.2 Class B Output Stage

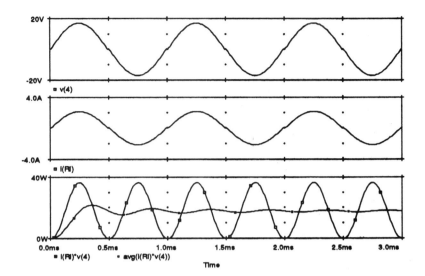

Figure 9.9 Several waveforms associated with the class B output stage shown in Fig. 9.7 when excited by a 17.9 V, 1 kHz sinusoidal signal. The upper graph displays the voltage across the load resistance, the middle graph displays the load current, and the lower graph displays the instantaneous and average power dissipated by the load.

into a quasi-constant steady-state value of approximately 17.7 W. This artifact is due to the PSpice algorithm used to compute the running average of a waveform.

The voltage and current waveforms supplied by the positive voltage supply $+V_{CC}$ are shown in the upper two graphs of Fig. 9.10. Shown in the lower-most graph are the instantaneous and average power supplied by $+V_{CC}$. Similar waveforms are found for the negative power supply $-V_{CC}$ except that current is supplied to the output stage in complementary time slots. The average power supplied by either $+V_{CC}$ or $-V_{CC}$ is found to be approximately 15 W, for a total supply power (P_S) of 30 W. The power conversion efficiency is then computed to be $\eta = P_L/P_S = 17.7/30 \times 100\% = 59\%$.

Another important quantity of concern in the design of an output stage is the amount of power dissipated by the two complementary transistors Q_N and Q_P. Figure 9.11 shows plots of the voltage, current, and power waveforms associated with transistor Q_P. Similar waveforms apply for Q_N. Both the voltage and current waveforms appear as expected. That is, the voltage waveform is sinusoid and the current waveform is a half-wave sinusoid. However, the instantaneous power waveform shown in the bottom-most graph of Fig. 9.11 has a rather strange shape. It appears periodic but is nonsinusoidal – almost pulse-like (less a few higher-order harmonic components). The reason for this is that the two transistors are being driven quite hard, and, although not apparent in the corresponding voltage and current

9 Output Stages And Power Amplifiers

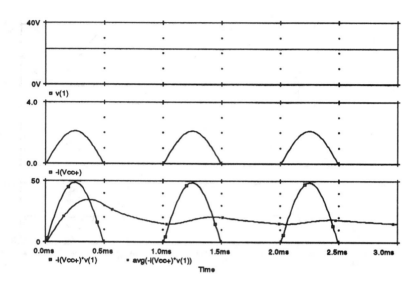

Figure 9.10 The voltage, current, and instantaneous and average power supplied by the positive voltage supply ($+V_{CC}$). The upper graph displays the voltage generated by $+V_{CC}$, the middle graph displays its current waveform and the lower graph displays the instantaneous and average power supplied to the rest of the circuit.

waveforms seen above, some distortion is present and is manifesting itself very clearly in the instantaneous power waveform. If the input voltage level of 17.9 V is reduce slightly, say to 17 V, then the ripple in the power waveform will disappear. The average power dissipated by transistor Q_P is seen to be approximately 6 W. Likewise, for transistor Q_N.

It is reassuring to confirm that the average power supplied by the two power supplies of 30 W equals (approximately) the sum of the power dissipated by the output stage of 12 W and that dissipated by the load of 17.7 W.

To compare the above results found through a Spice analysis with those calculated by a hand analysis, Table 9.1 was created for easy viewing. The accuracy of the hand calculated results as compared to those computed by Spice is also given as a relative error expressed in percent. As is evident, the hand calculated agree quite well with no more than an 8.3 per-cent error.

Transfer Characteristics And A Measure Of Linearity

The output signal of a class B output stage experiences crossover distortion as can be seen in the load voltage and current waveforms shown in Fig. 9.9, although barely due to the scale used there. Another way of investigating this crossover distortion is to plot the output voltage as a function of the input voltage level. To do this using Spice we simply

9.2 Class B Output Stage

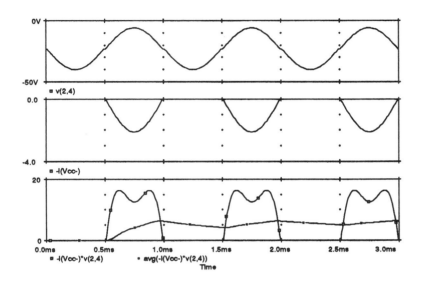

Figure 9.11 Waveforms of the voltage across, the current through, and the power dissipated in, the *pnp* transistor Q_P of the output stage shown in Fig. 9.7.

Power/Efficiency	Equation	Hand	Spice	Error %
P_S	$\dfrac{2}{\pi}\dfrac{\hat{V}_o}{R_L}V_{CC}$	31.2 W	30.0 W	4%
P_D	$\dfrac{2}{\pi}\dfrac{\hat{V}_o}{R_L}V_{CC} - \dfrac{1}{2}\dfrac{\hat{V}_o^2}{R_L}$	13.0 W	12.0 W	8.3%
P_L	$\dfrac{1}{2}\dfrac{\hat{V}_o^2}{R_L}$	18.2 W	17.7 W	2.8%
η	$\dfrac{P_L}{P_S}\times 100\%$	58.3%	59.0%	-1.2%

Table 9.1 The various power terms associated with the class B output stage shown Fig. 9.7 as computed by hand and Spice analysis. The rightmost column presents the relative error (in percent) between the values predicted by hand and Spice.

replace the transient analysis command given in the Spice input file listed in Fig. 9.8 with the following DC sweep command:

```
.DC Vi -10V +10V 50mV
```

We limit the swing of the input DC sweep level to ±10 V to highlight the deadband region of this amplifier. In addition, the following plot command should replace the transient plot

433

9 Output Stages And Power Amplifiers

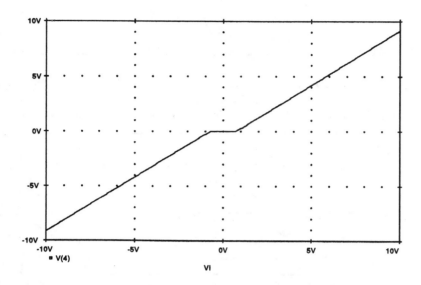

Figure 9.12 The transfer characteristics of a class B output stage constructed with National Semiconductor's NA51 and NA52 complementary transistors.

command given there in order to obtain a plot of the output stage transfer characteristic:

```
.Plot DC V(4)
```

Submitting the revised input file to Spice results in the transfer characteristic shown in Fig. 9.12. The slope of the line relating the input voltage to the output is very near unity and the deadband region is seen to extend between −0.72 V and +0.72 V.

Although the deadband effect is very visible in the amplifier transfer characteristic, engineers prefer to quantify the effect of this deadband region, and possibly other sources of distortion, in terms of a harmonic distortion measure. Spice provides this analysis capability through the *Fourier Analysis* command. This command simply decomposes a waveform generated through a transient analysis into it's Fourier series components. This includes the fundamental, a DC component, and the next eight harmonics. Spice will also compute the total harmonic distortion (THD) of the waveform

The syntax of the .FOUR command is given in Table 9.2. The keyword .FOUR specifies that a Fourier series decomposition is to be performed on each variable indicated in the variable list found after the field specifying the specified frequency of the fundamental, ie. *fundamental_frequency*. It is important to note that the Fourier analysis performed by Spice is not done on the entire waveform computed by the transient analysis. Instead it is performed only on the waveform defined between its end and one period of the fundamental

Analysis Requests	Spice Command
Fourier Analysis	.FOUR *fundamental_frequency* *output_variable_list*

Table 9.2 The general syntax of the Fourier Analysis command in Spice.

back from this end (ie. *1/fundamental_frequency* seconds). This implies that the transient analysis used in conjunction with a Fourier analysis must be at least *1/fundamental_frequency* seconds long.

The Fourier analysis command (.FOUR) does not require either a .PRINT or .PLOT command. The results of the analysis are automatically printed into the Spice output file in table form.

Returning to the Spice input file listed in Fig. 9.8, we can add a Fourier analysis command and compute the total harmonic distortion (THD) contained in the output voltage waveform. This is achieved simply by adding the following .FOUR command:

.FOUR 1kHz V(4)

where the fundamental frequency of 1 kHz is set equal to the frequency of the input sinusoidal signal.

The results of the Fourier analysis as computed by Spice are shown below:

```
FOURIER COMPONENTS OF TRANSIENT RESPONSE V(4)

DC COMPONENT =   9.648686E-05

HARMONIC   FREQUENCY    FOURIER      NORMALIZED    PHASE        NORMALIZED
   NO        (HZ)      COMPONENT     COMPONENT     (DEG)        PHASE (DEG)

    1      1.000E+03   1.683E+01     1.000E+00    -8.163E-05    0.000E+00
    2      2.000E+03   1.373E-04     8.162E-06    -9.067E+01   -9.067E+01
    3      3.000E+03   3.387E-01     2.013E-02    -1.800E+02   -1.800E+02
    4      4.000E+03   6.927E-05     4.117E-06    -8.964E+01   -8.964E+01
    5      5.000E+03   1.977E-01     1.175E-02    -1.800E+02   -1.800E+02
    6      6.000E+03   5.794E-05     3.443E-06    -8.846E+01   -8.846E+01
    7      7.000E+03   1.378E-01     8.188E-03    -1.800E+02   -1.800E+02
    8      8.000E+03   5.281E-05     3.139E-06    -8.740E+01   -8.740E+01
    9      9.000E+03   1.045E-01     6.213E-03    -1.800E+02   -1.800E+02

    TOTAL HARMONIC DISTORTION =    2.547618E+00 PERCENT
```

As is evident from above, the output voltage from the class B output stage is rich in odd-order harmonics, resulting in a rather high THD measure of 2.55%.

Generally, the distortion behavior of a class B output stage is much poorer than that which can be achieved with an emitter-follower circuit with the same input voltage level.

9   Output Stages And Power Amplifiers

To demonstrate this, let us return to the emitter-follower example of the previous section and compute its harmonic content when excited by a 17.7 V amplitude, 1 kHz input signal. Recall from Section 9.1 that the emitter-follower circuit given there could only handle input signals with peak values less than 2.7 V. Therefore, let us increase the load resistor from 1 kΩ to 5 kΩ and the level of the two power supplies to ± 23 V. This will increase the range of input signals that can be applied to the input of this amplifier before the output becomes distorted. (Computing the transfer characteristic of this revised voltage-follower circuit using Spice indicates that the linear region of this amplifier is between -21.4 V and $+23$ V).

The results of the Fourier analysis as computed by Spice for the voltage-follower circuit are as follows:

```
FOURIER COMPONENTS OF TRANSIENT RESPONSE V(4)

DC COMPONENT =  -6.772019E-01

HARMONIC   FREQUENCY   FOURIER      NORMALIZED    PHASE      NORMALIZED
  NO        (HZ)       COMPONENT    COMPONENT     (DEG)      PHASE (DEG)

   1       1.000E+03   1.752E+01    1.000E+00     2.912E-03   0.000E+00
   2       2.000E+03   5.782E-03    3.300E-04    -8.921E+01  -8.921E+01
   3       3.000E+03   2.596E-03    1.482E-04     1.971E+01   1.971E+01
   4       4.000E+03   1.738E-03    9.922E-05     8.523E+01   8.523E+01
   5       5.000E+03   8.973E-04    5.121E-05     9.995E+01   9.995E+01
   6       6.000E+03   7.882E-04    4.499E-05     7.459E+01   7.459E+01
   7       7.000E+03   9.442E-04    5.389E-05     7.217E+01   7.217E+01
   8       8.000E+03   9.679E-04    5.525E-05     7.415E+01   7.415E+01
   9       9.000E+03   9.591E-04    5.474E-05     7.316E+01   7.316E+01

       TOTAL HARMONIC DISTORTION =    3.928177E-02 PERCENT
```

Here we see that the THD for the emitter-follower circuit is 0.039%. This is obviously much less than that seen for the class B stage above at 2.55%.

9.3   Class AB Output Stage

Crossover distortion created by the output stage of an amplifier can be almost completely eliminated by a class AB configuration. In Fig. 9.13 we show a class AB output stage utilizing two diode-connected transistors (Q_1 and Q_2) for biasing. It is assumed that the two output transistors Q_N and Q_P are matched with $I_S = 10^{-13}$ A and $\beta = 50$. Further, it is assumed that the Q_N and Q_P have 3 times the junction area of Q_1 and Q_2. Under the conditions that $V_{CC} = 15$ V, $R_L = 100$ Ω, $I_{bias} = 3$ mA and an input sinusoidal signal having an amplitude of 10 V and a frequency of 1 kHz, we would like to determine the variation of current through Q_1 and Q_2 as a function of time. In addition, we also would like to observe the voltage across the two diode-connected transistors, designated as V_{BB}. This particular circuit was designed in Example 9.2 of Sedra and Smith where they selected the level of

9.3 Class AB Output Stage

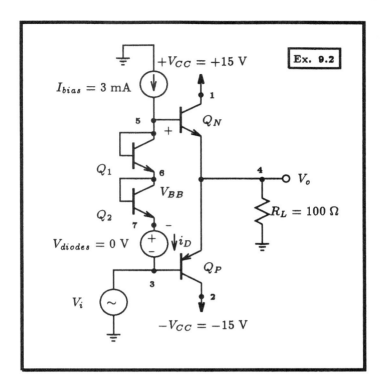

Figure 9.13 A class AB output stage utilizing diode-connected transistors for biasing.

I_{bias} so that the minimum current through Q_1 and Q_2 never dropped below 1 mA. We shall investigate whether this is indeed the case.

The Spice input file describing this class AB output stage is listed in Fig. 9.14. It is very similar to the input file used previously for the class B stage of the last section; the two major differences being that two diode-connected transistors are added, and the models for the two complementary transistors are greatly simplified. As output, the current supplied by $+V_{CC}$ is to be monitored which is equivalent to the collector current of Q_N. As well, the current flowing through the two diode-connected transistors is also to be monitored. A zero-valued voltage source V_{diodes} is place in series with these two diode-connected transistors in order to accomplish this. In addition, the output voltage V(4), and V(5,3) equal to V_{BB} are requested as output in the form of a plot.

On completion of Spice, the results of this analysis are shown in Fig. 9.15. The top graph displays the load voltage, and the graph below it illustrates the collector current of Q_N. Subsequently, the next graph displays the current flowing through the two diode-connected transistors Q_1 and Q_2. The bottom-most graph illustrates the biasing voltage V_{BB}.

Further review of these results reveals that the minimum current through the diode-

437

9   Output Stages And Power Amplifiers

Example 9.2: Class AB Output Stage

```
** Circuit Description **
* power supplies & current sources
Vcc+ 1 0 DC +15V
Vcc- 2 0 DC -15V
Ibias 0 5 DC 3mA
* input signal source
Vi 3 0 sin ( 0V 10V 1kHz )
* biasing diodes (transistors connected as diodes)
Q1 5 5 6 npn
Q2 6 6 7 npn
Vdiodes 7 3 0
* output buffer (junction area of Qn and Qp is 3 times Qd1 and Qd2)
Qn 1 5 4 npn 3
Qp 2 3 4 pnp 3
* load resistance
Rl 4 0 100
* simple transistor models
.model npn NPN  (Is=1e-13 Bf=50)
.model pnp PNP  (Is=1e-13 Bf=50)
** Analysis Requests **
.Tran 10us 3ms 0ms 10us
.OP
** Output Requests **
.Plot Tran i(Vcc+) i(Vdiodes)
.Plot Tran V(4) V(5,3)
.probe
.end
```

Figure 9.14 The Spice input file for computing the transient response of the class AB output stage shown in Fig. 9.13. A 10 V 1 kHz sinusoid is applied to the input.

connected transistors reaches 0.94 mA, a little less than the intended design value of 1 mA. Also, the biasing voltage V_{BB} varies between 1.25 V and 1.18 V during the time interval when Q_N is conducting, but remains constant at 1.25 V when it is not conducting. We also see that when the output voltage is at 0 V, Q_N is conducting a quiescent current of 8.3 mA and the two biasing transistors Q_1 and Q_2 are conducting a current of 2.8 mA. The former observation suggests that the quiescent power dissipated by Q_N is 8.3 mA × 15 V = 124.5 mW. Likewise, the same quiescent current must be passing through Q_P since the output voltage is at 0 V. Thus, it too is dissipating a quiescent power of 124.5 mW. Also, the total quiescent power dissipated by the two diode-connected transistors is 2.8 mA × 1.24 V = 3.5 mW.

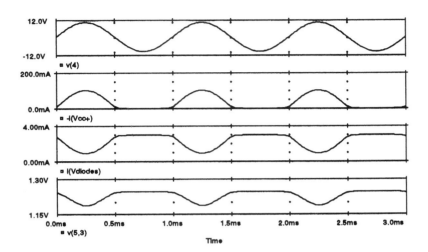

Figure 9.15 Voltage and current waveforms of the class AB output stage. Top graph: output load voltage; second graph: collector current of Q_N; third graph: current through diode-connected transistors Q_1 and Q_2; bottom graph: bias voltage V_{BB}.

9.4 Short-Circuit Protection

An important consideration of an output stage is its ability to recover from a direct short across its output terminals (ie. hot terminal and ground). This means that the amount of current either sourced or sinked by the output stage must be limited to a safe value in order not to exceed the power limitations of the two output transistors Q_N and Q_P. In Fig. 9.16 we illustrate a simple alteration to the class AB output stage which will limit the amount of current that can be sourced by the output stage through Q_N. This is achieved through the addition of transistor $Q_{protect}$, and resistors R_{E1} and R_{E2}.

To see the effectiveness of this approach, we shall compare the power dissipated by transistor Q_N with no current limiting protection to that dissipated by Q_N with current limiting when the output is shorted directly to ground. The input to the amplifier is assumed to be driven by a 10 V sinusoid of 1 kHz frequency. The junction area of $Q_{protect}$ is assumed to be equal to one-third that of Q_N and Q_P. All other transistor parameters are assumed to be the same as those used in the previous example in section 9.3.

The Spice input deck describing the class AB output stage with short-circuit protection is given in Fig. 9.17. A transient analysis is requested to compute both the collector current of Q_N and it's collector-emitter voltage. These two quantities are then multiplied together generating the instantaneous power dissipated by Q_N. The results of this analysis are shown

9 Output Stages And Power Amplifiers

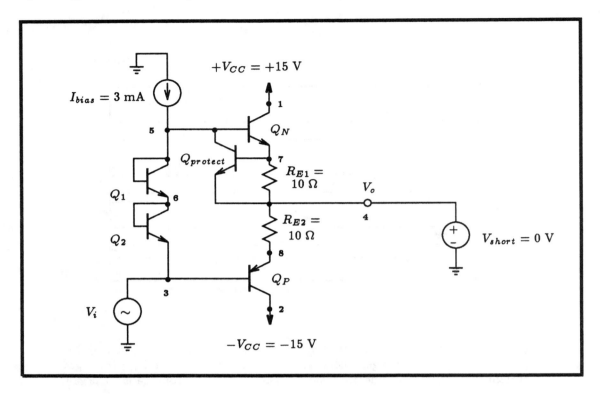

Figure 9.16 A class AB output stage with short-circuit protection. Note that the output terminal is shorted directly to ground.

in the top graph of Fig. 9.18. Also shown superimposed on this graph is the average power dissipated by Q_N. As is evident from these two curves, the peak instantaneous power is approximately 0.86 W, and the average power dissipated by Q_N is approximately 0.47 W.

It is interesting to compare the power dissipated by Q_N when it is not short-circuit protected. Consider removing $Q_{protect}$ from the Spice input file listed in Fig. 9.17. Everything else remains the same. The results are then computed and displayed in the bottom graph of Fig. 9.18. As is evident, the peak instantaneous power is approximately 2 W, and the average power dissipated by Q_N is approximately 1 W. Thus, the short-circuit protection circuit reduced the peak instantaneous power dissipated by Q_N by 60% and the average power dissipated by 56%. An even greater power reduction is possible by increasing the value of the two emitter resistors R_{E1} and R_{E2}. But, of course, this decreases the effective voltage swing of the output stage.

9.5 Spice Tips

- The current through a resistor is directly accessible in PSpice. No zero-valued voltage source is required to be placed in series with the resistor. This simplifies the creation of the input Spice deck.

```
Class AB Output Stage With Short Circuit Protection

** Circuit Description **
* power supplies
Vcc+ 1 0 DC +15V
Vcc- 2 0 DC -15V
Ibias 0 5 DC 3mA
* input signal source
Vi 3 0 sin ( 0V 10V 1kHz )
* biasing diodes (transistors connected as diodes)
Q1 5 5 6 npn
Q2 6 6 3 npn
* output buffer (junction area of Qn and Qp is 3 times Qd1 and Qd2)
Qn 1 5 7 npn 3
Qp 2 3 8 pnp 3
Re1 7 4 10
Re2 8 4 10
* short circuit protection transistor
Qprotect 5 7 4 npn
* short circuit output terminal
Vshort 4 0 0
* simple transistor models
.model npn NPN  (Is=1e-13 Bf=50)
.model pnp PNP  (Is=1e-13 Bf=50)
** Analysis Requests **
.Tran 10us 3ms 0ms 10us
** Output Requests **
.Plot Tran V(1,7) i(Vcc+)
.probe
.end
```

Figure 9.17 The Spice input file describing the class AB output stage shown in Fig. 9.16 having short-circuit protection.

- The Fourier Analysis command of Spice decomposes a time-domain waveform into its Fourier series components. This includes the fundamental, a DC component and the next eight harmonics. In addition, a total harmonic distortion measure (THD) is also provided.

- Fourier analysis is performed on the last cycle of a time-varying waveform computed by Spice. It is therefore important that by the final cycle the waveform has reached steady-state if the Fourier analysis results are to be interpreted correctly.

- Amplifier power efficiency can be computed using the results obtained through a transient analysis, ie. voltages and currents. The post-processing capabilities of Probe are helpful in generating graphical displays of the various power waveforms by multiplying different voltage and current signals together and by computing a running average.

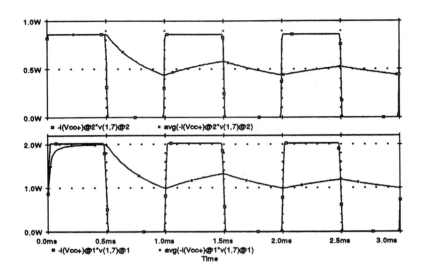

Figure 9.18 The top graph displays the instantaneous and average power dissipated by output transistor Q_N when short-circuit protected. The bottom graph displays the instantaneous and average power dissipated by Q_N with no short-circuit protection. As is evident, Q_N dissipates approximately 60% less power when protected against output short-circuit conditions.

9.6 Problems

9.1 A class A emitter follower, biased as in Fig. 9.1, uses $V_{CC} = 5$ V, $R = R_L = 1$ kΩ, with all transistors (including Q_3) identical. Assume β is very large. For linear operation, what are the upper and lower limits of the output voltage, and the corresponding inputs? How do these values change if the emitter-base junction area of Q_3 is made twice as big as that of Q_2. Half as big? Repeat for $\beta = 50$.

9.2 A source-follower circuit using enhancement NMOS transistors is constructed following the pattern shown in Fig. 9.1. All three transistors used are identical with $V_t = 1$ V and $\mu_n C_{OX} = 20$ mA/V^2. In addition, $V_{CC} = 5$ V, $R = R_L = 1$ kΩ. For linear operation, what are the upper and lower limits of the output voltage, and the corresponding inputs?

9.3 The BiCMOS follower shown in Fig. P9.3 uses devices for which $I_S = 100$ fA, $\beta = 100$, $\mu_n C_{OX} = 20$ mA/V^2, and $V_t = -2$ V. For linear operation, what is the range of output voltages obtained with $R_L = \infty$ as calculated by Spice? With $R_L = 100$ Ω? What is the smallest load resistance allowed for which a 1 V peak sine-wave output is available? What is the corresponding power-conversion efficiency?

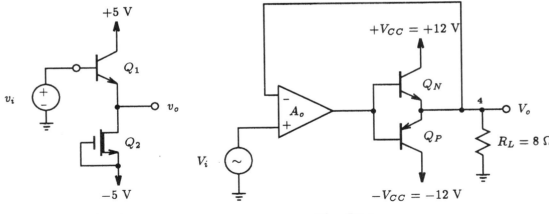

Fig. P9.3   Fig. P9.4

9.4 Consider the feedback configuration with class B output shown in Fig. P9.4. Let the amplifier gain $A_0 = 100$ V/V and the bipolar transistors be modeled after the NA51 and NA52 commercial transistors. Spice parameters for these devices can be obtain from Fig. 9.8. Compute the input-output voltage transfer characteristics using Spice. Compare these results to those generated by Spice for a corresponding class B output stage without feedback.

9.5 For the feedback amplifier shown in Fig. P9.4, with pertinent parameters described in Problem 9.4, compute the output voltage transient waveform for an input 5 kHz sinewave voltage signal of 6 V peak using Spice. Also, have Spice compute the Fourier Series coefficients of this output signal. What is the resulting Total Harmonic Distortion (THD) of this amplifier? Repeat the analysis on a corresponding class B output stage without feedback present. How do the THD results compare?

9.6 Consider the class B output stage using enhancement MOSFETs shown in Fig. P9.6. Let the devices have $|V_t| = 1$ V and $\mu_n C_{OX} = 2\mu_p C_{OX} = 200$ mA/V^2. With a 10 kHz sinewave input of 5 V peak and a high value of load resistance, what peak output results? What fraction of the sine-wave period does the crossover interval represent? For what value of load resistance is the peak output voltage reduce to half the input?

9.7 Consider the complementary BJT class B output stage constructed from 2N2222A commercial transistors. (See section 9.1 for Spice model parameters). For ±10 V power supplies and a 100 Ω load resistance, what is the maximum sine-wave output power available? What is the power conversion efficiency? For output signals of half this amplitude, find the output power, the supply power, and the power-conversion efficiency.

9 Output Stages And Power Amplifiers

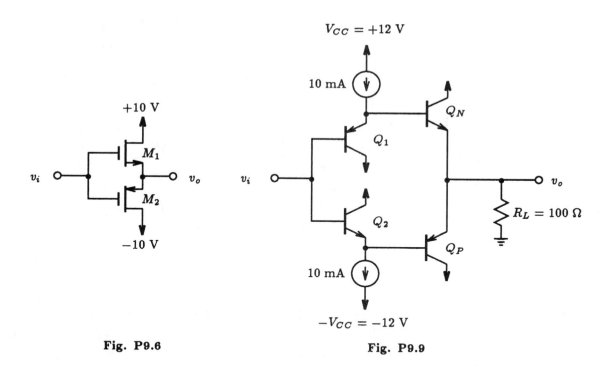

Fig. P9.6   Fig. P9.9

9.8 A class AB output stage using a two-diode bias network as shown in Fig. 9.13 utilizes diodes having the same junction area as the output transistors. For $V_{CC} = 10$ V, $I_{bias} = 0.5$ mA, $R_L = 100$ Ω and $\beta = 50$, what is the quiescent current computed by Spice when the output is at 0 V? How much power is dissipated by this output stage?

9.9 The circuit shown in Fig. P9.9 uses four matched transistors for which $I_S = 10$ fA and $\beta \geq 50$. What quiescent current flows in the output transistors? What bias current flows in the bases of the input transistors? Where does it flow? What is the net input current (the offset current) for a β mismatch of 10%? For a load resistance $R_L = 100$ Ω, what is the input resistance? What is the small-signal voltage gain?

9.10 Characterize a Darlington compound transistor formed from two npn BJTs modeled after the 2N3904 commercial transistor using Spice. For operation at 10 mA, what is the equivalent β_{eq}, V_{BEeq}, $r_{\pi eq}$ and g_{meq}?

9.11 The circuit shown in Fig. P9.11 operates in a manner analogous to that in Fig. 9.16 to limit the output current from Q_3 in the event of a short circuit or other mishap. With the aid of Spice, determine the value of R that causes Q_5 to turn on and absorb all of $I_{bias} = 2$ mA when the current being sourced reaches 150 mA. Assume that all devices can be described by $I_S = 10^{-14}$ A and $\beta = 100$. If the normal peak output current is 100 mA, find the voltage drop across R and the collector current in Q_5.

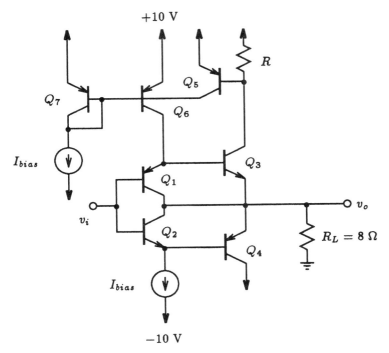

Fig. P9.11

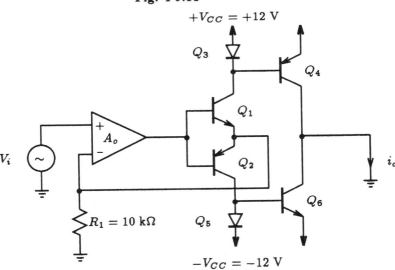

Fig. P9.12

9.12 For the current conveyor circuit shown in Fig. P9.12, assuming all transistors to have large β and $A_o = 10^6$, compute the output current i_o when the output is shorted directly to ground using Spice. Compare this to the situation where $\beta = 100$. By what percentage has i_o changed? The diodes are meant to be realized using a diode-connected transistor.

Chapter 10

Analog Integrated Circuits

In earlier chapters of this text the low-frequency behavior of differential pairs, current mirrors, output buffers, and other circuit building blocks was investigated using Spice. These were also combined, in a rather simplistic way, to form an operational amplifier whose large and small-signal static behavior was analyzed using Spice. This same approach is also used to realize other more sophisticated analog ICs. The very-popular bipolar 741 op amp circuit used in several of the examples of Chapter 2 is one such example. In this chapter we shall perform a detailed analysis of the 741 op amp using Spice. This will include an investigation into both its static and dynamic circuit behavior. We will also take a brief look at the expected thermal noise behavior of this op amp using the noise analysis capabilities of Spice. Following this, we shall investigate several CMOS and BiCMOS op amp designs that are finding important application in VLSI systems. Finally, we shall investigate the behavior of a D/A and an A/D converter circuits.

10.1 A Detailed Analysis Of The 741 Op Amp Circuit

A detailed circuit schematic of the 741 op amp is shown in Fig. 10.1. It consists of five main parts: (i) bias circuit, (ii) the input gain stage, (iii) the second gain stage, (iv) the output buffer, and (v) the short-circuit protection circuit. Each one of these stages, in one form or another, was briefly discussed and analyzed with Spice in previous chapters of this

10.1 A Detailed Analysis Of The 741 Op Amp Circuit

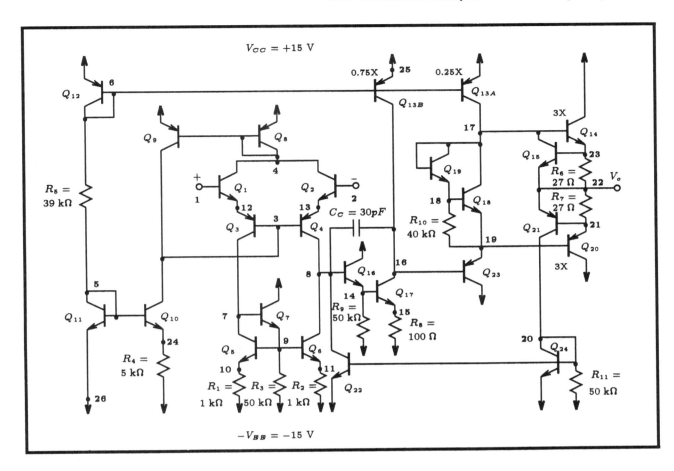

Figure 10.1 The 741 Op Amp Circuit.

text. For a detailed explanation of each section of the 741 op amp, see the discussion in Chapter 10 of Sedra and Smith.

In the following we shall demonstrate how one can utilize Spice to analyze the behavior of a complex analog circuit, such as the 741 op amp circuit shown in Fig. 10.1. Many of the procedures and set-ups used here would have been seen by our readers in early examples of this text, so our emphasis here serves more as a review of these methods; albeit, in a more complex setting. One exception to this is the Spice method used to compute the thermal noise behavior of the 741 op amp.

Assuming the typical set of device parameters for both the *npn* and *pnp* transistors shown in Table 10.1, we shall proceed to compute the DC operating point of the amplifier, its gain and frequency response, slew-rate and noise behavior. A reduced version of these parameters were used by Sedra and Smith in their hand analysis of the 741 op amp circuit in Chapter 10 of their text. Direct comparisons can be made between the results obtained through hand analysis and those obtained with Spice.

447

Parameter	NPN Transistor	PNP Transistor
β_F	200	50
β_R	2	4
V_{AF}	125V	50V
I_S	10fA	10fA
τ_F	0.35ns	30ns
r_b	200Ω	300Ω
r_c	200Ω	100Ω
r_e	2Ω	10Ω
C_{je}	1.0pF	0.3pF
V_{je}	0.7V	0.55V
M_{je}	0.33	0.5
C_{jc}	0.3pF	1.0pF
V_{jc}	0.55V	0.55V
M_{jc}	0.5	0.5
C_{js}	3.0pF	3.0pF
V_{js}	0.52V	0.52V
M_{js}	0.5	0.5

Table 10.1 Typical Spice parameters for integrated *npn* and *pnp* transistors [Gray and Meyer, 1984].

The Spice input file describing the circuit of the 741 op amp is listed in Fig. 10.2. The differential input is driven by the familiar multiple-input voltage-source arrangement discussed in section 6.1 and shown again in Fig. 10.3. This multiple source arrangement provides maximum flexibility when it comes to performing different types of analysis (eg. differential-mode and common-mode transfer characteristics) on a differential amplifier.

DC Analysis Of The 741

The first analysis that we will request that Spice perform is a DC operating point analysis. This will provide us with some insight into the DC behavior of all the transistors in the op amp circuit for grounded inputs — in case it's needed. In addition to this, we shall also request a DC sweep of the input differential voltage. This will indicate the boundaries of the high-gain region of the amplifier so that other analysis can be performed on the amplifier while it's biased around a known operating point; preferably a point within the linear region of the amplifier. Initially we set-up the DC sweep so that the input differential voltage is varied between the limits of the two power supplies. However, after the first submission to Spice, we refined the sweep range to vary between −400 μV and −200 μV in 10 μV increments so that we would obtain better resolution around the high-gain region of the amplifier.† During this analysis the input common-mode level is held at ground potential.

---

† To quickly converge on the location of the high-gain region of the op amp, one can connect the op

10.1 A Detailed Analysis Of The 741 Op Amp Circuit

```
The 741 Op-Amp

** Circuit Description **

* power supplies                        * short-circuit protection circuitry
Vcc 25 0 DC +15V                        Q15 17 23 22 npn
Vee 26 0 DC -15V                        Q21 20 21 22 pnp
                                        Q22 8 20 26 npn
* differential-mode signal level        Q24 20 20 26 npn
Vd 101 0 DC 0V                          R11 20 26 50k
Rd 101 0 1
EV+ 1 100 101 0 +0.5                    * biasing stage
EV- 2 100 101 0 -0.5                    Q10 3 5 24 npn
* common-mode signal level              Q11 5 5 26 npn
Vcm 100 0 DC 0V                         Q12 6 6 25 pnp
                                        R4 24 26 5k
* 1st or input stage                    R5 6 5 39k
Q1 4 1 12 npn
Q2 4 2 13 npn                           * transistor model statements
Q3 7 3 12 pnp                           .model npn NPN ( Bf=200 Br=2.0 VAf=125V Is=10fA Tf=0.35ns
Q4 8 3 13 pnp                           + Rb=200 Rc=200 Re=2 Cje=1.0pF Vje=0.70V Mje=0.33 Cjc=0.3pF
Q5 7 9 10 npn                           + Vjc=0.55V Mjc=0.5 Cjs=3.0pF Vjs=0.52V Mjs=0.5)
Q6 8 9 11 npn                           .model pnp PNP ( Bf=50 Br=4.0 VAf=50V Is=10fA Tf=30ns
Q7 25 7 9 npn                           + Rb=300 Rc=100 Re=10 Cje=0.3pF Vje=0.55V Mje=0.5 Cjc=1.0pF
Q8 4 4 25 pnp                           + Vjc=0.55V Mjc=0.5 Cjs=3.0pF Vjs=0.52V Mjs=0.5)
Q9 3 4 25 pnp
R1 10 26 1k                             ** Analysis Requests **
R2 11 26 1k                             .OP
R3 9 26 50k                             .DC Vd -400uV -200uV 10uV

* 2nd stage                             ** Output Requests **
Q13B 16 6 25 pnp 0.75                   .PLOT DC V(22)
Q16 25 8 14 npn                         .probe
Q17 16 14 15 npn                        .end
R8 15 26 100
R9 14 26 50k
Cc 8 16 30p

* output or buffer stage
Q13A 17 6 25 pnp 0.25
Q14 25 17 23 npn 3
Q18 17 18 19 npn
Q19 17 17 18 npn
Q20 26 19 21 pnp 3
Q23 26 16 19 pnp
R6 22 23 27
R7 21 22 27
R10 18 19 40k
```

Figure 10.2 The Spice input deck for analyzing the DC circuit behavior of the 741 op amp. Since the number of transistors exceed the limit imposed by the student version of PSpice, this particular Spice input deck is to be processed by the professional version of PSpice which has no transistor count limit.

On completion of Spice, the large-signal differential-mode transfer characteristics of the

---

amp in a unity-gain configuration with the positive terminal of the op amp connected to ground and compute the input-referred offset voltage of the op amp V_{OS}. Knowing this voltage we can infer that the differential transfer characteristics of the op amp will cross the horizontal voltage axis ($V_O = 0$ V) at $-V_{OS}$.

10 Analog Integrated Circuits

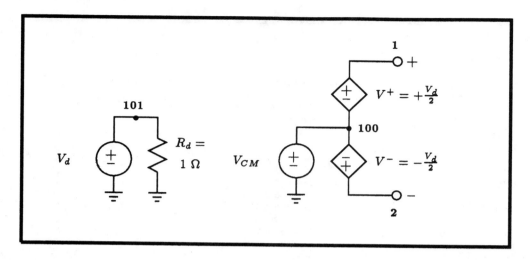

Figure 10.3 The multiple input voltage source used to drive the differential input of the 741 op amp.

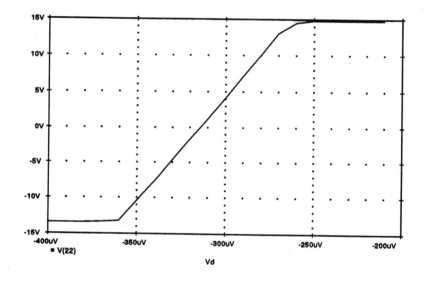

Figure 10.4 The large-signal differential input transfer characteristics for the 741 op amp circuit. The input common-mode voltage level is set to zero volts.

741 op amp, together with its DC operating point information, are calculated. The former is on display in Fig. 10.4. From this graph we observe several characteristics of the op amp. The first is that the linear region of the amplifier, as seen from the input terminals of the op amp, is bounded between -360 μV and -268 μV. Conversely, this corresponds to a maximum output voltage swing bounded in the negative direction by -13.2 V, and $+13.2$ V in the positive direction. This suggests that the small-signal DC gain of this amplifier is approximately $\frac{13.2-(-13.2)}{-268\mu-(-360\mu)} = +287$ kV/V. We also notice that the transfer characteristics

10.1 A Detailed Analysis Of The 741 Op Amp Circuit

of this amplifier cross the 0 V output axis somewhere between $-320\ \mu V$ and $-310\ \mu V$. A careful look using the Probe capability of PSpice indicates that the cross-over point occurs at -314.1 μV. Thus, this particular amplifier has an input-referred DC offset voltage of $+314.1\ \mu V$. This is the systematic offset of the op amp design and does not include offsets that result from component mismatches.

Reviewing the DC operating point information (which is not shown here) reveals that for grounded amplifier inputs, transistors Q_{13A} and Q_{13B} are saturated, causing transistors Q_{14}, Q_{18}–Q_{20}, and Q_{23} to cutoff. Except for these and the transistors of the protection circuit, the remaining transistors are operating in their active regions. Other DC information, such as input bias currents, static power dissipation, etc, was calculated through the .OP command; however, these results are obtained when the op amp is biased outside of it's normal operating region ie. linear region. Since the op amp is generally operated in it's linear region, it is more appropriate to repeat the DC operating point command with the op amp is biased in it's linear region. This is easily accomplished by applying a differential input offset voltage of $-314.1\ \mu V$ across the two inputs to the op amp. We can then repeat the DC analysis with the following element statement replacing the one listed in Fig. 10.2:

```
Vd 101 0 DC -314.1uV.
```

On completion of Spice, we obtain the following small-signal bias solution:

```
****     SMALL SIGNAL BIAS SOLUTION        TEMPERATURE =   27.000 DEG C
******************************************************************************

 NODE    VOLTAGE     NODE    VOLTAGE     NODE    VOLTAGE     NODE    VOLTAGE

(    1)-157.1E-06 (    2) 157.1E-06 (    3)   -1.0496 (    4)   14.4540
(    5)  -14.3510 (    6)   14.3420 (    7)  -13.9310 (    8)  -13.7450
(    9)  -14.4640 (   10)  -14.9920 (   11)  -14.9920 (   12)    -.5265
(   13)   -.5263  (   14)  -14.2900 (   15)  -14.9340 (   16)   -1.1909
(   17)    .5834  (   18)     .0348 (   19)    -.5789 (   20)  -15.0000
(   21)   -.0040  (   22) 592.6E-06 (   23)     .0052 (   24)  -14.9020
(   25)   15.0000 (   26)  -15.0000 (  100)    0.0000 (  101)-314.1E-06


    VOLTAGE SOURCE CURRENTS
    NAME         CURRENT

    Vcc         -1.841E-03
    Vee          1.841E-03
    Vd           3.141E-04
    Vcm         -6.897E-08

    TOTAL POWER DISSIPATION   5.52E-02  WATTS

****  VOLTAGE-CONTROLLED VOLTAGE SOURCES

    NAME         EV+         EV-
    V-SOURCE    -1.571E-04   1.571E-04
    I-SOURCE    -3.440E-08  -3.456E-08
```

10 Analog Integrated Circuits

Transistor	Hand	Spice	Transistor	Hand	Spice
Q_1	9.5	7.68	Q_{13B}	550	658
Q_2	9.5	7.71	Q_{14}	154	170
Q_3	9.5	7.59	Q_{15}	0	~ 0
Q_4	9.5	7.63	Q_{16}	16.2	17.1
Q_5	9.5	7.55	Q_{17}	550	644
Q_6	9.5	7.56	Q_{18}	165	198
Q_7	10.5	10.8	Q_{19}	15.8	16.2
Q_8	19	14.8	Q_{20}	154	168
Q_9	19	19.4	Q_{21}	0	~ 0
Q_{10}	19	19.6	Q_{22}	0	~ 0
Q_{11}	730	732	Q_{23}	180	213
Q_{12}	730	708	Q_{23}	0	~ 0
Q_{13A}	180	215			

Table 10.2 DC collector currents of the 741 circuit in μA as computed by hand analysis and by Spice.

From the above, we see that the 741 op amp dissipates a static power of 55.2 mW. We can also calculate the input bias and offset currents of the op amp from the currents passing through the two VCVS, EV+ and EV-. The input bias current I_B is $\frac{(-34.40n)+(-34.56n)}{2} =$ 34.48 nA and the offset current I_{OS} is $|(-34.40n) - (-34.56n)| = 0.16$ nA. This particular offset current is due solely to the systematic offset of the op amp design itself, as opposed to any random offset effect that can occur during manufacture.

Also included in the Spice output file is a detailed listing of the operating point of each transistor. Rather than list all of this information here, instead, in Table 10.2 we list the collector current (in μA) of each transistor found in the 741 op amp circuit. These are also compared with the current levels computed using hand analysis by Sedra and Smith. As is self-evident, there is reasonable agreement between the two sets of results.

To determine the input common-mode range (CMR) of the 741 op amp, we shall sweep the input common-mode level V_{CM} between the two power supply limits. To ensure that the op amp is biased in the linear region, we maintain the input differential offset voltage at -314.1 μV. To accomplish our task of varying the input common-mode level, we simply alter the .DC sweep command listed in the Spice deck of Fig. 10.2 according to:

```
.DC Vcm -15V +15V 0.1V.
```

No other alterations are necessary.

Re-submitting the revised input file to Spice results in the large-signal common-mode DC transfer characteristic displayed in Fig. 10.5. Here we see that linear operation is maintained over an input common-mode range (CMR) between -10.0 V and $+8.05$ V corresponding to

10.1 A Detailed Analysis Of The 741 Op Amp Circuit

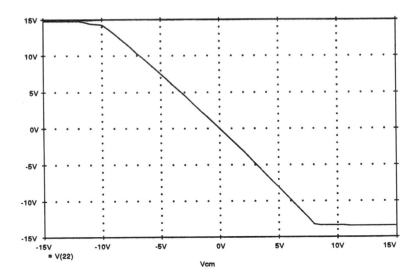

Figure 10.5 The large-signal common-mode DC transfer characteristics of the 741 op amp circuit shown in Fig. 10.1. An input differential offset voltage of $-314.1\ \mu V$ is applied to the input of the amplifier input to prevent any premature saturation.

an output voltage swing varying between -13.26 V and 14.27 V. The lower limit of the CMR is caused by Q_{13A} of the second stage saturating. Similarly, the upper limit of the CMR is caused by transistor Q_{17} also of the second stage saturating. We can also estimate the common-mode voltage gain in the linear region of the op amp to be $\frac{(-13.26)-14.27}{8.05-(-10.0)} = -1.53$ V/V.

We should note that these results are quite different from those normally encountered in an actual 741 op amp. Typically, the input CMR is determined by the transistors of the front-end stage leaving their linear regions rather than transistors of the second stage — as we have seen above. We attribute this difference to the fact that the 741 op amp consists of 6 different types of transistors (ie. small npn, large npn, lateral pnp, substrate pnp, dual-emitter pnp and a dual-collector pnp) each having different terminal characteristics. On the other hand, the above Spice simulation of the 741 op amp models the diversity of the different transistors with only two model types. Thus, some behavior of the actual 741 op amp is not properly captured in our simulations.

The above results are all examples of a large-signal analysis. Much of the analysis performed by Sedra and Smith is based on small-signal analysis. To obtain the small-signal parameters of the op amp we make use of the .TF command of Spice. For instance, the small-signal differential voltage gain and small-signal output resistance of the op amp can be computed by adding the following command to the Spice deck shown listed in Fig. 10.2:

.TF V(22) Vd

It is important to maintain the input differential offset voltage at -314.1 μV in order to keep the amplifier in its linear region. The input common-mode level is also held at ground potential. The results of Spice are listed below:

```
****     SMALL-SIGNAL CHARACTERISTICS

         V(22)/Vd =  2.946E+05

         INPUT RESISTANCE AT Vd =   1.000E+00

         OUTPUT RESISTANCE AT V(22) =   1.064E+02
```

Here we see that the small-signal differential voltage gain of $+294.6$ kV/V is quite close to the value $+287$ kV/V estimated from the slope of the op amp large-signal transfer characteristics shown in Fig. 10.4. The input resistance stated here is not the input differential resistance of the op amp. Rather, this is the 1 Ω resistance connected directly to V_d. To obtain an estimate of the input resistance of the op amp, we shall instead use the results of the AC analysis presented below. The output resistance of the op amp is seen to be quite low at 106.4 Ω.

We can repeat the above .TF analysis with respect to the input common-mode level around the operating point defined by $V_{CM} = 0$ V and $V_d = -314.1$ μV. One simply utilizes the command:

.TF V(22) Vcm

The results of the analysis are then found in the Spice output file as follows:

```
****     SMALL-SIGNAL CHARACTERISTICS

         V(22)/Vcm =  -1.555E+00

         INPUT RESISTANCE AT Vcm =   6.064E+08

         OUTPUT RESISTANCE AT V(22) =   1.064E+02
```

The common-mode voltage gain is -1.555 V/V, the input common-mode resistance is seen to be a very large 606.4 MΩ, and the output resistance is, as expected, the same as the previous calculation of 106.4 Ω.

Dividing the differential voltage gain of 2.946×10^5 V/V by the magnitude of common-mode voltage gain of 1.555 V/V we obtain the common-mode rejection ratio (CMRR) for the 741 op amp to be 105.6 dB.

Gain And Frequency Response Of The 741

We can carry the small-signal analysis over to the frequency domain and have Spice

10.1 A Detailed Analysis Of The 741 Op Amp Circuit

```
The 741 Op-Amp

** Circuit Description **

* power supplies                        * short-circuit protection circuitry
Vcc 25 0 DC +15V                        Q15 17 23 22 npn
Vee 26 0 DC -15V                        Q21 20 21 22 pnp
                                        Q22 8 20 26 npn
* DC & AC differential-mode signal level Q24 20 20 26 npn
Vd 101 0 DC -314.1uV AC 1V              R11 20 26 50k
Rd 101 0 1
EV+ 1 100 101 0 +0.5                    * biasing stage
EV- 2 100 101 0 -0.5                    Q10 3 5 24 npn
* common-mode signal level              Q11 5 5 26 npn
Vcm 100 0 DC 0V                         Q12 6 6 25 pnp
                                        R4 24 26 5k
* 1st or input stage                    R5 6 5 39k
Q1 4 1 12 npn
Q2 4 2 13 npn                           * transistor model statements
Q3 7 3 12 pnp                           .model npn NPN ( Bf=200 Br=2.0 VAf=125V Is=10fA Tf=0.35ns
Q4 8 3 13 pnp                           + Rb=200 Rc=200 Re=2 Cje=1.0pF Vje=0.70V Mje=0.33 Cjc=0.3pF
Q5 7 9 10 npn                           + Vjc=0.55V Mjc=0.5 Cjs=3.0pF Vjs=0.52V Mjs=0.5)
Q6 8 9 11 npn                           .model pnp PNP ( Bf=50 Br=4.0 VAf=50V Is=10fA Tf=30ns
Q7 25 7 9 npn                           + Rb=300 Rc=100 Re=10 Cje=0.3pF Vje=0.55V Mje=0.5 Cjc=1.0pF
Q8 4 4 25 pnp                           + Vjc=0.55V Mjc=0.5 Cjs=3.0pF Vjs=0.52V Mjs=0.5)
Q9 3 4 25 pnp
R1 10 26 1k                             ** Analysis Requests **
R2 11 26 1k                             .AC DEC 10 0.1Hz 100MegHz
R3 9 26 50k
                                        ** Output Requests **
* 2nd stage                             .PLOT  AC Vdb(22) Vp(22)
Q13B 16 6 25 pnp 0.75                   .PRINT AC Im(EV+) Im(EV-)
Q16 25 8 14 npn                         .probe
Q17 16 14 15 npn                        .end
R8 15 26 100
R9 14 26 50k
Cc 8 16 30p

* output or buffer stage
Q13A 17 6 25 pnp 0.25
Q14 25 17 23 npn 3
Q18 17 18 19 npn
Q19 17 17 18 npn
Q20 26 19 21 pnp 3
Q23 26 16 19 pnp
R6 22 23 27
R7 21 22 27
R10 18 19 40k
```

Figure 10.6 The Spice input deck for computing the frequency response of the 741 op amp.

compute the differential magnitude and phase response of the op amp. Several alterations must be made to the Spice input file shown listed in Fig. 10.2. For starters, the input differential excitation should be changed to include an AC voltage component. A one volt AC signal will be applied. The amplitude of this AC input is not important because an AC analysis is performed using a linearized model of the transistor circuit. A 1 volt amplitude is commonly chosen for the input signal because, in this way, the output voltage directly represents the transfer function of the circuit. We maintain the input DC offset voltage at −314.1 μV in order to maintain the op amp in its linear region. Thus the input excitation statement should be changed to read as follows:

```
Vd 101 0 DC -314.1uV AC 1V.
```

Following this, we shall request that Spice compute the frequency response of the amplifier over a frequency interval beginning at 0.1 Hz and ending at 100 MHz using the following statement:

```
.AC DEC 10 0.1Hz 100MegHz
```

A plot statement can be included in the input file to graphically display the output voltage results of the frequency response analysis. Its syntax would appear as follows:

```
.PLOT AC VdB(22) Vp(22).
```

We shall also include a .PRINT statement for the magnitude of the small-signal current flowing through the two VCVSs in series with the op amp input terminals. This will allow us to calculate the input differential resistance of the op amp knowing that the small-signal input differential voltage applied to the amplifier is 1 V. The command line for this is

```
.PRINT AC Im(EV+) Im(EV-).
```

The revised Spice input file for calculating the small-signal frequency response of the 741 op amp is listed in Fig. 10.6.

The results of the frequency-domain analysis are shown in Fig. 10.7. The low-frequency behavior is dominated by a single-pole response having a low-frequency gain of 109.4 dB and a 3 dB bandwidth of approximately 2.6 Hz. The unity-gain frequency f_t is read directly off the graph using Probe to be 0.652 MHz. We also find that this amplifier has a phase margin of 66.7°. The presence of a second pole is also evident. It is located at approximately 1.82 MHz.

The low frequency current drawn by the op amp input terminals was found by Spice to be as follows:

```
FREQ        IM(EV+)      IM(EV-)
1.000E-01   2.766E-07    2.732E-07
```

We see here that there is a small difference between the two currents drawn by the op amp input terminals. This difference is due to op amp systematic offset. To determine the input differential resistance to the amplifier under such asymmetric conditions, we shall work with the average of the two base currents. In this way, we can eliminate the presence of the offset current in the resistance calculation. Thus, we compute the average input base current to be 274.9 nA and therefore obtain the input differential resistance at 3.64 MΩ.

10.1 A Detailed Analysis Of The 741 Op Amp Circuit

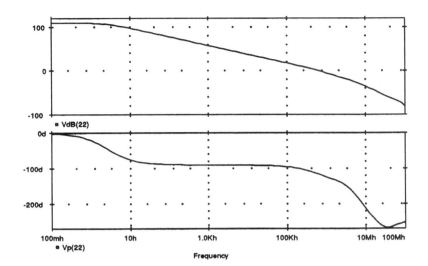

Figure 10.7 The differential magnitude and phase response of the 741 op amp. The low frequency response behavior of the 741 op amp is dominated by a single-pole roll-off beginning at 2.6 Hz. A second pole is also evident at 1.82 MHz.

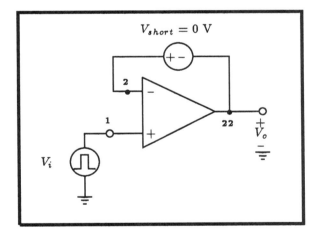

Figure 10.8 Circuit arrangement for computing op amp positive- and negative-going slew-rate limits.

Slew-Rate Limiting Of The 741

An important attribute of op amp behavior that usually limits the high-frequency operation of op amp circuits is its slew-rate limiting. Connecting the op amp in a unity-gain configuration and applying a large voltage pulse to its input will reveal both the positive-going and negative-going slew-rate of this op amp, as demonstrated in Fig. 10.8. The Spice input file depicting this arrangement is provided in Fig. 10.9. Here a zero-valued voltage

10 Analog Integrated Circuits

```
The 741 Op-Amp

** Circuit Description **

* power supplies
Vcc 25 0 DC +15V
Vee 26 0 DC -15V

* slew-rate limiting set-up
Vd 1 0 PWL ( 0,-5V 1ns,+5V 30000ns,+5V
+              30001ns,-5V 1s,-5V       )
Vshort 2 22 0

* 1st or input stage
Q1 4 1 12 npn
Q2 4 2 13 npn
Q3 7 3 12 pnp
Q4 8 3 13 pnp
Q5 7 9 10 npn
Q6 8 9 11 npn
Q7 25 7 9 npn
Q8 4 4 25 pnp
Q9 3 4 25 pnp
R1 10 26 1k
R2 11 26 1k
R3 9 26 50k

* 2nd stage
Q13B 16 6 25 pnp 0.75
Q16 25 8 14 npn
Q17 16 14 15 npn
R8 15 26 100
R9 14 26 50k
Cc 8 16 30p

* output or buffer stage
Q13A 17 6 25 pnp 0.25
Q14 25 17 23 npn 3
Q18 17 18 19 npn
Q19 17 17 18 npn
Q20 26 19 21 pnp 3
Q23 26 16 19 pnp
R6 22 23 27
R7 21 22 27
R10 18 19 40k

* short-circuit protection circuitry
Q15 17 23 22 npn
Q21 20 21 22 pnp
Q22 8 20 26 npn
Q24 20 20 26 npn
R11 20 26 50k

* biasing stage
Q10 3 5 24 npn
Q11 5 5 26 npn
Q12 6 6 25 pnp
R4 24 26 5k
R5 6 5 39k

* transistor model statements
.model npn NPN ( Bf=200 Br=2.0 VAf=125V Is=10fA Tf=0.35ns
+ Rb=200 Rc=200 Re=2 Cje=1.0pF Vje=0.70V Mje=0.33 Cjc=0.3pF
+ Vjc=0.55V Mjc=0.5 Cjs=3.0pF Vjs=0.52V Mjs=0.5)
.model pnp PNP ( Bf=50 Br=4.0 VAf=50V Is=10fA Tf=30ns
+ Rb=300 Rc=100 Re=10 Cje=0.3pF Vje=0.55V Mje=0.5 Cjc=1.0pF
+ Vjc=0.55V Mjc=0.5 Cjs=3.0pF Vjs=0.52V Mjs=0.5)

** Analysis Requests **
.TRAN 0.1ns 100us

** Output Requests **
.PLOT TRAN V(1) V(22)
.probe
.end
```

Figure 10.9 The Spice input deck for computing the positive- and negative-going slew-rate limit of the 741 op amp.

source described by

$$\text{Vshort 2 22 0}$$

is used to form a direct connection between the op amp output and its negative input terminal. This avoids having to renumber the negative input terminal of the op amp in the Spice deck with the same number used to describe the op amp output. Another voltage source is used to create the input pulse:

$$\text{Vd 1 0 PWL (0,-5V 1ns,+5V 30000ns,+5V 30001ns,-5V 1s,-5V)}.$$

10.1 A Detailed Analysis Of The 741 Op Amp Circuit

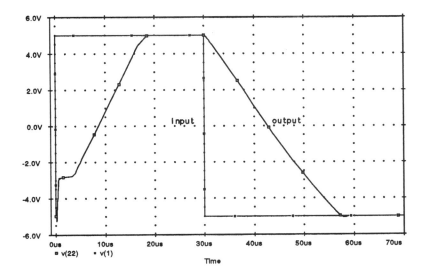

Figure 10.10 Input and output transient voltage waveforms of the 741 op amp circuit connected in a unity-gain configuration. Both the positive-going and negative-going slew-rate limits of the op amp is evident from these results.

Notice that this input pulse signal is described in a piece-wise linear fashion, beginning at a low level of −5 V and quickly rising to +5 V one ns after this, staying there for 30 μs and then returning to the −5 V level one ns later. It remains at this low level of −5 V for the rest of the duration of the pulse. A transient analysis is requested to compute the response of the op amp circuit arrangement over a 100 μs interval using a 0.1 ns sampling interval.

Submitting this input file to Spice results in the plot shown in Fig. 10.10. Both the input pulse to the amplifier and it's output response are shown. Here we see that the positive-going portion of the output signal has a very different shape than the negative-going portion. Instead of a gradual rise in the positive-going signal, initially there is small jump in the output voltage of 2.2 volts, followed by the output being held constant at −2.8 V for 3.6 μs, then rising linearly to +5 V in 18.2 μs. Thus, the positive-going slew-rate is estimated at +0.55 V/μs.

The negative-going response behaves more along the lines of what one expects, a steady decline from +5 V to −5 V. As a result the negative-going slew-rate is found to be −0.39 V/μs.

Noise Analysis Of The 741 Op Amp

As our final analysis, we shall utilize the noise analysis capability of Spice to estimate both the output-referred and input-referred thermal noise power spectral density $S_\eta(j\omega)$ of

10 Analog Integrated Circuits

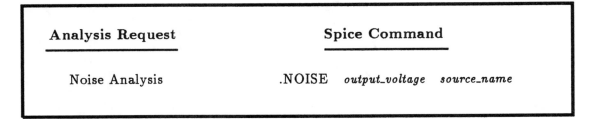

Table 10.3 The general syntax of the Noise Analysis command in Spice.

the 741 op amp circuit shown in Fig. 10.1. Analyzing the noise behavior of op amp circuits is a rather advanced topic and is not covered in the text by Sedra and Smith. It is included here because of its importance and to demonstrate yet another analysis capability of Spice.

Resistors and semiconductor devices generate various types of noise. Thermal noise is one important example of this noise, and the minimization of this noise component is an important consideration in the design of op amp circuits. The cause of thermal noise is the random movement of charge carriers in resistors or semiconductor devices. This movement, in turn, causes instantaneous voltage excursions to appear across the terminals of the device whose value, on average, is zero. In a circuit, these voltage variations also have an effect on the node voltages of the circuit. The total effect on any one particular node is simply the root-mean of the sum of the squares of each individual effect. The Noise Analysis capability of Spice performs exactly this computation in exactly this manner. In general, these noise fluctuations are functions of frequency and therefore are computed in conjunction with the .AC analysis of Spice. As a result, the noise level is reported in the output file in units of either $V/\sqrt{Hz}$ or $A/\sqrt{Hz}$, depending on its context.

The syntax of the Noise Analysis command is given in Table 10.3. The keyword .NOISE specifies that a noise analysis is to be performed on a given circuit over the frequency interval specified on the .AC command line. The next field *output_voltage* denotes a node voltage. The node(s) that this voltage appears across indicates the output port of the circuit. Spice will compute the effective noise voltage spectral density that appears at this port due to internal noise sources. Spice will not compute the effective output noise current spectral density associated with a short-circuit output. The subsequent field, indicated by *source_name*, specifies the name of an independent voltage or current source. The output noise voltage spectral density will be referred back to the port defined by this independent source as either an input-referred noise voltage or current spectral density depending on the input source type.

To access the noise information computed by Spice, one uses either a .PRINT or .PLOT command with reference to the information stored in the Spice variable *ONOISE* (output

10.1 A Detailed Analysis Of The 741 Op Amp Circuit

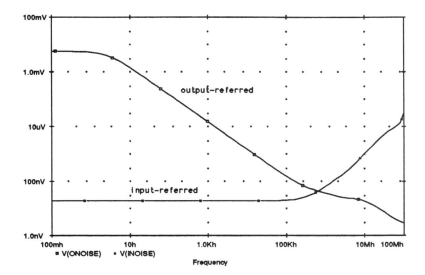

Figure 10.11 Output and input-referred noise voltage spectral density of the 741 op amp circuit. The vertical axis is in units of $V/\sqrt{Hz}$.

noise) or *INOISE* (equivalent input-referred noise).

Returning to the Spice input deck that we used to compute the frequency response of the op amp circuit, shown listed in Fig. 10.6, consider adding the following .NOISE command statement to the file, together with the statement specifying the output plot request:

```
.NOISE V(22) Vd
.PLOT NOISE ONOISE INOISE
```

Re-submitting the input file to Spice results in the output-referred and input-referred noise voltage spectral densities for the 741 op amp shown in Fig. 10.11. As a point of reference, we see that the output-referred noise voltage density at 1 kHz is 14.7 $\mu V/\sqrt{Hz}$. Conversely, the output noise power spectral density can be referred to the input of the op amp at this same frequency to find 19.0 $nV/\sqrt{Hz}$.

A Summary Of The 741 Op Amp Characteristics

We summarize the results that we have obtained through the application of Spice to the 741 op amp in Table 10.4. Some of these results are also compared to those computed by hand in Sections 10.1–10.6 of Sedra and Smith. For the most part, the results obtained through hand analysis agree quite well with those computed with Spice. This is especially true for the parameters obtained through the small-signal analysis. The same can not be said for many of the op amp's large-signal DC characteristics. We attribute this to the

Parameter	Units	Hand Calculation	Spice
Input-Referred Offset Voltage	μV	0	+314.1
Input Bias Currents	nA	47.5	34.48
Input Offset Current	nA	0	0.16
Quiescent Power Dissipation	mW	—	55.2
Differential-Mode Voltage Gain	kV/V	243.1	294.6
Common-Mode Voltage Gain	V/V	—	−1.55
Common-Mode Rejection Ratio	dB	—	105.6
Input Differential Resistance	MΩ	2.1	3.64
Input Common-Mode Resistance	MΩ	—	606.4
Output Resistance	Ω	∼ 75	106.4
Input Common-Mode Range	V	−12.6 to +14.4	−10.0 to +8.05
Output Voltage Swing	V	−13.5 to +14.0	−13.2 to +13.2
3dB Bandwidth	Hz	4.1	2.6
Unity-Gain Bandwidth	MHz	1	0.652
Phase Margin	degrees	—	66.7
Positive Slew Rate	V/μs	+0.63	+0.55
Negative Slew Rate	V/μs	−0.63	−0.39
Input-Referred Noise Voltage @ 1kHz	nV/$\sqrt{\text{Hz}}$	—	19.0
Output-Referred Noise Voltage @ 1kHz	μV/$\sqrt{\text{Hz}}$	—	14.7

Table 10.4 A comparison of the results of the analysis of the 741 op amp circuit shown in Fig. 10.1 by hand (Sedra and Smith) and Spice.

fact that the large-signal DC analysis performed by Sedra and Smith neglected the effect of transistor Early voltage, unlike that performed during the small-signal analysis. Of course, one could improve the accuracy of the hand analysis by including the effect of transistor Early voltage in the large-signal calculations; however, the additional complexity is probably not commensurate with the additional accuracy gained, especially when one could go directly to Spice.

10.2 A CMOS Op Amp

In Fig. 10.12 we show a two-stage CMOS op amp circuit with the device geometries seen listed in Table 10.4. The input differential stage of this amplifier was previously analyzed in Section 6.4 of this text for its large-signal differential-mode and common-mode transfer characteristics, among other things. Here an additional gain stage has been added in order to increase the gain of the amplifier, and a compensation network is included to maintain

10.2 A CMOS Op Amp

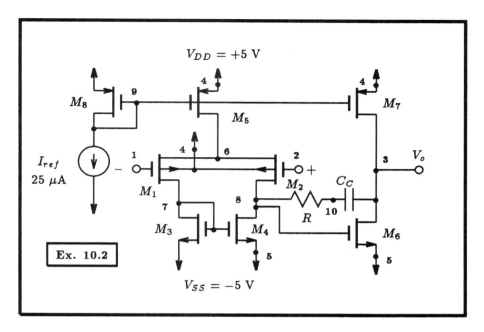

Figure 10.12 A two-stage CMOS op amp circuit with p-channel input transistors.

Transistor	M_1	M_2	M_3	M_4	M_5	M_6	M_7	M_8
W/L	120/8	120/8	50/10	50/10	150/10	100/10	150/10	150/10

Table 10.5 Transistor geometries for the CMOS op amp circuit shown in Fig. 10.12.

stability when the op amp is connected in a negative feedback loop. In Example 10.2 of Sedra and Smith this amplifier stage was analyzed by hand with rather simple transistor models where it was found to have a DC differential gain of +3125 V/V, an input common-mode range varying between −4.5 V to +3.0 V, and an output voltage swing ranging between −4.5 V and +4.4 V.

In the following we shall recalculate the above parameters of the CMOS amplifier with Spice using more realistic transistor models. In particular, we shall consider that each MOSFET is modeled after the transistors found in the 5-micron CMOS process at Bell Northern Research (BNR). We shall then compare these results with those predicted by the simple formulae presented in Sedra and Smith in Example 10.2 of their text using the DC operating point information provided by Spice.

The Spice description of the CMOS amplifier is shown in Fig. 10.12. The differential input is driven by the multiple source arrangement shown in Fig. 10.3. A level 2 MOSFET model of each transistor type is provided with a lengthly list of device parameters. These

Example 10.2: A CMOS Operational Amplifier (5um CMOS Models)

```
** Circuit Description **
* power supplies
Vdd 4 0 DC +5V
Vss 5 0 DC -5V
* differential-mode signal level
Vd 101 0 DC 0V
Rd 101 0 1
EV+ 2 100 101 0 +0.5
EV- 1 100 101 0 -0.5
* common-mode signal level
Vcm 100 0 DC 0V
* front-end stage
M1 7 1 6 4 pmos_transistor L=8u W=120u
M2 8 2 6 4 pmos_transistor L=8u W=120u
M3 7 7 5 5 nmos_transistor L=10u W=50u
M4 8 7 5 5 nmos_transistor L=10u W=50u
M5 6 9 4 4 pmos_transistor L=10u W=150u
* 2nd gain stage
M6 3 8 5 5 nmos_transistor L=10u W=100u
M7 3 9 4 4 pmos_transistor L=10u W=150u
* current source biasing stage
M8 9 9 4 4 pmos_transistor L=10u W=150u
Iref 9 5 25uA
* compensation network
Cc 8 10 10pF
R 10 3 10k
* 5um BNR CMOS transistor model statements
.MODEL nmos_transistor nmos ( level=2 vto=1 nsub=1e16 tox=8.5e-8 uo=750
+ cgso=4e-10 cgdo=4e-10 cgbo=2e-10 uexp=0.14 ucrit=5e4 utra=0 vmax=5e4 rsh=15
+ cj=4e-4 mj=2 pb=0.7 cjsw=8e-10 mjsw=2 js=1e-6 xj=1u ld=0.7u )
.MODEL pmos_transistor pmos ( level=2 vto=-1 nsub=2e15 tox=8.5e-8 uo=250
+ cgso=4e-10 cgdo=4e-10 cgbo=2e-10 uexp=0.03 ucrit=1e4 utra=0 vmax=3e4 rsh=75
+ cj=1.8e-4 mj=2 pb=0.7 cjsw=6e-10 mjsw=2 js=1e-6 xj=0.9u ld=0.6u )
** Analysis Requests **
.DC Vd -4mV +4mV 100uV
** Output Requests **
.PLOT DC V(3)
.probe
.end
```

Figure 10.13 The Spice input file describing the CMOS amplifier shown in Fig. 10.12. A Level 2 MOSFET model of each type of transistor is given.

were obtained through extensive measurements on transistors fabricated through the 5 μm CMOS process at BNR. The first analysis requested is a DC sweep of the input differential voltage V_d between the two supply limits with the input common-mode level V_{CM} set to 0 V.

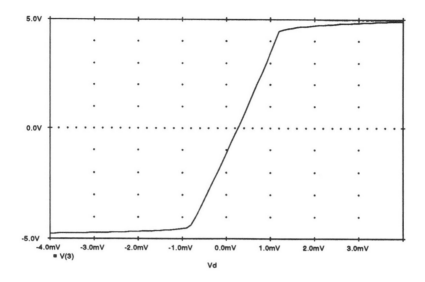

Figure 10.14 The large-signal differential-input transfer characteristics of the CMOS op amp circuit shown in Fig. 10.12. The input common-mode voltage level is set to zero volts.

The results of this DC sweep are shown plotted in Fig. 10.14. We see here that the linear region of the amplifier extends between the input voltage levels of −1.24 mV and +1.71 mV. Correspondingly, this gives rise to a maximum output voltage swing ranging between −4.40 V and +4.54 V. Thus, the small-signal differential gain can be estimated to be in the neighborhood of +3.031 kV/V. Also, we see that the input-referred offset voltage is approximately −220.0 μV.

Following this, we have performed a DC sweep of the input common-mode level with the differential input to the amplifier offset by +220.0 μV. To accomplish this, we replaced the DC sweep command given in Fig. 10.13 by the following one:

```
.DC Vcm -5V +5V 0.1V.
```

and modified the input differential voltage statement V_d according to:

```
Vd 101 0 DC +220.0uV.
```

The results of this analysis are displayed in Fig. 10.15. As is evident, the input common-mode range extends from the lower limit of the power supply V_{SS} to +3.1 V. We can further add commands .TF and .OP into the Spice deck and obtain the small-signal DC gain of the circuit and any relevant DC operating point information. These two new commands would appear in the Spice deck as follows:

10 Analog Integrated Circuits

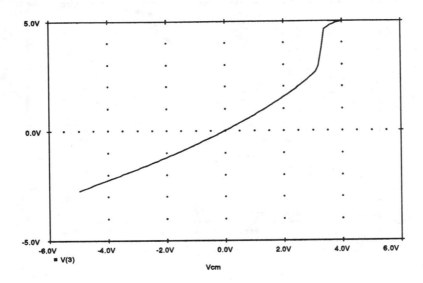

Figure 10.15 The large-signal common-mode DC transfer characteristics of the CMOS op amp circuit shown in Fig. 10.12. An input differential offset voltage of +220.0 µV is applied to prevent premature saturation.

```
                    .OP
                    .TF V(3) Vd.
```

One would then find in the output file the following information:

```
****     SMALL SIGNAL BIAS SOLUTION          TEMPERATURE =    27.000 DEG C
*****************************************************************************

    NODE   VOLTAGE      NODE   VOLTAGE      NODE   VOLTAGE      NODE   VOLTAGE

 (    1)-110.0E-06  (    2) 110.0E-06  (    3)    .0125  (    4)   5.0000
 (    5)  -5.0000   (    6)   1.8828   (    7)  -3.5404  (    8)  -3.5516
 (    9)   3.4242   (   10)    .0125   (  100)   0.0000  (  101) 220.0E-06


        VOLTAGE SOURCE CURRENTS
        NAME         CURRENT

        Vdd         -8.061E-05
        Vss          8.061E-05
        Vd          -2.200E-04
        Vcm          0.000E+00

        TOTAL POWER DISSIPATION   8.06E-04  WATTS

    NAME        M1               M2               M3               M4
    MODEL       pmos_transistor  pmos_transistor  nmos_transistor  nmos_transistor
    ID          -1.34E-05        -1.34E-05         1.34E-05         1.34E-05
    VGS         -1.88E+00        -1.88E+00         1.46E+00         1.46E+00
    VDS         -5.42E+00        -5.43E+00         1.46E+00         1.45E+00
    VBS          3.12E+00         3.12E+00         0.00E+00         0.00E+00
    VTH         -1.51E+00        -1.51E+00         9.52E-01         9.52E-01
    VDSAT       -3.18E-01        -3.18E-01         2.83E-01         2.83E-01
```

```
GM              7.18E-05           7.18E-05          5.35E-05          5.34E-05
GDS             7.85E-07           7.84E-07          6.05E-07          6.07E-07
GMB             8.53E-06           8.53E-06          3.88E-05          3.88E-05

NAME            M5                 M6                M7                M8
MODEL           pmos_transistor    nmos_transistor   pmos_transistor   pmos_transistor
ID              -2.69E-05          2.87E-05          -2.87E-05         -2.50E-05
VGS             -1.58E+00          1.45E+00          -1.58E+00         -1.58E+00
VDS             -3.12E+00          5.01E+00          -4.99E+00         -1.58E+00
VBS             0.00E+00           0.00E+00          0.00E+00          0.00E+00
VTH             -9.59E-01          9.38E-01          -9.54E-01         -9.64E-01
VDSAT           -4.56E-01          2.87E-01          -4.61E-01         -4.51E-01
GM              8.64E-05           1.14E-04          9.19E-05          8.08E-05
GDS             1.09E-06           6.85E-07          9.36E-07          1.42E-06
GMB             2.63E-05           8.17E-05          2.77E-05          2.48E-05

****    SMALL-SIGNAL CHARACTERISTICS

        V(3)/Vd =   3.603E+03

        INPUT RESISTANCE AT Vd =   1.000E+00

        OUTPUT RESISTANCE AT V(3) =   6.210E+05
```

Here we see that the small-signal DC gain of +3.603 kV/V has the same order of magnitude as our earlier estimate. We can check this value against the gain predicted by the formulae derived in Sedra and Smith; $A_0 = g_{m1}(r_{o2}||r_{o4})g_{m6}(r_{o6}||r_{o7})$. Substituting the appropriate values from the above Spice generated small-signal data we see that $A_0 = +3.630$ kV/V, which agrees quite closely with the value predicted by the .TF command of Spice.

We can also make use of the above DC and small-signal information to estimate the input CMR of this amplifier and its output voltage swing. The lower limit of the input CMR is determined by M_1 leaving the saturation region. This occurs when the voltage at the gate of M_1 drops below the voltage at its drain $(V(7) = -3.5404$ V) by a single threshold level. From the above Spice generated data, we see that the threshold voltage of M_1 is -1.51 V. We can then compute the lower limit of the input CMR to be -5.1 V. This limit seems to extend beyond V_{SS} of -5 V, unlike that calculated by hand where it was found that the lower CMR limit is -4.5 V. The reason for the difference is that in the hand calculation a threshold voltage of -1 V for M_1 was assumed neglecting its dependence on the source-substrate back-bias voltage (the body effect). If the hand calculation used the actual threshold voltage of M_1 at -1.51 V, we would obtain the exact same lower limit to the CMR found through Spice.

The upper limit of the input CMR is determined by M_5 leaving the saturation region. This occurs when the drain of M_5 rises one threshold voltage above its gate voltage of $+3.4242$ V. The threshold voltage of M_5 is -0.959 V. Hence, M_1 leaves saturation when its drain voltage exceed $+4.38$ V. To relate this voltage to the input of the amplifier, we simply subtract off the gate-source voltage of either M_1 or M_2. The result is the upper CMR limit of (4.38 V $-$ 1.88 V) or $+2.5$ V. This is somewhat lower than that predicted by hand analysis

10 Analog Integrated Circuits

at +3 V. It is also lower than that seen from the plot of the output voltage of the amplifier as a function of the input common-mode voltage shown in Fig. 10.15. There are two reasons for the discrepancy. The first is again due to the fact that the transistor body effect has been neglected. In addition, the current flowing through both M_1 and M_2 is somewhat reduced because of the dependence of the drain current of M_5 on its drain voltage. This, in turn, decreases the gate-source voltage of M_1 and M_2. Thus, a larger input common-mode voltage can appear at the input terminals of the amplifier before any internal saturation occurs.

Following similar reasoning, the maximum range of the output voltage is one threshold voltage above the gate voltage of M_7. Thus, it is simply (3.42 V + 0.954 V) or +4.37 V. Likewise, the minimum output voltage is one threshold voltage below the gate voltage of M_6 and is therefore −4.49 V. These results seem to be reasonably close to the results that are visible from Fig. 10.14 using Probe (ie. −4.40 V and +4.54 V) and those computed by hand (ie. −4.5 V and +4.4 V).

10.3 Two Different Technology Versions Of A Folded-Cascode Op Amp

In Fig. 10.16 we display two different versions of a folded-cascode op amp circuit; one realized using CMOS technology and the other using a combination of CMOS and Bipolar technology called BiCMOS. The dimensions of the various MOSFETs of the two circuits are provided in Table 10.6; the bipolar transistors are assumed all equal in size, except Q_6 and Q_7 which are three times the size of the other bipolar transistors.

In the following we shall compare the frequency response behavior of these two op amp circuits assuming that the MOSFETs are modeled with the complex BNR 5-micron CMOS models used in the previous section and the bipolar transistors are modeled using the complex set of parameters listed in Table 10.1. But before we begin this, we shall first compute the frequency response behavior of the two amplifiers by hand in order to obtain some sense of expected amplifier behavior. We shall assume that a simplified representation of the NMOS transistors are characterized by: $V_t = 1$ V, $\mu_n C_{OX} = 20$ μA/V^2 and $\lambda = 0.04$ V^{-1}; similarly, for the PMOS transistors: $V_t = -1$ V, $\mu_p C_{OX} = 10$ μA/V^2 and $\lambda = 0.04$ V^{-1}. A simplified model of the *npn* transistors of the BiCMOS circuit consists of $I_S = 10$ fA, $\beta = 200$ and $V_A = 125$ V.

According to the small-signal analysis performed by Sedra and Smith in Section 10.8 for a folded-cascode op amp, the voltage gain of either circuit is given by

$$A_0 = g_{m1} R_o \tag{10.1}$$

10.3 Two Different Technology Versions Of A Folded-Cascode Op Amp

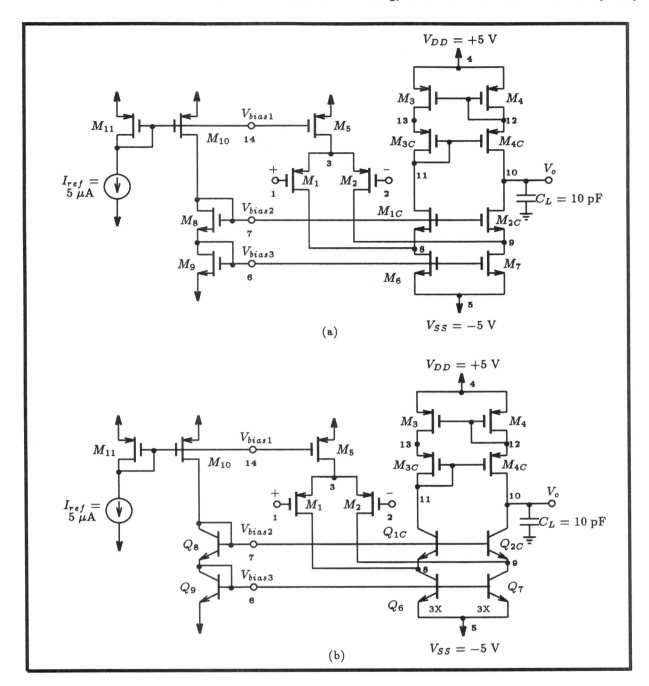

Figure 10.16 Folded-Cascoded Op Amp: (a) CMOS implementation (b) BiCMOS implementation.

Transistor W/L	M_1	M_2	M_3	M_4	M_5	M_6	M_7	M_8
	500/10	500/10	60/6	60/6	120/6	60/6	60/6	60/6
	M_9	M_{10}	M_{11}	M_{1C}	M_{2C}	M_{3C}	M_{4C}	
	20/6	30/6	30/6	60/6	60/6	60/6	60/6	

Table 10.6 Transistor geometries for the CMOS op amp circuit shown in Fig. 10.16.

I_D, I_C (A)	M_1	M_2	M_3	M_4	M_5	M_6 / Q_6	M_7 / Q_7	M_8 / Q_8
	10μ	10μ	5μ	5μ	20μ	15μ	15μ	5μ
	M_9 / Q_9	M_{10}	M_{11}	M_{1C} / Q_{1C}	M_{2C} / Q_{2C}	M_{3C} / Q_{3C}	M_{4C} / Q_{4C}	
	5μ	5μ	5μ	5μ	5μ	5μ	5μ	

Table 10.7 DC drain and collector currents of the CMOS and BiCMOS circuits shown in Fig. 10.16 as computed by hand analysis.

where R_o is the output resistance of the op amp, given by

$$R_o = [g_{m2C}r_{o2C}(r_{o7}\|r_{o2})][g_{m4C}r_{o4C}r_{o3}] \quad (10.2)$$

Moreover, the dominant pole of the amplifier is located at frequency

$$\omega_D = 2\pi f_D = 1/C_L R_o \quad (10.3)$$

and the corresponding unity-gain frequency is

$$\omega_t = 2\pi f_t = g_{m1}/C_L \quad (10.4)$$

Performing a DC hand analysis on the folded-cascoded CMOS amplifier shown in Fig. 10.16(a) with the above mentioned MOSFET device and geometry parameters, we obtain their corresponding drain currents listed in Table 10.7. To simplify the analysis, we have neglected the effect of channel-length modulation. This analysis is rather straightforward owing to the fact that the reference current I_{ref} is simply mirrored into the other branches of the op amp.

10.3 Two Different Technology Versions Of A Folded-Cascode Op Amp

Repeating the same DC analysis for the BiCMOS amplifier shown in Fig. 10.16(b) while neglecting the base currents of the bipolar transistors, one obtains exactly the same drain currents for each MOSFET as was found in the CMOS amplifier. The collector current of each npn transistor is equal to the drain current of the MOSFET that it replaces. Even though a MOSFET is replaced by a bipolar transistor, the same mirroring action is used in each circuit (ie. the same current gains).

Knowing the device operating currents, we can compute the small-signal parameters of the devices of interest, and thus, calculate expected frequency response behavior of the op amp. Below we list the transconductance and output resistance of the appropriate transistors of the CMOS and BiCMOS op amps. These can then be used to compute small-signal parameters A_0, R_o, f_D and f_t, as mentioned above.

Transistor	CMOS Circuit						BiCMOS Circuit					
	M_1	M_2	M_3	M_7	M_{2C}	M_{4C}	M_1	M_2	M_3	Q_7	Q_{2C}	M_{4C}
g_m	100μ	100μ	31μ	77.5μ	44.7μ	31.6μ	100μ	100μ	31μ	600μ	200μ	31.6μ
r_o	2.5M	2.5M	5M	1.67M	5M	5M	2.5M	2.5M	5M	8.3M	25M	5M

Substituting the above small-signal parameters into Eqns. (10.1)-(10.4), we obtain the following small-signal frequency response data for the two op amps:

Op-Amp	g_{m1}	R_o	A_o	f_D	f_t
CMOS	100 μA/V	174 MΩ	84.82 dB	91.2 Hz	1.59 MHz
BiCMOS	100 μA/V	678 MΩ	96.6 dB	23.4 Hz	1.59 MHz

As is evident from this table of values, the BiCMOS op amp has a larger low frequency gain but a lower bandwidth than the corresponding CMOS version. Interestingly enough, both op amps have identical unity-gain bandwidths.

To investigate realistic op amp frequency response behavior we make use of the two Spice decks listed in Figs. 10.17 and 10.18 where the transistors are modeled as mentioned earlier. A DC sweep command of the input differential voltage is performed first in order to locate the high-gain linear region of the amplifier. The common-mode input voltage level will be set at 0 V during this voltage sweep.

On completion of Spice, the large-signal differential characteristics of the CMOS and BiCMOS amplifiers are shown plotted in Fig. 10.15. As is evident, they have similar DC gains (ie. slope of the linear region), with the BiCMOS circuit having a slightly larger gain. From these results we have determined with the aid of Probe that the DC offset of the CMOS

```
A Folded-Cascode CMOS Op-Amp

** Circuit Description **
* power supplies
Vdd 4 0 DC +5V
Vss 5 0 DC -5V
* differential-mode signal level
Vd 101 0 DC 0V AC 1V
Rd 101 0 1
EV+ 1 100 101 0 +0.5
EV- 2 100 101 0 -0.5
* common-mode signal level
Vcm 100 0 DC 0V
* differential-pair steering control
M1 8 1 3 4        pmos_transistor L=10u W=500u
M2 9 2 3 4        pmos_transistor L=10u W=500u
M5 3 14 4 4       pmos_transistor L=6u  W=120u
* cascode stage
M3 13 12 4 4      pmos_transistor L=6u W=60u
M4 12 12 4 4      pmos_transistor L=6u W=60u
M3c 11 11 13 4    pmos_transistor L=6u W=60u
M4c 10 11 12 4    pmos_transistor L=6u W=60u

M1c 11 7 8 5      nmos_transistor L=6u W=60u
M2c 10 7 9 5      nmos_transistor L=6u W=60u
M6  8 6 5 5       nmos_transistor L=6u W=60u
M7  9 6 5 5       nmos_transistor L=6u W=60u
* biasing stage
M8 7 7 6 5        nmos_transistor L=6u W=60u
M9 6 6 5 5        nmos_transistor L=6u W=20u
M10 7 14 4 4      pmos_transistor L=6u W=30u
M11 14 14 4 4     pmos_transistor L=6u W=30u
Iref 14 5 5uA
* load capacitance
Cl 10 0 10pF
* transistor model statements
* 5u BNR CMOS transistor model statements
.MODEL nmos_transistor nmos ( level=2 vto=1 nsub=1e16 tox=8.5e-8 uo=750
+ cgso=4e-10 cgdo=4e-10 cgbo=2e-10 uexp=0.14 ucrit=5e4 utra=0 vmax=5e4 rsh=15
+ cj=4e-4 mj=2 pb=0.7 cjsw=8e-10 mjsw=2 js=1e-6 xj=1u ld=0.7u )
.MODEL pmos_transistor pmos ( level=2 vto=-1 nsub=2e15 tox=8.5e-8 uo=250
+ cgso=4e-10 cgdo=4e-10 cgbo=2e-10 uexp=0.03 ucrit=1e4 utra=0 vmax=3e4 rsh=75
+ cj=1.8e-4 mj=2 pb=0.7 cjsw=6e-10 mjsw=2 js=1e-6 xj=0.9u ld=0.6u )
** Analysis Requests **
.DC Vd -10mV +10mV 100uV
** Output Requests **
.PLOT DC V(10)
.probe
.end
```

Figure 10.17 The Spice input file describing the folded-cascode CMOS amplifier.

10.3 Two Different Technology Versions Of A Folded-Cascode Op Amp

and BiCMOS amplifiers are -857.1 μV and -500 μV, respectively. Also apparent, the output voltage swing of the BiCMOS amplifier is about one volt larger than its CMOS counterpart.

Offsetting the input to the two amplifiers with the appropriate DC voltage so that each amplifier is biased inside its linear region, we can compute the frequency response of each amplifier. This is accomplished by altering the input signal source statement in the Spice deck for the CMOS amplifier according to

```
Vd 101 0 DC -857.1uV AC 1V
```

and, likewise, for the BiCMOS amplifier with

```
Vd 101 0 DC -500.0uV AC 1V,
```

and adding the following AC analysis command in each input file

```
.AC DEC 10 1Hz 100MegHz,
```

we can then submit the revised Spice decks for analysis. The magnitude and phase behavior of the two amplifiers, as computed by Spice, are shown together in Fig. 10.20. Using Probe we were able to extract the following frequency response information from these results:

Op-Amp	A_o	f_D	f_t
CMOS	67.4 dB	794.3 Hz	1.82 MHz
BiCMOS	72.5 dB	400 Hz	1.58 MHz

Examining the above results, we see that the BiCMOS amplifier has a slightly larger low-frequency gain but a lower 3 dB bandwidth than the corresponding CMOS amplifier. This manifests itself into the CMOS amplifier having a larger unity-gain bandwidth product than the BiCMOS amplifier. In comparison with the results computed previously by hand, we find that the predicted DC gain and 3 dB bandwidth for the two amplifiers are quite different. The reason for the discrepancy can be traced back to the rather crude estimates of the small-signal model parameters used to compute them. There are two reasons for this: the simplified model parameters for the MOSFETs used in the hand calculations are not truly representative of actual transistor behavior, and secondly, the effect of transistor Early voltage and body-effect were neglected when the transistor bias levels were being computed. Using the small-signal parameters generated by Spice through the .OP command, hand analysis using Eqns. (10.1)-(10.4) will generate results similar to those computed by Spice. We leave this for the reader to confirm using Spice.

The final point that we want to discuss here is the location of the second or nondominant pole of each amplifier. According to the frequency response results shown in Fig. 10.20 the

10 Analog Integrated Circuits

```
A Folded-Cascode BiCMOS Op-Amp

** Circuit Description **
* power supplies
Vdd 4 0 DC +5V
Vss 5 0 DC -5V
* differential-mode signal level
Vd 101 0 DC 0V
Rd 101 0 1
EV+ 1 100 101 0 +0.5
EV- 2 100 101 0 -0.5
* common-mode signal level
Vcm 100 0 DC 0V
* differential-pair steering control
M1  8  1  3 4     pmos_transistor L=10u W=500u
M2  9  2  3 4     pmos_transistor L=10u W=500u
M5  3  14 4 4     pmos_transistor L=6u  W=120u
* cascode stage
M3  13 12 4  4    pmos_transistor L=6u W=60u
M4  12 12 4  4    pmos_transistor L=6u W=60u
M3c 11 11 13 4    pmos_transistor L=6u W=60u
M4c 10 11 12 4    pmos_transistor L=6u W=60u

Q1c 11 7 8 5    npn_transistor
Q2c 10 7 9 5    npn_transistor
Q6  8  6 5 5    npn_transistor 3
Q7  9  6 5 5    npn_transistor 3
* biasing stage
Q8  7  7 6 5    npn_transistor
Q9  6  6 5 5    npn_transistor
M10 7  14 4 4   pmos_transistor L=6u W=30u
M11 14 14 4 4   pmos_transistor L=6u W=30u
Iref 14 5 5uA
* load capacitance
Cl 10 0 10pF
* transistor model statements
* Bipolar transistor model statements
.model npn_transistor NPN ( Bf=200 Br=2.0 VAf=125V Is=10fA Tf=0.35ns
+ Rb=200 Rc=200 Re=2 Cje=1.0pF Vje=0.70V Mje=0.33 Cjc=0.3pF Vjc=0.55V
+ Mjc=0.5 Cjs=3.0pF Vjs=0.52V Mjs=0.5 )
* 5u BNR CMOS transistor model statements
.MODEL pmos_transistor pmos ( level=2 vto=-1 nsub=2e15 tox=8.5e-8 uo=250
+ cgso=4e-10 cgdo=4e-10 cgbo=2e-10 uexp=0.03 ucrit=1e4 utra=0 vmax=3e4 rsh=75
+ cj=1.8e-4 mj=2 pb=0.7 cjsw=6e-10 mjsw=2 js=1e-6 xj=0.9u ld=0.6u )
** Analysis Requests **
.DC Vd -10mV +10mV 100uV
** Output Requests **
.PLOT DC V(10)
.probe
.end
```

Figure 10.18 The Spice input file describing the folded-cascode BiCMOS amplifier.

10.3 Two Different Technology Versions Of A Folded-Cascode Op Amp

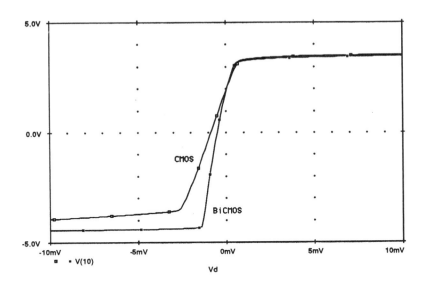

Figure 10.19 The large-signal differential transfer characteristics of the CMOS and BiCMOS amplifiers shown in Fig. 10.16.

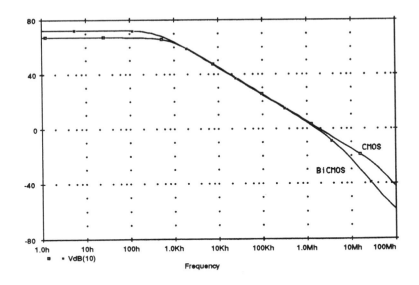

Figure 10.20 A comparison of magnitude and phase response of the CMOS and BiCMOS amplifiers shown in Fig. 10.16.

CMOS amplifier has a nondominant pole at 36 MHz. Whereas, the BiCMOS amplifier has its nondominant pole located at a much lower frequency of only 4.5 MHz. This result is rather contrary to what one would expected when a MOSFET is replaced by a corresponding bipolar transistor biased at the same current level. Since, after all, a bipolar transistor has

475

10 Analog Integrated Circuits

a higher transconductance than a corresponding MOSFET when biased at the same current level. However, this is not the entire story because the location of the nondominant pole is also determined by the parasitic capacitance associated with the transistor. As it turns out, in this particular case, the bipolar transistors used to replace the NMOS transistors have associated with it a much larger parasitic capacitance (this can be confirmed by viewing the small-signal model data generated by Spice for each transistor). Thus, what signal strength is gained from the increased transconductance is lost in the frequency domain to the added parasitic capacitance of the bipolar transistor. This is not necessarily what would happen in an actual design as the models for the MOSFETs and *npn* transistors used in this example were simply contrived and do not represent an actual BiCMOS technology.

10.4 Data Converters

Analog-to-digital (A/D) and digital-to-analog (D/A) converter circuits are important components of electronic systems. In the following we shall simulate the ideal behavior of a D/A and an A/D converter using Spice. More specifically, we shall simulate the operation of the conversion of a 4-bit digital word into a corresponding analog signal using a 4-bit R-2R ladder D/A converter circuit. Secondly, we shall also observe the conversion of an analog signal into a 3-bit digital signal using a flash converter. Our objective here is to investigate basic circuit operation as opposed to observing nonideal behavior.

R-2R Ladder D/A Converter

Figure 10.21 displays the circuit schematic of a 4-bit D/A converter utilizing an R-2R ladder network. The double throw switches are assumed ideal and are controlled by the appropriate voltage source whose level represents a single bit of a 4-bit digital number. It is precisely this digital number that is to be converted into an equivalent analog signal. The op amp is assumed pseudo-ideal and represented by the high-gain VCVS shown in Fig. 10.21(b). A Spice description of this circuit is shown listed in Fig. 10.22. Four digital signal generators are provided, generating the entire sequence of digital words possible with 4-bits over an 8 ms interval. A low logic level is represented by 0 V and a high logic level by +5 V.

Of special interest, and not seen by the reader until now, is the PSpice element statement used to represent a voltage-controlled switch. This element is not available in all versions of Spice, but greatly simplifies circuit modeling, and thus, is employed here to represent a single-pole single-throw switch.[†] The general syntax of the PSpice voltage-controlled switch

---

[†] An equivalent realization of a voltage-controlled switch can be described to Spice using a nonlinear

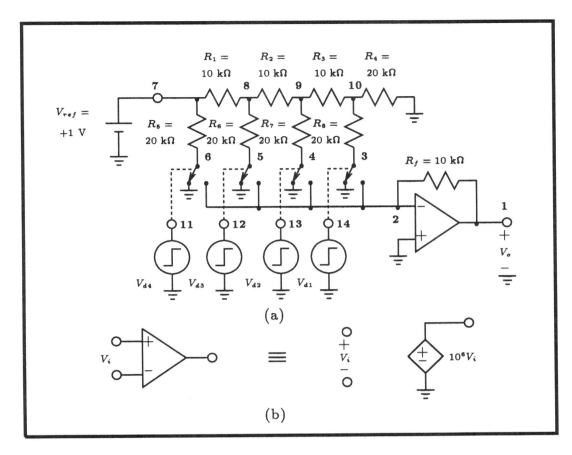

Figure 10.21 (a) A D/A converter utilizing an R-2R ladder network (b) op amp equivalent circuit.

statement is shown in Fig. 10.23. Two statements are necessary to describe a voltage-controlled switch to PSpice: a statement beginning with the letter S (for switch) specifying the connections made to the rest of the network and the name of the model used to characterize the switch. Another statement, cross-referenced to this element statement, contains a list of four parameters that describe the terminal behavior of the switch. These are: Von, the control voltage level required to turn on the switch; $Voff$, the voltage level that turns off the switch; and Ron and $Roff$, the on- and off-resistance of the switch, respectively.

In the PSpice description of the D/A converter seen listed in Fig. 10.22, the double-throw switch is represented by two complementary switches driven by the same digital control signal. The switches are complementary in the sense of their state of conduction. That is, one switch is conducting while the other is blocking.

VCCS statement having the following form: *Gswitch n+ n- poly(2) nc+ nc- n+ n- 0 0 0 0 1/Ron*. Nodes *n+* and *n-* are the terminals of the switch. Nodes *nc+* and *nc-* specify the two nodes whose voltage difference determines the state of the switch. The voltage appearing between nodes *nc+* and *nc-* divided by *Ron* specifies the conductance of the switch. If this voltage is zero, then the switch is open; otherwise, it is conducting with on-resistance *V(nc+,nc-)/Ron*.

10 Analog Integrated Circuits

```
R-2R Ladder D/A Converter

** Circuit Description **

* op-amp subcircuit
.subckt ideal_opamp 1 2 3
* connections:        | | |
*                output | |
*               +ve input |
*                    -ve input
Iopen1 2 0 0A
Iopen2 3 0 0A
Ecomp 1 0 2 3 1e6
.ends ideal_opamp

** Main Circuit **
* input digital signals
Vd1 14 0 PULSE (5 0 0 1ns 1ns 0.5ms 1ms)
Vd2 13 0 PULSE (5 0 0 1ns 1ns 1ms 2ms)
Vd3 12 0 PULSE (5 0 0 1ns 1ns 2ms 4ms)
Vd4 11 0 PULSE (5 0 0 1ns 1ns 4ms 8ms)
* reference voltage
Vref 7 0 DC 1V
* R-2R ladder network
R1 7 8 10k
R2 8 9 10k
R3 9 10 10k
R4 10 0 20k
R5 6 7 20k
R6 5 8 20k
R7 4 9 20k
R8 3 10 20k
* switches
S1 3 2 14 0 switch_model
S1c 3 0 14 0 complementary_switch_model
S2 4 2 13 0 switch_model
S2c 4 0 13 0 complementary_switch_model
S3 5 2 12 0 switch_model
S3c 5 0 12 0 complementary_switch_model
S4 6 2 11 0 switch_model
S4c 6 0 11 0 complementary_switch_model
* current-to-voltage converter
Rf 1 2 10k
Xopamp 1 0 2 ideal_opamp
* switch model
.model switch_model vswitch (Ron=1 Roff=1e6 Von=5V Voff=0V)
.model complementary_switch_model vswitch (Ron=1 Roff=1e6 Von=0V Voff=5V)
** Analysis Requests **
.TRAN 500us 8ms 0 500us
** Output Requests **
.PRINT TRAN V(1) V(11) V(12) V(13) V(14)
.probe
.end
```

Figure 10.22 The PSpice input file for analyzing the D/A converter shown in Fig. 10.21.

The results of the transient analysis are shown in Fig. 10.24. The top graph displays the sequence of digital signals applied as input to the D/A converter and the bottom graph displays the corresponding analog output signal. The glitches seen superimposed on the

10.4 Data Converters

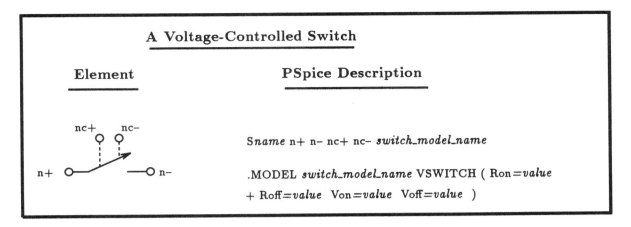

Figure 10.23 PSpice element and model description of a voltage-controlled switch. This element is not available in some versions of Spice, but can be realized in Spice using a nonlinear voltage-controlled current source; see footnote on page 477.

analog output signal are due to the very short interval when all the switches disconnect from the op amp negative input terminal on their way to their next position.

3-Bit A/D Flash Converter

In Fig. 10.25 we display a 3-bit A/D flash converter circuit. Seven comparators are required for 3-bit accuracy. A 10 kΩ resistor string is used to realize the reference voltages for the various comparators. An input voltage signal V_i is applied to the positive input terminal of each comparator, and if the input signal is larger than the reference voltage applied to the negative input terminal, the output of the comparator goes into its high logic state, otherwise it remains in its low state.

In the following we shall compute the output state of each comparator as a function of the input voltage level using Spice. We shall model the terminal characteristics of each comparator using the equivalent circuit shown in part (b) of Fig. 10.25. The front-end portion of the comparator is modeled as a high-gain amplifier with a VCVS. The output of the amplifier is maintained within the limits of the two power supplies by clamping the output voltage to one of the supply levels using the diode and DC voltage source limiter. The diodes will be assumed ideal, and made nearly so, by setting n small in the diode model statement (ie. $n = 0.01$).

The Spice file describing this circuit is shown listed in Fig. 10.26. Here we are requesting a DC sweep of the input voltage beginning at the lower power supply limit of −5 V and ending at the +5 V power supply level. In this way, we should be able to exercise all possible states of the A/D converter.

On completion of Spice, we have plotted the output voltage of each comparator as a

10 Analog Integrated Circuits

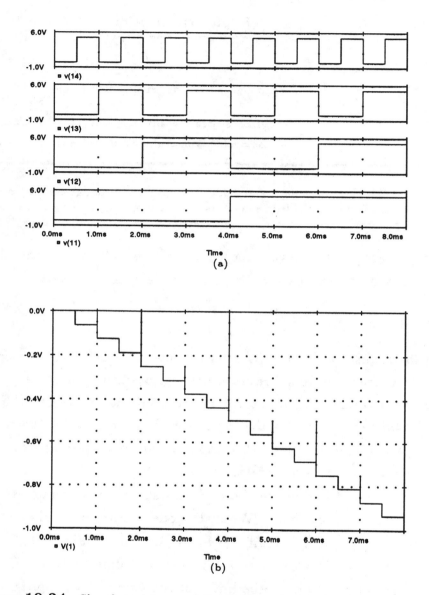

Figure 10.24 Signal waveforms of the D/A converter shown in Fig. 10.21: (a) switch control signals (b) analog output voltage signal as a function of time.

function of the input voltage level, as shown in Fig. 10.27. Clearly, for an input voltage level between −5 V and −3.75 V all comparator outputs are logically low. Subsequently, with the input increased to some level somewhere between −3.75 V and −2.5 V, comparator C_7 goes logically high with all other comparators remaining in the logic low state. Continuing, for the next input voltage interval between −2.5 V and −1.25 V, we see that comparator C_6 has now also gone into its high state. As we further increase the input voltage level, C_5 will

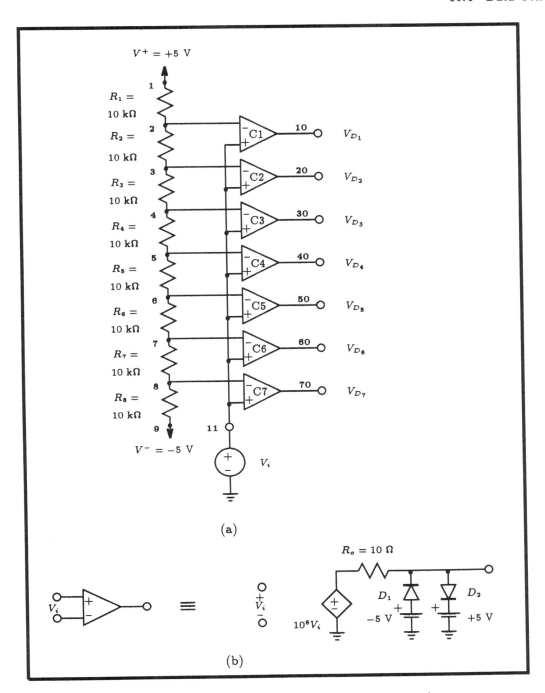

Figure 10.25 (a) A 3-Bit A/D flash converter utilizing 7 comparator circuits. (b) equivalent circuit representation of the comparator circuit.

be the next one that goes into its high state. This will be followed by C_4, C_3, etc., until all comparator outputs are logically high when the input level exceeds $+3.75$ V. Since the combined output of this A/D has 8 possible states, it can be represented by a 3-bit binary number. This conversion would be performed by a digital circuit that would follow the A/D

10 Analog Integrated Circuits

```
A 3-Bit A/D Flash Converter

** Circuit Description **

* comparator subcircuit
.subckt comparator  1 2 3
* connections:      | | |
*            output | |
*          +ve input |
*            -ve input
Iopen1 2 0 0A
Iopen2 3 0 0A
Ecomp 6 0 2 3 1e6
Rout 6 1 10
V+sat 4 0 DC +5V
V-sat 5 0 DC -5V
D1 1 4 ideal_diode
D2 5 1 ideal_diode
.model ideal_diode D (Is=100pA n=0.01)
.ends comparator

** Main Circuit **

* power supplies
Vcc 1 0 DC +5V
Vee 9 0 DC -5V
* input signal source
Vi 11 0 DC 0V
* reference levels for comparators
R1 1 2 10k
R2 2 3 10K
R3 3 4 10k
R4 4 5 10k
R5 5 6 10k
R6 6 7 10k
R7 7 8 10k
R8 8 9 10k
* comparators
Xc1 10 11 2 comparator
Rd1 10 0 1k
Xc2 20 11 3 comparator
Rd2 20 0 1k
Xc3 30 11 4 comparator
Rd3 30 0 1k
Xc4 40 11 5 comparator
Rd4 40 0 1k
Xc5 50 11 6 comparator
Rd5 50 0 1k
Xc6 60 11 7 comparator
Rd6 60 0 1k
Xc7 70 11 8 comparator
Rd7 70 0 1k
** Analysis Requests **
.DC Vi -5V +5V 100mV
** Output Requests **
.PLOT DC V(10) V(20) V(30) V(40) V(50) V(60) V(70)
.probe
.end
```

Figure 10.26 The Spice input file for analyzing the A/D converter shown in Fig. 10.25.

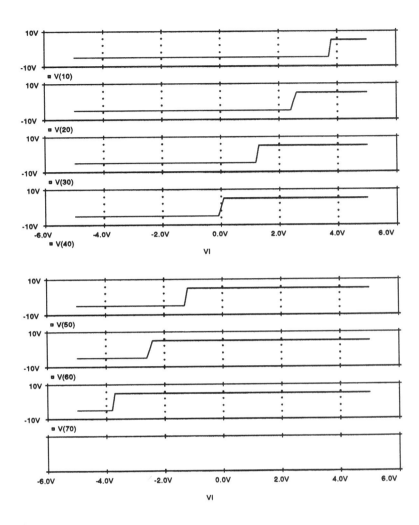

Figure 10.27 The output digital waveforms from the comparators of the A/D 3-bit flash converter shown in Fig. 10.25.

converter.

10.5 Spice Tips

- Inputs to differential amplifiers should consist of both a differential and a common-mode component. An interesting arrangement of several voltage sources was given in this chapter, illustrating how the differential and common-mode components can be independently adjusted.
- Performing a proper small-signal analysis on a high-gain amplifier circuit requires that the linear region of the amplifier be first located before any transfer function or frequency analysis is performed.
- The linear region of an op amp is located by first sweeping the input differential voltage

v_d between the limits of the power supplies with the common-mode voltage V_{CM} equal to 0 V. This analysis is repeated with a reduced sweep range centered more closely around the high-gain linear region of the amplifier until a smooth transition through the high-gain linear region is achieved. Following this, one must check to see whether the input common-mode voltage of 0 V keeps the amplifier in its linear region.

- To quickly converge on the location of the high-gain region of an op amp, one can connect the op amp in a unity-gain configuration with the positive terminal of the op amp connected to ground and compute the input-referred offset voltage of the op amp V_{OS}. Knowing this voltage we can infer that the differential transfer characteristics of the op amp will cross the horizontal voltage axis ($V_o = 0$ V) at $-V_{OS}$. This is usually a valid point within the high-gain region of the op amp.

- Once the linear region of the amplifier is located, the correct DC conditions can be applied to the inputs of the amplifier to establish the proper bias point about which the small-signal analysis of Spice can be performed.

- The small-signal input resistance of an op amp is computed using Spice by applying a known AC voltage across the input terminals of the differential amplifier and computing the AC currents that flow into the amplifier terminals. In many practical amplifier situations, these currents will not be equal. So, instead, the average of these two currents is used in the input resistance calculation.

- The output thermal noise behavior of electronic circuits can be estimated using the noise analysis capability of Spice.

- A voltage-controlled switch can be modeled in PSpice using a special element and model statement for the switch. An equivalent realization of a voltage-controlled switch can be described to Spice using a nonlinear VCCS statement having the following form: *Gswitch n+ n- poly(2) nc+ nc- n+ n- 0 0 0 0 1/Ron*.

10.6 Bibliography

P.R. Gray and R.G Meyer, *Analysis and Design of Analog Integrated Circuits*, 2nd ed. New York: Wiley, 1984.

10.7 Problems

10.1 Determine, with the aid of Spice, the differential-mode and common-mode voltage gain of the 741 op amp when the power supplies are reduced to $\pm 5V$.

10.2 Assume that the IC process used to fabricate the 741 op amp circuit undergoes some variation that causes the Early voltage of each device to decrease by 30%, but all other parameters remain invariant. Making the appropriate changes to the Spice parameters listed in Table 10.1, recalculate the large and small-signal parameters of the 741 op amp seen listed in Table 10.4 and compare them to those listed in this table.

10.3 One way to eliminate the input-referred offset voltage of an op amp such as the 741 is to connect what is known as a nulling resistor between the emitters of Q_5 and Q_6 of the first stage of the op amp seen in Fig. 10.1. With the aid of Spice determine the value of this nulling resistor that will cause the input offset voltage to be reduced to a value less than 10 μV. *Hint: Connect the op amp in a unity-gain configuration and adjust the value of the nulling resistor to eliminate the offset voltage.*

10.4 Through a processing imperfection, the β of Q_4 in Fig. 10.1 is reduced to 25 while the β of Q_3 remains at its regular value of 50. Find the input offset voltage that this mismatch introduces using Spice.

10.5 For a 741 op amp employing ± 7.5 V supplies, determine the maximum output voltage swing possible.

10.6 Determine the phase and gain margin of the 741 op amp when the compensation capacitor C_C is reduced from 30 pF down to 20 pF. Will the amplifier be stable when it is connected in a unity-gain configuration? Use the device parameters provided in Table 10.1.

10.7 This problem involves investigating the short-circuit protection capability of the 741 op amp. Consider connecting the op amp in a unity-gain configuration with the output terminal connected directly to ground. With the input to the voltage follower connected to +5 V, compare the power dissipated by the output stage with and without the internal short-circuit protection circuitry in place.

10.8 In a particular design of the CMOS op amp of Fig. 10.12 the designer wishes to investigate the effects of increasing W/L ratio of both M_1 and M_2 by a factor of 4. Assuming that all other parameters are kept unchanged:

(a) Find the resulting change in $(|V_{GS}| - |V_t|)$ and in g_m of M_1 and M_2.

(b) What change results in the voltage gain of the input stage? Also, how does this affect the overall voltage gain?

(c) What is the effect on the input offset voltage?

(d) If f_t is to remain the same as before the change to M_1 and M_2, what is the new value of C_C?

10.9 Consider a CMOS amplifier that is complementary to that in Fig. 10.12 in which each device is replaced by its complement of the same physical size with the supplies reversed.

Use the overall conditions as specified in Section 10.2. For all devices have Spice compute I_D and the small-signal parameters g_m and r_o. Compute the gain of the first and second stage, the overall amplifier voltage gain, the input common-mode range, and the output voltage range.

10.10 Using Spice determine the frequency response behavior of the CMOS folded-cascoded op amp shown in Fig. 10.16 for a load capacitance of 1 pF, 5 pF, 10 pF, 20 pF and 100 pF. How does the amplifiers phase margin vary?

10.11 The circuit in Fig. 10.21 can be used to multiply an analog signal by a digital one by feeding the analog signal to the V_{ref} terminal. In this case the D/A converter is called a multiplying DAC or MDAC. Given an input sine-wave signal of $0.1 \sin(\omega t)$ volts, use the circuit of Fig. 10.21 together with an additional op amp to obtain $v_o = 10D \sin(\omega t)$ where D is the digital word given by

$$D = \frac{b_1}{2^1} + \frac{b_2}{2^2} + \frac{b_3}{2^3} + \frac{b_4}{2^4}.$$

With the aid of Spice, compute the output transient voltage waveform of the DAC for the following values of D: (i) 0110 (ii) 1111, and (iii) 0000.

10.12 Repeat Problem 10.11 above, but this time, consider having the digital word D cycle through a digital pattern that resembles a symmetrical triangular waveform of 1/10-th the frequency of the input sine-wave. Using the transient analysis command of Spice, plot the output voltage waveform from the MDAC.

10.13 Using Spice compute the input resistance seen by V_{ref} in the circuit of Fig. 10.21 for all possible combinations of the switches.

10.14 With the aid of Spice, compare the transfer characteristics of the D/A converter shown in Fig. 10.21 with those obtained from an identical converter experiencing a +10 % increase in the value of resistor R_4.

Chapter 11

Filters And Tuned Amplifiers

In this chapter we shall use Spice and PSpice to investigate the frequency response of various types of active-RC filter circuits and an LC tuned amplifier. We shall also demonstrate how Spice (PSpice) can be used in the process of fine-tuning a design. This process is known as computer-aided design.

11.1 The Butterworth And Chebyshev Transfer Functions

Two filter functions commonly used to approximate low-pass transmission characteristics are the Butterworth and Chebyshev transfer functions. In Chapter 7 of this text we demonstrated how PSpice can be used to compute the frequency response of an arbitrary transfer function by specifying the gain of a dependent source as a Laplace Transform function. Adopting a similar approach, in the following we shall verify that the two filter functions calculated in Examples 11.1 and 11.2 of Sedra and Smith do indeed meet the required specifications.

The following 9th-order Butterworth transfer function,

$$T(s) = \frac{6.773 \times 10^4}{(s + 6.773 \times 10^4)} \frac{(6.773 \times 10^4)^2}{(s^2 + s1.8794 * 6.773 \times 10^4 + (6.773 \times 10^4)^2)} \times \qquad (11.1)$$

$$\frac{(6.773 \times 10^4)^2}{(s^2 + s1.5321 * 6.773 \times 10^4 + (6.773 \times 10^4)^2)} \frac{(6.773 \times 10^4)^2}{(s^2 + s1.0 * 6.773 \times 10^4 + (6.773 \times 10^4)^2)} \times$$

$$\frac{(6.773 \times 10^4)^2}{(s^2 + s0.3472 * 6.773 \times 10^4 + (6.773 \times 10^4)^2)}$$

11 Filters And Tuned Amplifiers

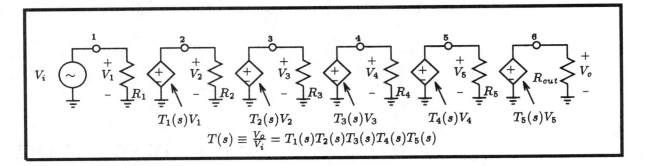

Figure 11.1 A cascade of several VCVSs whose gain is described by either a first-order or a second-order transfer function, forming an overall transfer function $T(s)$.

Example 11.1: 9th-Order Butterworth Filter Function

```
** Circuit Description **
* input signal
Vin 1 0 AC 1V
* a cascade of biquads forming a 9th-order Butterworth Filter Function
* first biquad
R1 1 0 100Meg
E1 2 0 Laplace {V(1)} = { (6.773e+4)/(s+6.773e+4) }
* 2nd biquad
R2 2 0 100Meg
E2 3 0 Laplace {V(2)} = { (6.773e+4*6.773e+4)/(s*s+1.8794*6.773e4*s+6.773e4*6.773e4) }
* 3rd biquad
R3 3 0 100Meg
E3 4 0 Laplace {V(3)} = { (6.773e+4*6.773e+4)/(s*s+1.5321*6.773e4*s+6.773e4*6.773e4) }
* 4th biquad
R4 4 0 100Meg
E4 5 0 Laplace {V(4)} = { (6.773e+4*6.773e+4)/(s*s+1.0*6.773e4*s+6.773e4*6.773e4) }
* 5th biquad
R5 5 0 100Meg
E5 6 0 Laplace {V(5)} = { (6.773e+4*6.773e+4)/(s*s+0.3472*6.773e4*s+6.773e4*6.773e4) }
* output
Rout 6 0 1k
** Analysis Requests **
.AC LIN 100 1Hz 20kHz
** Output Requests **
.PLOT AC VdB(6) Vp(6)
.probe
.end
```

Figure 11.2 The PSpice input deck for computing the magnitude response of the 9th-Order Butterworth transfer function given in Eqn. (11.1).

was calculated by Sedra and Smith in Example 11.1 so that it's magnitude response would satisfy the following filter specifications: $f_p = 10$ kHz, $A_{max} = 1$ dB, $f_s = 15$ kHz, $A_{min} = 25$ dB and dc gain $= 1$. An important constraint imposed by PSpice when specifying the gain of a dependent source as a Laplace transform expression, as demonstrated in section 7.1, is that the expression must fit on a single line in the PSpice input file (ie. limited to 131 characters). The transfer function specified in Eqn. (11.1) is very long and it could not be specified on a single line.

11.1 The Butterworth And Chebyshev Transfer Functions

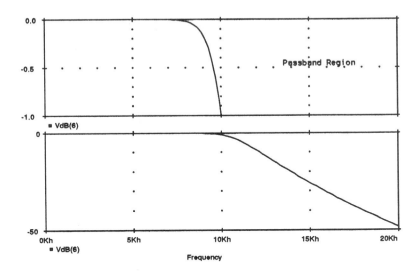

Figure 11.3 The magnitude response of the 9th-order Butterworth transfer function given in Eqn. (11.1) for $f_p = 10$ kHz, $A_{max} = 1$ dB, $f_s = 15$ kHz and $A_{min} = 25$ dB. The top graph displays an expanded view of the passband region and the bottom graph displays a view of both the passband and stopband regions.

To get around this problem, we simply break-up the transfer function into several first and second-order transfer functions and assign each one as the gain of a separate VCVS. The resulting set of VCVSs are then connected in cascade as shown in Fig. 11.1. The input file describing this circuit arrangement to PSpice is listed in Fig. 11.2. Here we are requesting that the transfer function be evaluated linearly over the frequency interval beginning at 1 Hz and ending at 20 kHz with 100 points collected.

On completion of Spice, the magnitude response of the 9th-order Butterworth transfer function is shown in Fig. 11.3. The top graph displays an expanded view of the passband region and the bottom graph displays a view of both the passband and stopband regions. On review of these results we see that at $f = f_p = 10$ kHz the gain is 1 dB below the dc value (and is monotonically decreasing). At $f = f_s = 15$ kHz we find that the gain is about 25 dB below the dc value (more precisely, using Probe we find at this frequency the gain is -25.8 dB). Thus, we can concluded that the transfer function of Eqn. (11.1) meets the required specifications.

Repeating the above analysis for the following 5th-order Chebyshev transfer function given in Example 11.2 of Sedra and Smith,

---

Example 11.2: 5th-Order Chebyshev Filter Function

```
** Circuit Description **
* input signal
Vin 1 0 AC 1V
* a cascade of biquads forming a 5th-order Chebyshev Filter Function
* first biquad
R1 1 0 100Meg
E1 2 0 Laplace {V(1)} = { (6.2832e+4/8.1408)/(s+0.2895*6.2832e+4) }
* 2nd biquad
R2 2 0 100Meg
E2 3 0 Laplace {V(2)} = { (6.2832e+4*6.2832e+4)/(s*s+0.4684*6.2832e+4*s+0.4293*6.2832e+4*6.2832e+4) }
* 3rd biquad
R3 3 0 100Meg
E3 4 0 Laplace {V(3)} = { (6.2832e+4*6.2832e+4)/(s*s+0.1789*6.2832e+4*s+0.9883*6.2832e+4*6.2832e+4) }
* output
Rout 4 0 1k
** Analysis Requests **
.AC LIN 100 1Hz 20kHz
** Output Requests **
.PLOT AC VdB(4) Vp(4)
.probe
.end
```

Figure 11.4 The PSpice input deck for computing the magnitude response of the 5th-Order Chebyshev transfer function given in Eqn. (11.2).

$$T(s) = \frac{6.2832 \times 10^4}{8.1408(s + 0.2895 * 6.2832 \times 10^4)} \frac{(6.2832 \times 10^4)^2}{(s^2 + s0.4684 * 6.2832 \times 10^4 + 0.4293 * (6.2832 \times 10^4)^2)} \times$$

$$\frac{(6.2832 \times 10^4)^2}{(s^2 + s0.1789 * 6.2832 \times 10^4 + 0.9883 * (6.2832 \times 10^4)^2)}, \quad (11.2)$$

we would like to verify that the magnitude of this transfer function meets the same specifications as did the above 9th-order Butterworth transfer function.

Adopting the same approach as for the above Butterworth transfer function, we can describe Chebyshev transfer function to PSpice with the input file shown listed in Fig. 11.4. Submitting this to PSpice results in the plot of the magnitude response shown in Fig. 11.5. Both an expanded view of the passband and a broader view of both the passband and stopband regions are shown. On inspection of these results, we see that the gain in the passband region (0 Hz to 10 kHz) oscillates between 0 and −1 dB. For frequencies above 10 kHz, we see that the gain rolls off monotonically with increasing frequency. At the stopband edge of $f = f_s = 15$ kHz, we see from the lower graph in Fig. 11.4 that the gain is −40 dB. Therefore, we can conclude that the lower-order Chebyshev filter function given by Eqn. (11.2) satisfies the same specifications as the higher-order Butterworth filter function described by Eqn. (11.1).

11.2 Second-Order Active Filters Based On Inductor Replacement

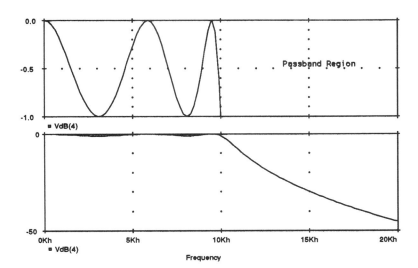

Figure 11.5 The magnitude response of the 5th-order Chebyshev transfer function given in Eqn. (11.2) for $f_p = 10$ kHz, $A_{max} = 1$ dB, $f_s = 15$ kHz and $A_{min} = 25$ dB. The top graph displays an expanded view of the passband region and the bottom graph displays a view of both the passband and stopband regions.

11.2 Second-Order Active Filters Based On Inductor Replacement

In Fig. 11.6 we display a circuit consisting of a cascade of two second-order simulated-LCR resonator circuits and a single first-order op amp RC circuit. The components of this circuit were selected such that it realizes the 5th-order Chebyshev function given in Eqn. (11.2). Using Spice we would like to verify this by comparing the magnitude response of this circuit with that computed directly from its transfer function as calculated above and shown in Fig. 11.5. We shall assume that the op amps are quasi-ideal and modeled after a VCVS with a gain of 10^6 V/V as shown in part (b) of Fig. 11.6. This step would normally be the first step after the circuit has been designed and one wants to verify that the synthesis procedure was correct.

The Spice input file describing this circuit is shown listed in Fig. 11.7. Since the same type of op amp is repeated many times in this circuit we have chosen to represent this quasi-ideal op amp with a single subcircuit named *ideal_opamp*. A detailed discussion of subcircuits was presented in Chapter 2 and will not be repeated here. An AC analysis is requested over the linear frequency interval beginning at 1 Hz and ending at 20 kHz. One-hundred points are to be computed. The input to the filter is driven by a 1 V AC signal.

Submitting the Spice input file listed in Fig. 11.7 to Spice, results in the magnitude

11 Filters And Tuned Amplifiers

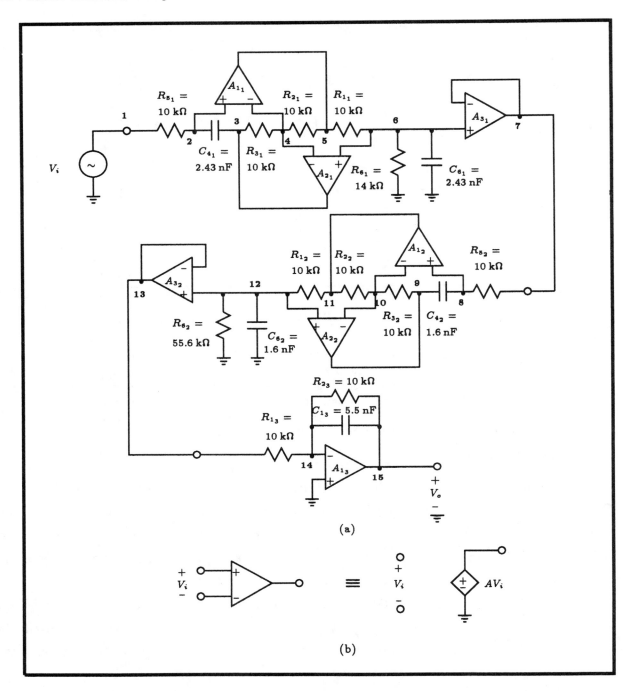

Figure 11.6 (a) A 5th-order Chebyshev filter circuit implemented as a cascade of two second-order simulated-LCR resonator circuits and a single first-order op amp-RC circuit. (b) a VCVS representation of an ideal op amp ($A = 10^6$).

response shown plotted in Fig. 11.8. Both an expanded view of the passband and a view of the passband and stopband regions are shown. When one compares these results with those computed directly from the transfer function given in Eqn. (11.2) above in Fig. 11.5 we see that they are, for all practical purposes, identical. Thus, we can conclude that we carried

11.2 Second-Order Active Filters Based On Inductor Replacement

```
Exercise D11.20: 5th-Order Chebyshev Filter Circuit (Fig. 11.6)

** Circuit Description **

* op-amp subcircuit
.subckt ideal_opamp 1 2 3
* connections:     | | |
*            output | |
*           +ve input |
*            -ve input
Eopamp 1 0 2 3 1e6
Iopen1 2 0 0A      ; redundant connection made at +ve input terminal
Iopen2 3 0 0A      ; redundant connection made at -ve input terminal
.ends ideal_opamp

** Main Circuit **
* input signal source
Vi 1 0 DC 0V AC 1V
* first biquad stage (Wo=41.17k rad/s   Q=1.4)
X_A1_1 5 2 4 ideal_opamp
X_A2_1 3 6 4 ideal_opamp
R1_1 5 6 10k
R2_1 4 5 10k
R3_1 3 4 10k
C4_1 2 3 2.43nF
R5_1 1 2 10k
C6_1 6 0 2.43nF
R6_1 6 0 14k
X_A3_1 7 6 7 ideal_opamp
* second biquad stage (Wo=62.46k rad/s   Q=5.56)
X_A1_2 11 8 10 ideal_opamp
X_A2_2 9 12 10 ideal_opamp
R1_2 11 12 10k
R2_2 10 11 10k
R3_2  9 10 10k
C4_2  8  9 1.6nF
R5_2  7  8 10k
C6_2 12  0 1.6nF
R6_2 12  0 55.6k
X_A3_2 13 12 13 ideal_opamp
* first-order stage
X_A1_3 15 0 14 ideal_opamp
R1_3 13 14 10k
R2_3 14 15 10k
C1_3 14 15 5.5nF
** Analysis Requests **
.AC LIN 100 1Hz 20kHz
** Output Requests **
.PLOT AC VdB(15) Vp(15)
.probe
.end
```

Figure 11.7 The Spice input deck for calculating the frequency response of the lowpass filter circuit shown in Fig. 11.6.

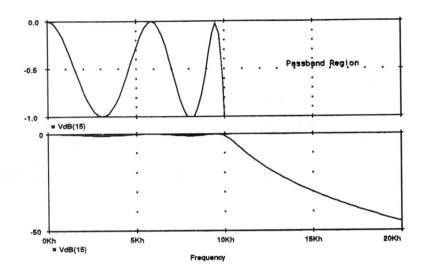

Figure 11.8 The magnitude response of the 5th-order lowpass filter circuit shown in Fig. 11.6. The top graph displays an expanded view of the passband region and the bottom graph displays a view of both the passband and stopband regions.

out the synthesis procedure correctly. The next step would be to investigate how the filter magnitude response is affected by the frequency characteristics of a real op amp. This we shall do in the next section on a Tow-Thomas biquad circuit.

11.3 Second-Order Active Filters Based On The Two-Integrator-Loop Topology

Another important class of biquadratic active-RC filter circuits are those that are formed by cascading two integrators in an overall feedback loop. In Fig. 11.9 we show the Tow-Thomas biquad based on this idea. The components were selected such that this circuit realized a second-order bandpass filter with $f_o = 10$ kHz, $Q = 20$ and unity center-frequency gain. Using Spice we would like to investigate the frequency response of this filter circuit assuming that each op amp is of the 741-type and compare it to its ideal frequency response (ie. one with ideal op amps).

One possible approach for evaluating the nonideal frequency response behavior of the bandpass filter shown in Fig. 11.9 is to describe the entire circuit to Spice including a detailed description of the 741 op amp at the transistor level. This would certainly lead to the most accurate representation of the bandpass filter circuit; however, it would require large amounts of computer time and storage to compute the frequency response. A more

11.3 Second-Order Active Filters Based On The Two-Integrator-Loop Topology

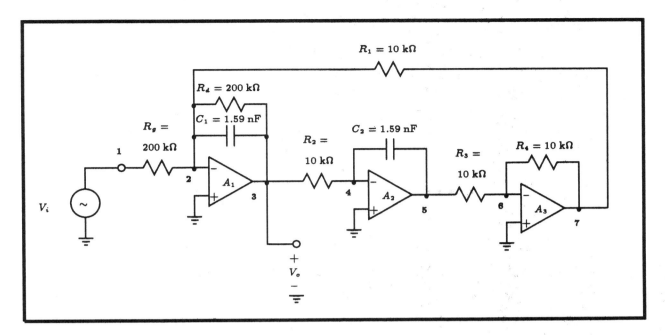

Figure 11.9 A second-order bandpass filter implemented with a Tow-Thomas biquad circuit with $f_O = 10$ kHz, $Q = 20$, and unity center-frequency gain. The op amps are assumed to be of the 741 type.

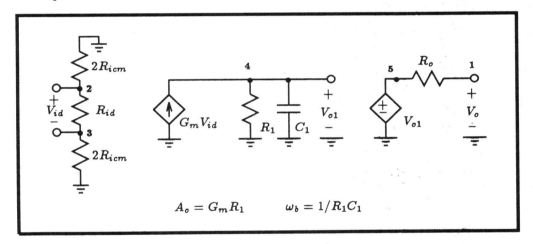

Figure 11.10 A one-pole equivalent circuit representation of an op amp operated within its linear region.

reasonable approach, and the one we will undertake here, is to utilize a *macromodel* of the 741 op amp.

To be more specific, we shall model the terminal behavior of the 741 op amp with the single-time constant linear network shown in Fig. 11.10. Other time-constants can also be included in this model by simple extension of the middle stage. One may also be tempted to add nonlinearities into this model; however, Spice linearizes a circuit about its operating point prior to the start of the AC analysis. Hence, the inclusion of any op amp nonlinearity

495

11 Filters And Tuned Amplifiers

in AC analysis serves no purpose and is ignored by Spice. Nonlinearities can, of course, be included in the model when performing transient analysis. It is a simple matter to derive the open-circuit transfer function of the equivalent op amp circuit shown in Fig. 11.10 and determine that it has a one-pole response given by

$$\frac{V_o}{V_{id}}(s) = \frac{G_m R_1}{1 + s R_1 C_1}. \tag{11.3}$$

The 741 op amp is an internally-compensated op amp which has a frequency response characterized by a one-pole frequency roll-off,

$$A(s) = \frac{A_o}{1 + s/\omega_b} \tag{11.4}$$

where A_o denotes the DC gain and ω_b is the 3-dB frequency. Nominally, the 741 op amp has a DC gain of 2.52×10^5 V/V and a 3 dB frequency of 4 Hz. Comparing the above two equations, we can write two equations in three unknowns. That is, $A_o = G_m R_1$ and $\omega_b = 1/R_1 C_1$. Assigning $C_1 = 30$ pF, we can solve for $G_m = 0.19$ mA/V and $R_1 = 1.323 \times 10^9$. We can also add some input and output resistances to account for the effects of loading. Here we shall choose $R_{id} = 2$ MΩ, $R_{icm} = 500$ MΩ and $R_o = 75$ Ω.

The Spice input file describing the Tow-Thomas biquad circuit of Fig. 11.9 is shown listed in Fig. 11.11. The op amp macromodel shown in Fig. 11.10 representing the small-signal nominal frequency response behavior of the 741 op amp is described by the subcircuit *nonideal_opamp* also included in this Spice listing. A 1 V AC signal is applied to the input of the filter, thus the output voltage of the filter will also represent the filter transfer function. The frequency response of the filter is computed between 8 kHz and 12 kHz using 100 points linearly separated from one another. We shall also concatenate another Spice deck on the end of this one with the nonideal subcircuit replaced with one containing a model of an ideal op amp, such as the one used in the previous example shown listed in Fig. 11.7, thus generating the "ideal" frequency response.

The results of the two circuit simulations, one assuming ideal op amps and another with op amps modeled after the 741-type, are shown collectively in Fig. 11.12. On comparison of their magnitude responses, we see that there are significant differences between them. In fact, we see that the center frequency of the circuit using the 741 op amp has shifted to the left by about 100 Hz and it's 3 dB bandwidth has decreased from 500 Hz to approximately 110 Hz. This, in effect, is an increase in the intended filter Q of 20 up to 90. In addition, the center frequency gain has also increased from 0 dB to 14 dB.

The above results serve to illustrate the adverse effects that actual op amps such as the 741-type have on the frequency response of a Tow-Thomas filter circuit. One possible

11.3 Second-Order Active Filters Based On The Two-Integrator-Loop Topology

---

Exercise D11.23: Second-Order Bandpass Filter Circuit (Nonideal Op-Amp)

```
** Circuit Description **

* op-amp subcircuit
.subckt nonideal_opamp 1 2 3
* connections:         | | |
*                output | |
*               +ve input |
*                -ve input
Ricm+ 2 0 500Meg
Ricm- 3 0 500Meg
Rid 2 3 2Meg
Gm 0 4 2 3 0.19m
R1 4 0 1.323G
C1 4 0 30pF
Eoutput 5 0 4 0 1
Ro 5 1 75
.ends nonideal_opamp

** Main Circuit **
* input signal source
Vi 1 0 DC 0V AC 1V
* Tow-Thomas Biquad
X_A1 3 0 2 nonideal_opamp
X_A2 5 0 4 nonideal_opamp
X_A3 7 0 6 nonideal_opamp
Rg 1 2 200k
R1 2 7 10k
R2 3 4 10k
R3 5 6 10k
R4 6 7 10k
Rd 2 3 200k
C1 2 3 1.59nF
C2 4 5 1.59nF
** Analysis Requests **
.AC LIN 100 8kHz 12kHz
** Output Requests **
.PLOT AC VdB(3) Vp(3)
.probe
.end
```

Figure 11.11 The Spice input deck for calculating the frequency response of the second-order bandpass filter circuit shown in Fig. 11.9. The op amp is assumed to have a DC gain of 252 kV/V and a 3 dB frequency of 4 Hz — much like the small-signal frequency response characteristics of a 741 op amp circuit.

approach for minimizing these effects is to add into the circuit what is known as a compensation capacitor. There are several ways in which to include the compensation capacitor in the circuit. Here we shall consider adding the compensation capacitor across resistor R_2 in

11 Filters And Tuned Amplifiers

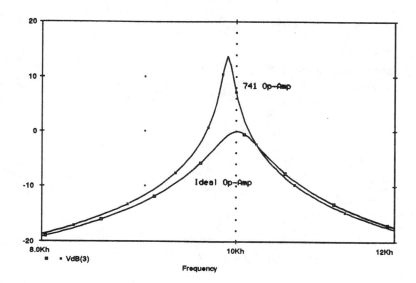

Figure 11.12 Comparing the magnitude response of the Tow-Thomas biquad circuit shown in Fig. 11.9 constructed with 741-type op amps to it's ideal magnitude response. These results illustrate the effect of the finite DC gain and bandwidth of the 741-opamp on the frequency response of the Tow-Thomas biquad circuit.

the circuit of Fig. 11.9. The actual value of the capacitor is not known at this time, so instead we shall search for the value of capacitance that improves the filter overall response. That is, we shall determine the value of capacitance that will make the magnitude response of the filter approach most closely to the desired or ideal response. To do this, we shall begin with a capacitance value of 20 pF and simulate the magnitude response of the filter. We shall compare this result to the ideal magnitude response and determine whether the added capacitance improves the filter response. If it does, we shall continue increasing the value of this capacitance until we no longer improve the filter response. Here we are utilizing Spice in the process of computer-aided *design* and not just analysis.

To demonstrate this design process, we have plotted in Fig. 11.13 the magnitude response that results from varying the value of the compensation capacitor from 0 pF to 80 pF in 20 pF increments. We see from these results that as the compensation capacitance increases from 0 pF, both the filter Q and the resonant peak of the filter response tend more closely towards the desired response. However, once the value of the compensation capacitor exceeds 60 pF, we begin to deviate away from the desired response. This suggests that the best result lies between 60 pF and 80 pF. More refinement of the same approach between these two values identifies that the best results occur when the compensation capacitor is set at 64 pF.

In Fig. 11.14 we compare the filter magnitude response with a compensation capacitor

11.3 Second-Order Active Filters Based On The Two-Integrator-Loop Topology

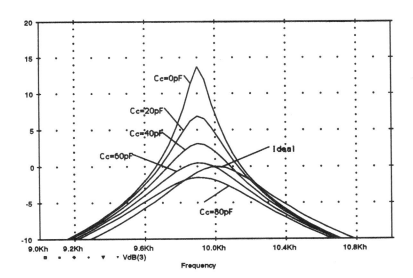

Figure 11.13 The magnitude response of the Tow-Thomas biquad circuit with different values of compensation capacitance. Also shown for comparison is the ideal response of the Tow-Thomas biquad circuit.

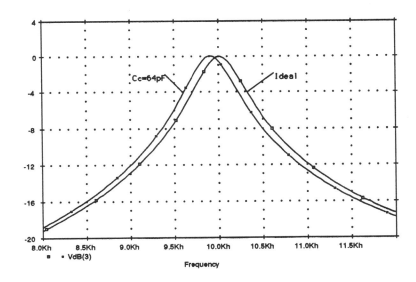

Figure 11.14 Comparing the magnitude response of the Tow-Thomas biquad circuit with a 64 pF compensation capacitor against the ideal response.

of 64 pF against the ideal response. As is evident, the actual filter magnitude response has about the same filter Q and similar center frequency gain as the ideal response, but have slightly different center frequencies. We can conclude from this computer-aided design

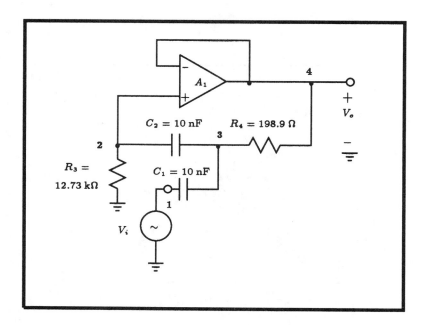

Figure 11.15 A single amplifier biquadratic circuit implementation of a highpass filter function with $f_O = 10$ kHz and $Q = 4$. The op amp is assumed to be of the 741-type.

process that much of the nonideal effects caused by the limited bandwidth of the 741 op amp has been eliminated by the addition of this single compensation capacitor. Finally, we should note that a closed form expression [Sedra and Brackett, 1978] exits for determining the required value of the compensation capacitor. This expression, however, relies on knowledge of the value of the op amp f_t. In practical situations, one would not know the exact value of f_t and the appropriate value of the compensation capacitance could be found by using a variable capacitor, in much the same way as done above with Spice.

11.4 Single-Amplifier Biquadratic Active Filters

Another important class of second-order filter circuits are the single-amplifier biquads or SABs. In Fig. 11.15 we display a filter circuit of the Sallen-and-Key type intended to realize a highpass filter function with $f_O = 10$ kHz and $Q = 4$. The op amp will be assumed to be of the 741 type – identical to the situation created in the previous problem. Using Spice we would like to calculate the magnitude response of this circuit over a frequency interval beginning at 1 kHz and ending at 10 MHz and compare it to the response of the same circuit when implemented with an ideal op amp. A logarithmic frequency sweep of ten points per decade will be chosen to be calculated.

The Spice input file describing the Sallen-and-Key filter circuit shown in Fig. 11.15 implemented with a 741 op amp is shown listed in Fig. 11.16. An identical Spice listing is

11.4 Single-Amplifier Biquadratic Active Filters

```
A Second-Order HP SAB Circuit (nonideal op-amp)

** Circuit Description **

* op-amp subcircuit
.subckt nonideal_opamp 1 2 3
* connections:        | | |
*                output | |
*              +ve input |
*                -ve input
Ricm+ 2 0 500Meg
Ricm- 3 0 500Meg
Rid 2 3 2Meg
Gm 0 4 2 3 0.19m
R1 4 0 1.323G
C1 4 0 30pF
Eoutput 5 0 4 0 1
Ro 5 1 75
.ends nonideal_opamp

** Main Circuit **
* input signal source
Vi 1 0 DC 0V AC 1V
* HP SAB Circuit
X_A1 4 2 4 nonideal_opamp
C1 1 3 10nF
C2 2 3 10nF
R3 2 0 12.73k
R4 3 4 198.9
** Analysis Requests **
.AC DEC 10 1Hz 100MegHz
** Output Requests **
.PLOT AC VdB(4)
.probe
.end
```

Figure 11.16 The Spice input deck for calculating the frequency response of the highpass SAB circuit shown in Fig. 11.15.

also appended at the end of this file with the nonideal op amp subcircuit replaced by an ideal op amp subcircuit. This will enable us to compare the frequency response of a real circuit implementation with the ideal response.

The results of the two simulations are shown plotted in Fig. 11.17. For frequencies less than 100 kHz, the two curves shown are very similar. Above this frequency, the magnitude response of the circuit implemented with a 741 op amp is beginning to deviate significantly from the ideal response. One might also be tempted to add that it appears more like a bandpass filter function than a highpass, with the exception that the magnitude response

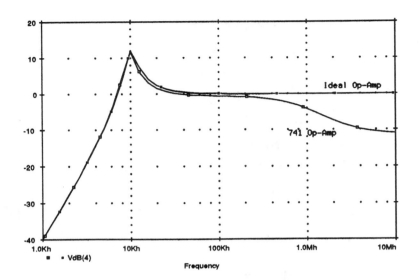

Figure 11.17 Comparing the magnitude response of the highpass SAB filter circuit shown in Fig. 11.15 constructed with an op amp modeled after the 741 type and one with an ideal op amp.

appears to be leveling off at about −11 dB. These results serve to illustrate the difficulty of realizing a highpass filter function using op amps that have an inherent lowpass frequency characteristic.

11.5 Tuned Amplifiers

In Fig. 11.18 we show a special kind of frequency selective network called the LC-tuned amplifier. Loading the drain of a typical resistor-biased transistor circuit with an LC tank circuit results in an amplifier with a highly selective amplitude response. This class of circuit finds extensive application in high-frequency communications systems such as radio. The tuned circuit portion of this amplifier was designed in Example 11.4 of Sedra and Smith such that this amplifier has a center-frequency of 1 MHz, a 3 dB frequency of 10 kHz and a center-frequency gain of −10 V/V. The resistor biasing network was added to bias transistor M_1 at $g_m = 5$ mA/V and $r_o = 10$ kΩ, much like that assumed by Sedra and Smith in their example. MOSFET M_1 is assumed to be 10 μm long and 1250 μm wide. It also has the following device parameters: $\mu_n C_{ox} = 100$ μA/V^2, $V_T = +2$ V and $\lambda = 0.1$ V^{-1}. Using Spice we would like to calculate the magnitude response of this circuit and compare it to what we are expecting.

The Spice input file describing this circuit is seen listed in Fig. 11.19. The input is

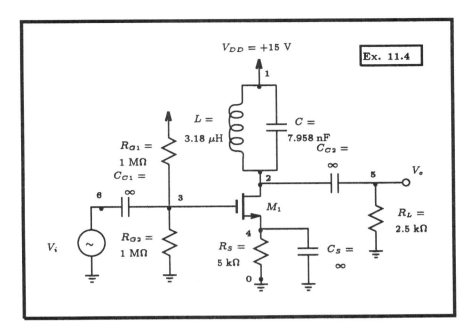

Figure 11.18 A single MOSFET tuned-amplifier with bias circuit.

driven by a 1 V AC signal and two types of analysis are requested. The first is simply an operating point calculation (.OP) which we shall use to compute the bias point of the MOSFET. These results will be used to determine whether the MOSFET has been biased correctly. Secondly, an AC command is specified to compute the frequency response of this amplifier in a frequency range beginning at 0.99 MHz and ending at 1.01 MHz with 100 points computed in a linear fashion between the two frequency end points.

Submitting the Spice input deck shown listed in Fig. 11.19 and collecting the results, we obtained the magnitude response of the tuned amplifier as shown in Fig. 11.20. Here we see that the center frequency of the magnitude response is almost exactly at the intended design value of 1 MHz. However, both the gain of the amplifier at its center frequency of 15.86 V/V and its 3 dB bandwidth of 9.01 kHz do not agree with their design values. To account for these differences, one only has to review the small-signal model for the MOSFET calculated by Spice through the .OP command, seen listed below:

```
****    OPERATING POINT INFORMATION    TEMPERATURE =   27.000 DEG C
*************************************************************************
**** MOSFETS

NAME       M1
MODEL      NMOS
ID         1.04E-03
VGS        2.29E+00
VDS        9.79E+00
VBS        0.00E+00
VTH        2.00E+00
VDSAT      2.90E-01
GM         7.18E-03
```

11 Filters And Tuned Amplifiers

Example 11.4: A Bandpass Tuned Amplifier

```
** Circuit Description **

* power supplies
Vdd 1 0 DC +15V
* input signal source
Vi 6 0 DC 0V AC 1V
* amplifier
Cc1 6 3 1uF
RG1 1 3 1Meg
RG2 3 0 1Meg
M1 2 3 4 4 NMOS L=10u W=1250u
Rs 4 0 5k
Cs 4 0 1uF
Cc2 2 5 1uF
* tuned circuit
L 1 2 3.18uH
C 1 2 7.958nF
* output load
Rl 5 0 2.5k
* mosfet model statement
.model NMOS nmos (kp=100u Vto=+2V lambda=0.1)
** Analysis Requests **
.OP
.AC LIN 100 0.98MegHz 1.02MegHz
** Output Requests **
.PLOT AC Vm(5)
.probe
.end
```

Figure 11.19 The Spice input deck for calculating the magnitude response of the tuned amplifier shown in Fig. 11.18.

```
GDS        5.27E-05
GMB        0.00E+00
```

Here we see that both the g_m and r_o (or 1/GDS) of the transistor has deviated from our initial design of 1 mA/V and 10 kΩ, respectively. Since, both the center frequency gain and 3 dB bandwidth of the tuned amplifier shown in Fig. 11.18 are functions of these two parameters (see Sedra and Smith) we would also expect them to deviate from the intended design values. To obtain better estimates of the center frequency gain and the 3 dB bandwidth, we can use the actual bias data generated by Spice. For instance, the equivalent resistance seen at the drain of M_1 is $R = r_o || R_L = (1/5.27 \times 10^{-5}) || 2.5 \times 10^3 = 2.208$ kΩ. Thus, the magnitude of the center frequency gain is $g_m R = 7.18 \text{ m} * 2.208 \text{ k} = 15.85$ V/V. Likewise, the 3 dB bandwidth is $B = 1/RC = 1/(2.208 \text{ k} * 7.958 \text{ n})$ or 9.057 kHz. Both these values are now

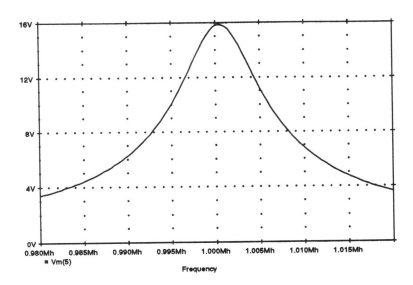

Figure 11.20 Frequency response of the tuned bandpass amplifier shown in Fig. 11.18.

much closer to what we obtained through simulation.

To demonstrate another example of a computer-aided design, let us consider adjusting the value of the load resistor R_L in the tuned amplifier shown in Fig, 11.18 such that the required 10 kHz bandwidth is achieved.

To make matters more realistic, we shall account for the MOSFET parasitic capacitances by including the following model statement for the device:

```
.model NMOS nmos (kp=100u Vto=+2V lambda=0.1 tox=100e-10 tt=100n
+                 cgso=100p cgdo=100p cgbo=50p cj=4e-4 cjsw=8e-10 )
```

The dc parameters of this MOSFET are identical to those used previously. The 1 kΩ source resistance will also be included in order to properly account for the effect of these capacitances on the overall circuit behavior. The Spice deck for this example is quite similar to that seen in Fig. 11.19 and is therefore not shown here.

To begin this computer-aided design process, we shall first determine the 3 dB bandwidth of this tuned amplifier with the MOSFET parasitic capacitances present and the load resistance set at its nominal value of 2.5 kΩ. This requires that we re-submit the revised Spice job for analysis and plot the magnitude response of this circuit. On doing so, we find that the 3 dB bandwidth is 9.06 kHz. In order to increase the bandwidth of this tuned amplifier to 10 kHz, we must decrease the value of the load resistance. How much we should decrease R_L is not known at this time; instead we shall perform a search using Spice. Begin-

11 Filters And Tuned Amplifiers

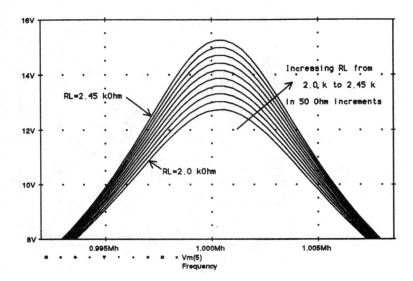

Figure 11.21 An expanded view of the magnitude response of the the tuned amplifier for various load resistances.

ning with R_L equal to 2.45 kΩ, we shall compute the 3 dB bandwidth and then repeat the same process while incrementally decreasing the value of R_L by, say, 50 Ω until we reach, say, 2.0 kΩ. Hopefully, we shall find that the amplifier 3 dB bandwidth has increased to 10 kHz, or more, with the range of load resistance that we selected above. If not, then we must continue the search process with values of R_L less than 2 kΩ.

In Fig. 11.21 we present an expanded view of the magnitude response of the tuned amplifier for the various load resistances. Although it is not directly evident, with the cursor facility of Probe, we were were able to determine the 3 dB bandwidth in each case. We found that for a load resistance of 2.20 kΩ the 3 dB bandwidth was slightly greater than 10 kHz (actually, 10.1 kHz). For a load resistance of 2.25 kΩ, we found that the 3 dB bandwidth is 9.966 kHz. We can therefore conclude that a 10 kHz bandwidth for this amplifier will occur with a load resistance somewhere between 2.20 kΩ and 2.25 kΩ. Using a more refined increment for the load resistance, we can repeat the same process stated above with the load resistance varied between 2.20 kΩ and 2.25 kΩ with a 10 Ω increment. On doing so, we find that the tuned amplifier will have a 10 kHz bandwidth when the load resistance is 2.24 kΩ. Any more refinement would probably not be worth the extra effort given that the resistor can not be specified any more accurately.

11.6 Spice Tips

- The element statement of a dependent source having a gain expressed in terms of a Laplace transform function can not be longer than 131 characters in length.
- The small-signal frequency response behavior of the 741 op amp can be modeled in Spice with a very simple macromodel. This macromodel consists of a single-time-constant linear network that represents the dominant pole behavior of the op amp.
- Spice is not only useful for analyzing a design, it is also useful for fine tuning a design.

11.7 Bibliography

A. S. Sedra and P. O. Brackett, *Filter Theory and Design: Active and Passive*, Portland, Ore.: Matrix, 1978.

11.8 Problems

11.1 A third-order lowpass filter has transmission zeros at $\omega = 2$ rad/s and $\omega = \infty$. Its natural modes are at $s = -1$ and $s = -0.5 \pm j0.8$. The dc gain is unity. With the aid of PSpice, plot the magnitude and phase of the transfer function for this filter.

11.2 With the aid of PSpice, prepare a plot of both the magnitude and phase of the following filter transfer functions capturing the most important frequency information:

(a) $T(s) = \dfrac{1}{(s^2 + \sqrt{2}s + 1)}$

(b) $T(s) = \dfrac{5725600}{(s + 125.3)(s^2 + 125.3s + 45698)}$

(c) $T(s) = \dfrac{0.2816(s^2 + 3.2236)}{(s + 0.7732)(s^2 + 0.4916s + 1.1742)}$

(d) $T(s) = \dfrac{0.083947(s^2 + 17.48528)}{(s^2 + 1.35715s + 1.55532)}$

(e) $T(s) = \dfrac{s^3}{(s^2 + 1000s + 10^6)(s + 1000)}$

11.3 Analyze the RLC network of Fig. P11.3 using Spice to determine the nature of its transfer function over the frequency range between 0.001 Hz and 10 Hz. Shunt the 2 H inductor with a 0.1 F capacitor. Observe the modified frequency response.

11.4 Design a Butterworth filter that meets the following lowpass specifications: $f_p = 10$ kHz, $A_{max} = 2$ dB, $f_s = 15$ kHz, and $A_{min} = 15$ dB. Use PSpice to confirm your design by plotting the magnitude of your transfer function.

11 Filters And Tuned Amplifiers

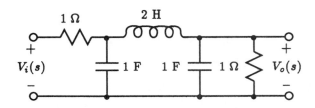

Fig. P11.3

11.5 Contrast the attenuation provided by a fifth-order Chebyshev filter at $\omega_s = 2\omega_p$ to that provided by a Butterworth filter of equal order. For both, $A_{max} = 1$ dB. Plot $|T(s)|$ for both filters on the same axes using PSpice.

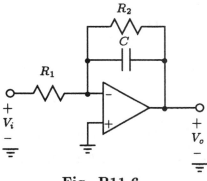

Fig. P11.6

11.6 Design the above lowpass op amp circuit in Fig. P11.6 to have a 3 dB frequency of 10 kHz, a dc gain of magnitude 10 and an input resistance of 10 kΩ. Verify your design using Spice. Assume a high-gain VCVS representation for the ideal op amp.

11.7 For the lowpass op amp circuit designed in the above problem, consider applying a 1 kHz sine-wave of 1 V amplitude to its input and observe its output steady-state time-response using Spice. Repeat for an input signal of the same amplitude but of a much higher frequency of 100 kHz. Now, combine these two sine-wave signals by connecting two voltage sources in series and apply it to the input of the filter. Compare the voltage signal appearing at the input to the filter with that appearing at the output. Comment on your results.

11.8 For the lowpass op amp circuit designed in Problem 11.7 above, consider applying a 1 V step to its input and observe its output step response using Spice.

11.9 Design the highpass op amp circuit shown in Fig. P11.9 to have a 3 dB frequency of 100 kHz, a high-frequency input resistance of 100 kΩ and a high-frequency gain magnitude of unity. Verify your design using Spice by plotting the magnitude response of the filter as a function of frequency. Assume a high-gain VCVS representation for the ideal op amp.

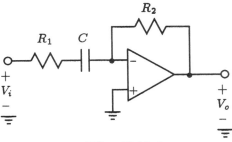

Fig. P11.9

11.10 For the highpass op amp circuit designed in the above problem, consider applying a 1 MHz sine-wave of 1 V amplitude to its input and observe its output steady-state time-response using Spice. Repeat for an input signal of the same amplitude but of much lower frequency at 10 kHz. Now, combine these two sine-wave signals by connecting two voltage sources in series and apply it to the input of the filter. Compare the voltage signal appearing at the input to the filter with that appearing at the output. Comment on your results.

11.11 For the highpass op amp circuit designed in Problem 11.10 above, consider applying a 1 V step to its input and observe its output step response using Spice.

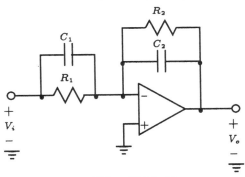

Fig. P11.12

11.12 For the first-order op amp circuit shown in Fig. P11.12, select the value of its components such that a transmission zero is formed at a frequency of 1 kHz, a pole at a frequency of 100 kHz, and has a dc gain magnitude of unity. The low-frequency input resistance is to be 1 kΩ. Plot the magnitude of the transfer function of the circuit that results using Spice.

11.13 By cascading a first-order op amp-RC lowpass circuit with a first-order op amp-RC highpass circuit one can design a wideband bandpass filter. Provide such a design for the case the midband gain is 12 dB and the 3 dB bandwidth extends from 100 Hz to 10 kHz. Select appropriate component values under the constraint that no resistors higher than 100 kΩ are to be used, and the input resistance is to be as high as possible. Verify your design by simulating the magnitude response of your circuit as a function of

11 Filters And Tuned Amplifiers

frequency.

11.14 Use two first-order op amp-RC all-pass circuits in cascade to design a circuit that provides a set of three-phase 60 Hz voltages, each separated by 120° and equal in magnitude. Use 1 μF capacitors. Verify your design by simulating the steady-state time-response of your circuit subject to a 1 Vrms 60 Hz sine-wave signal apply to its input. Plot your results using Spice.

11.15 Verify using Spice the design of a lowpass LCR resonator circuit that has natural modes with $\omega_o = 10^4$ rad/sec and $Q = 1/\sqrt{2}$. Utilize $C = 1$ μF.

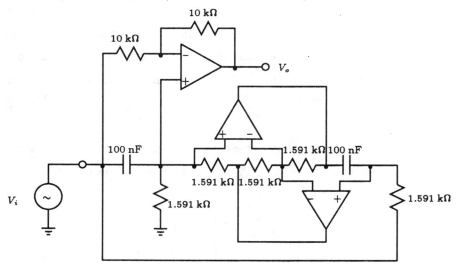

Fig. P11.16

11.16 The circuit of Fig. P11.16 has been designed to realize an allpass transfer function that provides a phase shift of 180° at 1 kHz and to have a $Q = 1$. Verify using Spice that this is indeed the case.

11.17 It is required to design a fifth-order Butterworth filter having a 3 dB bandwidth of 10^4 rad/s and a unity dc gain. Use a cascade of two second-order simulated-LCR resonator circuits and a single first-order op amp RC circuit. Verify your design using Spice.

11.18 The KHN circuit shown in Fig. P11.18 is used to realize a pair of complex poles located at $\omega = 10^4$ rad/s and $Q = 2$. With the aid of Spice, analyze the circuit to determine the transfer functions V_1/V_i, V_2/V_i and V_3/V_i. Classify each as either lowpass, bandpass, bandstop, etc.

11.19 For the KHN circuit shown in Fig. P11.18, compare the magnitude response $|V_2/V_i(j\omega)|$ between 1 Hz and 10 MHz for op amp dc gain A_o of 10^6, 10^4 and 10^2.

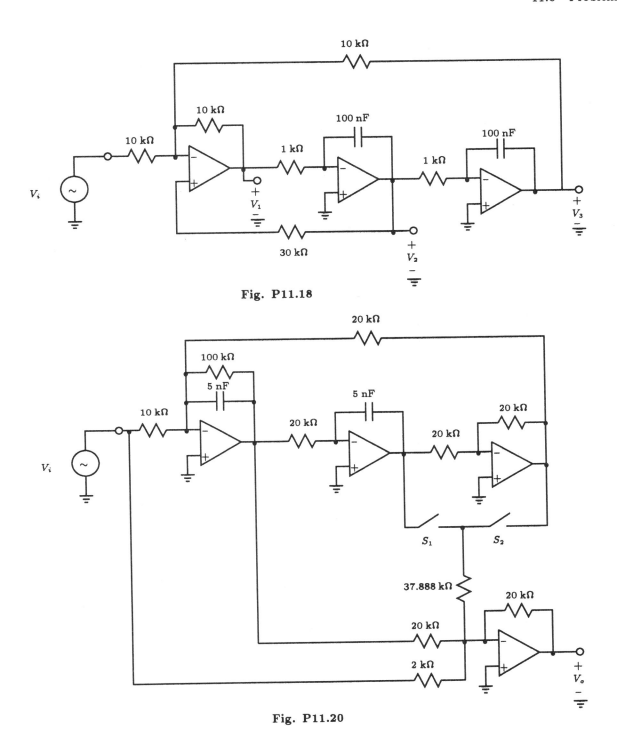

Fig. P11.18

Fig. P11.20

11.20 The filter circuit shown in Fig. P11.20 is a generalization of the Tow-Thomas biquad and commonly referred to as the "universal filter." The components of this filter circuit were chosen to realize a pair of complex poles described by $\omega_o = 10$ kHz and $Q = 5$, and a pair of transmission zeros, forming either a lowpass notch or highpass notch filter transfer function depending on which switch is closed. Using Spice, plot the magnitude of the output voltage for different switch positions. At what frequencies are the transmission zeros located at? Apply a 1 V AC signal to the input of the filter so that the output voltage represents the transfer function of the filter circuit. Also, assume that the op amps are almost ideal with a dc gain of 10^6.

11.21 The circuit of Fig. P11.21 has been designed to realize a lowpass notch filter with $w_o = 10^3$ rad/s, $Q = 10$, dc gain = 1, and $w_n = 1.2 \times 10^4$. Verify that this is indeed the case with Spice assuming that the op amps are almost ideal with a frequency independent voltage gain of 10^6 V/V.

11.22 Repeat Problem 11.21 above, but this time model each op amp after the small-signal frequency characteristics of the 741 op amp. Compare your results to those found previously in Problem 11.21.

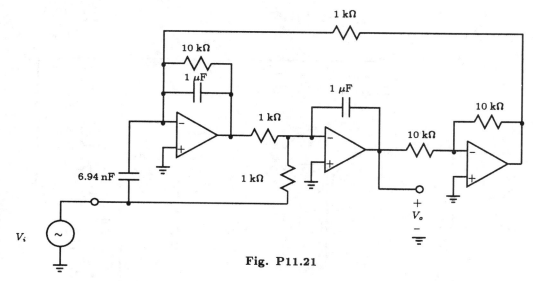

Fig. P11.21

11.23 Consider applying an input signal consisting of two 1 V peak sine-waves of 100 and 1909.86 Hz to the input of the notch filter shown in Fig. P11.21. Compare the voltage waveform appearing at the input of the filter with the voltage waveform appearing at its output.

11.24 Using Spice, plot the magnitude response $|V_2/V_i|$ of the KHN filter circuit shown in Fig. P11.18 assuming that each op amp has a unity-gain bandwidth of 10 MHz and a dc gain of 10^6 V/V. Compare these results to the ideal response (ie. ideal op amp case).

11.25 With the aid of Spice, compare the ideal or desired magnitude response of the 5th-order Chebyshev filter circuit shown in Fig. 11.6 with the magnitude response obtained when each op amp has a unity-gain bandwidth of 10^6 Hz.

11.26 Assuming that the resistors and capacitors used to realize the bandpass filter circuit in Fig. 11.9 have ±5% tolerances associated with them, select an arbitrary set of resistors and capacitors that lie within this tolerance, but are not the nominal value shown in the circuit schematic. Subsequently, with the aid of Spice, calculate the magnitude response $|V_o/V_i|$ of the modified filter circuit as a function of frequency. How does this frequency response compare to the nominal frequency response? Assume a high-gain VCVS representation for each op amp. (Some versions of Spice, such as PSpice, have a special analysis command called the Monte Carlo Analysis which allow the user to repeat this process of randomly selecting sets of component values according to a predefined distribution. Different analysis can then be compared and provide a useful sense of operation under normal manufacturing conditions).

11.27 In the example of section 11.3, the Tow-Thomas bandpass filter circuit shown in Fig. 11.9 was analyzed using Spice assuming that the op amps had a single-pole frequency response with unity-gain bandwidth of 1 MHz and found to deviate significantly from the ideal case. Consider compensating for the effect of op amp finite bandwidth by adding a capacitor of value $4C(\omega_o/\omega_t) = 63.6$ pF across resistor R_2 and re-analyzing the frequency response of the filter. How does the magnitude response compare with the ideal behavior? Is it better?

11.28 The magnitude response of the filter circuit shown in Fig. 11.9 deviates significantly from the desired or ideal frequency response, as demonstrated in section 11.3. One method used to decrease the deviation from the ideal response is to re-connect the noninverting input terminal of the op amp A_3 from ground to the inverting input terminal of op amp A_2. Using Spice, simulate this situation and compare the bandpass magnitude response of this modified filter circuit with its ideal response assuming that the op amps have a unity-gain bandwidth of 1 MHz.

11.29 For the highpass SAB circuit shown in Fig. 11.15, consider changing the resistors into capacitors and multiplying their value by a factor of 10^{12}. Likewise, change the capacitors into resistors and multiply their value by $1/10^{12}$. Plot the magnitude of the filter transfer function. What type of filter function results?

11.30 Design a fifth-order Butterworth lowpass filter with a 3 dB bandwidth of 5 kHz and a dc gain of unity using the cascade connection of two Sallen-and-Key circuits and a first-order section. Use a 10 kΩ value for all resistors. Verify your design using Spice.

11.31 To estimate the sensitivities of the bandpass SAB circuit shown in Fig. P11.31, consider varying, separately, each component of the circuit by +1% of its nominal value, and with the aid of Spice, compute the modified magnitude response of the filter transfer function. Then, using these results, together with the nominal magnitude response of the filter, calculate an estimate of the filter sensitivity as a function of frequency according to the formula:

$$S_x^{|T|}(j\omega) \cong \frac{1}{0.01} \frac{|T_{ACTUAL}(j\omega)| - |T_{IDEAL}(j\omega)|}{|T_{IDEAL}(j\omega)|}.$$

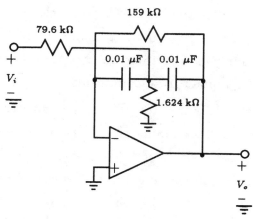

Fig. P11.31

11.32 Simulate the operation of the switched-capacitor integrator circuit shown in Fig. P11.32 and observe the voltage appearing at the output of the op amp for at least 10 μs. Initialize the feedback capacitor to 0 V. Model the op amp to have a unity-gain bandwidth of 10^6 Hz.

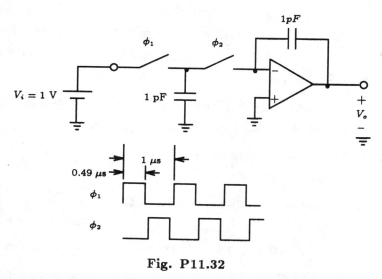

Fig. P11.32

11.33 A bandpass amplifier is shown in Fig. 11.18. Design the amplifier to produce a center frequency of 5 MHz with a 3 dB bandwidth of 50 kHz. What is the center frequency gain? Verify your design using Spice. Assume that the MOSFET has Spice parameters: $V_{to} = 1$ V, $k_p = 50$ μA/V^2 and $\lambda = 0.05$ V^{-1}.

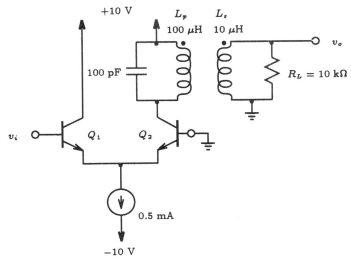

Fig. P11.34

11.34 A bandpass amplifier is shown in Fig. P11.34 whose output is coupled to the load resistor of 10 kΩ through a transformer. The transformer consists of a primary and secondary inductor, $L_p = 100$ μH and $L_s = 10$ μH, respectively, which have a coupling coefficient of 0.8. To specify this coupling, the following Spice statement is included in the Spice input file together with the two inductor statements: *K1 Lp Ls 0.8*. Using Spice determine the frequency response of this amplifier. What is the center frequency and bandwidth of this amplifier? Assume that the BJTs have model parameters $I_S = 10^{-16}$ A, $\beta_F = 100$ and $V_A = 75$ V.

Chapter 12

Signal Generators And Waveform – Shaping Circuits

In this chapter we shall utilize Spice to investigate the behavior of various types of signal generators and waveform-shaping circuits. Both sinusoidal and square-wave generation circuits will be considered. We shall first investigate the behavior of a Wien-bridge oscillator and an active-filter-tuned sinusoidal oscillator constructed with the 741 op amp. Subsequently, we will investigate the behavior of a crystal oscillator configured with a pair of complementary MOS transistors. Since these circuits are intended to be sinusoidal oscillators, the spectral purity of the signals generated from each circuit will be calculated using the Fourier series analysis capability of Spice. Following this, we shall investigate the behavior of several multivibrator circuits that produce square-waves. This will include the analysis of a bistable, astable, and monostable multivibrator circuits constructed with an op amp.

Another interesting circuit that we shall investigate with Spice is a nonlinear waveform-shaping circuit. With a triangular signal applied as input, this circuit shapes it into a sine-wave of the same frequency. This type of circuit is sometimes used in low-frequency function generators. We shall use Spice to investigate the quality of the sine-wave signal produced with this type of circuit.

The chapter concludes with the analysis of several precision rectifier op amp circuits commonly employed in instrumentation systems. This will include the analysis of a precision rectifier, a peak detector and a clamping circuit.

```
.subckt uA741           1 2 3 4 5
* connections:          | | | | |
*                       | | | | |
*   non-inverting input | | | | 
*        inverting input | | | 
*     positive power supply | | 
*       negative power supply | 
*                       output
*
*
  c1    11 12 8.661E-12
  c2     6  7 30.00E-12
  dc     5 53 dx
  de    54  5 dx
  dlp   90 91 dx
  dln   92 90 dx
  dp     4  3 dx
  egnd  99  0 poly(2) (3,0) (4,0) 0 .5 .5
  fb     7 99 poly(5) vb vc ve vlp vln 0 10.61E6 -10E6 10E6 10E6 -10E6
  ga     6  0 11 12 188.5E-6
  gcm    0  6 10 99 5.961E-9
  iee   10  4 dc 15.16E-6
  hlim  90  0 vlim 1K
  q1    11  2 13 qx
  q2    12  1 14 qx
  r2     6  9 100.0E3
  rc1    3 11 5.305E3
  rc2    3 12 5.305E3
  re1   13 10 1.836E3
  re2   14 10 1.836E3
  ree   10 99 13.19E6
  ro1    8  5 50
  ro2    7 99 100
  rp     3  4 18.16E3
  vb     9  0 dc 0
  vc     3 53 dc 1
  ve    54  4 dc 1
  vlim   7  8 dc 0
  vlp   91  0 dc 40
  vln    0 92 dc 40
.model dx D(Is=800.0E-18 Rs=1)
.model qx NPN(Is=800.0E-18 Bf=93.75)
.ends uA741
```

Figure 12.1 A Spice subcircuit description of a nonlinear macromodel of the 741 op amp. We shall rely on this macromodel of the 741 op amp in all examples of this chapter that contain op amps.

12.1 Op Amp-RC Sinusoidal Oscillators

In this section we shall investigate the behavior of a Wien-Bridge and an active-filter-tuned oscillator circuits assumed constructed with 741 op amps. Unlike examples of the previous chapter, where it was sufficient to represent the terminal behavior of the op amp with a linear network (see Section 11.3), the examples of this chapter rely on the nonlinear operation of the op amp (eg. output voltage saturation), and thus, its terminal behavior must be modeled more completely. Fortunately, many IC component manufacturers are

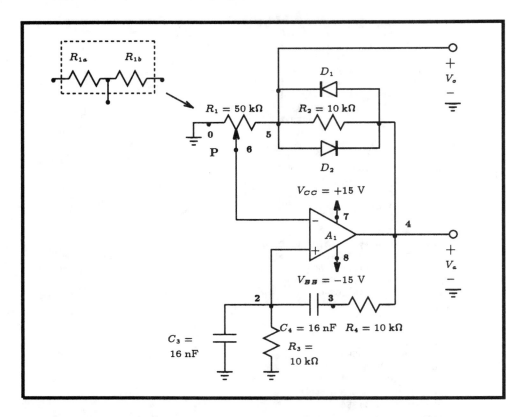

Figure 12.2 A Wien-bridge oscillator with a limiter used for amplitude control.

making available *macromodels* of their op amps in the form of a Spice subcircuit description, capturing much of the op amp's large-signal behavior in a much simpler circuit representation than a detailed transistor description [Texas Instruments, 1990]. This makes the task of circuit simulation more efficient without sacrificing accuracy. As an example, a Spice subcircuit description of the 741 op amp is shown listed in Fig. 12.1. We shall use this subcircuit description for the 741 op amp in all of the circuit simulations of this chapter that contain op amps.†

12.1.1 The Wien-Bridge Oscillator

In Fig. 12.2 we present a Wien-Bridge oscillator circuit with a diode limiter in its negative feedback path. The diode limiter serves to maintain the loop gain at unity and stabilizes the amplitude of the oscillations. The components were selected such that oscillations occur at 1 kHz. The potentiometer P is meant to be adjusted such that the oscillations just begin to grow. In the following we would like to investigate the behavior of this oscillator circuit for

---

† Those readers who have the PSpice student version will find that this op amp macromodel is contained in a file called NOM.LIB, along with other models of different electronic devices.

12.1 Op Amp-RC Sinusoidal Oscillators

```
A Wien-Bridge Oscillator With Amplitude Stabilization
* loop gain is 1.1 (R1a=15k R1b=35k)

** Circuit Description **

* op-amp subcircuit

+++++ place uA741 op-amp subcircuit here (see Fig. 12.1) +++++

** Main Circuit **
* power supplies
Vcc 7 0 DC +15V
Vee 8 0 DC -15V
* Wien-bridge oscillator
XAmp 2 6 7 8 4 uA741
R1a 6 0 15k
R1b 6 5 35k
R2 5 4 10k
R3 2 0 10k
C3 2 0 16nF IC=0V
R4 3 4 10k
C4 2 3 16nF IC=0V
* diode limiter circuit
D1 4 5 D1N4148
D2 5 4 D1N4148
* model statements
.model D1N4148 D (Is=0.1p Rs=16 CJO=2p Tt=12n Bv=100 Ibv=0.1p)
** Analysis Requests **
.OPTIONS itl5=0
.OP
.TRAN 200us 20ms 0ms 200us UIC
** Output Requests **
.PLOT TRAN V(4) V(5)
.probe v(4) v(5)
.end
```

Figure 12.3 The Spice input deck for computing the transient behavior of the Wien-bridge oscillator shown in Fig. 12.2.

different settings on the potentiometer. This is equivalent to selecting different values for R_{1a} and R_{1b}, such that $R_{1a} + R_{1b} = 50$ kΩ. Since oscillations begin when $(R_2 + R_{1b})/R_{1a} = 2$, or when $R_{1a} = 20$ kΩ and $R_{1b} = 30$ kΩ, we shall consider three possible settings: (a) $R_{1a} = 15$ kΩ, $R_{1b} = 35$ kΩ, (b) $R_{1a} = 18$ kΩ, $R_{1b} = 32$ kΩ, and (c) $R_{1a} = 25$ kΩ, $R_{1b} = 25$ kΩ. The first setting establishes a loop gain of 1.33 at the frequency of oscillation and should be more than enough for the oscillations to begin. The second setting reduces the loop gain to 1.1, just above the level that is necessary for oscillations to begin. Finally, the last setting creates a loop gain of 0.8 and should prevent the circuit from beginning or sustaining oscillations.

12 Signal Generators And Waveform – Shaping Circuits

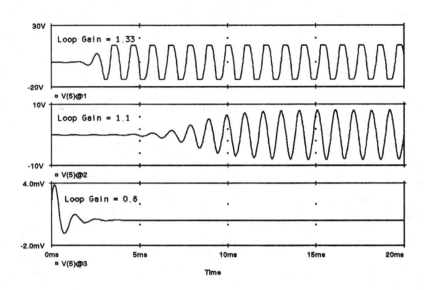

Figure 12.4 The startup transient behavior of the Wien-bridge oscillator shown in Fig. 12.2.

The Spice input file describing the Wien-bridge oscillator is shown in Fig. 12.2. The potentiometer is assumed to be set according to the setting described in (a) above. Notice that the voltage across each capacitor is initialized at 0 V. This is to demonstrate that this circuit will begin oscillations on its own without the need of any start-up circuit. The start of the oscillation is caused by the offset voltage of the op amp. To force Spice to utilize the supplied initial conditions, the .TRAN statement includes the keyword UIC (use initial conditions). The other two setting for the potentiometer are entered into separate Spice input file descriptions but attached to the end of the one shown in Fig. 12.3 before submission to Spice for execution. This will enable us to compare the output behavior for the three different circuit setups.

The results of the simulation are shown in Fig. 12.4. Here we see in the upper most graph the output oscillation for the Wien-bridge having a loop gain of 1.33. The frequency of oscillation is found to be 926.78 Hz slightly less than the intended value of 1 kHz. As is evident, the peaks of the output signal are clipped, indicating that the loop gain is too high. This is obviously not very useful as a sinusoidal generator because of the large distortion present. The middle graph illustrates the results of the oscillator with a reduced loop gain of 1.1 and demonstrates an undistorted sinusoidal waveform of 979.43 Hz frequency having a 8.165 V amplitude offset by −22.5 mV. The lowest-most graph confirms that oscillations will not begin when the loop gain is less than unity.

12.1 Op Amp-RC Sinusoidal Oscillators

It is interesting to compare the spectral purity of the waveform appearing at the output of the oscillator as marked in Fig. 12.2 with the waveform appearing at the op amp output (labeled as V_a). This is easily accomplished using the Fourier series analysis capabilities of Spice. Consider computing the first nine harmonics of the voltage waveform appearing across the output marked as V_o from which the total harmonic distortion (THD) can be computed. This requires that one add the following .FOUR command into the Spice file shown in Fig. 12.3 for a loop gain of 1.1:

```
.FOUR 979.43Hz V(5)
```

It is essential to use a close estimate of the oscillation frequency for the Fourier series analysis in Spice. A fractional error, say $\pm 0.1\%$, produces a noticeable change in the harmonic content in the output voltage waveform. The results of this analysis are then found in the output file as follows:

```
FOURIER COMPONENTS OF TRANSIENT RESPONSE V(5)

DC COMPONENT =  -1.159314E-02

 HARMONIC   FREQUENCY    FOURIER     NORMALIZED    PHASE      NORMALIZED
    NO        (HZ)      COMPONENT    COMPONENT    (DEG)      PHASE (DEG)

     1      9.794E+02    8.093E+00    1.000E+00    7.443E+00    0.000E+00
     2      1.959E+03    2.483E-01    3.068E-02    1.691E+01    9.469E+00
     3      2.938E+03    2.917E-01    3.604E-02   -4.998E+01   -5.742E+01
     4      3.918E+03    6.764E-02    8.358E-03   -1.021E+00   -8.464E+00
     5      4.897E+03    1.125E-01    1.390E-02   -1.121E+01   -1.866E+01
     6      5.877E+03    4.797E-02    5.927E-03   -2.027E+01   -2.772E+01
     7      6.856E+03    6.630E-02    8.193E-03    3.834E+01    3.090E+01
     8      7.835E+03    6.487E-02    8.016E-03    8.205E+00    7.612E-01
     9      8.815E+03    8.325E-03    1.029E-03    1.134E+02    1.059E+02

TOTAL HARMONIC DISTORTION =    5.168163E+00 PERCENT
```

Repeating this analysis for the output voltage waveform appearing at the op amp output using the command:

```
.FOUR 979.43Hz V(4)
```

we obtain the following results:

```
FOURIER COMPONENTS OF TRANSIENT RESPONSE V(4)

DC COMPONENT =  -1.196819E-02

 HARMONIC   FREQUENCY    FOURIER     NORMALIZED    PHASE      NORMALIZED
    NO        (HZ)      COMPONENT    COMPONENT    (DEG)      PHASE (DEG)

     1      9.794E+02    8.748E+00    1.000E+00    7.390E+00    0.000E+00
     2      1.959E+03    2.595E-01    2.967E-02    1.659E+01    9.195E+00
     3      2.938E+03    3.824E-01    4.371E-02   -2.690E+01   -3.429E+01
     4      3.918E+03    8.218E-02    9.394E-03    1.496E+00   -5.895E+00
     5      4.897E+03    1.669E-01    1.908E-02    4.072E+00   -3.318E+00
     6      5.877E+03    6.224E-02    7.115E-03   -1.151E+01   -1.890E+01
     7      6.856E+03    9.173E-02    1.049E-02    4.158E+01    3.419E+01
```

12 Signal Generators And Waveform – Shaping Circuits

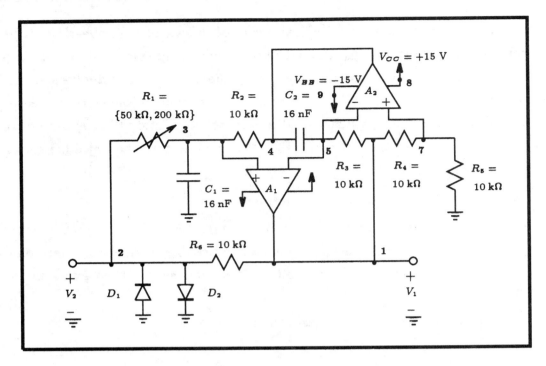

Figure 12.5 An active-filter tuned oscillator with the Q of the filter adjustable by R_1.

```
    8       7.835E+03    7.950E-02    9.088E-03    1.081E+01    3.416E+00
    9       8.815E+03    1.452E-02    1.660E-03    9.456E+01    8.717E+01

    TOTAL HARMONIC DISTORTION =    5.906799E+00 PERCENT
```

Comparison of the above two sets of results indicates that the voltage waveform appearing at the op amp output contains slightly more harmonic distortion than the voltage output marked V_o as indicated by the THD measures. But, in all fairness, the difference is very small and one may prefer to obtain the output from the low-impedance op amp output terminal instead of the high-impedance node that is presently marked as the output.

12.1.2 An Active-Filter-Tuned Oscillator

In Fig. 12.5 we display an active-filter-tuned oscillator. It consists of a high-Q bandpass filter connected in a positive feedback loop around a hard limiter. The filter circuit portion of this oscillator was designed to have a center frequency of 1 kHz and an adjustable Q determined by R_1. Using Spice we shall investigate the spectral purity of the output voltage waveform (V_1) by computing its harmonic content for different values of filter Q.

We shall begin our analysis with a filter circuit in the feedback loop having a moderate Q of 5. This is achieved with $R_1 = 50$ kΩ. The Spice description for this circuit arrangement is shown listed in Fig. 12.6. The op amps are modeled after the commercial 741 op amp

12.1 Op Amp-RC Sinusoidal Oscillators

```
The Active-Filter Tuned Oscillator

** Circuit Description **

* op-amp subcircuit

+++++ place uA741 op-amp subcircuit here (see Fig. 12.1) +++++

** Main Circuit **
* power supplies
Vcc 8 0 DC +15V
Vee 9 0 DC -15V
* high-Q filter circuit
Xopamp1 3 5 8 9 1 uA741
Xopamp2 7 5 8 9 4 uA741
R1 2 3 50k
C1 3 0 16nF IC=0V
R2 3 4 10k
C2 4 5 16nF IC=0V
R3 5 1 10k
R4 1 7 10k
R5 7 0 10k
* diode limiter circuit
D1 0 2 D1N4148
D2 2 0 D1N4148
R6 1 2 10k
* model statement
.model D1N4148 D (Is=0.1p Rs=16 CJO=2p Tt=12n Bv=100 Ibv=0.1p)
** Analysis Requests **
.OPTIONS itl5=0
.TRAN 20us 50ms 45ms 20us UIC
.FOUR 1kHz V(1)
** Output Requests **
.PLOT TRAN V(1) V(2)
.probe V(1) V(2)
.end
```

Figure 12.6 The Spice input deck for computing the transient behavior of the active-filter-tuned oscillator circuit shown in Fig. 12.5.

circuit and the diodes are modeled after the 1N4148 type. A transient analysis is requested beginning with the circuit initially at rest. This is achieved by setting the initial voltage across C_1 and C_2 to zero volts, as indicated by the keyword IC=0 V on the element statement for C_1 and C_2. In order to correctly compute the harmonic content of the output waveform, the circuit response must have reached its steady-state. As a lower bound on the amount of time required for this to happen, assume that the filter circuit within the feedback loop of the oscillator has an underdamped response characterized by a pair of complex poles described

12 Signal Generators And Waveform – Shaping Circuits

by ω_o and Q. As a result, its impulse response can be described by a damped sinusoidal having the following form:

$$Ke^{-\frac{\omega_o}{2Q}t}\cos(\omega_o\sqrt{1-\frac{1}{4Q^2}}\,t+\phi). \tag{12.1}$$

Clearly then, in order for the transient response to die out to less than 0.1% of its initial value, the exponential term in Eqn. (12.1) must reduce in value to something less than 0.001. The amount of time required for this to happen is then found from the following:

$$e^{-\frac{\omega_o}{2Q}T_S} \leq 0.001, \tag{12.2}$$

which, when rearranged, gives

$$T_S \geq -\frac{2Q}{\omega_o}\ln(0.001), \tag{12.3}$$

where we have denoted T_S by the time for the output response to settle into its steady-state behavior. Thus, for a filter Q of 5, we can expect the transient response to last for about 11 ms. Since the filter circuit is placed inside a positive feedback loop, the effective Q of the oscillator circuit will be much larger than the Q of the filter circuit. Thus, from Eqn. (12.3), we can expect that the time for the oscillator circuit to reach its steady-state will be much longer than 11 ms.

Through several iterations with Spice we found that the oscillator circuit shown in Fig. 12.5 for a filter Q of 5 required 50 ms to reach its steady-state. Furthermore, to get good coverage of the output waveform, mainly for plotting purposes, 10 points-per-period of oscillation was taken. Moreover, the results of this analysis revealed that the frequency of oscillation was precisely 1 kHz.

As the output request, only the output voltage and the voltage across the limiter are asked to be plotted. Also, we have requested that PROBE store only the results of these two voltages in order to reduce the amount of data that is stored on disk which can be quite enormous during a transient analysis.

Finally, a .FOUR command was added to the Spice input file to compute the harmonic content in the output voltage waveform having the following form:

.FOUR 1kHz V(1).

The results of the transient analysis are shown in Fig. 12.7. Here we display in the top graph the output voltage waveform of the oscillator and the lower graph displays the voltage signal appearing across the diode limiter. As is evident, the output signal appears undistorted, unlike that of the voltage appearing across the diode limiter. The frequency of

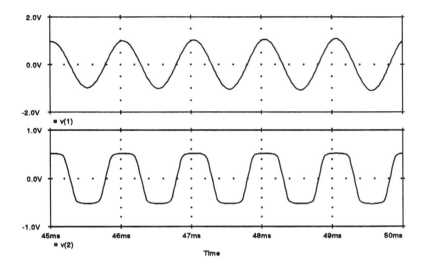

Figure 12.7 The steady-state transient response of the active-filter tuned oscillator shown in Fig. 12.5 for $Q = 5$.

oscillation of either waveform is seen to be very close to 1 kHz. The results of the Fourier series calculation reveals the following harmonic content in the output voltage waveform:

```
FOURIER COMPONENTS OF TRANSIENT RESPONSE V(1)

DC COMPONENT =  -6.624248E-03
```

HARMONIC NO	FREQUENCY (HZ)	FOURIER COMPONENT	NORMALIZED COMPONENT	PHASE (DEG)	NORMALIZED PHASE (DEG)
1	1.000E+03	1.270E+00	1.000E+00	4.450E+01	0.000E+00
2	2.000E+03	1.351E-02	1.063E-02	2.849E+01	-1.601E+01
3	3.000E+03	2.765E-02	2.177E-02	3.920E+01	-5.298E+00
4	4.000E+03	5.878E-03	4.627E-03	2.221E+01	-2.229E+01
5	5.000E+03	4.676E-03	3.681E-03	6.410E+01	1.960E+01
6	6.000E+03	3.573E-03	2.813E-03	2.499E+01	-1.952E+01
7	7.000E+03	2.726E-03	2.146E-03	2.529E+01	-1.921E+01
8	8.000E+03	2.635E-03	2.074E-03	2.553E+01	-1.897E+01
9	9.000E+03	2.277E-03	1.793E-03	3.218E+01	-1.232E+01

```
TOTAL HARMONIC DISTORTION =    2.533441E+00 PERCENT
```

As is evident, the total harmonic distortion (THD) is about 2.5%. Recall that for the Wien-bridge oscillator of the previous section the THD was approximately 2 times as large at 5%.

In order to repeat the above analysis for a filter Q of 20 (ie. $R_1 = 200$ kΩ), we should increase the time that the transient analysis is performed. According to Eqn. (12.3), if the filter Q is increased by a factor of 4, the transient time also increases by the same factor. Thus, we should increase the time that the transient is performed to 200 ms while maintaining

12   Signal Generators And Waveform – Shaping Circuits

the same time step. On doing so, we obtain the following Fourier series components for the output voltage waveform:

```
FOURIER COMPONENTS OF TRANSIENT RESPONSE V(1)

DC COMPONENT =    1.321303E-02

HARMONIC   FREQUENCY   FOURIER      NORMALIZED   PHASE        NORMALIZED
NO         (HZ)        COMPONENT    COMPONENT    (DEG)        PHASE (DEG)

    1      1.000E+03   1.268E+00    1.000E+00    -6.697E+01   0.000E+00
    2      2.000E+03   9.201E-03    7.254E-03    -5.044E+01   1.653E+01
    3      3.000E+03   5.859E-03    4.619E-03     2.491E+01   9.188E+01
    4      4.000E+03   2.674E-03    2.108E-03    -2.493E+01   4.204E+01
    5      5.000E+03   2.754E-03    2.171E-03    -3.491E+01   3.206E+01
    6      6.000E+03   1.652E-03    1.303E-03    -1.777E+01   4.920E+01
    7      7.000E+03   1.227E-03    9.673E-04    -1.296E+01   5.401E+01
    8      8.000E+03   1.162E-03    9.163E-04    -1.031E+01   5.666E+01
    9      9.000E+03   9.852E-04    7.767E-04    -9.472E+00   5.750E+01

TOTAL HARMONIC DISTORTION =    9.337391E-01 PERCENT
```

As expected the THD in the output waveform has decreased due to the increased selectivity of the filter circuit portion of the oscillator. In fact, we see that the THD decreased by almost the same factor that the filter Q was increased by.

12.2   Crystal Oscillators

In contrast to the active-RC oscillators circuit of the previous section, which can attain oscillation stabilities approaching 0.1%, crystal oscillators can obtain frequency stabilities many orders of magnitude greater, on the order of 1 in 10^6. Furthermore, at the heart of these oscillator circuits is a piezoelectric crystal that forms a resonant circuit with Q factor's on the order of 10,000 or more, resulting in highly selective oscillators.

An inexpensive crystal oscillator can be realized with a single crystal and a standard CMOS inverter, with several additional passive components, as shown in Fig. 12.8(a). In this particular case, the crystal is one that resonates at the TV color burst frequency of 3.579545 MHz and has a Q of 25,000. In the following we shall use Spice to compute the output waveform from this oscillator assuming that the two complementary transistors making up the CMOS inverter have parameters: $|V_T| = 1$ V, $\mu_n C_{OX} = 2\mu_p C_{OX} = 20$ μA/V^2 and $\lambda = 0.04$ V^{-1}. Furthermore, the lengths of each transistor are the same at 5 μm. The width of the NMOS 100 μm and the PMOS is 200 μm. Spice does not have any built-in model for piezoelectric crystals, so instead, we make use of its equivalent circuit representation shown in Fig. 12.8(b).

The Spice input file describing the crystal oscillator in Fig. 12.8(a) is seen listed in Fig. 12.9. The piezoelectric crystal is described by a subcircuit called *colorburst_crystal* in order

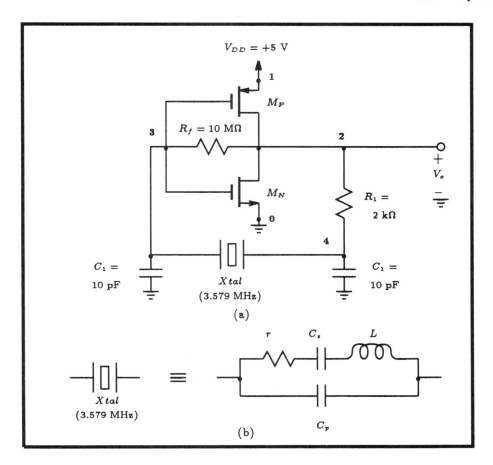

Figure 12.8 (a) A crystal oscillator utilizing a CMOS inverter as an amplifier. (b) An equivalent circuit representation of a piezoelectric crystal.

to make it easier to read the Spice input file. The model of this crystal was obtained from the PSpice library of device models supplied by the MicroSim Corporation. The inductor is initialized with a 1 μA current in order to assist in the transient analysis. More on this in a moment.

At this point we have to decide on how the transient analysis should be carried out. As it turns out, crystal circuits pose a difficult simulation problem for Spice owing to the unusually long transient behavior. From Eqn. (12.3) we see that the number of cycles of oscillation required to pass before the steady-state is reached is $\# \: of \: cycles = T_S/T = \ln(0.001)/\pi Q$ where T is the period of oscillation. Unlike earlier examples where the Q factors were on the order of tens, crystal oscillators have Q factors on the order of tens-of-thousands. As a result, incredibly long simulation times are necessary to obtain steady-state behavior.

Rather than perform the simulation on a personal computer, as were all other circuit simulation examples of this text, we have performed the above circuit simulation using a SUN 4 workstation running Spice version 2G6. In this way, we can obtain our results several order

12 Signal Generators And Waveform – Shaping Circuits

```
A Pierce Crystal Oscillator

** Circuit Description **

* 3.579545Mhz color burst, AT cut, parallel resonant, Q=25000,
.subckt colorburst_crystal 1 2
 l 1 11 0.0555779237 IC=1uA
 cs 11 12 3.56169600e-014
 r 12 2 50
 cp 1 2 8.90424001e-012
.ends

** Main Circuit **
* power supplies
Vdd 1 0 DC +5V
* oscillator circuit
Mp 2 3 1 1 pmos_transistor L=5u W=200u
Mn 2 3 0 0 nmos_transistor L=5u W=100u
Rf 2 3 10Meg
R1 2 4 75k
Xtal 3 4 colorburst_crystal
C1 3 0 10pF IC=2.5V
C2 4 0 10pF IC=2.5V
* model statement
.model pmos_transistor PMOS (kp=10u Vto=-1V lambda=0.04)
.model nmos_transistor NMOS (kp=20u Vto=+1V lambda=0.04)
** Analysis Requests **
.options limpts=100000 itl5=0 opts numdgt=7
.width out=80
* use the DC bias point as the initial conditions for the oscillator
.TRAN 20ns 20ms 0.0199972 20ns UIC
.FOUR 3.579MegHz
** Output Requests **
.PRINT TRAN V(2)
.end
```

Figure 12.9 The Spice input deck for computing the output voltage waveform of the crystal oscillator shown in Fig. 12.8(a).

of magnitude faster than what would be possible using a personal computer. In addition, to speed up the transition to the steady-state, several energy storage elements were initialized. The values used to initialized these elements were simply guessed at, keeping in mind values that would be typical of a low power oscillator circuit. Experience has shown that the quickest transition to steady-state was obtained when the inductor of the equivalent circuit representation of the crystal is initialized at some low current level as already suggested above.

The results of the simulation are shown plotted in Fig. 12.10. These results were

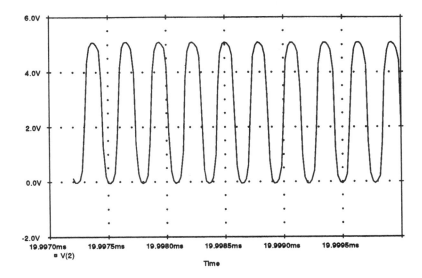

Figure 12.10 The steady-state output voltage waveform from the crystal oscillator shown in Fig. 12.8.

generated with Spice version 2G6 but plotted using the graphics capability available with the PSpice software package. As is evident from the results shown in Fig. 12.10 the output voltage waveform has reached steady-state with a frequency of oscillation of 3.579 MHz, set precisely by the crystal in the amplifiers feedback path. As is clearly evident, this output signal is highly distorted with a total harmonic distortion content of 20.9% as determined by the .FOUR command of Spice.

12.3 Multivibrator Circuits

In some applications, the need arises for square waveforms. Multivibrators are one such class of circuits that can be used to accomplish this. Multivibrators are conveniently classified as (1) bistable circuits, (2) astable circuits, or (3) monostable circuits. In Fig. 12.11 we present one type of each. In the following we shall analyze each with Spice assuming that the op amp in each circuit is modeled after the 741 type.

A Bistable Circuit:

In Fig. 12.11(a) we present an op amp circuit with a positive feedback loop that has two stable states. The state in which the circuit is in is dependent on the input level and the previous state that the circuit was in. As a result, this circuit exhibits a form of hysteresis. Using Spice we can deduce the hysteresis of the circuit shown in Fig. 12.11(a) from its

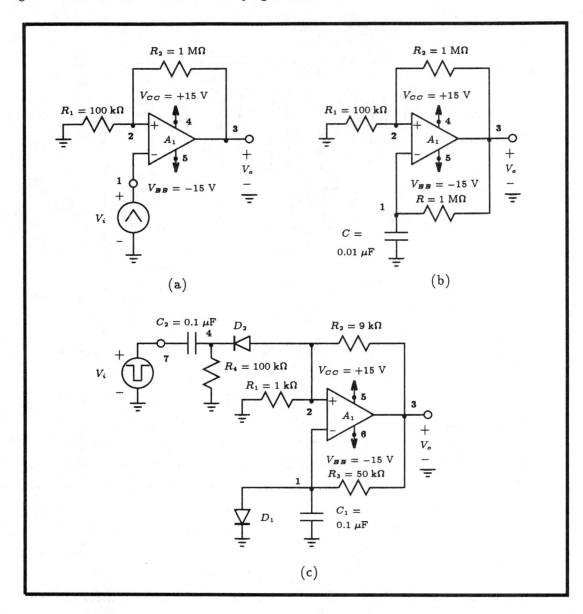

Figure 12.11 A collection of multivibrator circuits: (a) A bistable circuit. (b) An astable circuit. (c) A monostable circuit.

transfer characteristics. One might be tempted to perform a DC sweep of the input voltage V_i over the range supported by the two power supplies; however, two issues arise. The complete transfer characteristics of the bistable circuit requires that the input voltage level be swept from the negative supply level to the positive supply level, then returned back to the negative supply. This, therefore, requires two Spice runs; one for each sweep direction. Secondly, Spice has problems with convergence at the point where the circuit changes state and usually fails to complete the solution.

A better technique for obtaining the transfer characteristics of a regenerative circuit with

12.3 Multivibrator Circuits

```
The Bistable Circuit

** Circuit Description **

* op-amp subcircuit

+++++ place uA741 op-amp subcircuit here (see Fig. 12.1) +++++

** Main Circuit **
* power supplies
Vcc 4 0 DC +15V
Vee 5 0 DC -15V
* input triangular waveform
Vi 1 0 PWL ( 0,-15V 1s,+15V 2s,-15V )
* positive feedback op-amp circuit
Xopamp1 2 1 4 5 3 uA741
R1 2 0 100k
R2 2 3 1Meg
** Analysis Requests **
.OPTIONS itl5=0
.TRAN 10ms 2s 0ms 10ms
** Output Requests **
.PLOT TRAN V(1) V(3)
.probe
.end
```

Figure 12.12 The Spice input deck for computing the transfer characteristics of the bistable circuit shown in Fig. 12.11(a).

hysteresis is to apply a low frequency triangular waveform as input to the bistable circuit whose level varies between the two power supply levels. In this way, one mimics the same action one takes when measuring the transfer characteristics of a circuit in the laboratory. The frequency of the input signal is kept low to minimize the effects of op amp dynamics on the circuit transfer characteristics.

The Spice description of the bistable circuit shown in Fig. 12.11(a) is seen listed in Fig. 12.12. The input triangular waveform is described using a PWL transient source description beginning at −15 V and rising to +15 V in the first second and then decreasing back to −15 V in the next second. A transient analysis is requested to compute the output voltage over a two second interval. The results of this analysis are seen in Fig. 12.13. The top graph displays the input and output transient waveforms. The information contained in this figure is then translated into the transfer characteristics shown in the bottom graph. From this plot, we see that the lower threshold limit (V_{T_L}) equals −1.3266 V and the upper threshold limit (V_{T_H}) equals +1.3121 V. The width of the hysteresis is thus computed to be 2.639 V.

It is interesting to compare these threshold levels with those predicted by the theory

12 Signal Generators And Waveform – Shaping Circuits

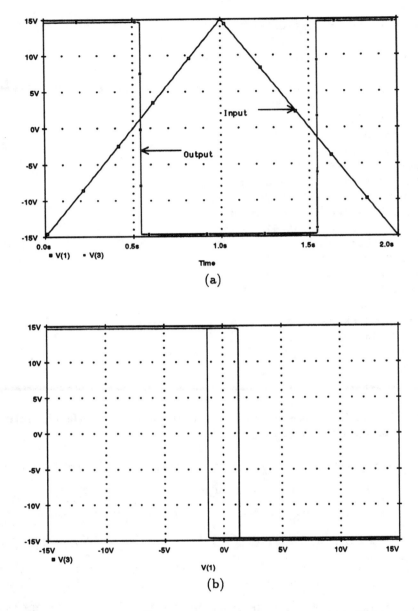

Figure 12.13 (a) The input and output waveforms of the bistable circuit shown in Fig. 12.11(a) as a function of time. (b) Transfer characteristics of the same bistable circuit.

presented by Sedra and Smith in Section 12.4. For instance, we see from Fig. 12.13 that the saturation level of the 741 op amp circuit is ±14.6 V. Since $\beta = R_1/(R_1 + R_2)$, we find with the appropriate values substituted that $\beta = 0.091$. Thus, we find: $V_{T_H} = -V_{T_L} = 1.33$ V, which are reasonably close to the threshold levels obtained through Spice simulation.

```
An Astable Multivibrator: Square-Wave Generator

** Circuit Description **

* op-amp subcircuit

+++++ place uA741 op-amp subcircuit here (see Fig. 12.1) +++++

** Main Circuit **
* power supplies
Vcc 4 0 DC +15V
Vee 5 0 DC -15V
* multivibrator circuit
Xopamp1 2 1 4 5 3 uA741
R1 2 0 100k
R2 2 3 1Meg
R  1 3 1Meg
C  1 0 0.01uF IC=0V
** Analysis Requests **
.OPTIONS itl5=0
.TRAN 500us 50ms 0ms 500us
** Output Requests **
.PLOT TRAN V(3) V(1)
.probe
.end
```

Figure 12.14 The Spice input deck for computing the transient behavior of the square-wave generator shown in Fig. 12.11(b).

Generation Of A Square-Wave Using An Astable Multivibrator:

A square-wave generator can be realized using a bistable multivibrator with an RC circuit in its feedback loop, as illustrated in Fig. 12.11(b). The components of the bistable circuit are the same as those used in the previous section, so its expected behavior is known. Since the upper and lower saturation limits of the bistable portion of this circuit are equal (ie. $L_+ = -L_- = 14.6$ V), as well, $\beta = 0.091$ and $\tau = RC = 10$ ms, the expected period of oscillation is then computed according to

$$T = 2\tau \ln(\frac{1+\beta}{1-\beta})$$

to give 3.65 ms.

To confirm that this circuit indeed oscillates with a period of 3.65 ms, we have created the Spice input file description of this circuit shown in Fig. 12.14 and submitted it to Spice. On completion of Spice, the output waveform is shown plotted in the top graph of Fig. 12.15. The bottom graph displays the almost-triangular voltage waveform appearing at the positive

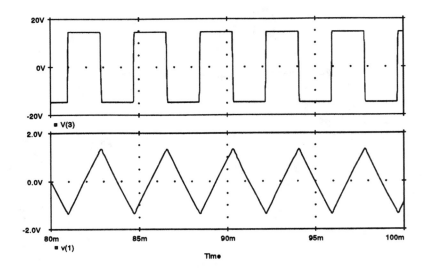

Figure 12.15 Several waveforms associated with the square-wave generator shown in Fig. 12.11(b) − top curve: output voltage waveform; bottom curve: voltage at input negative terminal of op amp.

input terminal to the op amp. The period of oscillation for each of these waveforms is found to be 3.74 ms, which agrees fairly well with that predicted by theory.

The Monostable Multivibrator:

A monostable is often used to create a pulse of a fixed duration that is triggered by another pulse of arbitrary duration. In Fig. 12.11(c) we present one such monostable that has been designed to generate a pulse of 755 μs duration when it receives a negative-going trigger pulse whose level varies between V_{CC} and ground, according to the formula:

$$T \cong C_1 R_3 \ln(\frac{0.7/L_- - 1}{\beta - 1}).$$

Using the Spice deck listed in Fig. 12.16 we have computed the response of the monostable when triggered with a negative-going pulse of a 50 μs duration. The width of this pulse was chosen to insure that the op amp has sufficient time to change state. Recall that the slew-rate of a 741 op amp is nominally 0.63 V/μs. For a complete change of state, the op amp output voltage undergoes a change of about 30 V. Thus, the time that it takes to change state is approximately 50 μs. The results of the simulation are shown in Fig. 12.17. In this diagram, the top graph displays the trigger pulse and the curve below it illustrates the output generated pulse. Here we see that the duration of this pulse is approximately

12.4 A Nonlinear Waveform-Shaping Circuit

```
The Monostable Multivibrator

** Circuit Description **

* op-amp subcircuit

+++++ place uA741 op-amp subcircuit here (see Fig. 12.1) +++++

** Main Circuit **
* power supplies
Vcc 5 0 DC +15V
Vee 6 0 DC -15V
* input trigger signal + circuit
Vtrig 7 0 PWL ( 0,+15V 10us,+15V 10.01us,0V 60us,0V 60.01us,+15V
+ 10ms, +15V 10.00001ms,0V 10.090ms,0V 10.09001ms,+15V 1s,+15V )
C2 7 4 0.1uF
R4 4 0 100k
D2 2 4 D1N4148
* monostable multivibrator circuit
Xopamp1 2 1 5 6 3 uA741
R1 2 0 1k
R2 2 3 9k
R3 1 3 50k
C1 1 0 0.1uF
D1 1 0 D1N4148
* model statements
.model D1N4148 D (Is=0.1p Rs=16 CJO=2p Tt=12n Bv=100 Ibv=0.1p)
** Analysis Requests **
.IC V(3)=+15V
.TRAN 1ms 2ms 0ms
** Output Requests **
.PLOT TRAN V(7) V(3) V(1)
.probe
.end
```

Figure 12.16 The Spice input deck for computing several time waveforms of the monostable op amp circuit shown in Fig. 12.11(c).

750 μs, as measured from the points where the output signal crosses the 0 V axis. This result seems to agree reasonably well with the value estimated using the above equation. Also evident, is the time that it takes the op amp output to change state which is found to be 67 μs. Finally, the bottom-most graph displays the voltage appearing at the negative input terminal of the op amp. Clearly, it takes this waveform 1441.1 μs to recover from the trigger pulse. Therefore, the maximum frequency of operation for this circuit is limited to 693 Hz.

12 Signal Generators And Waveform – Shaping Circuits

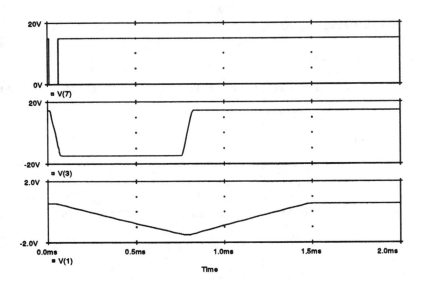

Figure 12.17 Several waveforms associated with the monostable op amp circuit shown in Fig. 12.11(c) – top curve: input trigger waveform; middle curve: output voltage waveform; bottom curve: voltage at input negative terminal of op amp.

12.4 A Nonlinear Waveform-Shaping Circuit

A three-segment sine-wave shaper with a triangular input signal is shown on display in Fig. 12.18. This type of circuit is used to shape the input triangular signal into a sinusoidal signal of the same frequency. Although this circuit has a very simple structure, it produces an output sinusoidal signal of relatively high spectral purity. In the following we shall use Spice to investigate the distortion content contained in the output signal created by the shaper circuit when a 10 V peak triangular signal of 1 kHz frequency is applied to its input.

Consider the Spice input description of the 3-segment shaper (Fig. 12.19). Here we have created the 10 V peak input triangular waveform of frequency 1 kHz using the pulse transient command of Spice. A transient analysis is requested using a sampling interval of 50 μs over a 4 ms time interval. Because the frequency of the output signal is known by virtue of the known input signal frequency, we can also include a Fourier series command at this time in the Spice input file.

The input and output signals as computed by Spice are shown in Fig. 12.20. As is evident, the output signal (V(7)) appears sinusoidal in shape with some distortion clearly present. The distortion components present in the output signal are shown below as computed by Spice:

12.5 Precision Rectifier Circuits

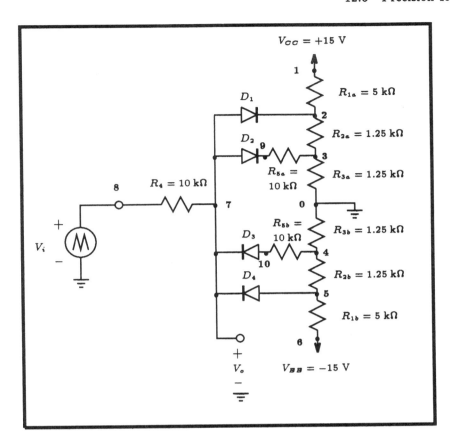

Figure 12.18 A three-segment sine-wave shaper.

```
FOURIER COMPONENTS OF TRANSIENT RESPONSE V(7)

DC COMPONENT =   1.217944E-02
```

HARMONIC NO	FREQUENCY (HZ)	FOURIER COMPONENT	NORMALIZED COMPONENT	PHASE (DEG)	NORMALIZED PHASE (DEG)
1	1.000E+03	5.892E+00	1.000E+00	-9.018E+01	0.000E+00
2	2.000E+03	7.432E-03	1.261E-03	-8.895E+01	1.237E+00
3	3.000E+03	6.693E-02	1.136E-02	8.939E+01	1.796E+02
4	4.000E+03	5.825E-04	9.886E-05	8.222E+01	1.724E+02
5	5.000E+03	1.643E-01	2.788E-02	-9.107E+01	-8.871E-01
6	6.000E+03	2.138E-03	3.628E-04	-8.672E+01	3.466E+00
7	7.000E+03	8.717E-03	1.480E-03	-8.918E+01	1.006E+00
8	8.000E+03	2.038E-03	3.459E-04	-8.494E+01	5.247E+00
9	9.000E+03	5.376E-02	9.125E-03	8.854E+01	1.787E+02

```
TOTAL HARMONIC DISTORTION =   3.152547E+00 PERCENT
```

Surprisingly, the total harmonic distortion is only 3.2%. When compared to the Wien-bridge sinusoidal oscillator of section 12.1.1, we see that the shaper circuit shown in Fig. 12.18 created an output signal with 2% less harmonic distortion.

12 Signal Generators And Waveform – Shaping Circuits

```
A 3-Segment Sine-Wave Shaper

** Circuit Description **
* power supplies
Vcc 1 0 DC +15V
Vee 6 0 DC -15V
* input triangular source
Vi 8 0 DC 0V PULSE (-10V 10V 0 0.5ms 0.5ms 1us 1ms)
* diode shaper circuit
R1a 1 2 5k
R1b 5 6 5k
R2a 2 3 1.25k
R2b 4 5 1.25k
R3a 3 0 1.25k
R3b 4 0 1.25k
R4 8 7 10k
R5a 3 9 10k
R5b 4 10 10k
D1 7 2 D1N4148
D2 7 9 D1N4148
D3 10 7 D1N4148
D4 5 7 D1N4148
* model statements
.model D1N4148 D (Is=0.1p Rs=16 CJO=2p Tt=12n Bv=100 Ibv=0.1p)
** Analysis Requests **
.FOUR 1kHz V(7)
.TRAN 50us 4ms 0ms 50us
** Output Requests **
.PLOT TRAN V(7) V(8)
.probe
.end
```

Figure 12.19 The Spice input deck for computing the total harmonic distortion of the three-segment sine-wave shaper shown in Fig. 12.18. A 10 V peak triangular waveform is applied to the input of this circuit.

12.5 Precision Rectifier Circuits

In this section we shall investigate the characteristics of several precision rectifier circuits used in different instrumentation applications. This will include Spice analysis of a precision half-wave rectifier, a peak detector and a clamping circuit, as shown in Fig. 12.21. We have purposely kept the input signal frequency low in order to minimize the dynamic effects (ie. slew rate) of the op amp on circuit operation. With higher input signal frequencies, these effects will begin to play a significant role in circuit operation. We encourage our readers to investigate this on their own.

12.5 Precision Rectifier Circuits

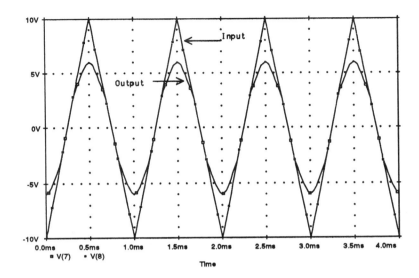

Figure 12.20 The input and output waveforms associated with the three-segment sine-wave shaper shown in Fig. 12.18. The output signal is very nearly sinusoidal with a total harmonic content of approximately 3.2%.

A Half-Wave Rectifier Circuit:

In Fig. 12.21(a) we present a precision half-wave rectifier circuit. The active components are made up of a single 741 op amp and two 1N4148-type diodes. Using a DC sweep of the input voltage level ranging between the two power supply levels we shall compute the input-output transfer characteristics of this circuit.

The Spice input file describing the circuit including the appropriate analysis requests is shown listed in Fig. 12.22. The resulting transfer characteristics are shown plotted in Fig. 12.23. The curve in part (a) of this figure illustrates the output voltage as a function of the input voltage when the latter is varied between the positive and negative supplies. For negative input signals, the output signal will be seen to be twice the size of the input signal with an 180° phase shift. For positive input signals, no signal appears at the output. Thus, half-wave rectification is performed. Figure 12.23(b) provides an expanded view of the transfer characteristic for signals in the range -50 μV to 200 μV. Here we see that the transfer characteristic does not go through the origin but instead has an output offset voltage of 211.9 μV. As well, we see from this curve that for positive input voltages larger than 100 μV, the output voltage does not go to zero but levels off at 20.9 μV.

12 Signal Generators And Waveform – Shaping Circuits

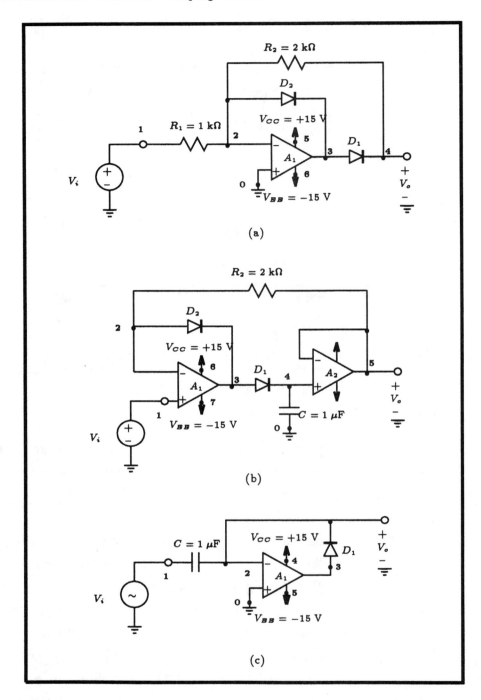

Figure 12.21 Precision rectifier circuits: (a) half-wave rectifier circuit. (b) peak detector circuit. (c) clamping circuit.

```
Precision Rectifier Circuit

** Circuit Description **

* op-amp subcircuit

+++++ place uA741 op-amp subcircuit here (see Fig. 12.1) +++++

** Main Circuit **
* power supplies
Vcc 5 0 DC +15V
Vee 6 0 DC -15V
* input signal source
Vi 1 0 DC 0V SIN( 0V 1V 1kHz )
* limiter circuit
Xopamp 0 2 5 6 3 uA741
R1 1 2 1k
R2 2 4 2k
D1 2 3 D1N4148
D2 3 4 D1N4148
* model statements
.model D1N4148  D (Is=0.1p Rs=6 CJO=1p Tt=3n Bv=100 Ibv=0.1p)
** Analysis Requests **
.DC Vi -15V +15V 0.5V
** Output Requests **
.PLOT DC V(4)
.probe
.end
```

Figure 12.22 The Spice input deck for computing the transient behavior of the peak detector shown in Fig. 12.21(a).

A Buffered Peak Detector:

In Fig. 12.21(b) we present a buffered precision peak detector. The purpose of this circuit is to hold the peak value of some input signal over long periods of time. To demonstrate the behavior of this circuit, we shall apply an input triangular signal with increasing peak values and observe the output signal from this detector. The Spice circuit description of this circuit is shown listed in Fig. 12.24, together with the piece-wise linear description of the input signal. A transient analysis is requested to compute the output signal over an 11 ms interval of the input signal.

The input and output signals for this experiment, as calculated by Spice, are shown in Fig. 12.25. We see here that this circuit does indeed hold the output level constant at the most recent peak level of the input signal.

12 Signal Generators And Waveform – Shaping Circuits

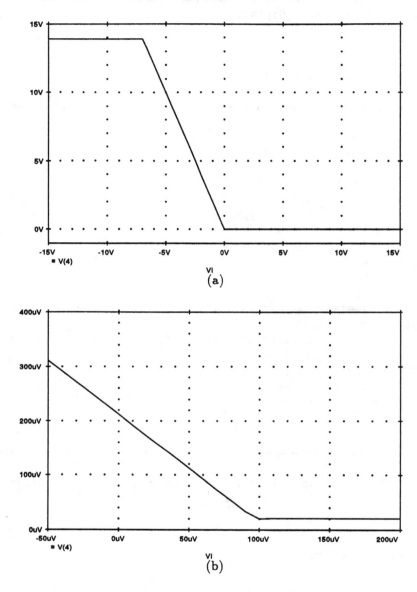

Figure 12.23 The input-output transfer characteristics of the precision half-wave rectifier circuit shown in Fig. 12.21(a): (a) Input voltage varied between the positive and negative supply rails. (b) Input voltage varied between −50 μV and 200 μV.

A Clamping Circuit:

As the final example of this section, we present in Fig. 12.21(c) a precision op amp clamping circuit whose function is to restore a DC level to some AC coupled input signal. Consider applying a 5 V peak input sinusoidal signal of 1 kHz frequency to the input of this clamping circuit with no DC offset. The Spice input file describing this setup is shown listed in Fig. 12.26. A transient analysis is requested, and the input and output signals are to be

```
A Buffered Precision Peak Detector

** Circuit Description **

* op-amp subcircuit

+++++ place uA741 op-amp subcircuit here (see Fig. 12.1) +++++

** Main Circuit **
* power supplies
Vcc 6 0 DC +15V
Vee 7 0 DC -15V
* input signal source
Vi 1 0 DC 0V PWL ( 0s,-15V 1ms,-7.5V 2ms,-15V 3ms,-15V 4ms,0V 5ms,-15V
+              6ms,-15V 7ms,+7.5V 8ms,-15V 9ms,-15V 10ms,+10V 11ms,-15V )
* limiter circuit
Xopamp1 1 2 6 7 3 uA741
Xopamp2 4 5 6 7 5 uA741
D1 3 4 D1N4148
D2 2 3 D1N4148
R 2 5 1k
C 4 0 1uF
* model statements
.model D1N4148 D (Is=0.1p Rs=16 CJO=2p Tt=12n Bv=100 Ibv=0.1p)
** Analysis Requests **
.OPTIONS itl5=0
.TRAN 0.5ms 11ms 0s 0.5ms
** Output Requests **
.PLOT TRAN V(1) V(3)
.probe
.end
```

Figure 12.24 The Spice input deck for computing the transient behavior of the peak detector shown in Fig. 12.21(b).

observed. The results of this analysis are shown in Fig. 12.27. Clearly, the output signal has been offset by the peak of the input signal and now varies between 0 V and +10 V. Also seen is that it takes less than one cycle of the input waveform to reach steady-state.

12.6 Spice Tips

- Various electronic manufacturers are making available Spice models of their components in the form of subcircuits that can be included directly into a Spice deck.
- The Fourier analysis command (.FOUR) of Spice requires a very good estimate of the frequency of the signal that is to be analyzed. A fractional error in this frequency estimate can produce noticeable change in the harmonic content of the output waveform.

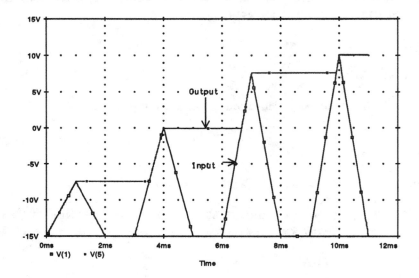

Figure 12.25 The input and output transient waveforms of the peak detector shown in Fig. 12.21(b).

- A Fourier analysis should only be performed on a periodic waveform that has reached steady-state. One must allow the transient in the output waveform to decrease to unappreciable level. As a rule-of-thumb, for a circuit dominated by a pair of complex conjugate poles described by ω_o and Q, the transient portion of the waveform will require $-\frac{2Q}{\omega_o}\ln(0.001)$ seconds for it to decrease to less than 0.1 per-cent of its initial value.
- The Q-factor associated with crystal oscillators are on the order of tens-of-thousands, and therefore these circuits require incredibly long simulation times before the steady-state is reached.
- Initializing the reactive elements of a circuit with values that are close to their steady-state current or voltage values can reduce the time required for a circuit to reach its steady-state.
- The transfer characteristics of regenerative circuits with hysteresis are best computed with Spice by applying a triangular waveform input whose level varies between the two power supply levels.

12.7 Bibliography

Staff, *Linear Circuits: Operational Amplifier Macromodels*, Data Manual, Texas Instruments, Dallas, Texas, 1990.

```
A Precision Clamping Circuit

** Circuit Description **

* op-amp subcircuit

+++++ place uA741 op-amp subcircuit here (see Fig. 12.1) +++++

** Main Circuit **
* power supplies
Vcc 4 0 DC +15V
Vee 5 0 DC -15V
* input signal source
Vi 1 0 SIN ( 0 5V 1kHz )
* limiter circuit
Xopamp1 0 2 4 5 3 uA741
D1 3 2 D1N4148
C 1 2 1uF IC=0V
* model statements
.model D1N4148 D (Is=0.1p Rs=16 CJO=2p Tt=12n Bv=100 Ibv=0.1p)
** Analysis Requests **
.OPTIONS itl5=0
.TRAN 50us 4ms 0s 50us UIC
** Output Requests **
.PLOT TRAN V(3) V(1)
.probe
.end
```

Figure 12.26 The Spice input deck for computing the transient behavior of the clamping circuit shown in Fig. 12.21(c).

12.8 Problems

12.1 Compare the amplitude of oscillation of the output voltage generated by the Wien-Bridge oscillator shown in Fig. 12.2 with and without the diode limiter in the op amp feedback path.

12.2 An oscillator is formed by loading a transconductance amplifier having a positive gain with a parallel RLC circuit and connecting the output to the input directly. Let the transconductance amplifier have an input resistance of 10 kΩ and output resistance of 10 kΩ. The LC resonator has $L = 10$ μH, $C = 1000$ pF, and $Q = 100$. For what value of transconductance G_m will the circuit oscillate? Confirm that this value of transconductance does indeed cause the circuit to oscillate. What is the resulting frequency of oscillation?

12 Signal Generators And Waveform – Shaping Circuits

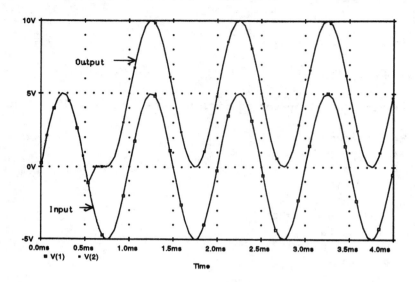

Figure 12.27 The input and output transient waveforms of the clamping circuit shown in Fig. 12.21(c).

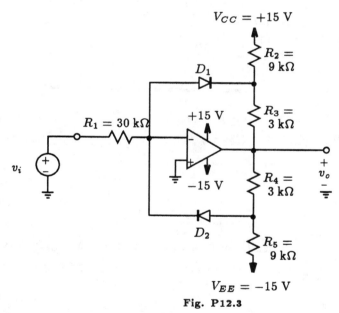

Fig. P12.3

12.3 Using Spice, determine the transfer characteristics of the comparator circuit shown in Fig. P12.3. Subsequently, connect a dc source $V_B = +1$ V to the virtual ground of the op amp through a resistor of $R_B = 10$ kΩ and observe that the transfer characteristics are shifted along the v_i-axis to the point $v_i = -(R_1/R_B)V_B$. Use the subcircuit description for the 741 op amp provided in Fig. 12.1 as the macromodel for the op amp in the comparator circuit. Assume that the two diodes have parameters $I_S = 10^{-14}$ A and $n = 1$.

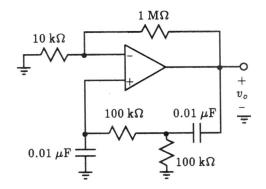

Fig. P12.4

12.4 For the circuit in Fig. P12.4, determine the loop transmission $L(j\omega)$ and the frequency for zero loop-phase. Verify that the circuit indeed oscillates at this frequency. Assume that the op amp is of the 741 type.

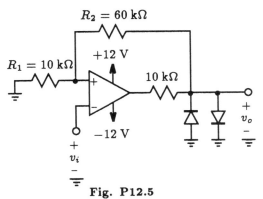

Fig. P12.5

12.5 For the circuit in Fig. P12.5, plot its input-output transfer characteristics v_o-v_i using Spice. Assume that the op amp is modeled after the 741 op amp and the two diodes after the commercial diode 1N4148. What is the maximum diode current?

12.6 Consider the circuit of Fig. P12.5 with R_1 eliminated and R_2 short-circuited. Plot the input–output transfer characteristics. Model the op amp after the 741 op amp and the two diodes after the commercial diode 1N4148.

12.7 Find the frequency of oscillation of the circuit in Fig. 12.11(b) for the case $R_1 = 10$ kΩ, $R_2 = 16$ kΩ, $C = 10$ nF, and $R = 62$ kΩ. Assume that the op amp is of the 741 type.

12.8 The circuit of Fig. P12.8 consists of an inverting bistable multivibrator with an output limiter and a noninverting integrator. Using equal values for all resistors except R_7 and a 0.5 nF capacitor, design the circuit to obtain a square wave at the output of the bistable multivibrator of 15 V peak-to-peak amplitude and a 10 kHz frequency. Plot the transient voltage waveform appearing at the integrator output. Model the op amp as a high-gain VCVS with a diode clamping circuit to limit the range of output voltage to ± 13 V and model the zener diodes with the equivalent circuit shown in Fig. 3.13.

12 Signal Generators And Waveform – Shaping Circuits

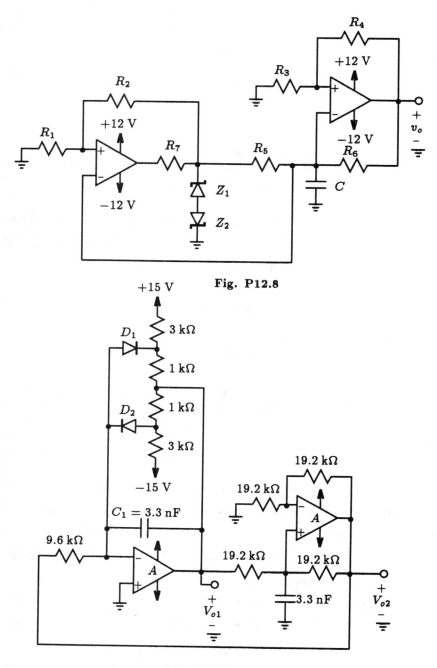

Fig. P12.8

Fig. P12.9

12.9 An oscillator circuit that generates two sinewave signals that are in quadrature is shown in Fig. P12.9. It oscillates at a frequency of 5 kHz. Using Spice, confirm that the two output signals are indeed in quadrature (ie. 90° apart) and that the frequency of oscillation is 5 kHz. Also, using the Fourier analysis capability of Spice, determine the total harmonic distortion present in the two output waveforms.

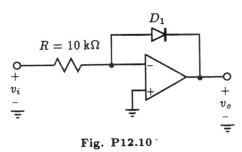

Fig. P12.10

12.10 Using Spice determine the transfer characteristics of the logarithmic amplifier circuit shown in Fig. P12.10. Assume that the op amp is of the 741 type and the diode has model parameters $I_S = 10^{-14}$ A and $n = 1.2$. Next, apply a 1 V, 1 kHz sine-wave to the circuits input and compute the output voltage waveform. Verify that the output voltage is related to the input voltage signal according to the expression:

$$v_o = -nV_T \ln(\frac{v_i}{I_S R}), \qquad v_i > 0$$

where V_T is the thermal voltage.

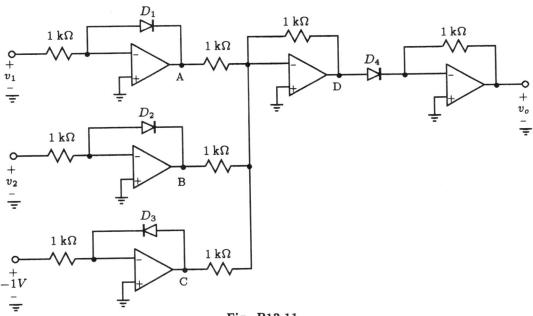

Fig. P12.11

12.11 The circuit of Fig. P12.11 implements the transfer characteristic $v_o = v_1 v_2$ for v_1 and $v_2 < 0$. Such a circuit is known as an analog multiplier. Using Spice, determine the transfer characteristics $v_o - v_1$ for the following values of v_2: 0.1, 0.5, 1.0, 2, 3, 5, 10 V. Assume all diodes to be identical, with 700 mV drop at 1 mA current and $n = 2$. Model the op amp as a high-gain VCVS with a diode clamping circuit to limit the range of output voltage to ± 12 V.

12.12 For the multiplier circuit shown in Fig. P12.11, use Spice to compute its output voltage waveform v_o assuming that the voltage waveform applied to input v_1 is described by $1 + 0.001\sin(2\pi 10^4)$ V and the voltage signal applied to input v_2 is described by $1 + 0.001\sin(2\pi 10^3)$ V.

12.13 Consider that a squarer circuit can be realized using the multiplier circuit shown in Fig. P12.11 by connecting the two inputs (v_1 and v_2) together, forming a single input. Apply a 1 volt peak sinewave of 1 kHz, offset by +5 V, to the input and compute the output voltage waveform for at least one complete period of the output signal. What happens to the output signal if the dc offset voltage is reduced to 0 V?

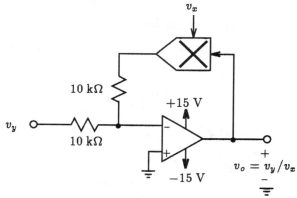

Fig. P12.14

12.14 A circuit that divides one signal by another can be realized using a multiplier circuit such as the one shown in Fig. P12.11 arranged in the feedback loop of an op amp as demonstrated in Fig. P12.14. The polygon with a cross in the center is meant to symbolize the multiplier circuit shown in Fig. P12.11. Verify the operation of this divider circuit by computing the transfer characteristics $v_o - v_x$ for the following values of v_y: -1, 0.1, 0.5, 1.0, 2, 3, 5 and 10 V.

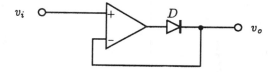

Fig. P12.15

12.15 Two superdiode circuits, such as the one shown in Fig. P12.15, are connected to a common load resistor of 20 kΩ and having the same input signal have their diodes reversed, one with cathode to the load, the other with anode to load. For a sinewave input of 10 V peak to peak and 1 kHz frequency, using Spice, plot the output voltage waveform and the current supplied by each superdiode circuit. Assume that the op amp and diode is modeled after the 741 and 1N4148, respectively.

12.16 The superdiode circuit of Fig. P12.15 can be made to have gain by connecting a resistor R_2 in place of the short circuit between the cathode of the diode and the negative input terminal of the op amp, and a resistor R_1 between the negative input terminal and ground. Design the circuit for a gain of 2. For a 10 V peak-to-peak input sinewave of 1 kHz frequency, determine, with the aid of Spice, the average voltage waveform appearing at the output. Assume that the op amp is modeled after the 741 op amp and the diode after the 1N4148.

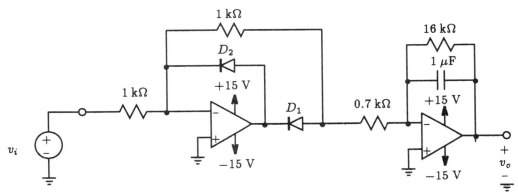

Fig. P12.17

12.17 The circuit shown in Fig. P12.17 is used as a voltmeter which is intended to function at frequencies of 10 Hz and above. It consists of precision half-wave rectifier followed by a first-order lowpass filter. Apply a 100 mV rms sinewave input signal of 10 kHz and observe the voltage signal appearing at the output. Does the output signal correspond to the average value of the input signal? Repeat with a triangular waveform and a square-wave as input. Model the op amp as a high-gain VCVS with a diode clamping circuit to limit the range of output voltage to ± 12 V. Assume that the diode has parameters $I_S = 10^{-14}$ A and $n = 1.6$.

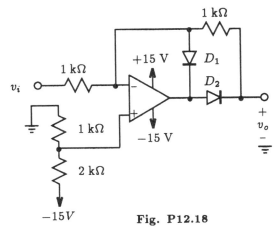

Fig. P12.18

12.18 Using Spice, plot the transfer characteristics of the circuit in Fig. P12.18. Assume that the op amp is modeled after the 741 op amp and the diode after the 1N4148.

12 Signal Generators And Waveform – Shaping Circuits

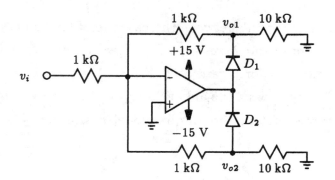

Fig. P12.19

12.19 With the aid of Spice, determine the transfer characteristics $v_{o1} - v_i$ and $v_{o2} - v_i$ of the circuit shown in Fig. P12.19. Assume that the op amp is modeled after the 741 op amp and the diode after the 1N4148.

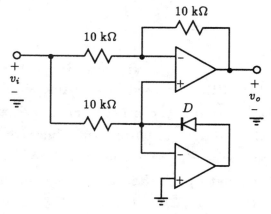

Fig. P12.20

12.20 Using Spice, plot the transfer characteristics of the circuit in Fig. P12.20. Assume that the op amp is modeled after the 741 op amp and the diode after the 1N4148.

Chapter 13

MOS Digital Circuits

In this chapter the basic operation of several types of simple MOS digital circuits will be investigated with the aid of Spice. Of particular interest are those digital circuits constructed in NMOS, CMOS or GaAs integrated circuit technologies. Our discussion here will mainly involve the static and dynamic analysis of several basic inverter circuits found in these three technologies, but some analyses of a two-input NOR gate and a D-type flip-flop will also be performed.

13.1 NMOS Inverter With Enhancement Load

As the first example of this chapter we have on display in Fig. 13.1 an NMOS inverter circuit with an enhancement load with the substrate connected to ground. Although, the source-to-body voltage (V_{SB}) of M_1 is zero, that of M_2 is equal to V_O. As a result, the threshold voltage of M_2 is no longer equal to the threshold voltage of M_1. For straightforward hand analysis, it is common to neglect the transistor body effect and assume that each transistor has the same threshold voltage. However, as we shall see from the following Spice circuit simulations of the inverter circuit shown in Fig. 13.1, neglecting the body effect results in substantial error in the large-signal input-output transfer characteristics. This is especially true for input signal levels near ground potential.

The Spice input description of the inverter circuit shown in Fig. 13.1 is seen listed in Fig. 13.2. The parameters associated with each transistor of the inverter are assumed as follows: $V_{to} = 0.7$ V, $W_1 = 9$ μm, $L_1 = 3$ μm, $W_2 = 3$ μm, $L_2 = 9$ μm, $\mu_n C_{OX} = 40$ μA/V^2, $2\phi_f = 0.6$ V. Each transistor is also assumed to have a body effect coefficient (γ) equal to 1.1 V$^{1/2}$.

13 MOS Digital Circuits

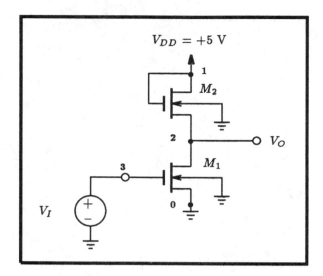

Figure 13.1 The enhancement-load inverter with substrate connected to ground.

```
An Enhancement-Load Inverter

** Circuit Description **
* dc supplies
Vdd 1 0 DC +5V
* input digital signal
Vi 3 0 DC 0V
* MOSFET circuit
M1 2 3 0 0 nmos_enhancement_mosfet L=3um W=9um
M2 1 1 2 0 nmos_enhancement_mosfet L=9um W=3um
* mosfet model statement (by default level 1)
.model nmos_enhancement_mosfet nmos (kp=40u Vto=0.7V phi=0.6V gamma=1.1)
** Analysis Requests **
.DC Vi 0V +5V 100mV
** Output Requests **
.Plot DC V(2)
.probe
.end
```

Figure 13.2 The Spice input file for calculating the input-output voltage transfer characteristics of the enhancement-load inverter circuit shown in Fig. 13.1.

The results of the computer simulation are shown in Fig. 13.3. Superimposed on the same graph are the input-output transfer characteristics of the inverter circuit when the body effect is eliminated (ie. set $gamma = 0$ on the transistor model statement in Fig. 13.2 and re-run the Spice analysis). Clearly, at low input levels, the two curves differ significantly, whereas, for higher input levels ($V_I > 2.5$ V), the two results agree quite closely.

It is interesting to compare the noise margins of these two situations. From Fig. 13.3, we see directly using Probe from the transfer characteristics accounting for the transistor body

13.1 NMOS Inverter With Enhancement Load

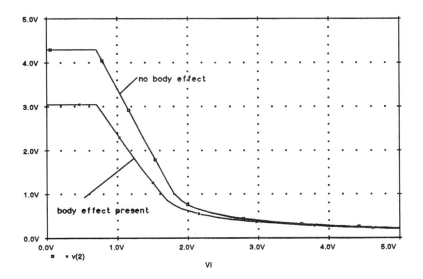

Figure 13.3 The input-output voltage transfer characteristics of the enhancement-load inverter circuit shown in Fig. 13.1 with and without the body effect present.

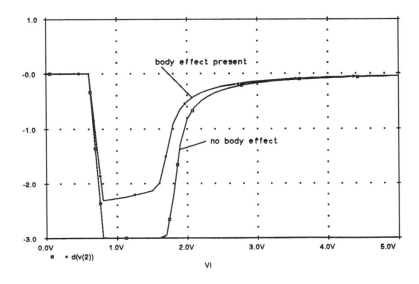

Figure 13.4 dV_O/dV_I as a function of the input voltage level V_I with and without the body effect present. These curves are used to determine the minimum inverter input voltage V_{IH}.

effect that $V_{IL} = 0.7$ V, $V_{OL} = 0.24$ V and $V_{OH} = 3.05$ V. To determine V_{IH}, it is necessary to determine the input level where the transfer characteristic has a slope of −1 in the vicinity of a logic low level. This is easy to do with the aid of the derivative procedure command

555

available in PROBE. Taking the derivative of the output voltage (V_O) with respect to the input voltage (V_I) for the above two cases and plotting it versus the input voltage level, we obtain the graphical results shown in Fig. 13.4. From this we see that the input level in the vicinity of a logic low level corresponding to a slope of −1 is 1.78 V. Thus, $V_{IH} = 1.78$ V. Therefore the two noise margins for the enhancement-load inverter with body effect included are:

$$NM_H = V_{OH} - V_{IH} = 3.05 - 1.78 = 1.27 \text{ V}$$
$$NM_L = V_{IL} - V_{OL} = 0.7 - 0.24 = 0.46 \text{ V}$$

Carrying out the above procedure for the characteristics of the enhancement-load inverter excluding the body effect we get the following two noise margins:

$$NM_H = V_{OH} - V_{IH} = 4.3 - 1.96 = 2.34 \text{ V}$$
$$NM_L = V_{IL} - V_{OL} = 0.7 - 0.26 = 0.44 \text{ V}$$

On comparison, we see that the body effect has decreased the high noise margin (NM_H) from 2.34 V to 1.27 V. The low noise margin (NM_L) remains relatively unchanged.

To complete our investigation of the static behavior of the enhancement-load inverter circuit, we shall compute its average static power dissipation assuming that the inverter spends equal time in each of its two states. We shall include the body effect in this analysis.

Returning to the Spice deck for the inverter, shown in Fig. 13.2, setting the input to the inverter to logic low, say $V_I = V_{IL} = 0.7$ V, will force the inverter into its high state. Replacing the DC sweep command with an .OP analysis command will have Spice compute the power dissipated by the inverter. The results of this analysis indicated that the power dissipated is very low at 5.56×10^{-11} W.

Repeating the above with the input at $V_I = 3.05$ V to correspond to an output voltage of V_{OL} for some preceding stage, we find that the static power dissipated by the inverter in its low state has increased substantially to a level of 462 μW. Thus, the average static power dissipated by the inverter with enhancement-load is 231 μW.

Dynamic Operation

We next consider the dynamic operation of the enhancement-load NMOS inverter. To facilitate our investigation we shall cascade two inverter circuits as shown in Fig. 13.5. The output of each inverter is assumed loaded with a 0.1 pF capacitor to represent the capacitance associated with the interconnect wiring between the two gates. A 0 to 5 V pulse of 100 ns duration is applied to the input of the first inverter circuit. To obtain realistic results, we shall model the NMOS transistors after a commercial 3 μm CMOS process. These model parameters are included in the Spice deck shown in Fig. 13.6. The static parameters remain

13.1 NMOS Inverter With Enhancement Load

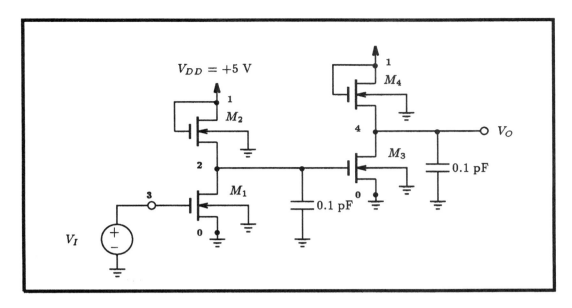

Figure 13.5 A cascade of two enhancement-load NMOS inverter circuits.

the same as those seen previously; however, it should be noted that the addition of the other parameters will alter the dc behavior of this circuit somewhat. A transient analysis is requested to be performed over a 200 ns interval using a 0.5 ns time step. The voltage at the output of each inverter will be plotted on completion of Spice.

The transient response of the cascade of inverters as calculated by Spice is presented in Fig. 13.7. The top graph displays the input voltage signal applied to the first inverter and the bottom graph displays the voltage waveform that appears at the output of each inverter circuit. As is evident, the rise time of each inverter is much slower than its fall time. To better estimate the rise and fall times of the inverter, as well as the propagation delay, an expanded view of the high-to-low and low-to-high output transitions of the first inverter is provided in Fig. 13.8.

From Fig. 13.8(a), together with the aid of Probe, we found that the 90% to 10% fall time t_{THL} is 1.063 ns. This is based on a logic low level V_{OL} of 0.128 V and a logic high value V_{OH} of 3.12 V. In addition, the high-to-low input-to-output delay t_{PHL} was found to be 0.524 ns. Repeating this for the low-to-high output voltage transition shown in Fig. 13.8(b), we find that the 10% to 90% rise time t_{TLH} is 45.9 ns. As well, the low-to-high input-to-output delay t_{PLH} is 7.71 ns. Averaging the above two input-to-output delays, we obtain the propagation time delay t_P for the NMOS enhancement-load inverter with a 0.1 pF load to be 4.12 ns.

It is interesting to note that the voltage waveform that appears at the output of the second inverter is somewhat different than that which appeared at the output of the first

13 MOS Digital Circuits

```
A Cascade Of Enhancement-Load Inverters (dynamic effects included)

** Circuit Description **
* dc supplies
Vdd 1 0 DC +5V
* input digital signal
Vi 3 0 PWL (0,0V 10ns,0V 10.1ns,5V 110ns,5V 110.1ns,0V 210ns,0V)
* 1st MOSFET inverter circuit
M1 2 3 0 0 MN L=3um W=9um
M2 1 1 2 0 MN L=9um W=3um
Cl1 2 0 0.1pF
* 2nd MOSFET inverter circuit
M3 4 2 0 0 MN L=3um W=9um
M4 1 1 4 0 MN L=9um W=3um
Cl2 4 0 0.1pF
* BNR 3um transistor model statements (level 3)
.MODEL MN nmos (level=3 vto=.7 kp=4.e-05 gamma=1.1 phi=.6
+ lambda=.01 rd=40 rs=40  pb=.7 cgso=3.e-10 cgdo=3.e-10
+ cgbo=5.e-10 rsh=25 cj=.00044  mj=.5 cjsw=4.e-10 mjsw=.3
+ js=1.e-05 tox=5.e-08 nsub=1.7e+16  nss=0 nfs=0 tpg=1 xj=6.e-07
+ ld=3.5e-07 uo=775  vmax=100000  theta=.11  eta=.05 kappa=1)
** Analysis Requests **
.TRAN 0.5ns 210ns 0ns 0.5ns
** Output Requests **
.Plot TRAN V(2) V(3) V(4)
.probe
.end
```

Figure 13.6 The Spice input file for computing the transient response for the NMOS inverter cascade shown in Fig. 13.5.

inverter. This, in turn, gives rise to different transient attributes which include t_r, t_f and t_P. Specifically, we find that the rise time is 38.6 ns and the fall time is 7.95 ns. Also, $t_{PLH} = 5.81$ ns and $t_{PHL} = 1.28$ ns which when combined will give a propagation delay t_P of 3.54 ns. The reason the second inverter has different transient attributes from the first inverter stems from to the fact that the signals that appear at the input to each inverter are different. The first inverter has an almost instantaneous voltage change at its input with a rise and fall time of about 0.1 ns, whereas the second inverter see an input voltage signal having an unsymmetrical rise and fall time of 38.6 ns and 7.95 ns, respectively.

At this stage of our investigation it would be instructive to check the formulae derived by Sedra and Smith in Section 13.2 of their text for estimating the low-to-high and high-to-low transition times t_{PLH} and t_{PHL}, respectively. According to their development,

13.1 NMOS Inverter With Enhancement Load

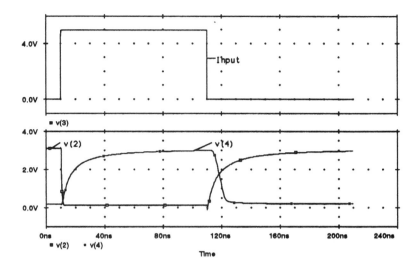

Figure 13.7 The transient response of a cascade of two enhancement-load NMOS inverter circuits. The top graph displays the input voltage signal and the bottom graph displays the two output voltage signals appearing at each inverter output.

$$t_{PHL} = \frac{C\left[V_{OH} - \frac{1}{2}(V_{OH} + V_{OL})\right]}{I_{HL}} \qquad (13.1)$$

and

$$t_{PLH} = \frac{C\left[\frac{1}{2}(V_{OH} + V_{OL}) - V_{OL}\right]}{I_{LH}} \qquad (13.2)$$

where I_{HL} and I_{LH} represent the average discharge and charge currents. These two currents are obtained from the static characteristics of each device. In terms of the development in Sedra and Smith, with the input V_I held high at $+5$ V (the same logic high level used to compute the transient response of the first inverter shown in Fig. 13.7), together with a simplified set of device parameters obtained from the level 3 NMOS model of the transistor given in the Spice deck of Fig. 13.6, ie. $k_p = 40 \ \mu A/V^2$, $V_t = 0.7$ V, $\gamma = 1.1$ $V^{1/2}$ and $\lambda = 0.01$ V^{-1}, we find that $I_{HL} = 861.2$ μA and $I_{LH} = 120.6$ μA.

Under the assumption that the effective capacitance seen by the inverter is dominated[†] by the interconnect capacitance of 0.1 pF, we can compute the transition times t_{PHL} and

---

[†] This is a reasonable assumption here because the parasitic capacitances associated with each MOSFET of the inverter are on the order of femto-farads. This was confirmed through a small-signal analysis of each transistor at each critical point of the inverter transfer characteristic.

13 MOS Digital Circuits

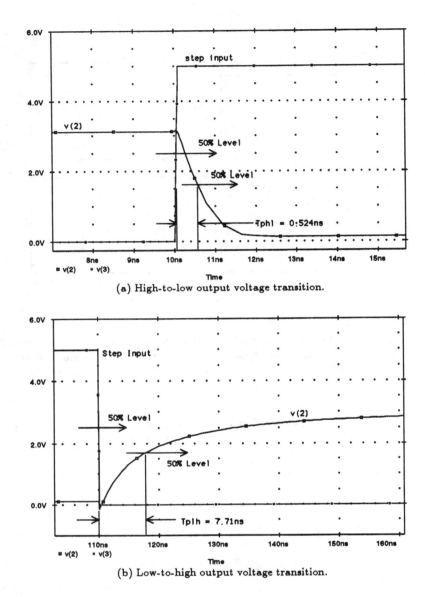

(a) High-to-low output voltage transition.

(b) Low-to-high output voltage transition.

Figure 13.8 An expanded view of the high-to-low and low-to-high output voltage transitions of the first inverter of the circuit shown in Fig. 13.5 when subjected to a step input.

t_{PLH} according to Eqns. (13.1) and (13.2) and obtain:

$$t_{PHL} = 0.174 \text{ ns} \qquad t_{PLH} = 1.24 \text{ ns}.$$

This can then be averaged to obtain an estimate of the propagation delay time at 0.71 ns. In comparison with the values obtained directly from Spice, $t_{PHL} = 0.524$ ns and $t_{PLH} = 7.71$ ns, we see that the above estimates are off by quite a large amount (for t_{PLH}, a factor of 6).

The reason for such a large discrepancy lies with the assumption that the static parameters of the MOSFET are those seen listed in the level 3 MOSFET model seen in Fig. 13.6.

13.1 NMOS Inverter With Enhancement Load

This is, unfortunately, not true. The value of many of the parameters in a level 3 Spice model of a MOSFET have no physical basis; they are simply empirical parameters made to fit the measured device characteristics using a computer optimization algorithm. As such, using a subset of the MOSFET device parameters can, and in this particular case, did, result in significant error. If one makes use of Spice, together with the level 3 MOSFET model, one will find the average discharge and charge currents, I_{HL} and I_{LH}, to be much less than those computed previously at 432 μA and 36 μA, respectively. Substituting these two results into Eqns. (13.1) and (13.2) above, we then obtain:

$$t_{PHL} = 0.347 \text{ ns} \qquad t_{PLH} = 4.16 \text{ ns}.$$

These two results are much closer to those read directly off the graph of the transient waveform for the first inverter computed by Spice in Fig. 13.8 (ie. $t_{PHL} = 0.524$ ns and $t_{PLH} = 7.71$ ns). Based on the accuracy of the above two results, we can conclude that Eqns. (13.1) and (13.2) are accurate to only 50% of the actual Spice computed values.

A similar result is obtained when the transition times for the second inverter is computed. By the two formulas given in Eqns. (13.1) and (13.2), together with the average discharge and charge current values of 198 μA and 36 μA, respectively, we find $t_{PHL} = 0.75$ ns and $t_{PLH} = 4.15$ ns. This, in turn, indicates a propagation delay of 2.45 ns. Comparing these results to those directly obtained from Spice (ie. $t_{PHL} = 1.28$ ns and $t_{PLH} = 5.81$ ns), we once again see that these estimates are within 50% of those found from the transient waveform computed by Spice.

Finally, the instantaneous power dissipated by each inverter circuit of Fig. 13.5 as computed by Spice is shown in Fig. 13.9. The top graph depicts the power dissipated by the first inverter and the bottom graph displays the power dissipated by the second inverter. From the top graph, during the first transition, we see that a sudden power glitch occurs, having a peak power of about 1.46 mW. This occurs as the inverter moves from the logic high to logic low state. After this, the power dissipated by this inverter settles into a steady-state value of 272.5 μW. Thus, we can state that the inverter dissipates a static power of 272.5 μW when in the logic low state. During the next transition, when the inverter is driven into the logic high state, we see that the power dissipated by this gate reduces to nearly zero. An infinitesimal leakage current of the reverse-biased pn junctions accounts for a very small power dissipation of 31 pW. The steady-state power dissipated by the second inverter in the logic low state is quite similar to that dissipated by the first inverter, a constant peak power of 263.6 μW. Also, when in the high state the power dissipated by this gate reduces to nearly zero. The difference in power dissipated by the two inverter circuits is due to the difference in the levels of their respective input signals.

13 MOS Digital Circuits

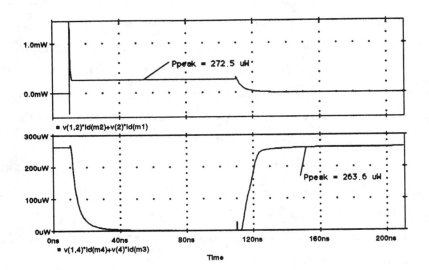

Figure 13.9 The instantaneous power dissipated by the first and second enhancement-load NMOS inverter cascade shown in Fig. 13.5. The top graph displays the power dissipated by the first inverter and the bottom graph displays the power dissipated by the second inverter.

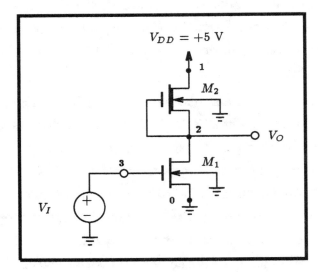

Figure 13.10 The depletion-load inverter with substrate connected to ground.

13.2 NMOS Inverter With Depletion Load

Replacing the enhancement load MOSFET in the inverter circuit of Fig. 13.1 by a depletion MOSFET will result in an inverter circuit with a sharper voltage transfer characteristic, and thus higher noise margins. In the following we shall compare the voltage transfer characteristic of the enhancement-load inverter circuit with the corresponding one

13.2 NMOS Inverter With Depletion Load

```
A Depletion-Load Inverter

** Circuit Description **
* dc supplies
Vdd 1 0 DC +5V
* input digital signal
Vi 3 0 DC 0V
* MOSFET circuit
M1 2 3 0 0 nmos_enhancement_mosfet L=3um W=9um
M2 1 2 2 0 nmos_depletion_mosfet L=9um W=3um
* mosfet model statements (by default level 1)
.model nmos_enhancement_mosfet nmos (kp=40u Vto=0.7V phi=0.6V gamma=1.1)
.model nmos_depletion_mosfet   nmos (kp=40u Vto=-3V  phi=0.6V gamma=1.1)
** Analysis Requests **
.DC Vi 0V +5V 20mV
** Output Requests **
.PLOT DC V(2)
.probe
.end
```

Figure 13.11 The Spice input file for calculating the input-output voltage transfer characteristics of the depletion-load inverter circuit shown in Fig. 13.10.

for depletion-load inverter circuit shown in Fig. 13.10. For a fair comparison we shall assume that the geometries of the corresponding devices of the two inverters are the same. The device parameters of both the enhancement and depletion-type MOSFET will be assumed the same as in first part of the previous example except that the threshold voltage of the depletion mode device will be set equal to –3 V.

The Spice input file describing the depletion-load inverter circuit of Fig. 13.10 is listed in Fig. 13.11. A DC sweep of the input voltage is requested between 0 V and 5 V. A plot of the output voltage is then requested. In order to compare the results of this analysis with those previously generated for the enhancement-load inverter we have concatenated the two Spice listings shown in Fig. 13.2 and Fig. 13.11 into one input file before submitting the file to Spice for execution.

The results of the two Spice analyses are shown collectively in Fig. 13.12. As is clearly evident, the depletion-load inverter has a steeper transition region than the enhancement-load inverter and therefore approaches more closely the ideal inverter characteristics. As a means of quantifying this we shall compare their respective noise margins.

From Fig. 13.12 we can directly obtain for the depletion-load inverter the output voltage levels V_{OL} and V_{OH} as 0.112 V and 5 V, respectively. To obtain the input threshold levels V_{IL} and V_{IH}, we must determine the input levels (V_I) that correspond to $dV_O/dV_I = -1$. Following the procedure outlined in the previous section, and with the aid of PROBE, we find

13 MOS Digital Circuits

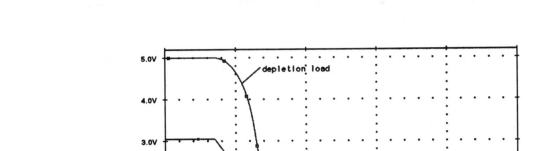

Figure 13.12 A comparison of the voltage transfer characteristics of depletion-load inverter with those of an equally-sized inverter having an enhancement load. The depletion-load inverter clearly has a steeper transition region and therefore approaches more closely to the ideal inverter characteristics.

$V_{IL} = 0.845$ V and $V_{IH} = 1.67$ V. Thus, the noise margins of the depletion-load inverter are $NM_L = 0.733$ V and $NM_H = 3.33$ V. Both of these values are larger than the corresponding values obtained for the enhancement-load inverter of the previous section; thus supporting the claim made that the depletion-load inverter behaves more ideally than the corresponding enhancement-load inverter.

13.3 The CMOS Inverter

For both custom and semicustom VLSI (very-large-scale-integration), CMOS is generally preferred over all other available technologies. Unlike NMOS technologies, CMOS digital circuits utilize both n-channel and p-channel enhancement-type MOSFETS. An example of a CMOS inverter is shown in Fig. 13.13. In the following we use Spice to compute its voltage transfer characteristics. The n-channel and p-channel MOSFETs will be assumed to be modeled after the same commercial 3 μm CMOS technology used in the previous examples of this chapter. Owning to the fact that the mobility μ for the n-channel device is about 3 times larger than that for the corresponding p-channel device, we shall make the p-channel device 3 times as wide as the n-channel device. Moreover, we shall make the dimensions of the n-channel device equal to the smallest transistor dimensions possible in this technology, ie. 3 μm by 3 μm.

13.3 The CMOS Inverter

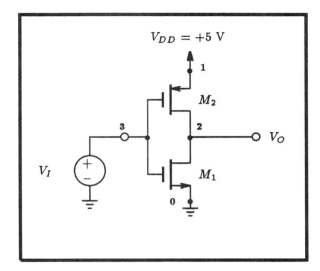

Figure 13.13 A CMOS inverter.

The Spice input file for the CMOS inverter is shown listed in Fig. 13.14. A sweep of the input voltage level is performed between ground and V_{DD}. The resulting voltage transfer characteristics computed by Spice are shown in Fig. 13.15. Here we see that the output voltage of this gate extends completely between the limits of the two power supply levels. Thus, $V_{OL} = 0$ V and $V_{OH} = +5$ V. Using the derivative feature of Probe, we find the input logic levels, V_{IL} and V_{IH}, as 2.03 and 2.825 V, respectively. In the case of V_{IL}, we find that the simulated result correlates very closely with that predicted by the simple formula derived in Section 14.5 of Sedra and Smith, ie. $V_{OL} = \frac{1}{8}(3V_{DD} + 2V_{tn}) = 2.05$ V. Similarly, the result predicted by $V_{IH} = \frac{1}{8}(5V_{DD} - 2V_{tp})$ of 2.93 V is quite close to that obtained by simulation. Combining the above results we find that the high and low noise margins are $NM_H = 2.175$ V and $MN_L = 2.03$ V, respectively. These noise margins are not equal owning to the fact that M_1 and M_2 are not exactly matched.

An important attribute of CMOS logic is the fact that the transistors do not conduct any current when settled into a particular logic state. To appreciate this, we plot in Fig. 13.16 the current that flows in the CMOS inverter circuit as a function of the input voltage level. These results indicate that insignificant current flows through the inverter when the input is either smaller than 0.7 V or greater than 4.2 V. We also see that a peak current of 43.8 μA flows through the inverter when the input signal level reaches the mid point between the voltage supply.

Dynamic Operation

The dynamic operation of a cascade of two CMOS inverters is considered next. The output of each inverter is assumed loaded with a 0.1 pF capacitor to represent the capacitance

13 MOS Digital Circuits

```
The CMOS Inverter

** Circuit Description **
* dc supplies
Vdd 1 0 DC +5V
* input digital signal
Vi  3 0 DC +5V
* MOSFET inverter circuit
M1 2 3 0 0 MN L=3um W=3um
M2 2 3 1 1 MP L=3um W=9um
* BNR 3um transistor model statements (level 3)
.MODEL MN nmos level=3 vto=.7 kp=4.e-05 gamma=1.1 phi=.6
+ lambda=.01 rd=40 rs=40  pb=.7 cgso=3.e-10 cgdo=3.e-10
+ cgbo=5.e-10 rsh=25 cj=.00044  mj=.5 cjsw=4.e-10 mjsw=.3
+ js=1.e-05 tox=5.e-08 nsub=1.7e+16  nss=0 nfs=0 tpg=1 xj=6.e-07
+ ld=3.5e-07 uo=775  vmax=100000  theta=.11  eta=.05 kappa=1
.MODEL MP pmos level=3 vto=-.8 kp=1.2e-05 gamma=.6 phi=.6
+ lambda=.03 rd=100 rs=100  pb=.6 cgso=2.5e-10 cgdo=2.5e-10
+ cgbo=5.e-10 rsh=80 cj=.00015  mj=.6 cjsw=4.e-10 mjsw=.6
+ js=1.e-05 tox=5.e-08 nsub=5.e+15  nss=0 nfs=0 tpg=1 xj=5.e-07
+ ld=2.5e-07 uo=250  vmax=70000  theta=.13  eta=.3 kappa=1
** Analysis Requests **
.DC Vi 0 5 50mV
** Output Requests **
.PLOT DC V(2) Id(M1)
.probe
.end
```

Figure 13.14 The Spice input file for calculating the input-output voltage transfer characteristics of the CMOS inverter shown in Fig. 13.13.

associated with the interconnect wiring between the two gates. The Spice deck for this particular circuit arrangement is provided in Fig. 13.18. A 0 to 5 V pulse of 20 ns duration is applied to the input of the first inverter circuit. A transient analysis is requested to be performed over a 50 ns interval using a 0.1 ns time step. The voltage at the output of each inverter will be plotted on completion of Spice.

The transient response of the cascade of CMOS inverters as calculated by Spice is presented in Fig. 13.19. The top graph displays the input voltage signal applied to the first inverter and the bottom graph displays the voltage waveform that appears at the output of each inverter circuit. Let us consider the behavior of the voltage signal that appears at the output of the first inverter. Here we see that both the rise and fall time of the output voltage signal are quite similar in duration. With the aid of the Probe facility, we find that the 90% to 10% fall time t_{THL} is 3.82 ns and the high-to-low input-to-output transition delay t_{PHL} is 2.44 ns. Repeating this for the low-to-high output voltage transition, we find that the

13.3 The CMOS Inverter

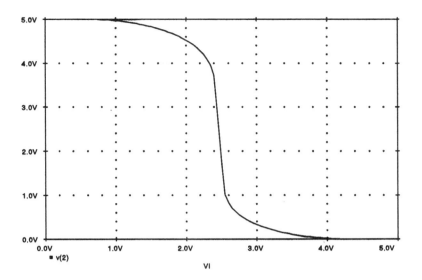

Figure 13.15 The input-output voltage transfer characteristic of the CMOS inverter shown in Fig. 13.13.

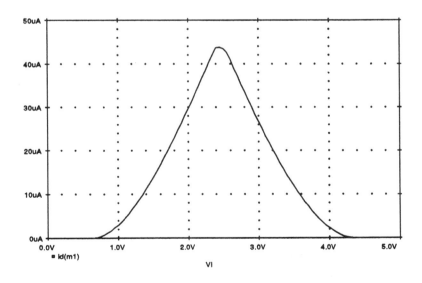

Figure 13.16 The current in the CMOS inverter versus the input voltage.

10% to 90% rise time t_{TLH} is 4.17 ns. As well, the low-to-high input-to-output delay t_{PLH} is 2.11 ns. Averaging the above two input-to-output delays, we obtain the propagation time delay t_P for the CMOS inverter with a 0.1 pF load to be 2.28 ns.

The voltage waveform appearing at the output of the second inverter is quite similar to

13 MOS Digital Circuits

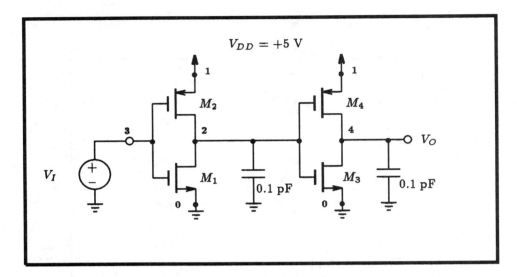

Figure 13.17 A cascade of two CMOS inverter circuits.

that which appears at the output of the first inverter. Using Probe we find $t_{THL} = 3.73$ ns, $t_{TLH} = 3.92$ ns, $t_{PHL} = 2.7$ ns and $t_{PLH} = 2.59$ ns. The propagation delay t_P is then 2.64 ns. These timing attributes are quite similar to those found for the output voltage waveform associated with the first inverter.

According to hand analysis (see Section 13.5 of Sedra and Smith), the expected transitional delays t_{PHL} and t_{PLH}, are found from the following two expressions:

$$t_{PHL} \approx \frac{0.8C}{\frac{1}{2}\mu_n C_{OX}(W/L)|_n V_{DD}} \tag{13.3}$$

and

$$t_{PLH} \approx \frac{0.8C}{\frac{1}{2}\mu_p C_{OX}(W/L)|_p V_{DD}} \tag{13.4}$$

Substituting the appropriate parameters obtainable from the Spice deck seen listed in Fig. 13.18, we find $t_{PHL} = 0.8$ ns and $t_{PLH} = 0.88$ ns. The propagation delay t_P is then 0.84 ns. Comparing these results with those computed by Spice, we find that the results predicted by the above equations underestimate the transition times by about a factor of 3. This is an unreasonably large error. Detailed investigation reveals that the theory used to derive Eqns. (13.3) and (13.4) assumed that the mobility of each MOSFET (μ_n and μ_p) was constant and independent of the voltages that appear at the terminals of the device. Experimental evidence indicates otherwise, and thus a more complicated relationship for the mobility factor has been included in the level 3 MOSFET model of Spice. Through additional simulations and some indirect calculations, it was found for an n-channel MOSFET over a 5 V range

13.3 The CMOS Inverter

```
A Cascade Of CMOS Inverters (dynamic effects included)

** Circuit Description **
* dc supplies
Vdd 1 0 DC +5V
* input digital signal
Vi 3 0 DC 1.237 PWL (0,0V 10ns,0V 10.1ns,5V 30ns,5V 30.1ns,0V 50ns,0V)
* 1st CMOS inverter
M1 2 3 0 0 MN L=3um W=3um
M2 2 3 1 1 MP L=3um W=9um
Cl1 2 0 0.1pF
* 2nd CMOS inverter
M3 4 2 0 0 MN L=3um W=3um
M4 4 2 1 1 MP L=3um W=9um
Cl2 4 0 0.1pF
* BNR 3um transistor model statements (level 3)
.MODEL MN nmos level=3 vto=.7 kp=4.e-05 gamma=1.1 phi=.6
+ lambda=.01 rd=40 rs=40  pb=.7 cgso=3.e-10 cgdo=3.e-10
+ cgbo=5.e-10 rsh=25 cj=.00044  mj=.5 cjsw=4.e-10 mjsw=.3
+ js=1.e-05 tox=5.e-08 nsub=1.7e+16  nss=0 nfs=0 tpg=1 xj=6.e-07
+ ld=3.5e-07 uo=775  vmax=100000  theta=.11  eta=.05 kappa=1
.MODEL MP pmos level=3 vto=-.8 kp=1.2e-05 gamma=.6 phi=.6
+ lambda=.03 rd=100 rs=100  pb=.6 cgso=2.5e-10 cgdo=2.5e-10
+ cgbo=5.e-10 rsh=80 cj=.00015  mj=.6 cjsw=4.e-10 mjsw=.6
+ js=1.e-05 tox=5.e-08 nsub=5.e+15  nss=0 nfs=0 tpg=1 xj=5.e-07
+ ld=2.5e-07 uo=250  vmax=70000  theta=.13  eta=.3 kappa=1
** Analysis Requests **
.TRAN 0.1ns 50ns 0ns 0.1ns
** Output Requests **
.Plot TRAN V(2) V(3) V(4)
.probe
.end
```

Figure 13.18 The Spice input file for computing the transient response for the CMOS inverter cascade shown in Fig. 13.17.

of $V_{GS} = V_{DS}$ that $\mu_n C_{OX}$ (k_p) varied between 10 μA/V^2 and 40 μA/V^2. Separately substituting these two extreme values of $\mu_n C_{OX}$ back into Eqn. (13.3) indicates that t_{PHL} can range between 0.8 ns and 3.2 ns. When compared to the Spice derived result of 2.7 ns we see that it lies between the two limits predicted by theory.

Similarly, for the p-channel MOSFET we find that $\mu_p C_{OX}$ varies between 4 μA/V^2 and 12 μA/V^2. Thus, we should expect that the Spice derived transition time t_{PLH} should lie somewhere between 0.86 ns and 2.66 ns. From above, we do indeed see that this is the case, ie. from Spice, $t_{PLH} = 2.59$ ns.

The instantaneous power dissipated by each CMOS inverter as computed by Spice is shown in Fig. 13.20. The top graph depicts the power dissipated by the first inverter

13 MOS Digital Circuits

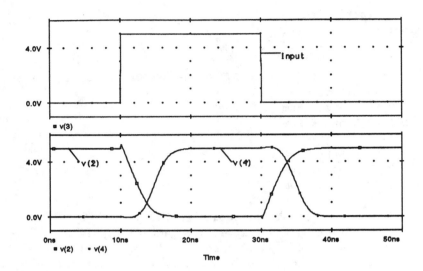

Figure 13.19 The transient response of a cascade of two CMOS inverter circuits. The top graph displays the input voltage signal and the bottom graph displays the voltage signals appearing at output of the two inverters.

and the bottom graph displays the power dissipated by the second inverter. The shapes of each power waveform is different owing to the different voltage signals that appear at their respective inputs.

13.4 A Two-Input CMOS NOR Gate

An important logic element of digital circuits is the two-input NOR gate shown in Fig. 13.21. Assuming that μC_{OX} of the p-channel MOSFETs is half that of the n-channel MOSFETs at 20 $\mu A/V^2$ and that the threshold voltage of the PMOS devices is equal but opposite to the V_t of the NMOS device at +2 V, in the following we shall compute the transfer characteristics of this simple gate under the following input conditions: (a) input terminal B connected to ground and (b) input terminals A and B joined together. This same circuit was analyzed by hand by Sedra and Smith in Example 13.3 of their text. Thus, one can compare the results computed here by Spice with those found by hand analysis. Note that the body effect is ignored in this analysis.

We begin our analysis with the Spice input file shown listed in Fig. 13.22. In this particular Spice listing, we have both input signals initially set to 0 V but the voltage at terminal A is varied with the DC sweep command between V_{DD} and ground. The substrate of each device is connected directly to their sources, thereby eliminating their respective

13.4 A Two-Input CMOS NOR Gate

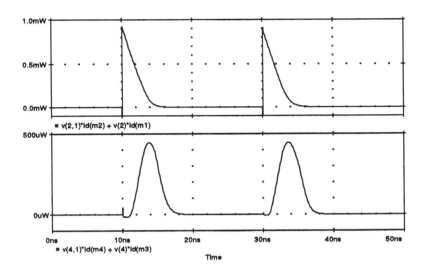

Figure 13.20 The instantaneous power dissipated by the first and second CMOS inverter cascade of Fig. 13.17. The top graph displays the power dissipated by the first inverter and the bottom graph displays the power dissipated by the second inverter.

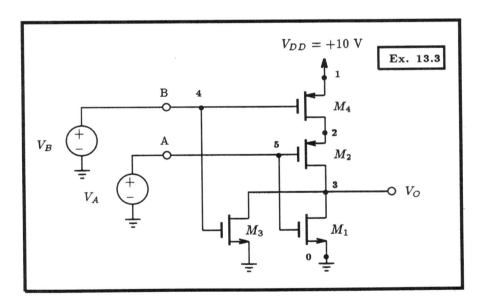

Figure 13.21 A two-input CMOS NOR gate.

body effects.

The results of the Spice simulation are shown in Fig. 13.23. The very sharp, or high gain, transition region is clearly visible with a threshold voltage V_{TH} occurring at 5.53 V. This compares very well with that calculated by hand in Sedra and Smith at 5.31 V.

13 MOS Digital Circuits

```
Example 13.3: A Two-Input CMOS NOR Gate

** Circuit Description **
* dc supplies
Vdd 1 0 DC +10V
* input digital signal
Va 4 0 DC 0V
Vb 5 0 DC 0V
* MOSFET circuit
M1 3 5 0 0 nmos_enhancement_mosfet L=3um W=3um
M2 3 5 2 2 pmos_enhancement_mosfet L=3um W=12um
M3 3 4 0 0 nmos_enhancement_mosfet L=3um W=3um
M4 2 4 1 1 pmos_enhancement_mosfet L=3um W=12um
* mosfet model statements (by default level 1)
.model nmos_enhancement_mosfet nmos (kp=20u Vto=+2V)
.model pmos_enhancement_mosfet pmos (kp=10u Vto=-2V)
** Analysis Requests **
.DC Va 0V +10V 50mV
** Output Requests **
.PLOT DC V(3)
.probe
.end
```

Figure 13.22 The Spice input file for calculating the input-output voltage transfer characteristics of the CMOS NOR gate shown in Fig. 13.21.

Also shown in Fig. 13.23 are the computed Spice results of the NOR gate when both input terminals are connected together and the input voltage is varied between V_{DD} and ground. Here we see that the threshold voltage V_{TH} has shifted down to a level of 4.48 V, which is almost identical to that computed by hand (4.5 V).

When the body effect is included in the analysis, we find very little difference in the results.

13.5 CMOS SR Flip-Flop

The simplest type of flip-flop is the set/reset (SR) flip-flop shown in Fig. 13.24. It consists of two cross-coupled NOR gates. The output of each gate is loaded with 0.1 pF so as to model the loading effect of other gates connected to the output terminals of the flip-flop. This provides a somewhat realistic situation in which to perform the simulation. In the following, with the aid of Spice, we shall investigate the dynamic behavior of this flip-flop with each NOR gate replaced by a CMOS NOR gate. Each transistor of this NOR gate will be modeled after the commercial 3 micron CMOS transistors used previously in Section 13.3. The Spice description of this flip-flop is provided in Fig. 13.25. Notice that

13.5 CMOS SR Flip-Flop

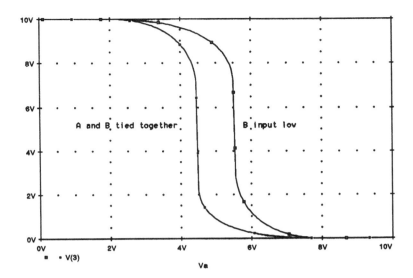

Figure 13.23 The input-output voltage transfer characteristic of the CMOS NOR gate shown in Fig. 13.21 under two different input conditions.

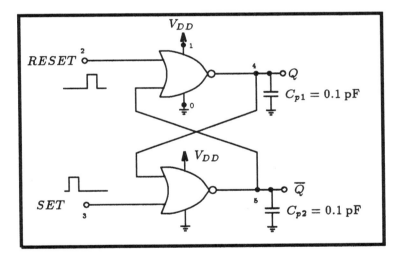

Figure 13.24 A set/reset (SR) flip-flop constructed from two cross-coupled NOR gates. Each output of the flip-flop is assumed loaded by a parasitic capacitance of 5pF.

here we have described the two-input CMOS NOR gate as a subcircuit. This simplifies the presentation, and, besides, reduces the amount of typing one has to do to create the Spice input file. By toggling the set and reset input terminals of the flip-flop between V_{DD} and ground, we can observe the storage capability of the flip-flop. This is achieved with the two input source PULSE statements.

The results of this analysis are shown in Fig. 13.26. The top two waveforms are the

```
A Set/Reset Flip-Flop

* CMOS NOR GATE subcircuit
.subckt NOR_GATE 4 5 3 1
* connections:     | | | |
*              input | | |
*                input | |
*                  output |
*                      Vdd
M1 3 5 0 0 MN L=3um W=3um
M2 3 5 2 2 MP L=3um W=18um
M3 3 4 0 0 MN L=3um W=3um
M4 2 4 1 1 MP L=3um W=18um
* BNR 3um transistor model statements (level 3)
.MODEL MN nmos level=3 vto=.7 kp=4.e-05 gamma=1.1 phi=.6
+ lambda=.01 rd=40 rs=40  pb=.7 cgso=3.e-10 cgdo=3.e-10
+ cgbo=5.e-10 rsh=25 cj=.00044  mj=.5 cjsw=4.e-10 mjsw=.3
+ js=1.e-05 tox=5.e-08 nsub=1.7e+16  nss=0 nfs=0 tpg=1 xj=6.e-07
+ ld=3.5e-07 uo=775  vmax=100000  theta=.11  eta=.05 kappa=1
.MODEL MP pmos level=3 vto=-.8 kp=1.2e-05 gamma=.6 phi=.6
+ lambda=.03 rd=100 rs=100  pb=.6 cgso=2.5e-10 cgdo=2.5e-10
+ cgbo=5.e-10 rsh=80 cj=.00015  mj=.6 cjsw=4.e-10 mjsw=.6
+ js=1.e-05 tox=5.e-08 nsub=5.e+15  nss=0 nfs=0 tpg=1 xj=5.e-07
+ ld=2.5e-07 uo=250  vmax=70000  theta=.13  eta=.3 kappa=1
.ends NOR_GATE

** Main Circuit **
* dc supplies
Vdd 1 0 DC +10V
* input digital signals
Vreset 2 0 PULSE (10 0 0 1us 100us 2ms 4ms)
Vset   3 0 PULSE (0 10 0 1us 100us 1ms 4ms)
* SR Flip-Flop
Xnor_gate1 2 5 4 1 NOR_GATE
Xnor_gate2 3 4 5 1 NOR_GATE
Cp1 4 0 0.1pF
Cp2 5 0 0.1pF
** Analysis Requests **
.TRAN 10us 3ms
** Output Requests **
.PLOT TRAN V(4) V(5)
.probe
.end
```

Figure 13.25 The Spice input file for investigating the different states of the SR flip-flop shown in Fig. 13.24.

SET and RESET voltage waveforms appearing at the inputs to the flip-flop. The bottom two waveforms Q and $\overline{Q}$ are the signals appearing at the flip-flop outputs. Clearly, when the SET input is high and the RESET input is low, the output Q goes high and $\overline{Q}$ goes

13.5 CMOS SR Flip-Flop

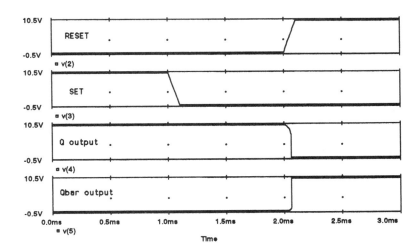

Figure 13.26 The input and output voltage waveforms for the SR flip-flop shown in Fig. 13.24.

low. This condition remains when both SET and RESET are low. However, we see from the simulated results that when RESET goes high and SET remains low, the two outputs change state.

As a matter of practical consideration when working with a SR flip-flop, one is sometimes faced with the question of what is the minimum time required to reset the flip-flop ($t_{RESETmin}$). This can easily be answered with the aid of Spice. Consider initializing the state of the flip-flop such that Q is logically high and $\overline{Q}$ is logically low. This is simply achieved by using the following initial condition (.IC) statement:

.IC V(4)=+10V V(5)=0V

Further, by adjusting the duration of a pulse applied to the RESET input of the flip-flop we can find the input condition which no longer resets the flip-flop. This will then indicate the minimum time required to reset the flip-flop. Let us begin our investigation with an input pulse of 5 ns duration, then check the state of the flip-flop. If the Q output of the flip-flop settles into the logic low state, then $t_{RESETmin}$ is probably something less than this value. We will then proceed to decrease this value by 0.5 ns and re-simulate the circuit to see whether the flip-flop has been reset. If, once again, this is not the case, then we keep repeating this process until we find the duration of the input pulse that will not reset the flip-flop.

Beginning our analysis with a pulse having a 5 ns duration was based on the time required

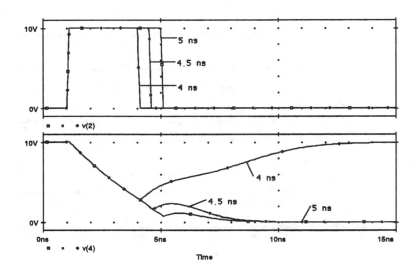

Figure 13.27 Determining the minimum time that is required to reset a SR flip-flop when initially in the logic high state. The top graph depicts the reset signal having various pulse widths. The bottom graphs shows the response of the flip-flop to the different inputs (Q output).

for a logic signal to propagate around a loop of two NOR gates. As a rough estimate of the time required for a signal to propagate through two NOR gates, let us consider that the propagation time of a NOR gate is about the same as that for an inverter circuit of similar dimensions. Recall that we found the propagation time for a CMOS inverter in Section 13.3 to be 2.28 ns. Thus, we can expect $t_{RESETmin}$ to be about 4.56 ns. Thus, to ensure that our starting point is one that resets the flip-flop, we shall begin our analysis with a pulse of 5 ns duration.

Our analysis will make use of the same Spice deck used previously for this flip-flop (seen listed in Fig. 13.25). We shall simply modify the two input signal sources as follows:

```
Vreset 2 0 PWL (0s,0V 1ns,0V 1.1ns,10V 5ns,10V 5.1ns,0V 15ns,0V)
Vset   3 0 DC 0
```

As well, we shall modify the analysis request as follows,

```
.TRAN 0.1ns 15ns 0 0.1ns.
```

In addition to a pulse input of 5 ns duration, we shall also concatenate together two other similar Spice decks having input pulses of 4.5 ns and 4 ns duration.

On completion of the Spice analysis, we find the Q output of the flip-flop shown in Fig. 13.27 for the three different input conditions. For an input reset signal of 5 or 4.5 ns duration, we see that the flip-flop changes state. But, for an input signal of 4 ns duration, we see that

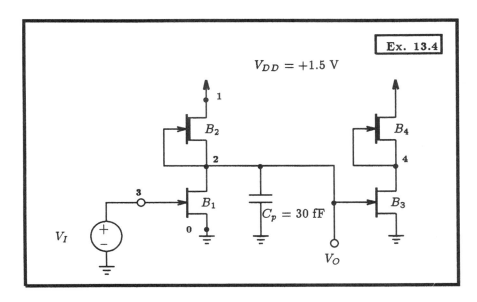

Figure 13.28 A cascade of two DCFL GaAs inverter circuits with the first gate loaded with 30 fF.

the flip-flop begins to change state, but then returns back to the high state. Thus, we can conclude that the minimum time to reset the flip-flop is greater than 4 ns but less than 4.5 ns. We could do more refinement of this search process if it was thought necessary, but we will, at this time, be content with $t_{RESETmin} = 4.5$ ns.

13.6 A Gallium-Arsenide Inverter Circuit

As an emerging IC technology, GaAs shows promise for high-speed electronic design. In Fig. 13.28 we present an inverter circuit in this technology, known as the direct-coupled FET (DCFL) inverter. Specifically, two inverter circuits are connected in cascade with a 30 fF lumped capacitor representing the distributed capacitance of the wiring that interconnects the two inverters. The lower two MESFETs of each inverter, B_1 and B_3, are enhancement MESFETs and their corresponding loads, B_2 and B_4, are depletion MESFETs. We shall assume that the parameters describing these two types of MESFETs per micron width are as follows: $L = 1$ μm, $V_{tD} = -1$ V, $V_{tE} = 0.2$ V, $\beta = 0.1$ mA/V^2, and $\lambda = 0.1$ V^{-1}. Moreover, in order to model the behavior of the Schottky diode that makes up the gate to source region of each MESFET, we shall set $n = 1.1$ and $I_S = 5 \times 10^{-16}$ A. The capacitive effects of each transistor will be modeled using Spice parameters $C_{GS} = 1.2 \times 10^{-15}$ F and $C_{GD} = 1.2 \times 10^{-15}$ F. Let the width of the input MESFET of each inverter be 50 μm, and let the width of the load MESFETs be 6 μm.

Using Spice we shall compute the voltage transfer characteristics of the first inverter with the second inverter present. As well, we shall compute the transient response of the

13 MOS Digital Circuits

Example 13.4: A DCFL GaAs Gate

```
** Circuit Description **
* dc supplies
Vdd 1 0 DC +1.5V
* input signal
Vi 3 0 PWL (0,0V 1ns,0V 1.1ns,700mV 3ns,700mV 3.1ns,0V 5ns,0V)
* amplifier circuit
B1 2 3 0 enchancement_n_mesfet 50
B2 1 2 2 depletion_n_mesfet 6
Cp1 2 0 30fF
B3 4 2 0 enchancement_n_mesfet 50
B4 1 4 4 depletion_n_mesfet 6
* mesfet model statements (by default, level 1)
.model enchancement_n_mesfet gasfet (beta=0.1m Vto=+0.2V lambda=0.1
+                     n=1.1 Is=5e-16 Cgd=1.2e-15 Cgs=1.2e-15 Rs=2500)
.model depletion_n_mesfet gasfet (beta=0.1m Vto=-1.0V lambda=0.1
+                     n=1.1 Is=5e-16 Cgd=1.2e-15 Cgs=1.2e-15 Rs=2500)
** Analysis Requests **
.DC Vi 0V 0.8V 10mV
.TRAN 0.1ns 5ns 0s 0.1ns
** Output Requests **
.PLOT DC V(2)
.PLOT TRAN V(2)
.probe
.end
```

Figure 13.29 The Spice input file for calculating the input-output transfer characteristics and the pulse response of the DCFL inverter circuit shown in Fig. 13.28.

first inverter for a 2 ns duration pulse input.

With the Spice deck provided in Fig. 13.29, the input-output voltage transfer characteristics for the DCFL inverter shown in Fig. 13.30 was calculated. Using Probe we find that $V_{OH} = 708$ mV, $V_{OL} = 247$ mV, and by taking the derivative of the output voltage with respect to the input voltage, we find $V_{IH} = 651$ mV and $V_{IL} = 491$ mV. Therefore the low and high noise margins are 245 mV and 56.7 mV, respectively. Here we see that the high noise margin appears quite low, a disadvantage of this type of logic gate. Nevertheless, these gates are presently receiving considerable attention in the development of GaAs digital circuits owing to their simplicity and low power dissipation.

The transient response of the DCFL inverter circuit is shown in Fig. 13.31 for a 2 ns duration pulse having a voltage swing between ground and V_{OH}. The input pulse has an assigned rise and fall time of 100 ps. The output response has an unsymmetrical rise and fall time of 623.5 ps and 337 ps, respectively. The two propagation delays t_{PHL} and t_{PLH} are found to be 198 ps and 211 ps, respectively. Thus, the average propagation delay of this

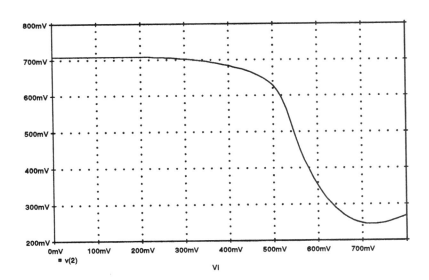

Figure 13.30 The input-output voltage transfer characteristics for the DCFL inverter circuit shown in Fig. 13.28.

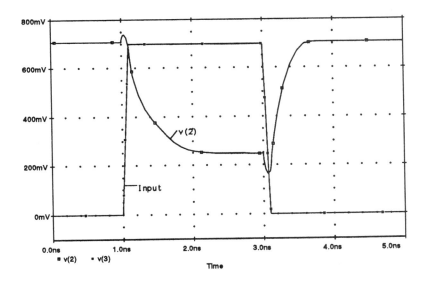

Figure 13.31 The transient response behavior of the DCFL inverter circuit shown in Fig. 13.28 subject to a 2 ns wide input pulse signal with 100 ps rise and fall times.

particular DCFL inverter gate is 204.5 ps.

13.7 Spice Tips

- The derivative dV_O/dV_I can be computed using the derivative command of Probe. A plot of V_O versus V_I must first be computed using the DC sweep command.

13.8 Problems

13.1 An n-channel enhancement MOS device, used in an inverter load operating from a 5 V supply, has V_{to} ranging from 1 to 1.5 and γ ranging from 0.3 to 1 $V^{1/2}$ and $2\phi_f = 0.6$ V. What is the lowest value of V_{OH} that will be found using transistors of this kind?

13.2 With the aid of Spice, determine the average propagation delay, the average power dissipation, and the delay-power product of an inverter for the following conditions:

$K_n(\mu A/V^2)$	$K_p(\mu A/V^2)$	$V_t(V)$	$V_{DD}(V)$	$C_L(pF)$
10	5	1	5	0.1
1	1	1	5	0.1
5	5	0.5	5	0.1
10	1	1	5	0.1
10	1	1	10	1.0
7.5	10	1	5	0.1

13.3 The circuit shown in Fig. P13.3 can be used in applications, such as output buffers and clock drivers, where capacitive loads require high current-drive capability. Let $V_{tE} = 1$ V, $V_{tD} = -3$ V, $V_{DD} = 5$ V, $\mu_n C_{OX} = 20$ $\mu A/V^2$, $W_1 = W_2 = 12$ μm, $L_1 = L_2 = L_3 = L_4 = 6$ μm. $W_3 = 120$ μm, and $W_4 = 240$ μm.

(a) Find V_{AH}, V_{OH}, V_{AL} and V_{OL} for v_i (high) = 5 V. Note that V_{AH} denotes the high level at node A, etc.

(b) Find the total static current with the output high and the output low.

(c) As v_o goes low, find the peak current available to discharge a load capacitance $C = 10$ pF. Also, find the discharge current for v_o at 50% output swing point and thus find the average discharge current and t_{PHL}.

(d) Repeat (c) for v_o going high, thus finding t_{PLH}.

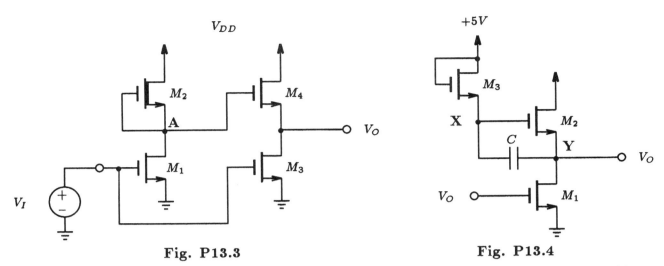

Fig. P13.3   Fig. P13.4

13.4 The circuit shown in Fig. P13.4, called a bootstrap driver, is intended to provide a large output voltage swing and high output drive current with a reasonable standing current. Let $V_{DD} = 5$ V, $V_t = 1$ V, $\mu_n C_{OX}\big|_1 = 40$ μA/V^2, $\mu_n C_{OX}\big|_2 = 10$ μA/V^2, $\mu_n C_{OX}\big|_3 = 1$ μA/V^2, and assume $C = 10$ pF. With the aid of Spice, answer the following questions:

(a) Find the voltages at node X and Y when v_i has been high (at 4 V) for some time.

(b) When v_i goes low to 0 V and Q_1 turns off, to what voltages do nodes X and Y rise?

(c) Find the current available to charge a load capacitance at Y as soon as v_i goes low and Q_1 turns off.

(d) As v_i goes high again, what current is immediately available to discharge the load capacitance at Y.

(e) What is the current shortly thereafter, when $v_Y = 4$ V?

13.5 A particular CMOS inverter uses n- and p-channel devices of identical sizes. If $\mu_n C_{OX} = 2\mu_p C_{OX} = 10$ μA/V^2, $|V_t| = 1$ V, and $V_{DD} = 5$ V, find V_{IL} and V_{IH} and hence the noise margins.

13.6 For a CMOS inverter having $V_{tn} = -V_{tp} = 1$ V, $V_{DD} = 5$ V, $\mu_n C_{OX} = 2\mu_p C_{OX} = 20$ μA/V^2, $(W/L)_n = 10$ μm/5 μm, and $(W/L)_p = 20$ μm/5 μm, determine using Spice, what the maximum current that the inverter can sink with the output not exceeding 0.5 V? Also find the largest current that can be sourced with the output remaining within 0.5 V of V_{DD}.

13.7 For an inverter with $\mu_n C_{OX} = 2\mu_p C_{OX} = 50$ μA/V^2, $V_{tn} = -V_{tp} = 0.8$ V, $V_{DD} = 5$ V, $(W/L)_n = 4$ μm/2 μm, and $(W/L)_p = 8$ μm/2 μm, using Spice, find the peak current drawn from a 5 V supply during switching.

13.8 If the inverter specified in Problem 13.7 is loaded with a 0.2 pF capacitance, determine,

581

13 MOS Digital Circuits

using Spice, the dynamic power dissipation when the inverter is switched at a frequency of 20 MHz. What is the average current drawn from the power supply?

13.9 For a CMOS inverter having $V_{tn} = -V_{tp} = 2$ V, $V_{DD} = 5$ V, $\mu_n C_{OX} = 2\mu_p C_{OX} = 50$ μA/V^2, and loaded with a 0.1 pF capacitance. If the inverter is to be clocked at 100 MHz, determine the values of (W/L) that will result in a delay-power product no larger than 0.1 pJ. Verify that this is indeed the case using Spice.

13.10 Consider a DCFL gate fabricated in GaAs technology for which $L = 1$ μm, $V_{tD} = -1$ V, $V_{tE} = +0.2$ V, and $\lambda = 0.1$ V^{-1}, and let $V_{DD} = 1.5$ V. Determine the widths of the two MESFETs that yields a $NM_H = 0.2$ V. Verify your choice using Spice. What is the resulting value of NM_L?

Chapter 14

Bipolar Digital Circuits

As the final chapter of this text we shall investigate the two dominant bipolar digital circuit families in use today: transistor-transistor logic (TTL) and emitter-coupled logic (ECL). The chapter concludes with a brief investigation into the behavior of a simple digital circuit that combines both bipolar and MOS technology into a single technology referred to as BiCMOS.

14.1 Transistor-Transistor Logic (TTL)

The basic circuit for a TTL NAND gate is shown in part (a) of Fig. 14.1. It consists of four *npn* transistors $Q_1 - Q_4$, one diode D_1, and several resistors. The input transistor Q_1 consists of a single BJT with two emitters and is referred to as a multi-emitter transistor. This particular multi-emitter transistor can be viewed as two *npn* transistors connected in parallel as shown in Fig. 14.1(b). If Q_1 has a single emitter then the NAND gate shown in Fig. 14.1(a) reduces to the basic TTL inverter circuit. With the aid of Spice, let us compute the voltage transfer characteristic of this gate from the A input to the NAND gate output. This is achieved by simply sweeping the dc voltage level at the A input between ground and the 5 volt supply. The B input will be connected to $V_{CC} = 5$ V throughout this simulation. In this arrangement, one emitter of the input transistor will be reverse biased and the NAND gate will behave exactly like an inverter circuit. The fanout associated with this NAND gate will be set to one. It is necessary to load the output of the NAND gate so that realistic results can be obtained. The circuit arrangement used in this simulation is depicted in Fig. 14.1(c). The Spice deck for this situation is provided in Fig. 14.2. Notice

14 Bipolar Digital Circuits

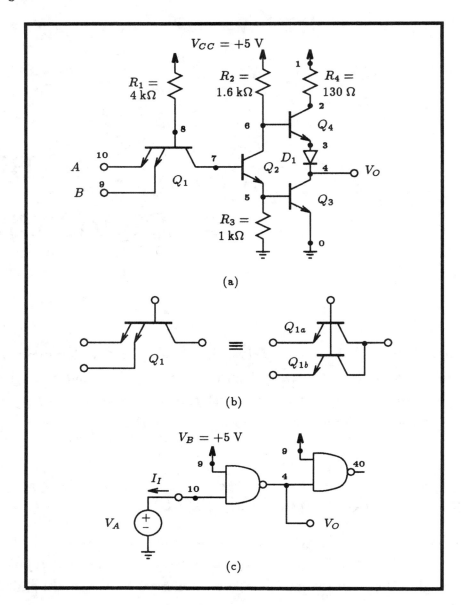

Figure 14.1 (a) A two-input TTL NAND gate. (b) A multi-emitter bipolar transistor and its equivalent representation for Spice simulation. (c) A cascade of two NAND gates.

that a subcircuit definition has been provided for the NAND gate. This is especially helpful when working with a large digital circuit consisting of many similar logic gates. Also, it is assumed that each resistor has a linear temperature coefficient of $TC = +1200$ ppm/°C. This will realistically model resistor behavior at different temperatures for the purpose of investigating circuit behavior versus temperature.

Running the Spice program on the above Spice deck results in the voltage transfer characteristic shown in Fig. 14.3. We see that the transfer characteristic consists of four

```
Voltage Characteristics Of A Two-Input TTL NAND Gate (Vb=logic 1)

* TTL Two-input NAND Gate
.subckt    NAND    10 9 4 1
* connections:     | | | |
*              inputA | | |
*                inputB | |
*                  output |
*                       Vcc
*
Q1a  7 8 10   npn_transistor
Q1b  7 8 9    npn_transistor
Q2   6 7 5    npn_transistor
Q3   4 5 0    npn_transistor
Q4   2 6 3    npn_transistor
QD1  3 3 4    npn_transistor
R1 1 8 4k     TC=1200u
R2 1 6 1.6k   TC=1200u
R3 5 0 1k     TC=1200u
R4 1 2 130    TC=1200u
* BJT model statement
.model npn_transistor npn (Is=1.81e-15 Bf=50 Br=0.02 Va=100
+                         Tf=0.1ns Cje=1pF Cjc=1.5pF)
.ends NAND

** Main Circuit **
* dc supplies
Vcc 1 0 DC +5V
* input digital signals
Va 10 0 DC  0V
Vb  9 0 DC +5V
* 1st NAND gate: inputB is held high
Xnand_gate1 10 9 4  1 NAND
* 2nd NAND gate: inputB is held high
Xnand_gate2  4 9 40 1 NAND
** Analysis Requests **
.DC Va 0V 5V 40mV
** Output Requests **
.Plot DC V(4) I(Va)
.probe
.end
```

Figure 14.2 The Spice input deck for calculating the voltage transfer characteristics of the two-input NAND gate shown in Fig. 14.1 with one input connected to logic high (+5 V).

main segments. With an input between 0 V and approximately 0.4 V, the output of the NAND gate is held constant at a level of 3.85 V. Beyond this input value of 0.4 V, but less than 1.2 V, the output voltage decreases at a linear rate of about −1.59 V/V. For inputs

14 Bipolar Digital Circuits

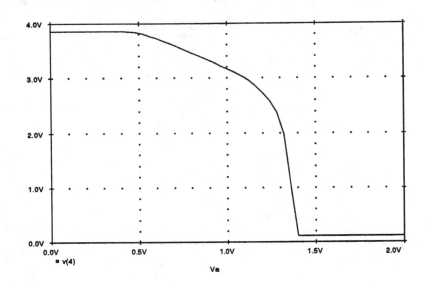

Figure 14.3 The voltage transfer characteristic of a two-input TTL NAND gate with one input connect to V_{CC}. The temperature of the circuit is assumed by default to be 27°C.

larger than 1.2 V, the output voltage deceases very quickly to a voltage level of about 0.111 V at an input voltage of 1.4 V. Above this value, the output voltage remains constant at 0.111 V. Based on this information, we can determine that $V_{OL} = 0.111$ V, $V_{OH} = 3.85$ V, $V_{IL} = 0.423$ V, and $V_{IH} = 1.4$ V. Thus, the low and high noise margins for this particular NAND gate with one input connected to V_{CC} are 0.312 V and 2.45 V, respectively.

Commercial TTL gates are specified to operate over a temperature range of 0 to 70°C. Using the Spice deck given in Fig. 14.2, we shall re-calculate the voltage transfer characteristics of the above NAND gate at three different temperatures: 0, 27 and 70°C. Adding the appropriate temperature statement into each Spice deck, such as

```
.TEMP 0C
```

for setting the temperature of the circuit at 0°C and repeating for the other two temperatures, we can obtain three different voltage transfer characteristics shown in Fig. 14.4. From these results, we summarize below the variation in the logic level parameters at the extreme limits of the above specified temperature range:

14.1 Transistor-Transistor Logic (TTL)

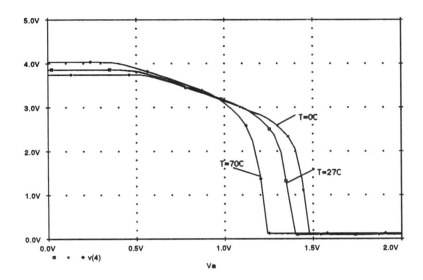

Figure 14.4 The temperature dependence of the voltage transfer characteristics of the TTL NAND gate shown in Fig. 14.1 when the B input is held at $V_{CC} = 5$ V.

Parameter	Variation (70°C/0°C)
V_{IL}	0.366 V / 0.493 V
V_{IH}	1.24 V / 1.48 V
V_{OL}	0.126 V / 0.101 V
V_{OH}	4.03 V / 3.75 V
NM_L	0.240 V / 0.392 V
NM_H	2.79 V / 2.27 V

Clearly, the low noise margins NM_L of this TTL gate decreases as the temperature of the circuit increases. In contrast, the high noise margins NM_H increases as the temperature of the circuit increases.

In addition to the voltage transfer characteristic, we have plotted in Fig. 14.5 the input current characteristics of the NAND gate as the input voltage is varied between the limits of the power supplies. This current is denoted by I_I in Fig. 14.1(c). Three different curves are provided corresponding to the three different circuit temperatures. If we consider the results obtained at a circuit temperature of 27°C, we see that for an input voltage level between 0 and 1.36 V, the current supplied by the input of the NAND gate decreases from 1.09 mA at 0 V to 0.722 mA at 1.36 V. When the input voltage goes above 1.36 V, the input gate current decreases rapidly and changes direction, thus flowing into the gate. The magnitude of this

14 Bipolar Digital Circuits

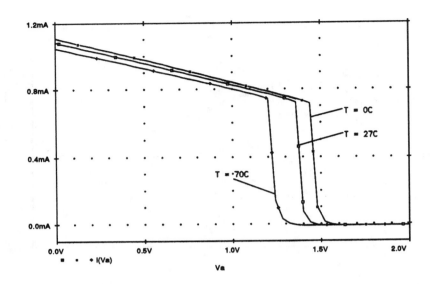

Figure 14.5 Input current characteristics of a single input of a TTL NAND gate as a function of the input voltage level.

current is quite small at about 7.4 μA. The level of this current remains constant for input voltages above 1.62 V. We can conclude from these results that the maximum input-low current I_{IL} is 1.09 mA and the maximum input-high current I_{IH} is 7.4 μA.

At other temperatures, current input characteristics are very similar, although the point where the input current changes direction decreases with increasing temperature. We can conclude over the 0 to 70°C temperature range that the maximum input-low current I_{IL} is 1.11 mA and occurs when the temperature of the circuit is at its lowest. In comparison, the input-high current level I_{IH} remains low with a maximum level of 7.4 μA over the two temperature extremes.

The low and high output characteristics of the NAND gate can be determined by varying the load. Consider the two circuit set-ups shown in Fig. 14.6. The circuit configuration in part (a) seen on the left hand side is used to determine the low state output voltage characteristic. The input to this gate is set equal to the minimum manufacturer's specified output voltage for a TTL gate established in the high state (ie. $V_A = V_{OH_{min}} = 2.4$ V). This is considered to be the worst-case input possible for establishing the output in the low state. The circuit shown on the right hand side in Fig. 14.6(b) is used to determine the high-state output voltage characteristic. The input to this gate is set equal to the maximum manufacturer's specified output voltage for a TTL gate set in the low state (ie. $V_A = V_{OL_{max}} = 0.4$ V). This is also a worst-case input situation.

14.1 Transistor-Transistor Logic (TTL)

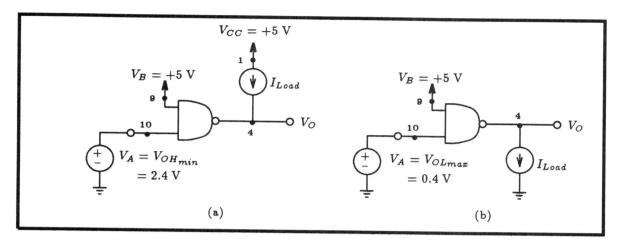

Figure 14.6 Circuit setup for simulating the output voltage characteristics of the TTL NAND gate under different load conditions: (a) low state (b) high state.

Let us consider the circuit setup in Fig. 14.6(a). Here the NAND gate is placed into the low state. With the Spice deck given in Fig. 14.7, we have requested that the load current source I_{Load} be swept between 0 and 150 mA and that the voltage at the output be plotted. This will be performed over three different temperatures: 0°C, 27°C and 70°C. We shall repeat this same experiment but with the NAND gate in the high state. The range of load current will be reduced to something between 0 and 20 mA. Several minor modifications will have to be made to the Spice deck seen listed in Fig. 14.7. We leave these to the reader to determine.

The results of these two Spice analyses are shown in Fig. 14.8. The top graph displays the low output voltage characteristics for different load currents at three different temperatures and the bottom graph displays the high output voltage characteristics. Let us first consider the low-state output voltage characteristics in Fig. 14.8(a) at room temperature. With no load, ie. $I_{Load} = 0$ mA, the output voltage is the offset voltage of about 100 mV. With increasing load current, the output voltage rises at a rate of about 1 mV/mA, for an equivalent output resistance of 1 Ω. With load currents exceeding 120 mA, the rate at which the output voltage changes increases dramatically to a near vertical slope. We note that at 27°C, the low state output voltage exceeds the manufacturer's maximum output voltage V_{OLmax} of 0.4 V at about a current of 120 mA. Similar behavior is seen to exist at different temperatures. At 0°C, the output resistance decreases to 0.85 Ω, and increases to 1.15 Ω at 70°C.

In the high state, the output voltage characteristics are shown in Fig. 14.8(b). With no load current, V_{OH} is about 4.4 V at 27° C but quickly reduces to 3.85 V with a very slight load current. For load currents increasing to about 5 mA, the output voltage changes by

14 Bipolar Digital Circuits

```
Fanout Behavior Of A NAND Gate

* TTL Two-input NAND Gate
.subckt    NAND     10  9  4  1
* connections:       |  |  |  |
*             inputA |  |  |  |
*                inputB |  |  |
*                   output |  |
*                         Vcc
*
Q1a  7  8  10   npn_transistor
Q1b  7  8  9    npn_transistor
Q2   6  7  5    npn_transistor
Q3   4  5  0    npn_transistor
Q4   2  6  3    npn_transistor
QD1  3  3  4    npn_transistor
R1   1  8  4k      TC=1200u
R2   1  6  1.6k    TC=1200u
R3   5  0  1k      TC=1200u
R4   1  2  130     TC=1200u
* BJT model statement
.model npn_transistor npn (Is=1.81e-15 Bf=50 Br=0.02 Va=100
+                          Tf=0.1ns Cje=1pF Cjc=1.5pF)
.ends NAND

** Main Circuit **
* dc supplies
Vcc 1 0 DC +5V
* input digital signals
Va 10 0 DC  +2.4; Va=Vohmin
Vb  9 0 DC  +5V
* 1st NAND gate: inputB is held high
Xnand_gate1 10 9 4  1 NAND
* current source load condition
Iload 1 4 0A
** Analysis Requests **
.Temp 27C
.DC Iload 0A 150mA 1mA
** Output Requests **
.Plot DC V(4)
.probe
.end
```

Figure 14.7 The Spice input deck for calculating the low-state output voltage characteristic of the NAND gate circuit setup shown in Fig. 14.6(a) at a circuit temperature of 27°C.

about 500 mV, thus the output resistance is 46 Ω. Beyond this current limit, we see that the output voltage changes at a new rate, with an output resistance of 118 Ω. At a load current of 13.3 mA, the output voltage reduces to the minimum specified manufacturer's

14.1 Transistor-Transistor Logic (TTL)

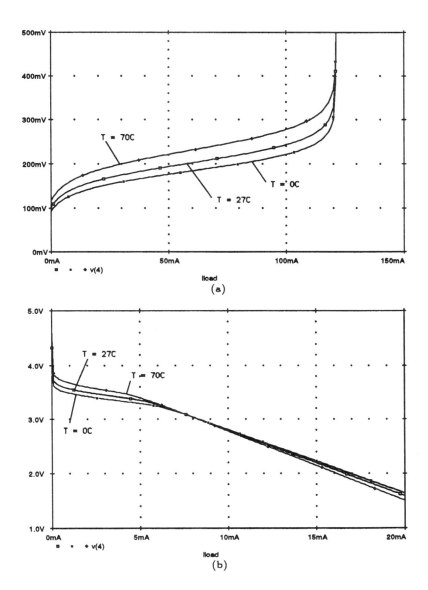

Figure 14.8 The output voltage characteristics of the TTL NAND gate in the two logic states under different load conditions: (a) low state (b) high state.

output voltage $V_{OH_{min}}$ of 2.4 V. Thus, the fanout of this particular gate is limited to loads less than this current. From the other curves seen in Fig. 14.8(b), we see that at large current values as the circuit temperature rises, the output voltage decreases, suggesting that the fanout of this gate decreases with increasing temperature.

To conclude our study of TTL logic gates, we have performed a transient analysis of the cascade of two NAND gates (connected as simple inverters) subject to a simple pulse input. The Spice deck seen listed in Fig. 14.2 was modified to perform this simulation. The dc source statement V_A was replaced by the following piecewise-linear source statement:

14 Bipolar Digital Circuits

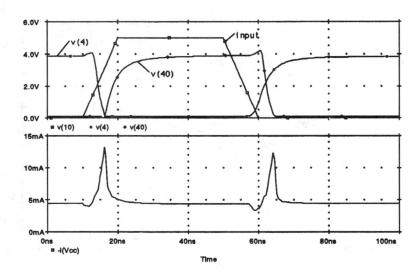

Figure 14.9 The transient response of a cascade of two TTL NAND gates (one input to each NAND gate is connected to logic high). The top graph displays the input voltage signal, and the voltage waveforms that appear at the output of of the two NAND gates. The bottom graph displays the current waveform associated with the power supply that powers each NAND gate.

```
Va 10 0 PWL( 0,0V 10ns,0V 20ns,5V 50ns,5V 60ns,0V 100ns,0V)
```

This source statement creates a logic pulse with a 5 V amplitude and a duration of 40 ns. The rise and fall times were both equal to 10 ns. A transient analysis was then requested over a time interval of 100 ns using the following statement

```
.TRAN 0.5ns 100ns 0s 0.5ns
```

Several results of the Spice analysis are shown in Fig. 14.9. In the top graph of this figure, the output voltage of the first and second NAND gates, together with the input voltage pulse, are plotted. Using Probe, we found that the transition delay times of the first inverter to be $t_{PHL} = 0.62$ ns and $t_{PLH} = 6.64$ ns. Thus, the propagation time t_P for this particular TTL gate is 3.63 ns. Similar results are also found for the second inverter. The lower graph in Fig. 14.9 displays the current supplied to the two NAND gates. Notice that a current spike with a peak of 13.24 mA occurs at the time that the two NAND gates change states. During nonswitching times, we see that the current drawn by the two NAND gates is constant at 4.41 mA.

14.2 Emitter-Coupled Logic (ECL)

Emitter-coupled logic, or ECL, is a nonsaturating family of bipolar digital circuits. Nonsaturating logic provides higher speed than digital circuits that incorporate transistors that

14.2 Emitter-Coupled Logic (ECL)

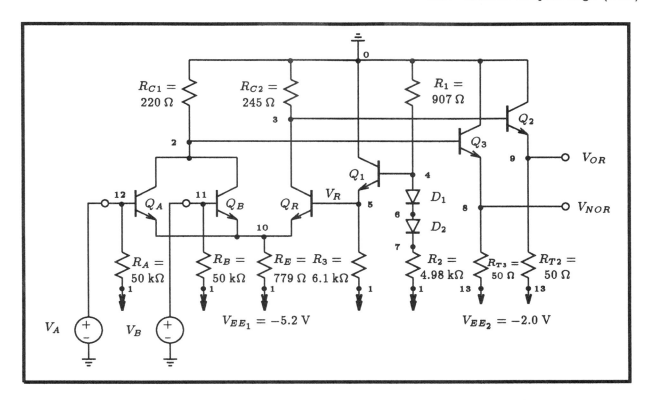

Figure 14.10 The two-input ECL OR gate with a complementary NOR output.

saturate during normal operation such as the TTL logic family. In the following we shall explore both the static and dynamic behavior of a simple ECL OR/NOR gate using Spice.

The basic schematic of an ECL OR/NOR gate is shown in Fig. 14.10. Unlike any of the previous logic families that we have encountered in this text, ECL logic has two outputs; an OR output and its complement, the NOR output. Both are connected to a –2 V level through two 50 Ω terminating resistors. The remaining transistors are biased using a –5.2 V supply. Using Spice, we shall calculate the voltage transfer characteristics of both the OR and NOR outputs of the ECL gate shown in Fig. 14.10 as a function of the input voltage V_A. We shall keep the B input to the ECL gate at a logic low level (ie. $V_B = -1.77$ V) throughout these calculations. We shall assume the following model parameters for each bipolar transistor: $I_S = 0.26$ fA, $\beta_F = 100$, $\beta_R = 1$, $\tau_F = 0.1$ ns, $C_{je} = 1$ pF, $C_{jc} = 1.5$ pF and $V_A = 100$ V. The diodes of this ECL gate will be realized by shorting the base and collector of an *npn* transistor together.

The Spice input file for the ECL gate is listed in Fig. 14.11. Here the level of the input voltage source V_A is varied between –2 V and 0 V. This represents the maximum voltage that can appear at the output of the ECL gate. The analysis request a plot of both the voltage at the OR and NOR outputs, and the voltage that appears at the base of Q_R (denoted V_R in Fig. 14.10).

14 Bipolar Digital Circuits

```
Two-Input ECL OR Gate With Complementary NOR Output

** Circuit Description **
* dc supplies
Vee1 1  0 DC -5.2V
Vee2 13 0 DC -2.0V
* input digital signals
Va 12 0 DC 0V
Vb 11 0 DC -1.77V
* ECL Gate
Qa 2 12 10 npn_transistor
Qb 2 11 10 npn_transistor
Qr 3  5 10 npn_transistor
Q2 0  3  9 npn_transistor
Q3 0  2  8 npn_transistor
Ra 12 1 50k TC=1200u
Rb 11 1 50k TC=1200u
Re 10 1 779 TC=1200u
Rc1 0 2 220 TC=1200u
Rc2 0 3 245 TC=1200u
Rt2 9 13 50 TC=1200u
Rt3 8 13 50 TC=1200u
* temperature-compensated voltage reference circuit
Q1  0 4 5 npn_transistor
QD1 4 4 6 npn_transistor
QD2 6 6 7 npn_transistor
R1  0 4 907   TC=1200u
R2  7 1 4.98k TC=1200u
R3  5 1 6.1k  TC=1200u
* BJT model statement
.model npn_transistor npn (Is=0.26fA Bf=100 Br=1
+                          Tf=0.1ns Cje=1pF Cjc=1.5pF Va=100)
** Analysis Requests **
.TEMP 27C
.DC Va -2V 0V 10mV
** Output Requests **
.Plot DC V(8) V(9) V(5)
.probe
.end
```

Figure 14.11 The Spice input deck for calculating the input-output voltage transfer characteristics of the OR and NOR outputs of the ECL gate shown in Fig. 14.10.

The results of the Spice analysis are shown in Fig. 14.12 for inputs between −2 and 0 V. Within the linear region of the gate, the two transfer characteristics are seen to be symmetrical about an input voltage of −1.32 V. With inputs larger than −0.42 V, the NOR transfer characteristic no longer behaves in the usual manner. Instead, we see that the NOR output begins to increase. It is therefore imperative that the input voltage to the ECL gate

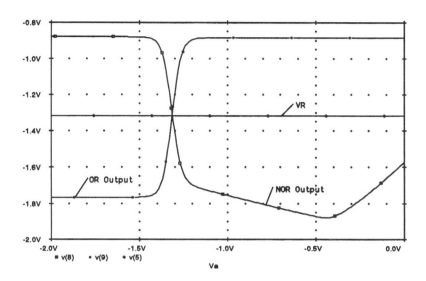

Figure 14.12 The input-output voltage transfer characteristics of the OR and NOR outputs of the ECL gate shown in Fig. 14.10.

be limited to levels less than −0.42 V.

In the case of the transfer curve for the OR output, we find with the help of Probe that $V_{OL} = -1.77$ V, $V_{OH} = -0.884$ V, $V_{IL} = -1.41$ V and $V_{IH} = -1.22$ V. Thus, the low and high noise margins are $NM_H = 0.339$ V and $NM_L = 0.358$ V.

From the transfer curve for the NOR output, we find that $V_{OL} = -1.78$ V, $V_{OH} = -0.879$ V, $V_{IL} = -1.41$ V and $V_{IH} = -1.22$ V. Thus, the corresponding low and high noise margins are $NM_H = 0.345$ V and $NM_L = 0.377$ V. In both cases, the NOR and OR outputs have noise margins that are nearly equal to one another (a maximum difference of 32 mV).

An important part of ECL logic is the voltage reference circuit. In Fig. 14.10, this consists of transistor Q_1, diodes D_1 and D_2, and resistors R_1, R_2 and R_3. This circuit is responsible for setting the dc voltage at the base of transistor Q_R. At a room temperature of 27°C, a Spice analysis reveals that the voltage appearing at the base of Q_R is −1.32 V. This is very nearly equal distance away from the high and low output voltage levels of the NOR and OR outputs. At different temperatures, the level of this voltage will change. However, the goal of a well designed ECL gate is that the output voltage levels will remain at equal distance from the voltage generated by the reference voltage circuit V_R over a wide range of temperatures. To demonstrate that the voltage reference circuit incorporated into the circuit of Fig. 14.10 accomplishes this goal, we shall compare the voltage transfer characteristics of an ECL OR/NOR gate with a temperature-compensated bias network in place (Fig. 14.10)

14 Bipolar Digital Circuits

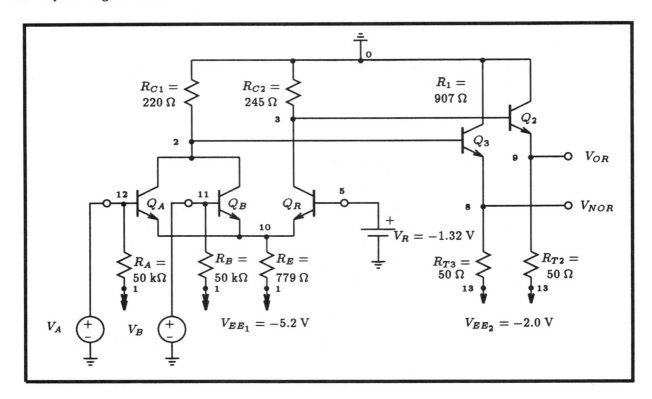

Figure 14.13 Removing the voltage reference bias circuit of the ECL OR/NOR gate shown in Fig. 14.10 and replacing it with a temperature independent voltage source V_R.

against a corresponding gate with only a dc voltage source representing the bias network. The latter circuit is shown Fig. 14.13. Our comparison will be based on computing the OR and NOR voltage characteristics of the gate at two different temperature extremes, 0°C and 70°C.

The results of a Spice analysis for these two different ECL circuits are shown in Fig. 14.14. The graphs in part (a) are the OR and NOR voltage transfer characteristics for the ECL gate with a temperature-compensated bias network. Also shown is the reference voltage V_R appearing at the base of Q_R. The graphs in (b) are the corresponding results without temperature compensation. Reviewing the results seen in the top figure, we see that regardless of the temperature of the circuit the transfer characteristics are seen to be symmetrical about the reference voltage V_R. In contrast, the uncompensated results in Fig. 14.13(b) show that as the temperature changes the voltage characteristics are less symmetrical about V_R. To quantify these observations, we tabulate below in Table 14.1 the output voltages levels V_{OL} and V_{OH}, their corresponding average $V_{avg} = \frac{V_{OL}+V_{OH}}{2}$ and the reference level V_R for each case. One measure of the symmetry of the output voltage characteristics is the closeness of V_{avg} with respect to V_R. Thus, we added another row at

14.2 Emitter-Coupled Logic (ECL)

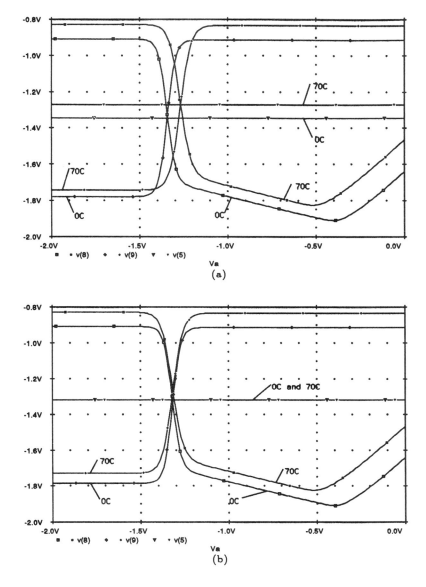

Figure 14.14 Comparing the input-output voltage transfer characteristics of the OR and NOR outputs of the ECL gate shown in Fig. 14.10 with: (a) a temperature-compensated bias network (b) a temperature-independent voltage source.

each temperature in Table 14.1 that indicates $V_{avg} - V_R$. On comparison of these results, we see that over a 70°C temperature change, V_{avg} for either the OR or NOR output never deviates more than 23 mV from V_R in the temperature compensated ECL gate but deviates as much as 38 mV in the uncompensated ECL circuit; thus, indicating the advantages of incorporating a temperature-compensated bias circuit in an ECL gate.

Our final investigation involving ECL gates concerns its dynamic operation. Let us consider the following situation: Two ECL gates are required to communicate over a distance

14 Bipolar Digital Circuits

Temperature	Parameter	Temp. Compensated		No Temp. Compensated	
		OR	NOR	OR	NOR
0°C	V_{OL}	−1.779 V	−1.799 V	−1.786 V	−1.799 V
	V_{OH}	−0.9142 V	−0.9092 V	−0.9142 V	−0.9092 V
	$V_{avg} = \frac{V_{OL}+V_{OH}}{2}$	−1.3466 V	−1.3541 V	−1.3501 V	−1.3541 V
	V_R	−1.345 V	−1.345 V	−1.32 V	−1.32 V
	$|V_{avg} - V_R|$	1.6 mV	9.1 mV	30.1 mV	34.1 mV
70°C	V_{OL}	−1.742 V	−1.759 V	−1.729 V	−1.759 V
	V_{OH}	−0.8338 V	−0.8285 V	−0.8338 V	−0.8285 V
	$V_{avg} = \frac{V_{OL}+V_{OH}}{2}$	−1.288 V	−1.294 V	−1.2814 V	−1.294 V
	V_R	−1.271 V	−1.271 V	−1.32 V	−1.32 V
	$|V_{avg} - V_R|$	17 mV	23 mV	38 mV	26.2 mV

Table 14.1 Characteristics of an ECL gate with and without temperature compensation at two different temperatures.

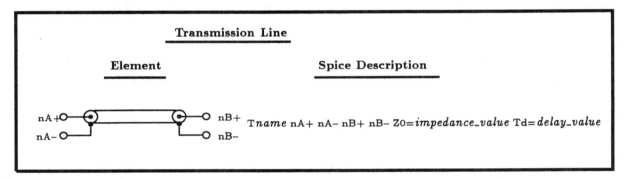

Figure 14.15 The Spice element description for a lossless transmission line.

of 1.5 meters. A coaxial cable having a characteristic impedance (Z_O) of 50 Ω is used to connect them. Based on the manufacturer's data, signals propagate along this cable at about half the speed of light, or 15 cm/ns. Thus, we can expect a signal injected at one end of the coaxial cable to take 10 ns before it reaches the other end. If we assume that the transmission is lossless (ie. ignore the skin effect), then we can model the behavior of the coaxial cable using the *Transmission Line* element statement of Spice. This element of Spice has not yet been introduced to the reader, so we shall do that next.

The general syntax of a lossless transmission line statement is provided in Fig. 14.15. The input port terminals of this transmission line are denoted $nA+$ and $nA-$, and the output port terminals are denoted $nB+$ and $nB-$. The characteristic impedance Z_O of the transmission line, as well as its propagation delay time T_d, are specified by the two fields marked Z0=*impedance_value* and Td=*delay_value*, respectively. Initial conditions at

14.2 Emitter-Coupled Logic (ECL)

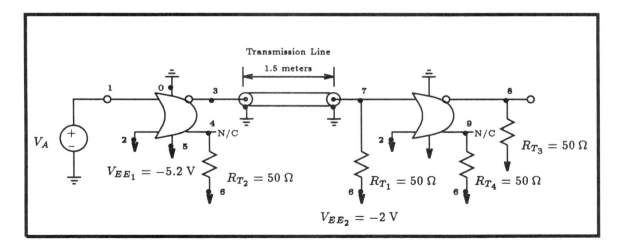

Figure 14.16 Circuit arrangement depicting a cascade of two ECL gates interconnected by 1.5 meters of coaxial cable.

the input and output ports can also be specified, but these are purposely left off the diagram of Fig. 14.15. Further details on the transmission line can be obtained from any Spice reference guide, see for instance [Vladimirescu, 1981].

Returning to the problem at hand, consider the circuit arrangement for the two interconnected ECL gates shown in Fig. 14.16. Here the two gates are interconnected with 1.5 meters of coaxial cable. It is assumed that the outputs of each gate are open collectors, thus each must be terminated with 50 Ω. Notice that the output of the ECL gate that is driving the coaxial cable is terminated by a 50 Ω resistor at the input to the second gate. This is necessary to minimize the potential of line reflections. With the aid of Spice, let us simulate this situation and determine the time that it takes for a logic change at one end, ie. at the input to the first gate, to appear at the output of the second gate. The Spice deck describing this circuit arrangement is provided in Fig. 14.17. A voltage step beginning at −1.77 V and rising to −0.884 V in 1 ns is applied as input to the first gate. A transient analysis over a 30 ns interval is requested. The inputs to each gate, together with their outputs, are to be plotted when the analysis is completed.

The results of the Spice analysis are illustrated in Fig. 14.18. We see here that the output change of the first ECL gate appears at the input to the second gate after 10 ns. The result of this input change to the second gate causes a new logic state to appear at the output of the second gate 1.06 ns later. Thus, we see that the overall propagation delay from the input to the first gate to the output of second gate is 12.15 ns.

To complete our transient analysis of two ECL gates interconnected by a 1.5 meter coaxial cable having a characteristic impedance of 50 Ω, consider the effect of replacing this coaxial cable with one that has a characteristic impedance of 300 Ω. With the terminating

14 Bipolar Digital Circuits

```
A Cascade Of Two ECL Gates Interconnected By A Lossless Transmission Line

* ECL two-input OR/NOR gate
.subckt OR/NOR_ECL   12 11 8 9 1
* connections:        |  |  | | |
*                inputA  |  | | |
*                   inputB  | | |
*                      NORoutput | |
*                         ORoutput |
*                               Vee1
*
* ECL Gate
Qa 2 12 10 npn_transistor
Qb 2 11 10 npn_transistor
Qr 3  5 10 npn_transistor
Q2 0  3  9 npn_transistor
Q3 0  2  8 npn_transistor
Ra 12 1 50k TC=1200u
Rb 11 1 50k TC=1200u
Re 10 1 779 TC=1200u
Rc1 0 2 220 TC=1200u
Rc2 0 3 245 TC=1200u
* temperature-compensated voltage reference circuit
Q1  0  4 5 npn_transistor
QD1 4  4 6 npn_transistor
QD2 6  6 7 npn_transistor
R1  0 4 907    TC=1200u
R2  7 1 4.98k TC=1200u
R3  5 1 6.1k  TC=1200u
* BJT model statement
.model npn_transistor npn (Is=0.26fA Bf=100 Br=1
+                          Tf=0.1ns Cje=1pF Cjc=1.5pF Va=100)
.ends OR/NOR_ECL

** Main Circuit **
* dc supplies
Vee1 5 0 DC -5.2V
Vee2 6 0 DC -2.0V
* input digital signals
Va 1 0 PWL (0,-1.77V 2ns,-1.77V 3ns,-0.884V 30ns,-0.884V)
Vb 2 0 DC -1.77V
* 1st OR/NOR gate: inputB is held low
Xnor_gate1 1 2 3 4 5 OR/NOR_ECL
* 2nd OR/NOR gate: inputB is held low
Xnor_gate2 7 2 8 9 5 OR/NOR_ECL
* transmission line interconnect + terminations
Tinterconnect 3 0 7 0 Z0=50 Td=10ns
Rt1 7 6 50
Rt2 4 6 50
Rt3 8 6 50
Rt4 9 6 50
** Analysis Requests **
.TRAN 0.05ns 30ns 0s 0.05ns
** Output Requests **
.Plot TRAN V(1) V(3) V(7) V(8)
.probe
.end
```

Figure 14.17 The Spice input deck for calculating the transient response of a cascade of two ECL gates interconnected by 1.5 meters of coaxial cable as shown in Fig. 14.16.

14.2 Emitter-Coupled Logic (ECL)

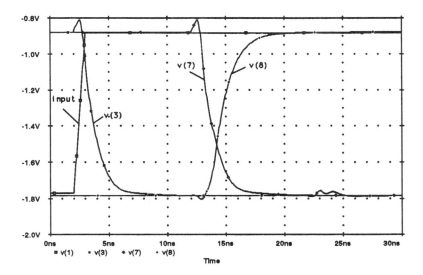

Figure 14.18 The transient response of a cascade of two ECL gates interconnected by a 1.5 meter coaxial cable having a characteristic impedance of 50 Ω.

resistance R_{T_1} remaining at 50 Ω, the transmission line is no longer terminated properly. We should then expect signal reflections to appear along the transmission line. Using Spice, let us re-simulate this situation and determine what effect it has on the overall propagation delay from the input to the first gate to the output of the second gate. We can use the same Spice deck as in the previous case except that we should replace the transmission line element statement by the following one:

```
Tinterconnect 3 0 7 0 Z0=300 Td=10ns.
```

Also, we shall modify the .TRAN command so that the transient analysis is performed over a longer time interval. A time interval of 400 ns was found sufficient for our purposes here.

The results of the Spice analysis are shown in Fig. 14.19. These results are very different than the results seen previously for the situation when the transmission line was terminated properly. Most notable is the voltage signal that appears at the input to the second gate (V(7)). This signal seems to change in small increments beginning at −0.879 V and decreasing to −1.16 V, followed by another small step change to −1.35 V, and again to −1.48 V, and so on until it reaches the logic low level of −1.77 V. The 90% to 10% fall time associated with this decreasing signal is found using Probe to be 102 ns. Such a slow falling signal at the input to the second gate gives rise to this gate changing state 33.8 ns after the initiation of the level change at the input of the first gate; a rather long propagation delay. Moreover, the rise time of the second gate is also very slow at 20.7 ns. Our conclusion here is that two

14 Bipolar Digital Circuits

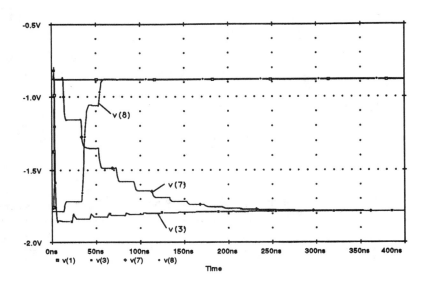

Figure 14.19 The transient response of a cascade of two ECL gates interconnected by a 1.5 meter coaxial cable having a characteristic impedance of 300 Ω.

ECL gates interconnected by a transmission line not properly terminated can give rise to unpredictably long propagation delays.

14.3 BiCMOS Digital Circuits

We conclude our study of digital circuits with a look at a new technology that combines the low power, high packing density and high input impedance of CMOS with the faster operation and higher drive capability of bipolar, called BiCMOS. We shall compare the transient behavior of a two-input NAND gate implemented with 5-micron[†] BiCMOS and CMOS technology. The advantages of a BiCMOS technology over a CMOS technology should become evident from this example. The circuit schematics of the two NAND gates are shown in Fig. 14.20. The BiCMOS NAND gate shown in Fig. 14.20(a) consists of both n-channel and p-channel MOSFETs, *npn* bipolar transistors, and two 20 kΩ resistors R_1 and R_2 that provide the gate with rail-to-rail voltage swing once the bipolar transistors turn off. Typically, these two resistors would be implemented using MOSFETs, however, to keep the ideas simple we shall stick to using resistors. The logic gate shown in Fig. 14.20(b) is the familiar two-input CMOS NAND gate. Both circuits of Fig. 14.20 are loaded at the output with a 5 pF capacitor.

[†] BiCMOS technologies tend to be in the submicron region. We use a 5-micron technology simply because its models are readily available to us.

14.3 BiCMOS Digital Circuits

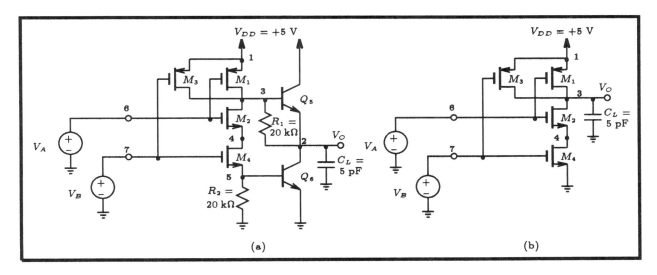

Figure 14.20 A logic NAND gate: (a) BiCMOS implementation (b) CMOS implementation.

Comparing the BiCMOS NAND gate to its corresponding CMOS implementation in Fig. 14.20 we see that the BiCMOS circuit is made up of a CMOS NAND gate with a bipolar output stage. It is precisely the introduction of these bipolar transistors that increase the drive capability of the BiCMOS gate. To demonstrate this, we shall have Spice compute the output voltage and output current waveform of both types of gates. Direct comparison can then be made.

The Spice input file describing the BiCMOS NAND gate is provided in Fig. 14.21. The A input is driven by a voltage signal generator whose output produces a 5 volt symmetrical square-wave signal of 500 kHz frequency. The B input is to be maintained at 5 V so that we can observe the effect of a single change at the input on the output. Initially, the A input to the NAND gate is logic low so that the output will begin in the logic high state. We may want to begin the transient analysis with the load capacitor initialized to 5 V; however, we must be aware that the internal nodes of the gate will not be properly initialized. Thus, it is important that we allow at least one cycle of the square-wave input to complete before we conclude that we have a typical output waveform. Therefore, we shall request that the transient analysis be performed over two periods of the input signal. We will then focus our attention on the high-to-low output transition of the gate.

The results of the Spice analysis are shown in Fig. 14.22. The top graph depicts the input and output voltage waveform of the BiCMOS gate circuit between 1.8 μs and 2.3 μs. We can extract from these two waveforms that the high-to-low transitional delay t_{PHL} is about 25.2 ns. The waveform in the bottom-graph represents the current that flows into the gate from the 5 pF load capacitor. As is evident, the discharge current waveform simply

14 Bipolar Digital Circuits

```
A BiCMOS Two-Input Nand Gate

** Circuit Description **
* dc supplies
Vdd 1 0 DC +5V
* input digital signals
Va 6 0 PULSE (0V 5V 0 10ns 10ns 1us 2us)
Vb 7 0 DC +5V
* CMOS input stage
M1 3 6 1 1 pmos_transistor L=5u W=30u
M2 3 6 4 0 nmos_transistor L=5u W=15u
M3 3 7 1 1 pmos_transistor L=5u W=30u
M4 4 7 5 0 nmos_transistor L=5u W=15u
* Bipolar output stage
Q5 1 3 2 npn_transistor
Q6 2 5 0 npn_transistor
* High-Swing resistors
R1 5 0 20k
R2 2 3 20k
* Load capacitance
CL 2 100 5pF IC=+5V
* Capacitor discharge current meter
VCL 100 0 0
* MOS and BJT model statements
.MODEL nmos_transistor nmos ( level=2 vto=1 nsub=1e16 tox=8.5e-8 uo=750
+ cgso=4e-10 cgdo=4e-10 cgbo=2e-10 uexp=0.14 ucrit=5e4 utra=0 vmax=5e4 rsh=15
+ cj=4e-4 mj=2 pb=0.7 cjsw=8e-10 mjsw=2 js=1e-6 xj=1u ld=0.7u )
.MODEL pmos_transistor pmos ( level=2 vto=-1 nsub=2e15 tox=8.5e-8 uo=250
+ cgso=4e-10 cgdo=4e-10 cgbo=2e-10 uexp=0.03 ucrit=1e4 utra=0 vmax=3e4 rsh=75
+ cj=1.8e-4 mj=2 pb=0.7 cjsw=6e-10 mjsw=2 js=1e-6 xj=0.9u ld=0.6u )
.model npn_transistor npn (Is=10fA Bf=100 Br=1 Tf=0.1ns Cje=1pF Cjc=1.5pF Va=100)
** Analysis Requests **
.TRAN 100ns 2.3us 1.8us 100ns UIC
** Output Requests **
.PLOT TRAN V(2) V(6) I(VC1)
.probe
.end
```

Figure 14.21 The Spice input file used to calculate the discharging current waveform of the BiCMOS NAND gate shown in Fig. 14.20(a) when a square wave is applied to its A input. The B input is held logical high throughout this example. A zero-valued voltage source VCL is placed in series with the load capacitor in order to monitor the capacitor discharge current.

rises and falls in accordance with the output voltage waveform. Using Probe we find that the peak value is 1.75 mA.

In contrast, the voltage and current waveforms associated with the CMOS NAND gate are shown in Fig. 14.23. Here we see that the output voltage has a much longer high-to-low transitional delay t_{PHL} of 43.5 ns; a factor of 1.6 increase over the delay associated

14.3 BiCMOS Digital Circuits

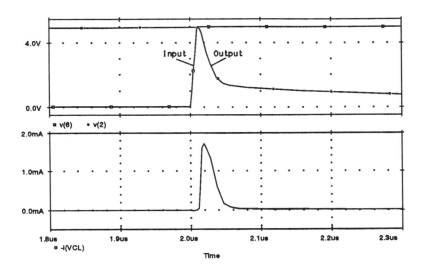

Figure 14.22 Voltage and current waveforms associated with BiCMOS NAND gate shown in Fig. 14.20. The top graph illustrates the input and output voltage of the gate circuit. The bottom graph depicts the load capacitor C_L discharge current waveform.

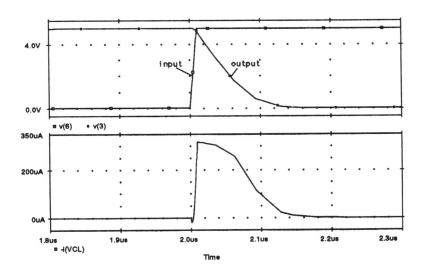

Figure 14.23 Voltage and current waveforms associated with CMOS NAND gate shown in Fig. 14.20. The top graph illustrates the input and output voltage of the gate circuit. The bottom graph depicts the load capacitor C_L discharge current waveform.

with the BiCMOS gate. In addition, the discharge current waveform shown in the bottom graph does not simply increase to a peak level and fall back down, as was the case for the BiCMOS gate. Instead, we see that the discharge current reaches a level of 317 μA and stays

14 Bipolar Digital Circuits

there for about 30 ns. In other words, the current available to discharge the load capacitor saturates. As a result, the output voltage of the gate begins to slew and the delay-time increases significantly. This is clearly evident in the middle waveform shown in Fig. 14.23. Similar results also occur during the charging of the load capacitor.

In conclusion, the BiCMOS gate provides extra drive capability, preventing the gate from slew-rate limiting. As a result, it has a shorter propagation delay than a corresponding CMOS gate.

14.4 Bibliography

A. Vladimirescu, K. Zhang, A. R. Newton, D. O. Pederson, and A. Sangiovanni-Vincentelli, "SPICE Version 2G6 User's Guide," Dept. of Electrical Engineering and Computer Sciences, University of California, Berkeley, CA, 1981.

14.5 Problems

14.1 A BJT for which $\beta_F = 100$ and $I_S = 10$ fA operates with a constant base current of 1 mA but with the collector open. Using Spice, determine the saturation voltage V_{CEsat} for this transistor.

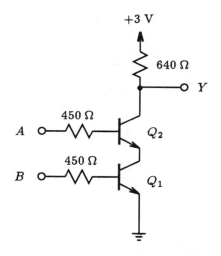

Fig. P14.2

14.2 Determine the logic function implemented by the circuit shown in Fig. P14.2 by cycling through the 4 possible inputs using Spice.

14.3 Consider the circuit in Fig. P14.3. If $V_{CC} = 5$ V, $R_C = R_B = 1$ kΩ, with the aid of Spice, determine the voltage levels at Y and $\overline{Y}$ when the set and reset inputs are inactive but following an interval in which S was high while R was low? Assume that the transistors have device parameters: $I_S = 10$ fA, $\beta_F = 100$, $\beta_R = 1$, $V_A = 50$ V, $\tau_F = 0.1$ ns, $C_{je} = 1$ pF and $C_{jc} = 1.5$ pF.

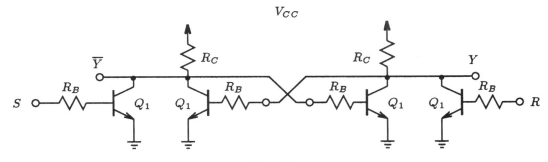

Fig. P14.3

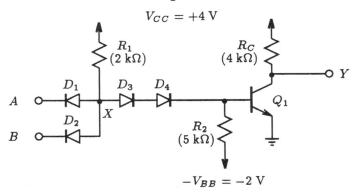

Fig. P14.4

14.4 For the diode-transistor logic (DTL) gate shown in Fig. P14.4, use Spice to determine the logic levels: V_{OH}, V_{OL}, V_{IL} and V_{IH}. Then, calculate the corresponding noise margins. Assume that the transistor has device parameters: $I_S = 10$ fA, $\beta_F = 100$, $\beta_R = 1$, $V_A = 50$ V, $\tau_F = 0.1$ ns, $C_{je} = 1$ pF and $C_{jc} = 1.5$ pF. Further, assume a base-collector shorted transistor for the *pn* junction diodes.

14.5 Repeat Problem 14.4 above, but for a fan-out of 5.

14.6 For the DTL gate of Fig. P14.4, with the aid of Spice, determine the total current in each supply and also the gate power dissipation for the two cases: v_Y high and v_Y low. Then, find the average power dissipation in the DTL gate. Use the same transistor parameters given in Problem 14.4.

14.7 Consider the DTL gate of Fig. P14.7 when the output is low and the gate is driving N identical gates. Assume that the transistors have Spice parameters $I_S = 10$ fA, $\beta_F = 50$, $\beta_R = 0.1$, and $V_A = 50$ V. Further, assume a base-collector shorted transistor for the *pn* junction diodes. Use Spice to determine:

(a) the output voltage for $N = 0$, and

(b) the maximum allowed fan-out N under the constraint that the output voltage does not exceed twice the value found in (a).

14 Bipolar Digital Circuits

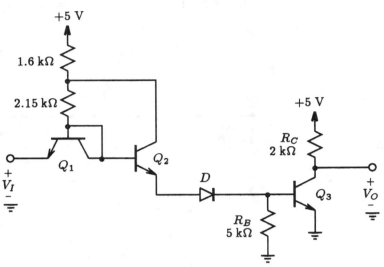

Fig. P14.7

14.8 For the DTL gate of Fig. P14.7 use Spice to plot the base current supplied to Q_3 when v_i goes high and the reverse current that flows out of its base when v_i goes low. What is the storage time t_s associated with Q_3. Assume that the transistors have device parameters $I_S = 10$ fA, $\beta_F = 100$, $\beta_R = 1$, $V_A = 50$ V, $\tau_F = 10$ ns, $C_{je} = 1$ pF and $C_{jc} = 1.5$ pF. The diode is realized by short circuiting the base to the collector of an npn transistor.

14.9 A variant of the T^2L gate shown in Fig. 14.1 is being considered in which all resistances are tripled. For both inputs high, use Spice to determine all node voltages and branch currents. Assume that the transistors have Spice parameters $I_S = 1.81$ fA, $\beta_F = 50$, $\beta_R = 0.1$, and $V_A = 100$ V.

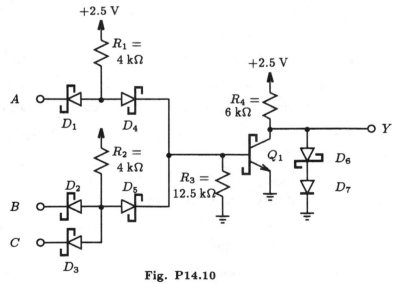

Fig. P14.10

14.10 For the BJT circuit shown in Fig. P14.10, use Spice to determine the following:
  (a) What logic function is performed?

(b) Determine V_{OL} and V_{OH}?

(c) Find V_{IL} and V_{IH} at the A input when both the B and C inputs are held at ground potential.

(d) What are the noise margins?

(e) Determine the average power dissipated by this gate when both input B and C are held at ground.

Assume that the transistors have device parameters $I_S = 10$ fA, $\beta_F = 100$, $\beta_R = 1$, $V_A = 50$ V, $\tau_F = 0.1$ ns, $C_{je} = 1$ pF and $C_{jc} = 1.5$ pF. Further, assume a base-collector shorted transistor for all pn junction diodes. For the Schottky-diodes assume: $I_S = 1$ pA, $n = 1$, $C_{j0} = 0.2$ pF, $\phi_o = 0.7$ V and $m = 0.5$.

14.11 For the ECL circuit shown in Fig. P14.11, determine, with the aid of Spice, the following:

(a) Find V_{OH} and V_{OL}.

(b) For the input at B sufficiently negative for Q_B to be cut off, what voltage at A causes a current of $1/2$ mA to flow in Q_R.

(c) Repeat (b) for a current in Q_R of 0.99 mA.

(d) Repeat (c) for a current in Q_R of 0.01 mA.

(e) Use the results of (c) and (d) to specify V_{IL} and V_{IH}.

(f) Find NM_H and NM_L.

Assume that the transistors have Spice parameters $I_S = 10$ fA, $\beta_F = 50$, $\beta_R = 0.1$, and $V_A = 75$ V.

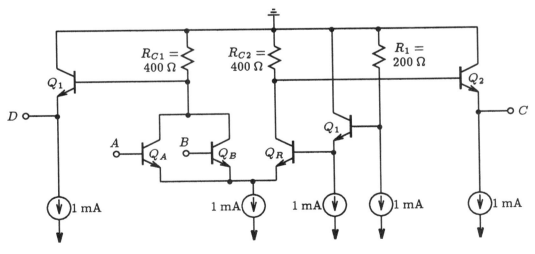

Fig. P14.11

14.12 For the BiCMOS and CMOS NAND gates shown in Fig. 14.20, with the aid of Spice, compute the propagation delay time for the following capacitive loads: 10 fF, 100 fF, 1 pF and 10 pF. The propagation delay time is defined as the time between the midway points of the input and output voltage waveforms. Sketch the *delay-time* versus *load capacitance* for both gates. What can you conclude from these results? Use the same transistor models and dimensions as was used in section 14.3.

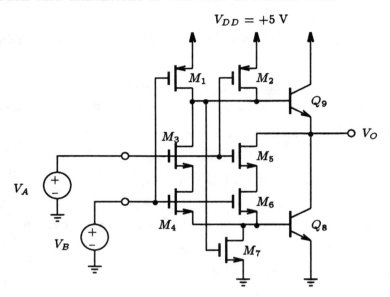

Fig. P14.13

14.13 Shown in Fig. P14.13 is an example of a practical BiCMOS NAND gate. With the aid of Spice, compute the voltage and current waveform associated with a 1 pF load capacitor. Use the same transistor models and dimensions as was used in section 14.3.

Appendix A.

Device Model Parameters

Each built-in semiconductor device of Spice is modeled with a sophisticated set of mathematical equations that describe the static and dynamic terminal behavior of the device. To allow the user to customize a particular device to their application, a set of parameters can be specified on a .MODEL statement according to the following Spice syntax:

.MODEL   *model_name*   *type*   (*parameter_list* ....)

Here the field labeled *model_name* is the model name, and the field labeled *type* is one of the following eight device types:

Type	Description
D	Diode model
NPN	NPN BJT model
PNP	PNP BJT model
NJF	N-channel JFET model
PJF	P-channel JFET model
NMOS	N-channel MOSFET model
PMOS	P-channel MOSFET model
GASFET	N-channel MESFET model

The next field, labeled *parameter_list*, describes a set of parameters that characterize the semiconductor device. A discussion of these parameters is the focus of this appendix.

In the following we shall outline all the parameters that make up the various models of Spice. This includes a description of the model parameters for semiconductor diodes, bipolar transistors (BJTs) and junction and metal-oxide semiconductor field-effect-transistors (JFETs and MOSFETs). Finally, we conclude with a description of the model parameters that make up the metal semiconductor field-effect-transistor (MESFET) found in PSpice.

A.1 Diode Model

The DC characteristics of the diode are determined by the parameters IS and N. A resistor RS accounts for the series ohmic resistance of the diode. Charge storage effects are

Appendix A.

Spice Name	Model Parameter	Units	Default	Example
IS	Saturation current	A	1×10^{-14}	1.0×10^{-14}
RS	Ohmic resistance	Ω	0	10
N	Emission coefficient	–	1	1.0
TT	Transit-time	sec	0	0.1 ns
CJO	Zero-bias junction capacitance	F	0	2 pF
VJ	Junction potential	V	1	0.6
M	Grading coefficient	–	0.5	0.5
EG	Activation energy	eV	1.11	1.11 Si, 0.69 Sbd, 0.67 Ge
XTI	Saturation-current temperature coefficient	–	3.0	3.0 pn-junction, 2.0 Sbd
KF	Flicker noise coefficient	–	0	
AF	Flicker noise exponent	–	1	
FC	Coefficient for forward-bias depletion capacitance formula	–	0.5	
BV	Reverse-bias breakdown voltage	V	∞	40.0
IBV	Reverse-bias breakdown current	A	1×10^{-10}	1.0×10^{-10}

Table A.1 Semiconductor diode model parameters.

modeled by a transit time TT and a nonlinear depletion-layer capacitance. The parameters that effect the depletion layer capacitance are CJO, VJ, M and FC. The temperature dependence of the saturation current is defined by the parameters EG and XTI. The flicker noise behavior of the diode is defined by the parameters KF and AF. Reverse breakdown is modeled by an exponential increase in the reverse diode current and is determined by the parameters BV and IBV.

The parameters used to model a semiconductor diode in Spice are listed in Table A.1.

A.2 BJT Model

The bipolar junction transistor model in Spice is an adaptation of the integral charge control model of Gummel and Poon. The model will automatically simplify to the Ebers-Moll model when certain parameters are not specified.

The forward static current gain characteristic of the BJT is defined by the parameters IS, BF, NF, ISE, IKF and NE. Correspondingly, the reverse current gain characteristic of the BJT is defined by the parameters IS, BR, NR, ISC, IKR and NC. The output conductance of the forward and reverse regions of the transistor is determined by VAF and VAR, respectively. Resistors RB, RC and RE represent an ohmic resistance in series with each terminal of the

BJT. The current dependence of RB can be modeled by the parameters IRB and RBM. Base charge storage is modeled by forward and reverse transit times, TF and TR, and the nonlinear depletion-layer capacitances are determined by CJE, VJE, and MJE for the base-emitter junction, CJC, VJC and MJC for the base-collector junction, and CJS, VJS and MJS for the collector-substrate junction. The temperature dependence of the saturation current IS is determined by parameters EG and XTI. Additionally, base current temperature dependence is modeled by parameter XTB. The flicker noise behavior of the diode is defined by the parameters KF and AF.

The BJT parameters used in the modified Gummel-Poon model are listed in Table A.2. There are 40 parameters associated with this model.

A.3 JET Model

The JFET model in Spice is derived from the FET model of Shichman and Hodges. The DC characteristics are defined by the parameters VTO, BETA, LAMBDA and IS. Two ohmic resistances RD and RS are included in series with the drain and source terminals of the JFET. Charge storage is modeled by a nonlinear depletion-layer capacitance for both gate junctions using parameters CGS, CGD, PB and FC. The flicker noise behavior of the JFET is defined by the parameters KF and AF.

The JFET model parameters are listed in Table A.3.

A.4 MOSFET Model

Spice provides three MOSFET device models that have different large-signal $i\text{-}v$ characteristics. The variable LEVEL specifies the model that is to be used to represent a particular MOSFET:

LEVEL=1 => Shichman-Hodges
LEVEL=2 => MOS2, an analytical model (as described in [A. Vladimirescu, 1981])
LEVEL=3 => MOS3, a semi-empirical model (see [A. Vladimirescu, 1981])

The DC characteristics of the MOSFET are defined by the device parameters VTO, KP, LAMBDA, PHI and GAMMA. These parameters are computed by Spice if process parameters (NSUB, TOX, ...) are given and user-specified values are not given instead. Two ohmic resistances RD and RS are included in series with the drain and source terminals of the MOSFET. Charge storage is modeled by a nonlinear thin-oxide capacitance, several nonlinear depletion-layer capacitances and overlap capacitances. There are two built-in models of the

Appendix A.

Spice Name	Model Parameter	Units	Default	Example
IS	Transport saturation current	A	1×10^{-16}	1.0×10^{-15}
BF	Ideal maximum forward beta	–	100	100
Nf	Forward current emission coefficient	–	1	1.0
VAF	Forward Early voltage	V	∞	200
IKF	Corner for forward beta high current roll-off	A	∞	0.01
ISE	B-E leakage saturation current	A	0	1.0×10^{-13}
NE	B-E leakage emission coefficient	–	1.5	2.0
BR	Ideal maximum reverse beta	–	1	0.1
NR	Reverse current emission coefficient	–	1	1.0
VAR	Reverse Early voltage	V	∞	200
IKR	Corner for reverse beta high current roll-off	A	∞	0.01
ISC	B-C leakage saturation current	A	0	1.0×10^{-13}
NC	B-C leakage emission coefficient	–	2	1.5
RB	Base ohmic resistance	Ω	0	100
IRB	Current where base resistance falls halfway to its min value	A	∞	0.1
RBM	Minimum base resistance at high currents	Ω	RB	10
RE	Emitter resistance	Ω	0	1
RC	Collector resistance	Ω	0	10
CJE	B-E zero-bias depletion capacitance	F	0	2 pF
VJE	B-E built-in potential	V	0.75	0.6
MJE	B-E junction exponential factor	–	0.33	0.33
TF	Ideal forward transit time	sec	0	0.1 ns
XTF	Coefficient for bias dependence of TF	–	0.75	0.6
VTF	Voltage describing VBC dependence of TF	V	∞	
ITF	High-current parameter for effect on TF	A	0	
PTF	Excess phase at $freq = 1/(2\pi TF)$	degree	0	
CJC	B-C zero-bias depletion capacitance	F	0	2 pF
VJC	B-C built-in potential	V	0.75	0.5
MJC	B-C junction exponential factor	–	0.33	0.5
XCJC	Fraction of B-C depletion capacitance connected to internal base node	–	1	
TR	Ideal reverse transit time	sec	0	10 ns
CJS	Zero-bias collector-substrate capacitance	F	0	2 pF
VJS	Substrate junction built-in potential	V	0.75	
MJS	Substrate junction exponential factor	–	0	0.5
XTB	Forward and reverse beta temperature exponent	–	0	

Table A.2 BJT model parameters .... continued next page ....

Spice Name	Model Parameter	Units	Default	Example
EG	Energy gap for temperature effect on IS	eV	1.11	
XTI	Temperature exponent for effect on IS	–	3	
KF	Flicker-noise coefficient	–	0	
AF	Flicker-noise exponent	–	1	
FC	Coefficient for forward-bias depletion capacitance formula	–	0.5	

Table A.2 BJT model parameters.

Spice Name	Model Parameter	Units	Default	Example
VTO	Threshold voltage	V	−2.0	−2.0
BETA	Transconductance parameter	A/V^2	1×10^{-4}	1×10^{-3}
LAMBDA	Channel length modulation parameter	1/V	0	1×10^{-4}
RD	Drain ohmic resistance	Ω	0	100
RS	Source ohmic resistance	Ω	0	100
CGS	Zero-bias G-S junction capacitance	F	0	5 pF
CGD	Zero-bias G-D junction capacitance	F	0	1 pF
PB	Gate junction potential	V	1	0.6
IS	Gate junction saturation current	A	1×10^{-14}	1×10^{-14}
KF	Flicker noise coefficient	–	0	
AF	Flicker noise exponent	–	1	
FC	Coefficient for forward-bias depletion capacitance formula	–	0.5	

Table A.3 JFET model parameters.

charge storage effects associated with the thin-oxide. The flag/coefficient XQC determines which of the two models will be used; a voltage-dependent or a charge-controlled capacitance model [A. Vladimirescu, 1981]. Other parameters of the MOSFET model that determine the charge storage effects are CBD, CBS, CJ, CJSW, MJ, MJSW, PB and FC. The overlap capacitances are set by the parameters CGSO, CGDO and CGBO. The flicker noise behavior of the diode is defined by the parameters KF and AF.

The MOSFET parameters used for the three different MOSFET models in Spice are listed in Table A.4. There are 42 parameters associated with the three models of the MOSFET.

Appendix A.

Spice Name	Model Parameter	Units	Default	Example
LEVEL	Model index (eg. 1,2 or 3)	—	1	
VTO	Zero-bias threshold voltage	V	0	1.0
KP	Transconductance parameter	A/V^2	2.0×10^{-5}	3.1×10^{-5}
GAMMA	Bulk threshold parameter	$V^{1/2}$	0	0.37
PHI	Surface potential	V	0.6	0.65
LAMBDA	Channel-length modulation (LEVEL 1 and 2 only)	1/V	0	0.02
RD	Drain ohmic resistance	Ω	0	1.0
RS	Source ohmic resistance	Ω	0	1.0
CBD	Zero-bias B-D junction capacitance	F	0	20 fF
CBS	Zero-bias B-S junction capacitance	F	0	20 fF
IS	Bulk junction saturation current	A	1.0×10^{-14}	1.0×10^{-15}
PB	Bulk junction potential	V	0.8	0.87
CGSO	Gate-source overlap capacitance per meter channel width	F/m	0	4.0×10^{-11}
CGDO	Gate-drain overlap capacitance per meter channel width	F/m	0	4.0×10^{-11}
CGBO	Gate-bulk overlap capacitance per meter channel length	F/m	0	2.0×10^{-10}
RSH	Drain and source diffusion sheet resistance	Ω/sq.	0	10.0
CJ	Zero-bias bulk junction bottom cap. per sq-meter of junction area	F/m^2	0	2.0×10^{-4}
MJ	Bulk junction bottom grading coeff.	—	0.5	0.5
CJSW	Zero-bias bulk junction sidewall cap. per meter of junction perimeter	F/m	0	2.0×10^{-9}
MJSW	Bulk junction sidewall coefficient	—	0.33	
JS	Bulk junction saturation current per sq-meter of junction area	A/m^2	1.0×10^{-8}	
TOX	Oxide thickness	meter	1.0×10^{-7}	1.0×10^{-7}
NSUB	Substrate doping	$1/cm^3$	0	4.0×10^{15}
NSS	Surface state density	$1/cm^2$	0	1.0×10^{10}
NFS	Fast surface state density	$1/cm^2$	2×10^{-5}	1.0×10^{10}
TPG	Type of gate material: +1 op. to substrate −1 same as substrate 0 Al gate	—	1	
XJ	Metallurigical junction depth	meter	0	1.0 μm
LD	Lateral diffusion	meter	0	0.8 μm
UO	Surface mobility	$cm^2/(V \cdot s)$	600	700
UCRIT	Critical field for mobility degradation (LEVEL 2 only)	V/cm	1×10^4	1.0×10^4
UEXP	Critical field exponent in mobility degradation (LEVEL 2 only)	—	0	0.1

Table A.4 MOSFET model parameters .... continued next page ....

Spice Name	Model Parameter	Units	Default	Example
UTRA	Transverse field coefficient (mobility) (deleted for LEVEL 2)	–	0	0.3
VMAX	Maximum drift velocity of carriers	m/s	0	5.0×10^4
NEFF	Total channel charge (fixed and mobile) coefficient (LEVEL 2 only)	–	1	5.0
XQC	Thin-oxide capacitance model flag and coefficient of channel charge share attributed to drain (0-0.5)	–	1	0.4
KF	Flicker-noise coefficient	–	0	1.0×10^{-26}
AF	Flicker-noise exponent	–	1	1.2
FC	Coefficient for forward-bias depletion capacitance formula	–	0.5	
DELTA	Width effect on threshold voltage (LEVEL 2 and LEVEL 3)	–	0	1.0
THETA	Mobility modulation (LEVEL 3 only)	1/V	0	0.1
ETA	Static feedback (LEVEL 3 only)	–	0	1.0
KAPPA	Saturation field factor (LEVEL 3 only)	–	0.2	0.5

Table A.4 MOSFET model parameters

A.5 MESFET Model

The MESFET model that we describe here is that provided in the PSpice program by MicroSim Corporation. Spice version 2G6 does not have a device model for the MESFET.

PSpice provides three MESFET device models that have different large-signal i-v characteristics. The variable LEVEL specifics the model that is to be used to represent a particular MESFET:

$$\text{LEVEL}=1 => \text{Curtice}$$
$$\text{LEVEL}=2 => \text{Raytheon}$$
$$\text{LEVEL}=3 => \text{TriQuint}$$

The MESFET is modeled as an intrinsic JFET with resistances RD, RS and RG in series with the drain, source and gate, respectively. The DC characteristics are defined by the parameters VTO, BETA, ALPHA, LAMBDA, IS, N and M. Charge storage is modeled by a nonlinear depletion-layer capacitance for both gate junctions using parameters CGS, CGD, PB and FC. A capacitance between drain and source CDS can also be declared. The flicker noise behavior of the diode is defined by the parameters KF and AF. Effects of temperature can also be modeled using parameters EG, XTI, VTOTC, BETATCE, TRG1, TRD1 and TRS1.

Appendix A.

PSpice Name	Model Parameter	Units	Default	Example
LEVEL	Model index (eg. 1,2 or 3)	–	1	2
VTO	Threshold voltage	V	-2.5	-2.0
ALPHA	Saturation voltage parameter	V^{-1}	2.0	1.0
BETA	Transconductance parameter	A/V^2	0.1	1×10^{-2}
B	Doping tail extending parameter (LEVEL=2 only)	V^{-1}	0.3	
LAMBDA	Channel length modulation parameter	V^{-1}	0	
GAMMA	Static feedback parameter (LEVEL=3 only)	–	0	
DELTA	Output feedback parameter (LEVEL=3 only)	$(A \cdot V)^{-1}$	0	
Q	Power-law parameter (LEVEL=3 only)	–	2	
TAU	Conduction current delay time	sec	0	
RG	Gate ohmic resistance	Ω	0	
RD	Drain ohmic resistance	Ω	0	100
RS	Source ohmic resistance	Ω	0	100
IS	Gate junction saturation current	A	1×10^{-14}	1×10^{-14}
N	Gate junction emission coefficient	–	1	
M	Gate junction grading coefficient	–	0.5	
VBI	Gate junction potential	V	1	
CGS	Zero-bias G-S junction capacitance	F	0	1 pF
CGD	Zero-bias G-D junction capacitance	F	0	1 pF
CDS	D-S capacitance	F	0	1 pF
FC	Coefficient for forward-bias depletion capacitance formula	–	0.5	
EG	Bandgap voltage	eV	1.11	
XTI	IS temperature exponent	–	0	
VTOTC	VTO temperature coefficient	V/°C	0	
BETATCE	BETA exponential temp. coefficient	%/°C	0	
TRG1	RG temperature coefficient (linear)	°C^{-1}	0	
TRD1	RD temperature coefficient (linear)	°C^{-1}	0	
TRS1	RS temperature coefficient (linear)	°C^{-1}	0	
KF	Flicker noise coefficient	–	0	
AF	Flicker noise exponent	–	1	

Table A.5 MESFET model parameters.

The MESFET parameters used to describe the three different PSpice models are listed in Table A.5.

A.6 Bibliography

A. Vladimirescu, K. Zhang, A. R. Newton, D. O. Pederson, and A. Sangiovanni-Vincentelli,

Appendix A.

"SPICE Version 2G6 User's Guide," Dept. of Electrical Engineering and Computer Sciences, University of California, Berkeley, CA, 1981.

P. Antognetti and G. Massobrio, *Semiconductor Device Modeling with SPICE,* New York: McGraw-Hill, 1988.

A. Vladimirescu and S. Liu, *The simulation of MOS integrated circuits using SPICE2,* Memorandum no. M80/7, February 1980, Electronic Research Laboratory, University of California, Berkeley.

Staff, *PSpice Users' Manual,* MicroSim Corporation, Irvine, California, Jan. 1991.

Appendix B.

Spice Options

Spice allows the user to reset program control and specify user options for various simulation purposes. This is accomplished using an .OPTIONS statement in the Spice input file. The syntax of the .OPTIONS statement has the following form:

.OPTIONS   *list_of_options*

Table B.1 lists the various options available in Spice version 2G6. Similar options are also available in PSpice; one should refer to the PSpice User's Manual for the exact details. Any combination of the options listed in Table B.1 may be included in the *list_of_options* and in any order. There are two kinds of options: flags that initiate specific action and flags that re-assign a value to a specific parameter. The variable x associated with these types of flags shown in this table represents some positive number.

Appendix B.

OPTIONS	EFFECT
ACCT	causes accounting and run time statistics to be printed.
LIST	causes the summary listing of the input data to be printed.
NOMOD	suppresses the printout of the model parameters.
NOPAGE	suppresses page ejects.
NODE	causes the printing of the node table.
OPTS	causes the options values to be printed.
GMIN=x	sets the value of GMIN, the minimum conductance allowed by the program. The default value is 1.0×10^{-12}.
RELTOL=x	resets the relative error tolerance of the program. The default value is 0.001.
ABSTOL=x	resets the absolute current error tolerance of the program. The default value is 1 picoamp.
VNTOL=x	resets the absolute voltage error tolerance of the program. The default value is 1 microvolt.
TRTOL=x	resets the transient error tolerance. The default value is 7.0. This parameter is an estimate of the factor by which Spice overestimates the actual truncation error.
CHGTOL=x	resets the charge tolerance of the program. The default value is 1.0×10^{-14}.
PIVTOL=x	resets the absolute minimum value for a matrix entry to be accepted as a pivot. The default is 1.0×10^{-13}.
PIVREL=x	resets the relative ratio between the largest column entry and an acceptable pivot value. The default value is 1.0×10^{-3}.
NUMDGT=x	is the number of significant digits printed for output variable values. The variable x must satisfy the relation $0 < x < 8$. The default is 4. Note: this option is independent of the error tolerance used by Spice (ie. if the values of options RELTOL, ABSTOL, etc., are not changed then one may be printing numerical "noise" for NUMDGT > 4.
TNOM=x	resets the nominal temperature. The default value is 27°C (300°K).
ITL1=x	resets the DC iteration limit. The default is 100.
ITL2=x	resets the DC transfer curve iteration limit. The default is 50.
ITL3=x	resets the lower transient analysis iteration limit. The default is 4.
ITL4=x	resets the transient analysis timepoint iteration limit. The default is 10.
ITL5=x	resets the transient analysis total iteration limit. The default is 5000. Set ITL5=0 to omit this test.
ITL6=x	resets the DC iteration limit at each step of the source stepping method. The default is 0 which means not to use this method.
CPTIME=x	is the maximum cpu-time in seconds allowed for this job.
LIMTIM=x	resets the amount of cpu-time reserved by Spice for geneating plots should a cpu-time limit cause job termination. The default value is 2 seconds.
LIMPTS=x	resets the total number of points that can be printed or plotted in a DC, AC, or transient analysis. The default value is 201.
LVLTIM=x	if x is one (1), the iteration timestep control is used. If x is two (2), the truncation-error timestep is used. The default value is 2. If METHOD= GEAR and MAXORD > 2 then LVLTIM is set to 2 by Spice.
METHOD= name	sets the numerical integration method used by Spice. Possible names are GEAR or TRAPEZOIDAL. The default is trapezoidal.

Table B.1: Options available in Spice .... continued next page ....

Appendix B.

OPTIONS	EFFECT
MAXORD=x	sets the maximum order for the integration method if Gear's variable-order method is used. The variable x must be between 2 and 6. The default is 2.
DEFL=x	resets the value for MOS channel length; the default is 100.0 micrometer.
DEFW=x	resets the value for MOS channel width; the default is 100.0 micrometer.
DEFAD=x	resets the value for MOS drain diffusion area; the default is 0.0.
DEFAS=x	resets the value for MOS source diffusion area; the default is 0.0.

Table B.1: Options available in Spice.

Index

A

AC analysis; *See* Analysis: AC
A/D converter, 479-483
Active devices:
    bipolar junction transistor (BJT), 143-148
    junction field-effect transistor (JFET), 212-218
    metal-oxide-semiconductor FET (MOSFET), 194-203
    metal-semiconductor FET (MESFET), 247-251
    semiconductor diode (D), 94-97
Active filter, 491-502
Ammeter, 263
Analysis:
    AC (.AC), 16-17
    DC (.DC), 16-17
    Fourier (.FOUR), 434-435
    Noise (.NOISE), 459-461
    Operating Point (.OP), 16-17
    Sensitivity (.SENS), 168-169
    Temperature (.TEMP), 101-102
    Transfer Function (.TF), 48-49
    Transient (.TRAN), 16-18
    types, 5-6
Astable multivibrator, 533-534

B

Beta:
    β_{ac}, 179, 329-330
    β_{dc}, 179, 329-330
    β_F, 146, 329-330
    β_R, 146
Biasing point analysis, *See* Analysis: operating point
BiCMOS:
    digital circuits, 583-602, 602-606
    folded cascode, 468-476
    op amp, 468-476
Bipolar:
    current mirror, 279-287
    digital circuits, 583-602
    junction transistor, 143-148
Biquad:
    HKN, 510-511
    single-amplifier, 500-502
    Tow-Thomas, 494-500
    two-integrator-loop, 494-500
Bistable multivibrator, 529-532
BJT (bipolar junction transistor):
    amplifier, 160-167
    capacitances, 326-330
    characteristics, 149-150
    Ebers-Moll model, 612-613
    Gummel-Poon model, 612-613
    model parameters, 612-613
    model statement, 145-146
    modes of operation, 151-154
    small-signal parameters, 160-161
    statement, 143-146
Bode plot, 398-399
Breakdown:
    diode, 97
    BV, 97
    IBV, 97

C

Capacitance:
    BJT, 326-330
    diode, 326-330
    MOS, 326-330
    overlap, 328
Capacitor:
    element line, 11
    initial conditions, 11

Index

Circuit description, 9-16
Class A output stage, 423-428
Class AB output stage, 436-438
Class B output stage, 428-436
Clipping, 427
CMOS:
    amplifier, 239-242
    gate circuits, 570-577
    op amp, 462-468
CMOS logic:
    flip-flop, 572-577
    NOR gate, 570-572
    transfer characteristic, 567-573
    transmission gate, 244
CMRR (common-mode rejection ratio), 77, 297
Coefficient of coupling, 115
Comment line, 6
Common-mode,
    gain, 77
    input resistance, 271
    range, 296, 311, 452, 467-468
    rejection ratio, 77, 296, 297, 454
    voltage, 263
Comparator, 479
Compensation, 410-413
Continuation element line, 7
Controlled sources, *See* Dependent sources
Convergence, Spice, 305
Converter, 476-483
    A/D, 479-483
    D/A, 476-479
Crossover distortion, 428-432
Crystal oscillator, 526-529
Current-controlled current source, 14
Current-controlled voltage source, 14
Current convention, 9, 11
Current conveyor, 445
Current mirror, 279-287
    BJT, 280,
    cascode, 286
    compensated, 280
    MOS, 316-318
    multiple-output, 319
    output resistance, 282-285
    Wilson, 280

Current source, 287-290
    controlled/dependent, 15
    independent, 12
    element line, 12, 15

D

D/A converter, 476-479
Damped integrator, 53-57
DC analysis, 5, 16-17
DC path to ground, 84, 119
DC sensitivity, 168-172
DC sweep, 5, 6
.DC Spice statement, 16-17
Deadband, 433, 434
DEC, *See* Analysis: AC
Dependent sources, 14-15
Device models, 94-97, 143-146, 194-198, 212-214, 247-251, 611-618
Differential amplifier
    BJT, 261-279
    frequency response, 344-352
    MESFET, 298-304
    MOSFET, 290-297
Digital circuit
    bipolar, 583-602
    BiCMOS, 602-606
    GaAs, 577-579
    MOSFET, 553-577
    noise margins, 554-556
Diode
    breakdown, 97
    clamping circuit, 542-543
    element line, 95
    emission coefficient (n), 97
    ideal, 102-104
    model line, 95
    model parameters, 95-97, 611-612
    peak detector, 541
    rectifier, 115-127, 537-539
    regulator, 104-108, 109-114, 127-132
    shaper, 536-537
    temperature dependence, 101-102, 516-517, 611-612
    voltage doubler, 135-136
    zener, 108-109

Distortion
    crossover, 428, 432, 436
    harmonic, 434
Dot commands, 16
    .AC, 17
    .DC, 16
    .END, 6
    .ENDS, 60
    .FOUR, 434
    .IC, 18
    .MODEL, 144, 194, 95
    .NOISE, 460
    .NODESET, 305
    .OP, 16
    .OPTIONS, 128, 525, 620
    .PLOT, 19
    .PRINT, 19
    .PROBE, 25
    .SENS, 168
    .SUBCKT, 60
    .TF, 48
    .TRAN, 17

E

Early voltage, 146
Eber-Moll BJT model, 612
ECL (emitter coupled logic), 592-602
Element:
    lines, continued, 7
    lines, rule for, 7
    name, number of characters, 9
    types, 9
End, input file (.END), 6
End, subcircuit (.ENDS), 60
Engineering suffixes, 8
Error messages, 58

F

Feedback
    Bode plots, 398-399
    effects,
    factor β, 358
    frequency compensation, 410-413
    loop gain, 388-394
    Nyquist plot, 398-399
    resistance, 360
    series-series, 367-372
    series-shunt, 359-367
    shunt-series, 380-388
    shunt-shunt, 374-380
    stability, 394-408
    transfer function, 360
File:
    input, 6
    output, 21
Filter
    Butterworth, 487
    Chebyshev, 489-490
    inductance simulation, 491
    second-order, 491-502
    two-integrator-loop, 494-500
Flash converter, 479-483
Flip-flop
    set-reset, 572-577
Folded-cascode, 468-476
Fourier analysis, *See* Analysis: Fourier
Frequency response analysis, *See* Analysis: AC

G

GaAs (gallium arsenide), 247-255
    amplifier, 298-304
    digital circuits, 577-579
Gain
    common-mode, 77-80, 267, 271, 273, 274, 275, 276, 290, 291, 304, 454
    current, 33
    differential, 267, 272, 273, 290, 291, 297, 454, 467
    power, 35
    small-signal, 48
    voltage, 33
Gallium-arsenide MESFET
    model parameters, 250-252, 617-618
    model statement, 249
    small-signal model, 251-252

Index

g-parameters, 388
Graphics post-processor, Probe, 25-27
Gummel-Poon BJT model, 612

H

Half-wave rectifier, 115-119
Harmonics, *See* Analysis: Fourier
h-parameters, 367
Hybrid π-model, 160-161, 326-330
Hysteresis, 529

I

Independent source, 10-14
    current convention, 11
    piecewise linear, 12
    pulse, 12
    sinusoidal, 12
    triangular, 164
Inductor coupling, 115-117
Inductance/inductor
    coefficient of coupling, 115
    dot convention, 116
    element line, 11
    mutual, 116
Input
    bias currents, 81, 267, 277-279, 307, 451
    common-mode range, 296, 302, 305, 311, 452, 465
    current range, 285-286
    offset current, 267, 451, 456
    offset voltage, 275, 276, 293
    resistance, 174, 178, 184, 225, 227, 228, 229, 267, 268, 270, 271, 274, 311, 454, 456
Instrumentation amplifier, 59-63
Integrator,
    damped, 53-57, 81-83
    Miller, 50-53, 81-83
    switched-capacitor, 514
Initial transient conditions
    for capacitor, 9
    for inductor, 9
    .IC statement, 18

J

JFET (junction field-effect transistor), 212-218
    element line, 212
    model line, 212
    model parameters, 213-214, 613
    small-signal model, 225

K

K, *See* coefficient of coupling
k_p, MOSFET transconductance coefficient, 197
KHN biquad, 510

L

λ, channel-length modulation, 197
Library files, 118
Limiter circuit, 132
Line regulation, 111
LIN, *See* Analysis: AC
Logic
    BJT, 583-606
    circuits, 553-606
    CMOS, 570-577
    emitter-coupled (ECL), 592-602
    inverter, 553-564
    noise margins, 554-556
    transistor-transistor (TTL), 583-592
Loop
    gain, 388-394
    transmission, 394

M

Magnitude response, 324, 330, 331, 334, 338, 341, 346, 352
MESFET (metal semiconductor field-effect transistor)
    amplifier, 253, 298-304
    element line, 249
    logic, 577-579
    model line, 249, 250-251
    model parameters, 251
    small-signal model, 251-252

Miller
- integrator, 50-53
- theorem, 332

Models
- BJT, 145-146, 612-613
- DIODE, 95-97, 611-612
- JFET, 213-214, 613
- MESFET, 250-251, 617-618
- MOSFET, 196-198, 613-615

Monostable multivibrator, 534-535

MOSFET (metal-oxide semiconductor field-effect transistor)
- amplifier, 225-230, 290-297, 235-242, 462-476
- body effect parameter, 197
- capacitances, 326-330
- characteristics, 200-203
- differential amplifier/pair, 235-242, 290-297, 462-468
- digital circuits, 553-577
- element line, 195, 196-198
- inverters, 553-564
- model line, 195
- model parameters, 613-615
- op amp, 462-476
- output resistance, 201
- resistor, 242
- saturation region, 204-207, 207-209
- small-signal model, 221-225
- switch, 242-246
- transmission gate, 244
- triode region, 209-211

Mutual coupling coefficient, 115

N

n, emission coefficient, 97

Negative feedback, 358-413

Nodes
- circuit, 8
- floating, 84
- ground, 8

.NODESET statement, 305

Noise analysis, *See* Analysis: Noise

Nonlinear Circuits:
- clamping, 542-543
- comparator, 529-532
- limiter, 132-133
- multivibrator, 529-535
- peak detector, 541
- rectifier, 115-132, 537-543
- dc restorer, 133-135
- voltage doubler, 135-136
- waveform generator, 536-537

Nyquist plot, 394, 396, 398, 399

O

OCT, *See* Analysis: AC

Offset:
- current, 81, 277-279, 451, 456,
- input/output, 279-282
- reduction/nulling, 485
- systematic/random, 308, 311, 451
- voltage, 81, 275-277, 451

On-resistance, 242

Op amp
- bandwidth, 65-67, 454-456, 468-476
- bias/offset current, 81, 451
- BiCMOS, 468-476
- buffer/follower, 57-59
- CMOS, 462-468, 468-476
- common-mode rejection, 62, 77-80, 454
- compensation, 410-413
- DC problems, 81-83
- filters, 487-515
- frequency response, 65-67, 454-456, 468-476
- gain, 454-456
- integrator, 50-57
- internal structure, 68
- large-signal operation, 73-77
- macromodel,
   - large-signal, 67-77
   - small-signal, 65-67
- resistances, 80-81
- RC oscillator, 522-526
   - active-filter tuned, 522-526
   - Wien-bridge, 518-522
- saturation, 71
- slew-rate, 75-77, 457-459
- virtual ground, 49

Index

Op amp 741 type
    common-mode gain, 454
    DC analysis, 448-454
    frequency response, 454-456
    input resistance, 456
    noise, 459-461
    output resistance, 454
    slew-rate, 457-459
Operating/quiescent point analysis, *See* Analysis: Operating Point
.OPTIONS control line, 128, 620
    ITL5, 128
    all, 620-622
Oscillator
    active-filter tuned, 522-526
    amplitude stabilization, 518
    CMOS, 526-529
    Colpitts, 526-529
    crystal, 526-529
    distortion, 520-522, 524-526, 528-529, 536-537
    Pierce, 526-529
    sinusoidal, 517-529
    square wave, 529-534
    Wien-bridge, 518-522
Output Stage
    class A, 423-428
    class AB, 436-438
    class B, 428-436
Overdrive factor, 186
Overlap capacitance, 328

P

Phase
    angle, 11, 19
    margin, 403, 407, 408, 409, 410, 411
Piecewise linear (PWL) source, 12
Plot (.PLOT) control statement, 19
    Probe, 25-27
Plus (+) sign, 7
Polar plot, 394, 396, 398, 399
Post-processor, graphics, 25
Power
    amplifier, 423
    conversion efficiency, 429, 431
    transformer, 115

Power dissipation
    instantaneous, 33, 428, 429, 431, 432, 439, 440, 442
    static, 16, 307
Precision circuits, 537-543
Print (.PRINT) control statement, 18
Probe (.PROBE) control statement, 25
PSpice, 3
PULSE, 12
PWL, 12

Q

Quiescent operating point, 16

R

Rectifier, 115-132
    clamped capacitor, 542-543
    full-wave, 119-127
    half-wave, 115-119
    peak, 541-543
    power supply, 127-132
    precision, 537-543
    ripple, 107
    transformer-coupled, 116
Regulator/regulation, 104-108
Resistor
    element line, 11
    temperature coefficient, 173
Ripple voltage, 121-127, 130-131
Rise/fall time, 578

S

Scale factor abbreviations, 7
Schottky-barrier diode (SBD), 250
Second-order filters:
    biquad, 491-502
    sensitivity, 514
    single-amplifier biquad, 500-502
Sensitivity
    active, 168-172, 231-235
    passive, 168-172, 231-234
    temperature, 172-174

Index

Sensitivity analysis, *See* Analysis: Sensitivity
Series:
    -series feedback, 367-372
    -shunt feedback, 359-367
Set/reset (SR) flip-flop, 572-577
Shunt:
    -series feedback, 380-388
    -shunt feedback, 374-380
SIN independent signal source, 12
Signal generators:
    astable multivibrator, 533-534
    bistable multivibrator, 529-532
    crystal oscillator, 526-529
    monostable multivibrator, 534-535
    op amp-RC oscillator, 517-526
    sinusoidal oscillator, 517-526
Single-amplifier biquad, 500-502
Slew rate, 72, 75-77, 457-459
Small-signal:
    analysis
        DC, 16-17
        AC, 17
    DIODE model, 96
    BJT model, 160
    JFET model, 223
    MESFET model, 251-252
    MOSFET model, 223
    transfer function, 48-49
Source:
    current-controlled current, 14
    current-controlled voltage, 14
    dependent, 14
    independent current, 10-14
    independent voltage, 10-14
    piece-wise linear, 12
    pulse, 12
    sinusoidal, 12
    voltage-controlled current, 14
    voltage-controlled voltage, 14
Spice:
    acronym, 1
    analyses with, 5-6
    circuit description, 9-16
    input file, 6-21
    output file, 21-25

Square wave generator:
    pulse generator, 12
    piece-wise linear generator, 12
SR flip-flop, 572-577
Stability
    Bode plot, 394-408
Subcircuit, 59-62
    external nodes, 61
    internal nodes, 60
    .ENDS element line, 60
    .SUBCKT element line, 60
    Subcircuit call (X), 61
Suffixes
    element dimensions, 8
    scale factor abbreviations, 8
Switch
    MOSFET, 242-246
    voltage-controlled, 476
Symmetrical excitation, 261-266, 448

T

Temperature:
    coefficient, 173
    sweeping, 172-174
    .TEMP statement, 101-102
Temperature analysis, *See* Analysis: temperature
Thermal noise, 459
Threshold voltage, 197
Title line, 6
Total harmonic distortion (THD), 434
Tow-Thomas biquad, 494-500
Transfer function analysis, *See* Analysis: Transfer Function
Transformer, 115, 116
Transient response
    .TRAN control statement, 17-18
    UIC (use initial conditions), 17-18
Transistor
    bipolar junction, *See* BJT
    field-efect, *See* FET; JFET; MOSFET
    -transistor logic, 583-592
Transmission:
    gate, 242-246
    line, 597-602
    loop, 392-394

Index

Triangular waveforms, 164-165
TTL, 583-592
Tuned amplifier, 502-506
Two-integrator-loop biquad, 494-500

U

Unity-gain follower, 57-59
Universal active filter, 512
UIC (use initial conditions), 17-18

V

Variables
    AC analysis, 19
    DC analysis, 19
    noise analysis, 461
    transient analysis, 19
Virtual ground, 49
Voltage-controlled current source, 14
Voltage-controlled voltage source, 14
Voltage-controlled switch, 476

W

Waveforms:
    square, 12
    triangular, 164-165
Wien-bridge oscillator, 518-522
Wilson current source/mirror, 280

X

X, subcircuit call, 61

Y

y-parameter, 380

Z

z-parameter, 372
Zener:
    diode, 108-109
    voltage regulator, 109-114